D1346347

SHEEP BREEDING

STUDIES IN THE AGRICULTURAL AND FOOD SCIENCES

STUDIES IN THE AGRICULTURAL
AND FOOD SCIENCES

Sheep Breeding

SECOND EDITION

Edited by
G.L. Tomes, D.E. Robertson and R.J. Lightfoot
Muresk and Perth, Western Australia

Revised by WILLIAM HARESIGN
University of Nottingham, England

BUTTERWORTHS

LONDON   BOSTON
Sydney   Wellington   Durban   Toronto

| United Kingdom | **Butterworth & Co (Publishers) Ltd** |
| London | 88 Kingsway, WC2B 6AB |

Australia	**Butterworths Pty Ltd**
Sydney	586 Pacific Highway, Chatswood, NSW 2067
	Also at Melbourne, Brisbane, Adelaide and Perth

| Canada | **Butterworth & Co (Canada) Ltd** |
| Toronto | 2265 Midland Avenue, Scarborough, Ontario, M1P 4S1 |

| New Zealand | **Butterworths of New Zealand Ltd** |
| Wellington | T & W Young Building, 77–85 Customhouse Quay, 1, CPO Box 472 |

| South Africa | **Butterworth & Co (South Africa) (Pty) Ltd** |
| Durban | 152–154 Gale Street |

| USA | **Butterworth (Publishers) Inc** |
| Boston | 10 Tower Office Park, Woburn, Massachusetts 01801 |

Originally published in 1976 by the Western Australian Institute of Technology

© Butterworth & Co (Publishers) Ltd, 1979

ISBN 0 408 10633 6

British Library Cataloguing in Publication Data

Sheep breeding. – 2nd ed. – (Studies in the
  agricultural and food sciences)
  1. Sheep breeding
  I. Tomes, G.J. II. Robertson, D.E. III.
  Lightfoot, R.J. IV. Haresign, William.
  V. International Sheep Breeding Congress,
  Muresk and Perth, Western Australia, 1976.
  VI. Series
  636.3′08′2   SF376.2   78–41188

  ISBN 0–408–10633–6

Printed in Great Britain by
Billing & Sons Limited
Guildford, London and Worcester

PREFACE

It was fitting that a book on Sheep Breeding should be produced to mark the 50th Anniversary of Muresk, Western Australia's principal Agricultural College. Sheep and wheat have always been the mainstays of Western Australian and Australian agriculture, and Muresk lies in the wheat belt where 75 per cent of Western Australia's 32,000,000 sheep are located. In maintaining sheep breeding and research throughout its history, Muresk has handsomely fulfilled its founder's expectations. Besides educating students and farmers in all aspects of the sheep industry, the College has been associated with a wide variety of research and with the distribution of quality stock to farmers.

Muresk farm was used in many sheep-breeding experiments when under the control of the Department of Agriculture. Since 1969, when Muresk joined the Western Australian Institute of Technology, research has been concentrated on artificial insemination, co-operative breeding, nutrition and reproduction.

The material content of the book covers the term Sheep Breeding in its widest context, incorporating sections on the structure and objectives of national sheep industries from many of the major sheep-producing areas of the world, genetic selection and breed improvement, stud breeding and co-operative breeding schemes, reproduction in the ewe and finally male reproduction together with artificial insemination. The style of writing differs between chapters in that some are designed as general review articles covering various principles of sheep breeding, whilst others report important recent research findings in specific areas of interest.

The wide range of nationalities amongst the contributors attests to the importance of this subject in many countries, and it is hoped that the book will further the progress of sheep breeding—production and research— throughout the world.

CONTENTS

I

NATIONAL SHEEP INDUSTRIES—
STRUCTURE, BREEDS AND OBJECTIVES

1

THE SHEEP AND WOOL INDUSTRY OF AUSTRALIA — A BRIEF OVERVIEW

V.R. SQUIRES
CSIRO, Deniliquin, N.S.W., Australia

INTRODUCTION

Australia has an immense sheep industry, the largest in the world (Chapman, Williams and Moule 1973). The gross annual value of the products has exceeded $U.S.1500 million. More than three-quarters of the sheep are bred and kept for apparel wool. Most wool is exported, thereby contributing significantly to Australia's export income. In 1970 the total number of sheep shorn and the total annual wool production in Australia reached record levels of 180 million sheep and 923,000 tonnes of greasy wool. Sheep numbers have since decreased with a corresponding decline in wool production.

Australia produces about 30% of the world's wool, twice as much as U.S.S.R. (the second largest producer), and wool is one of the most important of rural exports. Ninety percent of wool produced is exported—95% if semi-processed wools are included. Japan (34%), Italy (9%), France (8%), United Kingdom (8%) and the Federal Republic of Germany (7%) are the principal buyers, (Bureau of Census and Statistics 1970).

DEVELOPMENT OF THE SHEEP AND CATTLE INDUSTRY IN INLAND AUSTRALIA

Commercial livestock grazing in Australia has been shaped, over a period of more than a century and a half, by three interrelated factors: environment, profit and cultural pressure. Geared essentially to distant export markets, its development reflects the abundance of land and scarcity of labour and capital on which Australia's economic growth was initially based.

The rapid increase in numbers of livestock following their introduction into Australia, and the attendant development of an industry based on use of extensive land areas, calls for critical examination.

What are the factors required for development of a pastoral industry in any new, remote and undeveloped country? It appears that at least four conditions must be co-existent in order to encourage an enterprise on an extensive scale:

1. Availability of great areas of essentially free land producing suitable forage.
2. Demand for animal products, locally or in a foreign market.

3. A low requirement for labour.
4. Suitable animals for foundation stock.

In inland Australia there were for all practical purposes, unlimited areas of free land; its vegetation was dominated by perennial plants, including grasses, forbs and palatable shrubs. The forage was of such a nature that it could be utilized for all classes of domestic livestock—horses, sheep, cattle and goats.

The native pastures were co-extensive with conditions highly favourable for both forage growth and livestock production in good seasons. Winter temperatures permitted year-long grazing of livestock and did not require that shelter be provided. Lack of competition from a large native wildlife population, relatively few predatory animals, and, especially at first, freedom from diseases and pests.

The last point may have been of particular significance (McDonald 1959). When livestock are introduced to an entirely new area they take with them only the diseases with which those particular animals happen to be infected at the time—the others to which their species are subject are left behind. As a net result livestock taken to new areas commonly suffer fewer diseases for a considerable number of years. Another point is that there were no native ungulates in Australia from which pests and diseases could be contracted. For these reasons, domestic livestock introduced into Australia, and spreading through it as colonization proceeded, were singularly free from disease.

Throughout Australia in the early part of the colony's existence the demands were for animal products for local subsistence. This period was ephemeral merging into that in which there was an active market in the early gold mining districts. Sale of animal products to mining communities was more important in Victoria and New South Wales than in other colonies. In addition to edible products, important in the local economy, domestic animals produced items of even greater significance—wool, hides and tallow. These items were non-perishable and compact, withstood shipment for long distances, and had a high value per unit of weight which enabled them to compete successfully for space on cargo vessels against virtually all other commodities. There is abundant documentation of the fact that throughout much of the colonial era the value of the wool commonly represented the entire worth of the sheep (Barnard 1962).

In inland Australia, wool was a highly marketable commodity from the beginning, continuing to be the principal incentive to sheep production until recent times (Barnard 1962).

Livestock raising on the pastoral lands of Australia required a small amount of labour, none of which needed to be highly skilled. With land which was ample in area and initially abundant in forage production, animals could be turned loose to roam freely for most of the year. Each flock of sheep needed a shepherd, and the wool had to be shorn. Labour to meet these requirements was readily available except for periods during the goldrushes or in times of war. Following initial introduction of sheep into the coastal regions of New South Wales, reservoirs of animals accumulated rapidly. Importations from England and from Spain supplemented the natural increase. Selected from the hardy stock which had survived the long sea voyage and by breeding they readily adapted to the environment of their new pastures. Their inherent vigour, coupled with relative freedom from disease already discussed, fitted them admirably to their role as foundation stock for Australia. As Moule (1962) and McDonald (1959) point out there was a considerable amount of breeding and selection among sheep once the inland pastoral areas were opened up.

In summary, all major factors favouring a pastoral industry in a remote new and undeveloped region were found in Australia. Moreover, these conditions were in such a favourable combination that stock raising became a very important industry within a short time (Barnard 1962; Butlin 1962).

Livestock increased at an astonishing rate when simply left to run relatively wild on the grazing lands of the new colonies. It was a cause of wonder even among contemporary observers (Butlin 1962). Some indication of the rapidity with which herds and flocks could build up is afforded by the rates of unimpeded increase in livestock. For the present purpose two years has been used as the minimum breeding age for both cattle and sheep (in practice it often has been nearer one year, especially under the conditions prevailing a century or more ago). These rates of unimpeded increase apply only to ideal conditions of course (Table 1). But with ample forage, a favourable climate, relatively few diseases and essentially no competition or serious predators, the grazing lands of inland Australia approached the ideal in no small degree during several periods (Butlin 1962).

TABLE 1

UNIMPEDED INCREASE IN LIVESTOCK NUMBERS

Years	Total Number of Animals	
	Cattlet	Sheep*
1	2	2
5	7	8
10	30	54
15	142	406
20	675	3076
25	3209	—

†Assuming on average one young per birth
*Assuming on average 1.5 young per birth

Periodic droughts caused great fluctuations in livestock numbers (McDonald 1959; Moule 1968; Perry 1967, 1968) but over periods of 10-20 years at a time numbers rose rapidly. The data for sheep numbers, as an example, shown on a logarithmic scale (Fig. 1) provide a simple growth curve which resembles in broad outline, the growth curve of a typical bacterial culture. The initial growth from 1820-1890 was typical of the growth of a new organism in a suitable environment—the numbers increased to 20 million, thereafter increasing steadily to 106 million by 1891 when most of the areas suitable for sheep had been occupied. Then from the year 1891 to the present, the curve shows a general tendency toward a stationary phase in which there are violent fluctuations in numbers. These fluctuations in the arid zone livestock population are thought to be drought related (McDonald 1959; Chapman *et al.* 1973). Excellent accounts of the history of the development of Australia's pastoral industries have been published by Peel (1973), Barnard (1962) and by Alexander and Williams (1973); there are detailed accounts relating to past and current status of each particular livestock industry.

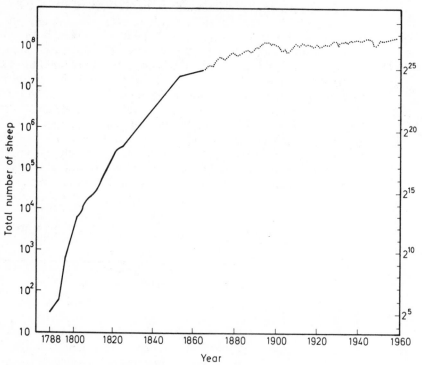

Figure 1—Growth in the Australian sheep population

THE INDUSTRY STRUCTURE

The last 25 years have been a dynamic period for the Australian sheep and wool industry in that there have been great changes in population and production characteristics. During this time span the industry has experienced mixed fortunes, ranging from the relatively buoyant conditions of the early 1950s to those at present where it is being subjected to strong economic pressures.

The sheep industry is dispersed throughout inland Australia and encompasses a wide range of environmental conditions with annual rainfall ranging from less than 250 mm to more than 1000 mm (Fig. 2). Even though there is increasing diversity within the woolgrowing enterprise as annual rainfall increases, wool is always the major product.

The Bureau of Agricultural Economics in its surveys of the Australian sheep industry (Lawrence 1971), divided the sheep grazing regions into three broad zones: the pastoral zone, the wheat-sheep and the high rainfall zones (Fig. 3).

The largest zone in terms of area, the pastoral zone, consists of the arid and semi-arid regions which run through the Northern Territory, Queensland, New South Wales and Western Australia. This zone is characterized by almost no agricultural activity especially towards it inner boundaries. Large sheep flocks are a feature of the sheep grazing areas and the main activity on these sheep properties is wool growing, using Merino-type sheep. Some beef cattle are also run on what are predominantly sheep growing properties.

The wheat-sheep zone is bounded by the pastoral zone towards the centre of the continent and the high rainfall zone at the coastal side. It is the second largest zone in terms of area. Properties in this zone typically combine sheep and/or beef cattle raising with a wheat or other cereal growing enterprise. The majority of the national sheep flock is carried in this zone and many of the flocks produce non-Merino sheep for either direct slaughter or for fat lamb breeding programmes.

The high rainfall zone is the smallest of the three zones in terms of total area but the second largest in respect to sheep numbers carried. This zone lies on the seaboard edge of the continent in each of the States and covers all of Tasmania. In the high rainfall zone sheep are generally carried in conjuction with beef cattle and, as in the wheat-sheep zone, many properties produce prime lambs in the sheep enterprise.

Although the boundaries of these zones are not clear and one zone merges into the next, the surveys reveal appreciable differences in production and land-use between the zones. Of the sheep population 23% are in the pastoral zone, 44% in the wheat-sheep zone and 33% in the high rainfall zone (Lawrence 1971). The boundaries of the pastoral zone coincide with the semi-arid and arid zone as defined by Perry (1967).

Even within the pastoral zone there are regional differences in sheep productivity. Brown and Williams (1970) examined the geographical aspects of the distribution of sheep and their productivity in the pastoral zone.

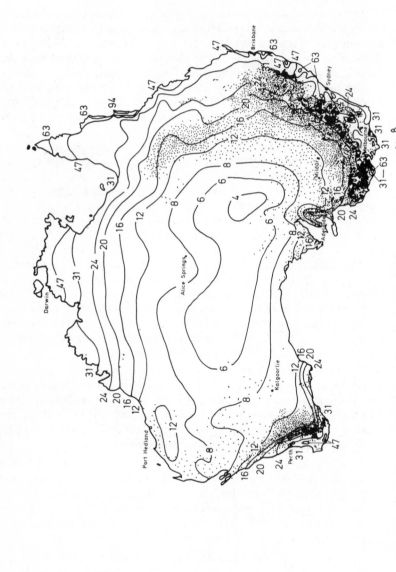

Figure 2—Sheep distribution in Australia (*After* Atlas of Australian Resources 1970). Each dot represents 50,000 sheep and each isohyet represents 50 percentile rainfall in inches per year

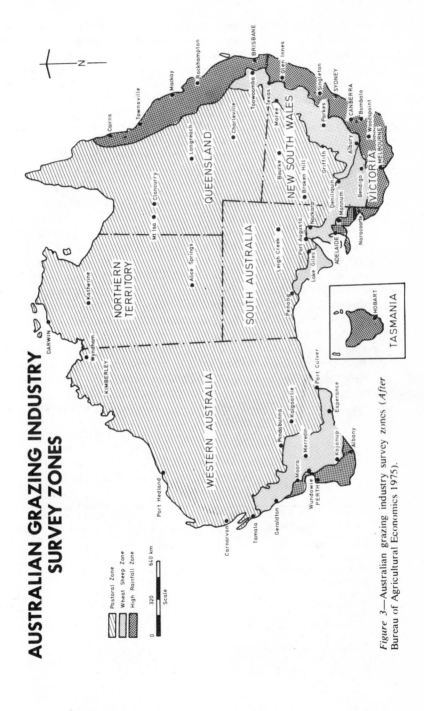

Figure 3—Australian grazing industry survey zones (*After* Bureau of Agricultural Economics 1975).

The average greasy fleece weight of sheep in the northern sub-tropical areas of eastern and western Australia were considerably lower than those of sheep in the temperate semi-arid southern areas. Similar north-south trends in marking percentages were apparent in eastern Australia but were not so clearly seen in western Australia. The low productivity of sheep in northern Australia must impose a severe economic penalty on pastoralists in these areas where there are about 6.5 million sheep which produce less than 2.5 kg of clean wool each year.

Three-quarters of Australia's sheep are Merinos (73% in 1970). Other pure breeds include Corriedales (6%) and Polwarths (2%). Cross-bred sheep (over 12%) and Merino comebacks (less than 3%) comprise the remainder.

Sheep are now grazed on more than 85,000 properties that vary greatly in area and size of flock (Tables 2 and 3). Most of the sheep in the inland are run on large properties where capital (Table 4) and labour components (Table 2) are low by overseas standards (Campbell 1965; Waring 1969). The industry covers one-third of the continent (Figure 2) but directly employs or supplies a living for less than 5% of the population (Chapman *et al* 1973). The Merino flocks are found throughout the sheep grazing areas of Australia and although wool is their major product some of these flocks provide ewes that form the foundation of the prime lamb industry.

The history of sheep breeding in Australia is dominated by the development of a number of strains of the Merino sheep, well adapted to the production of medium and fine wool in most of the many and varied parts of the country. There are various detailed authoritative accounts of the introduction of sheep by the early colonists and of the early breeding and husbandry methods (McDonald 1959). The earliest sheep were natives of

TABLE 2

SUMMARIZED STATISTICS OF
AUSTRALIAN SHEEP ENTERPRISES†

| | | Zone | |
Attribute	Pastoral	Sheep/Wheat	High Rainfall
Area of holding (ha)	15,725	1,144	546
Sheep/holding	3,900	1,511	1,185
Sheep units/ha	0.18-0.58*	2.1	5.3
Wool/sheep (kg)	4.0	4.1	3.9
Labour units (man-years)	2.8	2.7	1.8
Sheep units/labour unit	2,124	973	1,383

†Summarized values for all States. Values for individual States may differ markedly from the averages shown.
*Extremely variable and hence a range of values is given

Source: Bureau of Agricultural Economics (1976)

TABLE 3

AVERAGE FLOCK SIZES BY ZONES FOR
PROPERTIES PRODUCING MERINO APPAREL WOOL

Flock Size	Pastoral	Zone Sheep/Wheat	High Rainfall
Number of sheep		% of properties	
200 — 500	Nil	15.7	21.9
500 — 1,000	1.8	25.2	12.4
1,000 — 2,000	16.8	31.9	28.9
2,000 — 5,000	47.7	23.3	29.4
5,000 — 10,000	25.3	3.5	6.3
10,000 — 20,000	7.1	0.4	1.1
>20,000	1.3	Nil	Nil

Source: Bureau of Agricultural Economics (1976)

TABLE 4

AVERAGES PER PROPERTY FOR SELECTED ATTRIBUTES
PERTAINING TO SHEEP AND WOOL PRODUCTION
IN THE MAJOR ZONES

Item	Pastoral	Zone Sheep/Wheat	High Rainfall
Total capital invested ($)*	193,024	145,473	150,213
Sheep and lambs shorn	5,244	1,801	2,112
Wool produced (kg)	22,957	7,123	7,932
Ewes mated as % of flock	49.2	50.7	45.8
Lambing %	50.2	66.8	79.2
Wool cut/sheep shorn (kg)	4.4	3.9	3.8
Returns from wool ($)*	39,063	11,877	16,635
Sheep trading gain ($)	5,803	3,994	5,295

Source: Bureau of Agricultural Economics (1976)
* 1972-73 figures

South Africa and Bengal, transported from the Cape and Calcutta. It was not until the introduction of Spanish Merinos, also from the Cape, in 1797 that the foundations were laid for the development of an adapted breed for Australian conditions. The Merino thrived in the new environment, producing superfine wool for the English market. Later, the combined forces of changing markets and the discovery of the inland grazing areas of Australia stimulated the creation of development of two new Merino types which became important as the Australian Merino. The two strains were the medium-wool Peppin and the South Australian strong-wool. The medium-wool Peppin (64's count) was developed by the Peppin brothers at their "Wanganella" property in the Riverina of south-west New South Wales. South Australia breeders favoured a more robust type of sheep with dense, long-stapled wool of 60's count. The Merino sheep has, through selection over the past 150 years, developed varieties suited to hot-dry and cool-wet environments. In hot-dry summer rainfall country the Merino is low in production (3-5 kg greasy wool) and in fertility (30-50% of lambs a year) (Macfarlane 1968a). Chapters in both Barnard (1962) and Alexander and Williams (1973) provide additional information on aspects of the development of the Merino sheep in Australia.

Sheep and wool are part of the Australian tradition. However, the importance of wool has declined; wool prices fell to marginal levels in 1971, stimulating interest in sheep meat production at the expense of wool and in substitution of cattle for sheep (McCarron 1975).

THE OUTLOOK FOR THE FUTURE

In his recent review of the future prospects for the pastoral industries of Australia, Ferguson (1973) cites alternative uses of grazing lands, possibilities of improving ruminant productivity, and decreasing costs of production as questions which will have bearing. Competitions from man-made products (synthetics) could also play a major role. Already man-made fibres have made a major inroad into the market for wool. Vegetable protein could have a bearing on the future market for red meat.

Reduction in the demand for wool stimulated a switch of resources to production of beef cattle. Sheep and cattle have always shared approximately equal proportions of Australia's major photosynthetic resource. The ratio of sheep to cattle numbers has fluctuated in a cyclic fashion reflecting the response of producers to changing relative returns from these enterprises. Sheep and cattle share much of the same rangeland, except in the far north where sheep have been replaced almost entirely. Marked seasonal and year-to-year variations in pasture growth in the inland have favoured the production of wool, since wool can be as efficiently produced on a fluctuating as on a constant feed intake (Ferguson 1962). Wool growth continues on feed intakes insufficient to maintain body weight, while meat production would be negative under such conditions.

Pastoral industries in the inland are non-competitive for land-use with cropping, due to unsuitable climate, soil or terrain. Proclamation of some areas within the arid zone as recreation areas and as National Parks represents some conflict of interest but the actual extent of these special reserves as proportion of the total land area available is small (Costin and Mosely 1969).

Sound ecological reasons have been advanced for removing livestock from much of the arid zone. Unless more conservative stocking is adopted, production is likely to continue to decline in the pastoral zone as desert formation increases.

The grazing industries appear safe from displacement by food crops in the pastoral zone but there is likely to be conflict of interest in land use in high rainfall and wheat-sheep zones as pressure for food grain production increases.

Despite a seemingly inevitable doubling of the world's population by the end of the century, the future prospects for wool are clouded by the uncertaintities of its competitive position with the fibre products of the petrochemical industry. However, the exponentially increasing depletion of the world's supplies of fossil fuels and mineral resources may give the competitive advantage to wool. The energy cost of harvesting the photosynthetic resource of inland Australia is small, the ruminant serving as a self-fuelled, self-propelling, self-servicing and self-reproducing harvesting machine. It is also a food or fibre processing factory (Ferguson 1973; Macfarlane 1968).

During one-and-a-half centuries of occupation by graziers their pastoral activities have brought about development of transport and communications systems throughout thousands of square kilometres of arid areas. These facilities are now being utilized in the search for minerals, oil and gas which no doubt will yield far greater returns than wool or meat in years to come. And although in future the relative significance of pastoral production may decline, it must be remembered that in past years, sheep and cattle played a major role both in terms of development and in yielding revenue. In fact for more than 100 years pastoral production has been virtually the only major source of revenue for about four-fifths of inland Australia.

REFERENCES

Alexander, G. and Williams, O.B. (Editors) (1973)—*The Pastoral Industry of Australia— practice and technology of sheep and cattle production.* Sydney University Press: Sydney.

Barnard, A. (Editor) (1962)—*The Simple Fleece—Studies in the Australian Wool Industry.* Melbourne University Press: Melbourne.

Brown, G.D. and Williams, O.B. (1970)—Geographical distribution of the productivity of sheep in Australia. *Journal of the Australian Institute of Agricultural Science* 36: 182-98.

Bureau of Agricultural Economics (1976) *The Australian sheep industry survey 1970-71 to 1972-73.* Government Printer, Canberra pp. 270.

Bureau of Census and Statistics (1970)—*Official Year Book of the Commonwealth of Australia,* Government Printer: Canberra.

Butlin, N.G. (1962)—Distribution of the sheep population: preliminary statistical picture, 1860-1957. In: Barnard, A. (Editor) *The Simple Fleece—Studies in the Australian Wool Industry.* Melbourne University Press: Melbourne.

Campbell, K.O. (1965)—Problems of adaptation of pastoral businesses in the arid zone. *Journal of the Australian Agricultural Economics Society* 10: 14-26.

Chapman, R.E., Williams, O.B. and Moule, G.R. (1973)—The Wool Industry. In: Alexander, G. and Williams, O.B. (Editor) *The Pastoral Industries of Australia.* Sydney University Presss: Sydney.

Costin, A.B. and Mosely, J.G. (1969)—Conservation and recreation in arid Australia. In: Slatyer, R.O. and Perry, R.A. (Editors) *Arid Lands of Australia.* Australian National University Press: Canberra.

Ferguson, K.A. (1962)—The relation between the responses of wool growth and body weight to changes in food intake. *Australian Journal of Biological Sciences* **15** : 720-31.

Ferguson, K.A. (1973)—Future prospects for the pastoral industries. In: Alexander, G. and Williams, O.B. (Editors) *The Pastoral Industries of Australia.* Sydney University Press: Sydney.

Lawrence, M.J. (1971)—The Australian sheep industry survey: 1967-68 to 1969-70. Preliminary results. *Quarterly Review of Agricultural Economics* **24** : 217-35.

Macfarlene, W.V. (1968a)—Cropping Australia's desert with ruminants. *Span* **11** : 150-54.

Macfarlene, W.V. (1968b)—Protein from the wasteland: water and the physiological economy of ruminants. *Australian Journal of Science* **31** : 20-30.

McCarron, A.C. (1975)—The changing structure and economic situation of the Australian sheep and beef cattle grazing industry. *Quarterly Review of Agricultural Economics* **28** : 152-76.

McDonald, I.W. (1959)—Merino sheep in Australia. In: Keast, A., Croker, R.L., and Christian, C.S. (Editors) *Biogeography and Ecology of Australia.* Junk: Hague, pp. 487-99.

Moule, G.R. (1962)—The ecology of sheep in Australia. In: Barnard, A. (Editor) *The Simple Fleece—Studies in the Australian Wool Industry.* Melbourne University Press: Melbourne.

Moule, G.R. (1968)—Sheep and wool production in semi-arid pastoral Australia. *World Review of Animal Production* **4**(17) : 40-49, (18) : 46.

Peel, Lynette J. (1973)—History of the Australian pastoral industries to 1960. In: Alexander, G. and Williams, O.B. (Editors) *The Pastoral Industries of Australia.* Sydney University Press: Sydney.

Perry, R.A. (1967)—The need for rangelands research in Australia. *Proceedings of the Ecological Society of Australia* **2** : 1-14.

Perry, R.A. (1968)—Australia's arid rangelands. *Annals of the Arid Zone* **7** : 243-49.

Waring, E.J. (1969)—Some economic aspects of the pastoral industry in Australia. In: Slatyer, R.O. and Perry, R.A. (Editors) *Arid Lands of Australia.* Australian National University Press: Canberra.

2

AUSTRALIAN SHEEP BREEDING PROGRAMMES — AIMS, ACHIEVEMENTS AND THE FUTURE

K.A. FERGUSON
CSIRO, Canberra, Australia

INTRODUCTION

Australia does not have a sheep breeding programme in the sense of a planned programme that has been laid down either by Government Departments of Agriculture, by research organizations such as CSIRO, or by organizations of sheep breeders. Thus I have interpreted the title given to me—Australian Sheep Breeding Programmes—to mean existing practices that have evolved since the wool industry was established in Australia at the beginning of last century.

Australia was colonized at a time of rapid population growth during the industrial revolution in Europe with a consequent increased demand for wool. Australia provided 800 million hectares of land sparsely occupied by about 600,000 Aborigines. 65% of the area proved to be suitable for grazing and much of this was unsuitable for competing uses such as crop production and forestry. Today only $2\frac{1}{4}\%$ of Australia's land is used for cropping because of the restraints imposed by soil, climate and terrain (Nix, 1973, 1975; Gifford *et al* 1975).

Both sheep and cattle spread into Australia's empty grazing lands but prior to the development of refrigerated transport there was no overseas market for beef. Wool had a high unit value which could withstand the transport cost to European markets. Ultimately the sheep and cattle distributions over the grazing lands of Australia were determined by historical trial and error influenced by the suitability of sheep and cattle for particular environments.

The sheep are distributed in three broad zones (Fig. 1). The pastoral zone is too dry for cropping. The wheat-sheep zone encompasses the restricted area available for cropping and the high rainfall zone is mostly too steep for cultivation. Sheep proved unsuitable for the wet tropics and for the humid eastern coastal strip where grazing land is largely used for dairying.

The variability of rainfall in Australia produces frequent periods of pasture shortage unsuitable for meat production. Wool growth continues on sub-maintenance intakes while bodyweight is being lost. Such conditions favour Merino wool production since the limited arable land precludes the widespread adoption of supplementary feeding.

The stocking rate is much lower in the pastoral zone and property sizes are correspondingly larger (Table 1). In the higher and more reliable rainfall conditions of the wheat-sheep and high rainfall zones there is an increase in

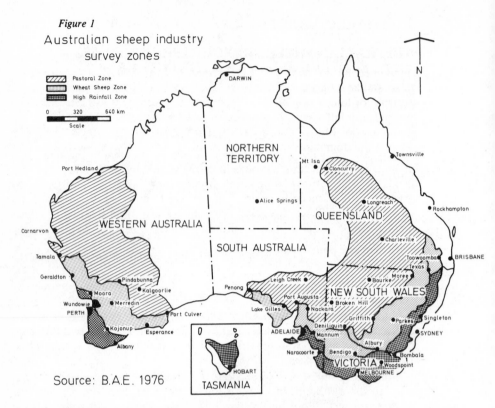

Figure 1

Australian sheep industry survey zones

Pastoral Zone
Wheat Sheep Zone
High Rainfall Zone

Source: B.A.E. 1976

production of lamb for meat with a corresponding rise in the non-Merino sheep population, largely Merino × Border Leicester ewes which are mated to Dorset rams to produce lambs for market. Flocks of Corriedales and Polwarths make up most of the remaining non-Merino flocks.

TABLE 1

SHEEP DISTRIBUTION IN AUSTRALIA 1972-73

Zone	No. of properties	Average property size Ha	Sheep per property	Sheep equivalents per hectare	% Merino	% Crossbred	% Corriedale Polwarth
Pastoral	6,459	24,103	4,947	0.30	97.5	1.9	
Wheat sheep	44,400	1,161	1,605	2.9	79.9	13.4	4.7
High rainfall	28,762	639	1,814	6.0	60.9	18.6	16.5

Source: B.A.E. 1976.

STRUCTURE OF AUSTRALIA'S SHEEP BREEDING PROGRAMME

The expansion of the Merino into the grazing lands of Australia was accompanied by the evolution of different strains of Merino promoted by geographical isolation, different environmental conditions, and increased demand for longer stapled wool. The short stapled fine wools established in the high rainfall zone proved less suitable for the pastoral zone and the medium wool Merino developed by the Peppin brothers in the Riverina became the dominant type in New South Wales. In South Australia a distinctive type of strong wool Merino was developed which spread to Western Australia (Pattie 1973).

A hierarchical three-tiered structure developed in which a group of closed stud flocks called parent studs supplied rams to a second tier of daughter and general studs, the daughter studs drawing outside blood only from their parent stud and the general studs drawing outside blood from more than one parent stud (Short and Carter 1955). In turn, the three kinds of stud supply more than half the rams required by the third tier of commercial flocks (Table 2).

TABLE 2

RAMS SOLD BY SHEEP STUDS IN AUSTRALIA 1971

	No.	%
Parent Studs	13,616	10.8
Daughter Studs	46,915	37.3
Family Groups	60,531	48.1
General Studs	64,968	51.9
All Studs	125,499	100.0

Source: Roberts, Jackson and Phillips 1975.

The number of daughter and general studs has increased with the growth of the sheep population, the number of parent studs by definition remaining restricted. The family groups deriving from each parent stud have preserved some degree of genetic diversity within the fine, medium and strong wool strains of Australian Merino. However, the different family groups have unequal influences on the genetic composition of the Merino population (Fig. 2).

The majority of fine wool rams are now supplied by the Merryville family group, while the Peppin medium wool family groups, Uardry and Haddon Rig, dominate the medium wool ram market. Most of the strong wools are now supplied by the Collinsville family group.

An increasing proportion of daughter and general studs are now breeding their own replacement rams (Connors and Reid 1976), tending to dilute the genetic influence of the parent studs and establishing a two-tier structure. A proportion of the larger commercial flocks (the exact number is unknown) are also breeding their own rams.

In recent years an additional feature of sheep breeding programmes in Australia as in New Zealand is the emergence of the co-operative nucleus breeding systems, the largest scheme being in Western Australia. The co-operating sheep breeders measure performance objectively and contribute their best ewes to a central nucleus flock which in turn supplies rams to the contributing flocks. Ewes from the entire population can thus be drawn on to produce rams rather than only the 2% of the population represented by the traditional stud system. However, the relative efficiency of the two systems probably depends on the extent to which objective measurement of performance is employed and the culling percentage used.

At present the number of rams tested for clean fleece weight and mean diameter by the Merino studs is less than 10% of the number of rams sold.

Figure 2

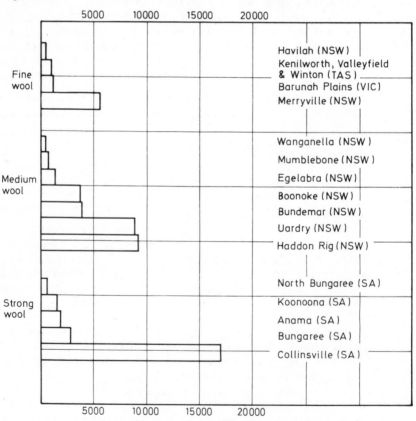

Rams sold by family groups 1971

Source: Roberts, Jackson & Phillips 1975

Rams produced by the nucleus flock in the co-operative breeding systems are not only selected on measured performance but are produced from ewes also selected on measured performance. Traditionally, sheep studs have relied on visual methods of appraising performance and appear to remain unconvinced of the merits of objective measurement. The percentage of rams culled by the studs is generally low.

CRITERIA OF PERFORMANCE

In Australia the amount and kind of wool per sheep dominates estimates of value with little attention being given to fertility or bodyweight unlike the practice in New Zealand where environmental conditions are more favourable for meat production.

Traditional selection criteria which are still largely utilized include some characteristics of little value and others which are negatively correlated with wool weight such as crimp frequency. However, the introduction of measured clean wool yield and mean fibre diameter as a basis for wool selling has brought about changes in wool classing which will be reflected in sheep classing. Separate quality count and yield lines within a flock are now combined since crimp frequency, the major determinant of quality count, has a variable relation to mean fibre diameter (A.W.C. 1973). Tops made from the bulked lines have been shown not to be inferior in variability of fibre diameter and length (Andrews and Rottenbury 1975).

The price margin for finer wool has in the past been balanced by the lower fleece weights of finer wool traditionally selected by high crimp frequency. However, more widespread recognition that the positive genetic correlation between fleece weight and fibre diameter is small may lead to a lessening or a reversal of the present trend towards coarser wool and a reduction of the price margin for decreased diameter.

The worsening of the terms of trade for the wool grower over the past 25 years is reflected in a marked decrease in hired labour on sheep properties despite an increase in sheep numbers (B.A.E. 1976). The shortage of labour has focused attention on genetic solutions of problems previously coped with by more labour-intensive non-genetic means. For example, the elimination of skin folds easily achieved by selection reduces shearing time and susceptibility to blowfly strike.

GENETIC PROGRESS

Over the past 30 years greasy fleece weights of sheep in Australia have increased by 0.02 kg or 0.4% per year without any change in yield of clean wool (Figs. 3 and 4). This small increase can be accounted for by a decrease in the proportion of high quality count wool grown (Fig. 5) and by an increase in the area of sheep properties fertilized with superphosphate (Waring and Morris 1974). If correction is made for these influences, levels of wool production appear to have reached a plateau with the existing breeding system.

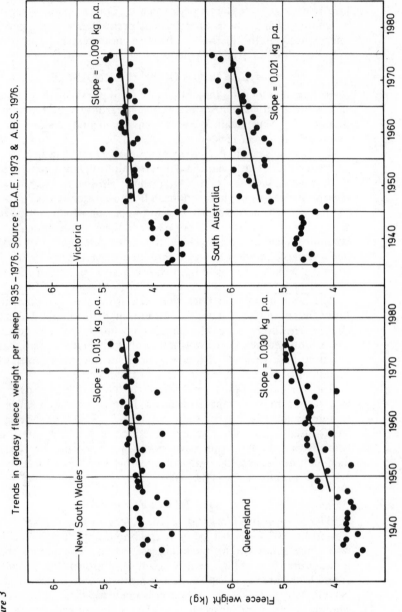

Figure 3   Trends in greasy fleece weight per sheep 1935–1976. Source: B.A.E. 1973 & A.B.S. 1976.

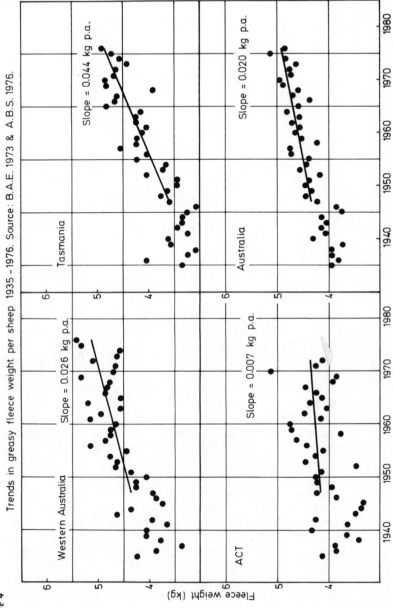

Figure 4

Trends in greasy fleece weight per sheep 1935 - 1976. Source: B.A.E. 1973 & A.B.S. 1976.

Figure 5

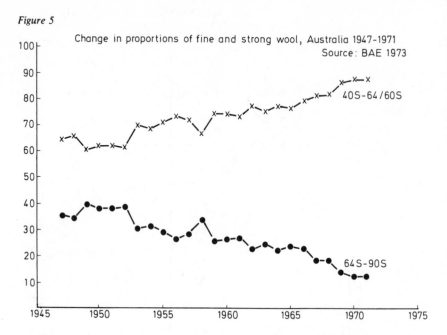

Change in proportions of fine and strong wool, Australia 1947–1971
Source: BAE 1973

It has been generally assumed that further genetic progress will be achieved if the Merino studs adopt fleece measurement programmes which they have been exhorted to do for the past twenty years by State Government extension services. However, I am less confident than formerly that fleece measurement will result in continued progress now that the results of experimental programmes initiated during the past 25 years are becoming available.

In a CSIRO experiment, selection for clean fleece weight with maintenance of mean fibre diameter and exclusion of sheep with excessive skin wrinkles resulted in the predicted increase of clean fleece weight of 2.3% per year (Turner and Young, 1969) for nine years after which further progress appears to have slowed (Fig. 6).

Another CSIRO experiment in which there was no restriction on fibre diameter or skin wrinkles was duplicated in two environments simultaneously. The initial responses of 6 and 8% per year were in excess of the responses expected on the basis of heritability, selection intensity and variance in clean fleece weights, possibly because the base flock had been unselected for many years (Fig. 7), (Dunlop 1976, Personal communication). However, a plateau appears to have been reached after nine years.

Again, in the N.S.W. State Department of Agriculture Experimental Station, selection for clean fleece weight resulted in the expected response for seven years after which progress has declined (Fig. 8).

Figure 6—Response to selection (1950–1974) for clean fleece weight with restriction on fibre diameter increase and high wrinkle score in Peppin Merino (Source: Turner, Dolling and Kennedy 1968 and unpublished data).

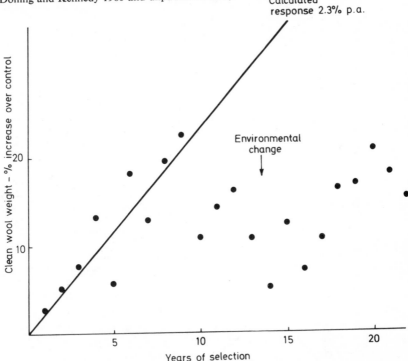

Unfortunately, we have no extensive data on the relative productivity of the sheep in these selection experiments compared with those in stud flocks. However, there is no evidence that the flocks selected on objective measurement have reached fleece weights in excess of those reached in stud flocks (Saville and Robards 1972). Thus there is no guarantee that the use of objective fleece measurements by the studs will increase their fleece weights and thereby fleece weights in the Merino sheep population at large.

Alternatively it may be argued that the apparent plateau in clean fleece weights reached in the selection experiments was due to the small size and restricted genetic diversity of the foundation flocks. The results achieved by co-operative nucleus breeding systems, if adequately documented, will throw light on this question. Both in the national flock and in the selection experiments progress may be limited by nutritional restraints.

THE FUTURE

The future of sheep breeding programmes in Australia depends of course on the economic viability of wool as a textile fibre. Analysis of factors influencing wool prices (Table 3) indicate the dominant effects of synthetic

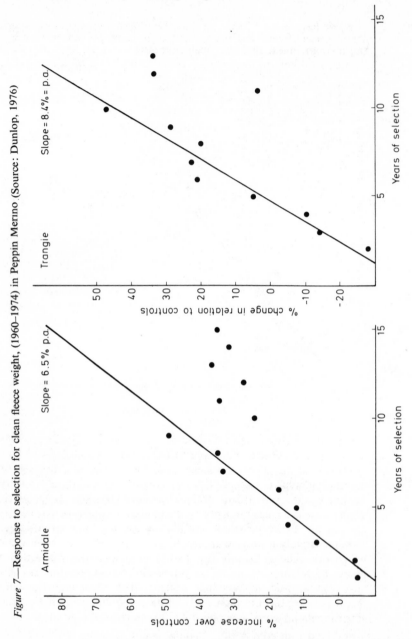

Figure 7—Response to selection for clean fleece weight, (1960–1974) in Peppin Merino (Source: Dunlop, 1976)

fibre prices and the exchange rate in recent years. Unfavourable influences leading to a depression of sheep numbers and diversification to other forms of production will also lead to a further reduction of cost inputs for wool production on sheep properties in Australia. Sheep studs will be disinclined to invest in objective fleece measurement and commercial properties are likely to increase the trend to breeding their own rams.

Figure 8—Response to selection for clean fleece weight (1952–1974) in Peppin Merino. (Source: Pattie & Barlow 1974 and unpublished data)

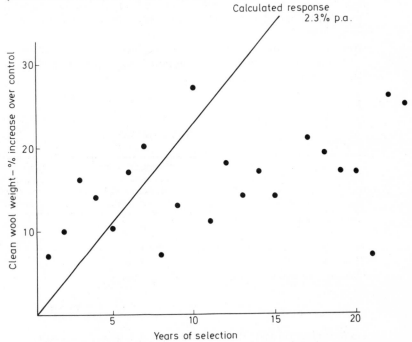

Unfavourable economic conditions for wool production also lead to a reduction of funds for wool research since in the past these funds have been largely supplied by a levy on wool income. However, assuming more optimistic projections for wool prices and wool production, one can speculate on possible research which might assist the wool industry to increase the genetic efficiency of wool production.

In the existing strains and family groups of the Australian Merino, different characteristics favourable for wool production are present to different degrees (Carter and Clarke 1957; Dun and Hayward 1962; Dunlop 1962, 1963; Jackson and Roberts 1970; Saville and Robards 1972). The South Australian Merino is outstanding in clean fleece weight associated with greater fibre diameter, staple length, and body size. It also has least skinfolds and face cover and has the greatest ability to increase wool production under good nutritional conditions. The Merryville, Egelabra and Haddon Rig family groups are outstanding in density of fibres per unit area, a characteristic not only favourable for increased wool production but probably also for exclusion of water, dirt and vegetable matter from the fleece. The Egelabra and Collinsville family groups are noted for improved colour of the fleece, probably reflecting a decreased suint content. Susceptibility to blowfly strike may be reduced by a low suint of the fleece and by increased density. Although data are limited there are suggestions that different strains vary in other important characteristics such as the maintenance energy requirements and susceptibility to infection by internal parasites. There has been a dearth of research on the interbreeding of strains to increase genetic diversity as a basis for more effective selection based on objective measurement although some general studs have already been established on this tenet.

TABLE 3

TREND IN WOOL PRICES: 1973 TO 1971 AND 1971 TO 1975
(CENTS PER KG)

Change	Change in trend price of wool between 1963 and 1971	Change in trend price of wool between 1971 and 1975
Growth in population and real income per head	+ 24	+ 14
Inflation and exchange rate movements	+ 20	+ 31
Changes in the supply of wool	− 17	+ 12
Changes in synthetic fibre prices	− 134	+ 51
Net effect	− 107	+ 108

Source: Dalton 1976.

The selection intensity possible for any characteristic is lowered by the number of characteristics for which selection is practised. Rather than select for all important characteristics simultaneously, general studs can introduce rams known to have particular characteristics developed well above the range in their flock and so achieve a rapid gain to a new level. In this respect research flocks in State Departments of Agriculture and CSIRO may be able to play a role in producing sheep superior in characteristics which are difficult for the stud breeder to measure such as resistance to internal parasites, high fertility and particular wool follicle characteristics.

Improvements in non-genetic aspects of sheep nutrition may reduce non-genetic variance in fleece characteristics and also provide conditions which enhance genetic variation. Thus it is known that Merino sheep provided with optimum diets are capable of producing clean fleece weights of 7-9 kg (Ferguson 1972, 1975) which compares with the present Australian trend average of 2.8 kg. Agronomists and plant breeders have seldom sought potential new pasture plants in terms of their wool promoting capacity and the principal feedstuffs used for supplementary feeding on sheep properties — hay, silage and cereal grains — are not capable of producing maximum rates of wool growth. Improvements in sown pastures and supplementary feedstuffs could thus provide nutritional conditions leading to the expression of increased genetic variation in wool characteristics. The development of supplements which provide increased amounts of cystine should also improve conditions for selection (Ferguson 1975).

I have painted a broad canvas and leave it to my colleagues to develop aspects in more detail. No doubt they will also express their disagreement with some of the generalizations I have made.

REFERENCES

Andrews, M.W. and Rottenbury, R.A. (1975) *Wool Technology and Sheep Breeding* **22** : 23.

Australian Bureau of Statistics (1976) Sheep Numbers, Shearing and Wool Production Forecast, 1975-76 Season. Ref. No. 10.56

Australian Wool Corporation (1973) Technical Report of the Australian Wool Board's Objective Measurement Policy Committee. Government Printer, Canberra.

Bureau of Agricultural Economics, Canberra (1973) Statistical Handbook of the Sheep and Wool Industry.

Bureau of Agricultural Economics, Canberra (1976) The Australian Sheep Industry Survey 1970-71 to 1972-73.

Carter, H.B. and Clarke, W.H. (1957) *Australian Journal of Agricultural Research* **8** : 91.

Connors, R.W. and Reid, R.N.D. (1976) Proceedings of Australian Society of Animal Production, Vol. XI, P.9

Dalton, M.E. (1976) Wood Situation and Outlook. Bureau of Agricultural Economics, Canberra. P.28.

Dun, R.B. and Hayward, L.T. (1962) Proceedings of Australian Society of Animal Production, Vol. IV. P.178.

Dunlop, A.A. (1962) *Australian Journal of Agricultural Research* **13** : 503.

Dunlop, A.A. (1963) *Australian Journal of Agricultural Research* **14** : 690.

Ferguson, K.A. (1972) Proceedings of Australian Society of Animal Production, Vol. IX : 314.

Ferguson, K.A. (1975) Proceedings of IV International Symposium on Ruminant Physiology, Sydney, August, 1974, P.448.

Gifford, R.M., Kalma, J.D., Aston, A.R., Millington, R.J. (1975) *Search* **6**(6) : 212.

Jackson, N. and Roberts, E.M. (1970) *Australian Journal of Agricultural Research* **21** : 815.

Nix, H.A. (1973) Proceedings Symposium on Land Use Planning in North Queensland. A.I.A.S., Innisfail, P.1.

Nix, H.A. (1975) "Climate and Rice". Proceedings of a Conference at International Rice Research Institute, Los Banos, Philippines.

Pattie, W.A. (1973) "The Pastoral Industries of Australia", Ed. G. Alexander and O.B. Williams, Chapter 10. Sydney University Press.

Pattie, W.A. and Barlow, R. (1974) *Australian Journal of Agricultural Reserach* **25** : 643.

Roberts, E.M., Jackson, N. and Phillips, J.M. (1975) *Wool Technology and Sheep Breeding* **22** : 6.

Saville, D.G. and Robards, G.E. (1972) *Australian Journal of Agricultural Research* **23** : 117.

Short, B.F. and Carter, H.B. (1955) CSIRO Bulletin No. 276.

Turner, H.N., Dolling, C.H.S. and Kennedy, J.F. (1968) *Australian Journal of Agricultural Research* **19** : 79.

Turner, H.N. and Young, S.S.Y. (1969) *Quantitative Genetics and Sheep Breeding*, Macmillan of Australia.

Waring, E.J. and Morris, J.G. (1974) *Quarterly Review of Agricultural Economics* **27** : 39.

3

SHEEP BREEDING IN FRANCE

P. MAULEON
I.N.R.A., Station de Physiologie de la Reproduction 37380, Nouzilly, France

SUMMARY

Sheep production in France is characterized by the dominant demand for a particular type of carcass (18-21 kg) of suckling lambs from a great variety of sheep farming methods. The flocks are small and there are numerous breeds. Many English blood lines have been introduced into the local hardy ewes giving seasonal variation in the possibilities of reproduction. So, the programmes of technical improvements are cross-breeding for meat production, increasing prolificacy by selection or hybridization with Romanov and Finnish breeds and the control of oestrus and ovulation.

INTRODUCTION

There are 7 million ewes aged more than 1 year in France. There has been a tendency towards a gradual increase in this figure since the Second World War (16.6% in the last 10 years).

This chapter will be divided into 3 parts: (1) Characteristics of sheep production; (2) The structure of flocks, zones of sheep production, physiological definitions of the breeds; (3) Important technical advances of recent years.

PRODUCTION

Meat represents 90% of the income from sheep farming. Production has fluctuated around 130,000 tonnes over the past few years, representing a slow but regular increase of 26% in 10 years. However, the production is not sufficient for the national consumption, which is continually increasing (3.3 kg/person) and the deficit is of the order of 60,000 tonnes, or 45% of the national production.

The type of carcass most in demand is that of suckling lambs which represent 77% of the total head slaughtered, and 67% of the net carcass weight; the remainder is composed of cast animals. Carcass weights of lambs slaughtered now range between 18-21 kg at $2\frac{1}{2}$ to $4\frac{1}{2}$ months throughout France, although previously those from the South were lighter, and those from the North, heavier.

Milk is produced by 800,000 dairy ewes. It is destined for the production of quality cheese of which 10% is exported.

Wool represents 4% of the receipts from sheep production. The interest in wool has diminished with decreasing prices.

There has appeared an overall demand for quality meat and milk for cheese, such that bidding prices for these products are high (3 times higher than world prices).

A solid economic organization of the production exists for the collection and sale of these products: almost all of the milk produced, 35% of the wool and 15-20% of the meat. This last being the most recent, continues to increase.

VARIOUS ASPECTS OF SHEEP BREEDING IN FRANCE

(a)   Structure of flocks and zones of production

The natural and economic areas of sheep farming are very varied. The number of farms where sheep are bred has decreased by 18% in 7 years, and the mean number of ewes per farm has increased from 30 to 40: 47.5% of ewes are found in 15,500 farms (9% of farms) where there are more than 100 ewe-mothers, 80% of farms have less than 50 mothers. French sheep farms are thus very small.

TABLE 1

DIVISION OF FLOCKS BY TYPE OF FARMING

	% Nos.
Indoor sheep-folds	41.6
Semi-out-door	33.4
Out-door	10.9
Transhumance	11.2
Diverse	2.9

The national sheep population tends to be concentrated in regions of traditional sheep farming. These are:

Mountain areas: the animals winter in folds, but have external grazing areas either next to the folds, or a long way away (transhumance). Flock management is based on the exploitation of poor pastures during the gestation period, when there is least need. As soon as nutritional requirements increase intensified areas are used, or feed reserves from spring. The lambs are produced in folds. Sheep are often the only possible form of production and although in these regions, flocks are in excess of 50 mothers, the number is rarely sufficient to ensure a reasonable income.

The mid-west is becoming a specialized area. The number of sheep has increased by 45% over the past 10 years. Sheep are bred entirely on pasture, together with grassland culture which is increasing, and recently the culture of maize has been begun. The intensification necessitates more productive livestock, nutritional supplementation of young on pasture and the destruction of parasites. In this region farmers seek to spread the sale of lambs over a maximum period in order to maintain prices.

The milking area (Causses du Sud—Massif Central) is one of the regions with the greatest density of sheep. The system of breeding is highly evolved, to allow milking machines to be used. Lambing takes place in December and lactating ewes benefit from food reserves grown in spring, mainly lucerne. In summer, the ewes use the extensive pastoral zone and the lambs, after early weaning are fattened in the neighbouring regions. In the Pyrénées Occidentales, where the climate is moist and mild, winter is spent on pasture and the animals move to the mountains in summer. The organization of selection for this production is very efficient.

Sheep breeding may also constitute a supplementary production in other areas:

The pastures of the Atlantic coast on the clay soil surrounding the North of the Massif Central are regions where sheep co-exist with cattle. The principle of this type of breeding is to satisfy the needs of the ewes with lambs at the time of abundant pasture growth from March to July and to sell the lambs before the dry period to leave pasture for the cattle. Flocks are small, usually less than 20 animals.

The industrial cereal area of the Parisian Basin have possibilities for intensive and abundant nutrition in autumn and winter. They produce lambs in the non-breeding season in folds. The possibilities for intensification are greatest, if qualified labour can be sufficiently well paid by a high productivity.

(b) Animal populations

Populations should be capable of producing a uniform type of lamb carcass (18-21 kg) from very different areas.

TABLE 2

PRINCIPAL FRENCH BREEDS

		% of total effective
1. *Group of littoral breeds*	Texel, Bleu du Maine, Contentin Avranchin	4
2. *Group of breeds highly crossed with English blood* (areas of extensive culture or pasture, area of mid-West)	Ile-de-France, Berrichon du Cher, Charmoise, Blackhead, Southdown	42
3. *Group highly crossed with Merinos* (areas of extensive culture or mountain, with transhumance)	Merinos précoce, Merinos d'Arles, Merinos de l'Est (wool)	10
4. *Group of rustic breeds*		
Mountain breeds	Limousine, Caussenardes, Préalpes Tarasconaise, Blanc de Lozère, Bizet, Rava	24
Milking breeds	Lacaune, Pyrénéenne, Corse	20

The origin of the breeds is very varied. There are a few imported breeds which have undergone selection towards a French type. Many breeds were created by hybridization between local breeds and English sires, above all Leicester. Others are local rustic breeds improved by selection. During the 1st Empire sheep populations underwent infusions of Merino blood.

A very great effort has been expended during recent years to determine the characteristics of reproduction, in particular seasonal variations in sexual activity.

TABLE 3

CHARACTERISTICS OF REPRODUCTION IN THE
PRINCIPAL FRENCH BREEDS

Breed	Mean date of the last oestrus of sexual season	Mean date of the first oestrus of sexual season
Ile-de-France	20 January	15 August
Préalpes	1 March	25 July
Romanov (Russian import)	20 February	20 September
Limousine	9 January	5 July

(From Thimonier 1975)

The periods of anoestrus are thus relatively long but the percentage of animals which go into deep anoestrus, with no ovarian activity is very variable. Ewes of the Ile-de-France breed have 2 months of the year during which 80-100% of females show total inactivity of the ovaries (Thimonier and Mauleon 1969).

PRINCIPAL NEW PROGRAMMES OF TECHNICAL IMPROVEMENTS

There is still a large scope for progress of sheep farming in France in classic techniques, principally in nutrition and disease problems. In fact it would be interesting to explain the motivation behind research programmes which may lead to modifications in sheep production.

(a) Butchering potential of "male" breeds used for large-scale cross-breeding for meat

Following a recent survey, it appears that 46% of farmers use at least one breed of male different from their female breed.

The increase in weight at slaughter of lambs is an objective imposed by the demand for heavier lambs. The French market demands that an increase in weight is not accompanied by an increase in fatty tissue.

Rams of the Berrichon du Cher breed are the most widely used of the rustic breeds in double crossings along with the Romanov.

There is no one breed which is absolutely better than another but genetic combinations may be better adapted to particular farming conditions.

(b)   Increase in prolificacy

More than one third of breeders consider that an increase in prolificacy has priority among their breeding problems.

On the average, every female aged more than 1 year only produces 1 lamb per year. It has been estimated that half the breeders could rapidly reach a production of 1.5 lambs per year.

The level of prolificacy varies by 100% among the French breeds, the most prolific being the littoral breeds.

Genetic improvement based on the control of performance has allowed an increase in growth potential. Existing structures are used for the collective organization of selection for prolificacy in the main French breeds.

However, the efficiency of selection remains limited and the search for new criteria means that the discrimination of prolific strains is a major research objective.

The Romanov and Finnish breeds constitute a genetic potential which could rapidly increase the productivity of flocks either by single or double crossing, or by creating synthetic strains. These breeds are beginning to be used in all systems of farming but the economic conclusions cannot yet be drawn and no strategy for their use has been defined.

(c)   Oestrus and ovulation control

(i)   Induction of pregnancy regardless of season

Since most of the French breeds have a period of anoestrus, this possibility is used for the following reasons:

to mate non-pregnant ewes during spring or autumn, or to have the major production during the non-breeding season: the latter is the most widely used, with excellent results of 70-80% fertility at a single induced oestrus, and 160-180% prolificacy.

an increase in lambing rhythm: the degree of intensification is limited to 3 lambings in 2 years, with 5 months of gestation, 2 months lactation and 1 month weaning followed by mating; that is, 8 months between 2 births.

a planned and regular lamb production enabling contracts for their sale to be organized in advance.

early reproduction in lambs: mating is spread over more than 1 ½ months in 70% of young females leading to poor preparation of females for lambing. Lambings are spread out, making their surveillance difficult, and leading to high mortality in the newborn lambs.

A first mating and reproduction at 8-9 months with 60-75% fertility is possible, but so far, this aspect remains limited to the milking sheep of the Roquefort region.

Introduction of lambs or ewe-lambs into the normal mating system to produce lambs for intensive breeding: young females are normally mated in August-September during the mating period of non-pregnant adults. But the spread of births in February-March interferes with the next, and principal mating period in April-May. By advancing and reducing the mating period of the young females, early lambing in January can be obtained with a second mating in April-May.

Extensive mating system for lambs: in open-air breeding, parturition occurs in February-March, even April and the new lambs are 7-8 months old in October-November, during the mating period; they are often insufficiently developed, and poor fertility results. If they can be mated during the non-breeding season, when they are 12 months old, this allows a second mating in autumn with the rest of the flock, giving higher fertility in the young females. This method is at present being extensively developed.

(ii)   Grouping of oestrus and lambing periods

In milking flocks, it is important to have homogeneous groups for weaning of lambs at 4 weeks, and a precise time for starting to milk the lactating females, since industrial dairies close at a fixed date.

In transhumant flocks, there should be no births in the mountain pastures from June to October, and the spring lambs should be old enough to follow the rest of the flock. Thus, mating must take place at a precise time, either in April-May, or October-November. During the former period, results are variable, but the technique of control allows them to be improved.

In flocks of sheep for milk or meat production, artificial insemination permits the development of selection programmes (large-scale crossing with prolific breeds) or diffusion of the best progenitors. This technique, without detection of oestrus, and perfect synchronization, is rapidly developing—actually about 200,000 animals. It is indispensable for this objective in small flocks.

(iii)   Increase in prolificacy

This is a consequence of the use of PMSG, which is administered systematically with all progestagen treatments. The increase in lambing rate is 40 additional lambs for every 100 ewes lambing.

In fact, this presupposes, in order not to have a consequent loss of lambs, that this treatment is done by a good breeder (previous lambing averages 130 lambs for 100 ewes lambing) and that he is well informed of the necessity for increased surveillance of births and of saving the lambs by artificial rearing.

CONCLUSION

In the six-country European Common Market, French sheep production represented more than two thirds of the community production. In the enlarged Europe, French production represents less than one third of the total production and follows a long way behind British production (230,000 t). The rules of the Community game have not yet been defined.

ACKNOWLEDGEMENT

The skilful assistance of Dr. Meredith Lemon for the English correction is gratefully acknowledged.

REFERENCES

Anonyme (1973) Bulletin de l'Institut Technique d'Elevage ovin et caprin, Paris.
Thimonier, J. (1975) *Journées de la Recherche ovine et caprine en France*, Paris, 18.
Thimonier, J. and Mauleon, P. (1969) *Annales de Biologie animale, Biochimie. Biophysique* **9** : 233.

4

NATIONAL SHEEP BREEDING PROGRAMMES NEW ZEALAND

A.L. RAE

Massey University, Palmerston North, New Zealand

SUMMARY

The first national sheep breeding programme established in New Zealand was the National Flock Recording Scheme which was started in 1967. In 1976, as a result of experience gained, a revised system called Sheeplan has been introduced. It contains a greater number of options to meet the requirements of different breeds. The traits recorded are: number of lambs born for each ewe, weaning weight of lambs, later liveweights at 6, 9 and 12 months of age, hogget fleece weight and other fleece traits. The breeder must record number of lambs born but may choose none or any combination of the remaining traits. The main outputs to the breeder are: (i) Two-tooth ewe and ram selection lists containing breeding values, a selection index and production information for the traits recorded; (ii) a summary of the performance of each ewe in the flock updated annually and (iii) summaries of the performance of the progeny of each sire used in the flock.

INTRODUCTION

Breeding plans formulated on a national scale can be thought of as encompassing two main ingredients. The first requirement is a recording service which has as its objective the measurement or assessment of the traits of economic importance on individual animals and along with this, the processing and presentation of the records in a way which will assist in identifying genetically superior animals. The second need is for a plan whereby the genetic merit of the selected individuals may be spread effectively through the population. In New Zealand, in common with many other countries, the main endeavour on a national scale has been devoted to the first requirement: the development of an effective recording scheme. This development is the main theme of this chapter.

SOME BACKGROUND INFORMATION

The way in which a flock recording scheme is organized and the traits recorded depend to a considerable degree on the structure of the sheep industry it is expected to serve.

The stratification of breeds in the sheep industry in New Zealand is essentially that of a two-tier system. The basic breeding-ewe breeds, such as

35

the New Zealand Romney, Perendale, Coopworth and Corriedale are maintained as pure-bred flocks breeding their own replacements on the hill or poorer pasture areas. Surplus young ewes and cast-for-age ewes from these flocks are bought by farmers on lower country and are mated to rams of the specialized meat breeds for prime lamb production, both the ewe and wether lambs being slaughtered. A large number of breeds of sire is available for this purpose including Southdown, Suffolk, Dorset Down, Poll Dorset, Dorset Horn, Hampshire, Border Leicester and breeds such as the South-Suffolk and South-Dorset Down which are derived from crosses with the Southdown.

Partly as a consequence of the two-tier stratification, improving reproductive rate is of high economic importance in the basic breeding-ewe breeds. Wool production is also important but with varying emphasis from breed to breed. It has been found convenient to classify the breeds into groups: fine apparel wool such as Merino and Corriedale, bulky apparel wool such as fine Perendale, general purpose wool such as Romney and Coopworth and specialty carpet wool such as the Drysdale. The selection criteria for these groups have been discussed by the New Zealand Society of Animal Production Study Group (1974).

In the specialized meat breeds, it is clear that high growth rate, low lamb mortality in crosses, reduced fat and increased muscling in the carcass are important requirements. There is some variation between breeds in the relative emphasis to be placed on these traits, the contrast being greatest between larger breeds such as the Suffolk and the smaller breeds such as the Southdown.

Several studies (quoted in Rae, 1964) have shown that the registered ram breeding flocks of the breeds studied (Romney, Southdown and Corriedale) form a typical hierarchical structure, with a small number of nucleus flocks supplying most of the sires for a larger group of multiplying flocks. Hence, a recording scheme has to be able to supply the needs of flocks of varying size and status in the traditional ram breeding structure as well as coping with the specialized requirements of co-operative breeding schemes.

FLOCK RECORDING SCHEMES IN NEW ZEALAND

In the past, several systems of record keeping were designed to assist breeders in recording the information required for pedigree purposes and for a variety of productive traits. Probably the most widely used in the period 1930-1950 was a card index system designed by Waters at Massey Agricultural College (Waters, 1939). This system or adaptations of it is still being used by many breeders. It is a satisfactory system for a small flock. Its disadvantages are the substantial amount of work required in keeping the cards up-to-date and the fact that none of the information is processed to assist in selection.

(a)   The National Flock Recording Scheme

The first sheep recording scheme to operate on a national scale in New Zealand was started in 1967 by the Department of Agriculture (now Ministry of Agriculture and Fisheries). It was introduced on the basis of a report prepared by E.A. Clarke and A.L. Rae and the scheme has been described in detail by Clarke (1967).

The scheme catered for two distinct classes of sheep: dual-purpose breeds for which reproductive rate and wool production are important selection objectives, and breeds producing sires for cross-bred prime lamb production where growth rate is of major importance.

In the option for dual-purpose breeds, the traits recorded were: number of lambs reared, weaning weight of the lamb, greasy fleece weights for both ram and ewe hoggets and, at the breeder's option, visual assessments of quality number, grade of the fleece and comments on fleece faults. The measure of reproductive rate used was weight of lamb weaned by the ewe at each lambing.

The inputs supplied by the breeder to the Flock Recording Office were: (i) a mating list giving tag numbers of the ewes mated to each sire; (ii) a lambing list giving tag numbers of lambs born to each ewe and their date of birth, sex, birth and rearing rank; (iii) a weaning weight list and (iv) a hogget shearing list for both ewe and ram hoggets in which fleece weights and fleece observations were recorded.

The most important output was the two-tooth ram and ewe selection lists which were designed to aid the breeder in his selection of replacement rams and ewes and to assist buyers in selecting rams. These lists, in numerical order of tag number, contained: (i) the sire and dam of each individual along with its birth and rearing rank; (ii) the individual's own records for weaning weight (adjusted for birth and rearing rank, age of dam, sex, and age at weaning), hogget fleece weight and fleece traits, and any coded remarks made by the breeder; (iii) a breeding value for weight of lamb weaned based on the average performance of the dam over all her available lambing; (iv) a selection index combining weight of lamb weaned and hogget fleece weight.

In addition to the selection lists, a ewe summary updated annually was produced. It included the lambing performance of each ewe in the flock and served as a basis for the culling of the ewe flock.

The prime lamb sire breed's option was based on recording weaning weight of the lamb which after adjustment for known non-genetic factors was returned to the breeder in a selection list.

At its peak in 1974, about 170,000 ewes from 630 flocks were entered in the National Flock Recording scheme. This represented about 30% of the stud ewes in the country and about 15% of the flocks. Experience with the procedures in practice, comment and criticism from breeders along with further information from research suggested that: (i) The scheme did not have sufficient flexibility to cope with the variety of needs of the different breeds within the two broad options and of the individual breeders within each breed; (ii) It was regarded by some breeders as being too complex and there appeared to be a need for a simple limited option; (iii) There was a strong demand from breeders for sire summaries, an aspect of the scheme which had been planned but never put into operation; (iv) Research had indicated the need to include a wider range of traits (particularly liveweights at various ages). The steps which were taken to initiate a re-examination of the scheme have been detailed by Wallace (1974) and Dalton and Callow (1975) and will not be recounted here. It is sufficient to state that many individual breeders, the New Zealand Federation of Livestock Breeding Groups, Farm Advisory Officers (Animal Husbandry) and Sheep and Beef Officers of the Ministry of Agriculture and Fisheries, research and university personnel have contributed to the formulation of the revised recording scheme called Sheeplan.

(b) Sheeplan

The traits recorded in Sheeplan are: (i) number of lambs born or reared for each lambing of a ewe; (ii) weaning weight; (iii) later weights of ewe and ram hoggets—liveweights taken in the autumn, winter and spring; (iv) hogget greasy fleece weight, and an assessment of quality number, or fibre diameter and a fleece grade along with remarks on faults and other aspects of the fleece.

Number of lambs born or reared must be recorded by all members of Sheeplan. The remaining traits are optional and each breeder may choose the combination of traits which suits his particular breed and the objectives of improvement in his flock. Thus, for a Romney flock, a common combination of traits would be number of lambs born, weaning weight, a winter liveweight for ram hoggets and a spring liveweight for ewe hoggets, fleece weights and other fleece information for ram and ewe hoggets. The simplest scheme (for which a demand was predicted but not yet realized) is the number of lambs born on its own. This supplies all the information required by the breeder to meet the requirement of the breed societies for registration and records at least the economically most important trait.

For breeders of prime lamb sire breeds, a common option would be, in addition to the compulsory number of lambs born, a weaning weight and an autumn liveweight.

(1) Inputs

The input lists supplied by the breeder are, with few exceptions, similar to those described for the National Flock Recording Service. The major difference is the elimination of the mating list and the introduction of fate lists to allow the breeder to enumerate the animals which have either left or have entered the flock. Lambing, weaning and hogget shearing lists follow the same general pattern as described earlier. Additional lists are required for each of the later liveweight options.

(2)   Outputs

The underlying principles which have been applied in organizing the information to be supplied to the breeder are:—

(i) That, in addition to the records of each animal adjusted, where necessary, for non-genetic effects, predicted breeding values for the important traits should be supplied.

(ii) That all the available information, including information from correlated traits should be used in predicting the breeding value of each trait. These breeding values may then be combined following the method given by Henderson (1963) by weighting by their relative economic values to give a selection index.

(iii) That provision be made in the processing of the records for differing sets of relative economic values, correction factors, genetic and phenotypic parameters to be used for each breed. Indeed, in the long run, there is no reason why each breeder should not have his own set of relative economic values for the traits he is recording.

(iv) That sufficient pedigree information be supplied so that the breeder does not have to maintain further records to meet the registration requirements of the breed societies.

Thus, the two-tooth selection lists contain, in addition to the sire and dam (and her lambing records), records of adjusted weaning weight, adjusted later liveweight and fleece weight all expressed as deviations from the means of all progeny in the flock of the same sex. They also present breeding values for number of lambs born (or reared), weaning weight, latest liveweight and hogget fleece weight if these have been recorded by the breeder. For number of lambs born (or reared), the primary information is the average number of lambs born (or reared) per lambing of the dam of the individual. The selection index found by using the appropriate set of relative economic values is also included.

As noted earlier, provision is being made for each breed or group of breeds to have its own set of genetic and phenotypic parameters for computing the breeding values of the traits.   At present, estimates of these parameters are available for the Romney, Southdown and Perendale breeds. As new information comes to hand, it will be used to update and expand the sets of parameters in use.

For the meat breeds, the selection lists supply breeding values for weaning weight and for autumn liveweight, in both cases using all the weights which the breeder has recorded to increase the accuracy of prediction of the breeding values.

The selection lists are prepared in numerical order of tag number but the breeder can request an additional list presented in descending order of the size of the selection index.

(3) Ewe Summary

This list is produced annually to assist the breeder in making selection decisions about the ewes in the flock. It gives for each ewe the sire, dam and sire of dam, fleece weight and fibre diameter, and the progeny records for each lambing. A breeding value for number of lambs born is also included.

Where the weaning weight option has been taken, two other breeding values are provided: (i) Mothering ability includes two components which indicate the ability of the ewe to rear her lambs successfully. They are lamb survival to weaning expressed as the ratio of number of lambs weaned to number of lambs born and average adjusted weaning weight of the lambs weaned. Selection index procedures have been used to combine these two traits into a score for each lambing of the ewe. These annual scores are then averaged over all the lambings which the ewe has had and this average is used to predict the breeding value of the ewe for mothering ability. When available, the average annual score for mothering ability of the dam of the ewe is also included in predicting the ewe's breeding value for this trait. (ii) Lamb production is an assessment of the total lamb production of the ewe for a particular lambing. The aggregate genotype involved is considered to be made up of the number of lambs weaned and their average adjusted weaning weight, the two traits being weighted by their relative economic values. In predicting this aggregate genotype, a selection index is constructed using number of lambs born, number of lambs weaned and the average adjusted weaning weights. As with mothering ability, this index is worked out for each lambing of the ewe, and the average of all the ewe's lambings is used to predict her breeding value for lamb production. Information on the lamb production of the dam of the ewe is also included in the prediction of breeding value.

A further output related to the Ewe Summary is the Closed Ewe File. This is put out annually and includes all ewes which have died or been culled. It contains the same information as the most recent Ewe Summary for each of these ewes.

(4) Sire Summary

Interpretation of the average performance of the progeny of the sires used in the flock gives rise to several difficulties. These arise because: (i) particularly with traits measured later in the life of the progeny (e.g., later liveweights and especially two-tooth lambing performance of the daughters) the data may be biased by the culling which has taken place; (ii) non-random matings of various types may have taken place in the flock so that the average genetic merit of ewes mated to the different sires may be different;

(iii) the numbers of progeny of the sires used may differ markedly.

The last of these difficulties is overcome by the usual techniques of predicting breeding values of sires. The effect of non-random mating would appear to be not of major significance and could be allowed for if necessary. The effect of selection is more difficult to account for and the techniques suggested by Henderson (1975) are being investigated. In the meantime, the procedure being adopted is to highlight in each sire summary the extent to which culling has taken place in the data.

Sire summaries in Sheeplan are produced at the option of the breeder and as a general principle are printed out at the time the particular input data on the progeny are being processed.

DISCUSSION

It should be noted that the sheep recording programmes which have been developed in New Zealand have been primarily directed towards genetic improvement in the sheep industry. Little attention has been given to recording as an aid to flock management. Consequently, the recording systems have emphasized the performance of the individual animal within the flock rather than the flock as a whole.

In addition, a considerable amount of emphasis has been placed on processing and presenting the records to assist the breeder in making his selection decisions. While this adds to the value of the records in increasing the effectiveness of selection, it requires an extra extension effort to ensure that the records are understood and used properly by both the breeder and the purchaser of rams. The importance of an adequate advisory service closely involved in the organization and operation of the scheme and assisting both the breeder and the buyer cannot be overemphasized.

Finally, New Zealand experience would stress the importance of a recording scheme remaining flexible so that it can accommodate the changing needs of breeders and take account of new research results. Since a great deal of information relevant to improving the effectiveness of the scheme will come from the analysis of records already collected, there is need for a continuing organization to undertake the work involved.

REFERENCES

Clarke, E.A. (1967). *Proceedings of the New Zealand Society of Animal Production* **27** : 29.
Dalton, D.C. and Callow, C. (1975). *Sheepfarming Annual* 1975: 187.
Henderson, C.R. (1963). In *"Statistical Genetics and Plant Breeding"*. (NAS-NRC 982) p. 141.
Henderson, C.R. (1975). *Biometrics* **31** : 432.
New Zealand Society of Animal Production Study Group (1974). *"Guidelines for wool production in New Zealand (Editorial Services Ltd: Wellington)."*
Rae, A.L. (1964). *Proceedings of the New Zealand Society of Animal Production* **24** : 111.
Wallace, L.R. (1974). *Sheepfarming Annual* 1974:150.
Waters, R. (1939). *Proceedings of the Eighth Annual Meeting of Sheep Farmers.* (Massey Agricultural College) 1939: 96.

5

SHEEP BREEDING PROGRAMMES IN SOUTH AFRICA

G.J. ERASMUS
S.A. Fleece Testing Centre, Middelburg, Cape, South Africa

SUMMARY

Facilities for breed improvement programmes based on measured performance were provided by the Department of Agricultural Technical Services in 1964. The two performance recording schemes provide free performance recording and fleece measurement facilities as well as an advisory and educational service to sheep breeders.

The limited participation in these schemes is largely attributed to the fact that the traditional stud breeding industry is often not compatible with selection based on measured performance. The greatest support for fleece measurement comes from a newly formed breed society for Merinos where selection is based on an index incorporating clean fleece weight, body weight, wrinkle score and fibre diameter. Between-flock evaluation is done by comparing home-bred progeny to the progeny of rams from a control flock mated to a random sample of ewes and indexes are corrected accordingly to make them comparable. The control flock is kept reasonably genetically stable so that breeding improvement can be measured.

Results of inter-flock progeny tests are given, illustrating that phenotypic differences between flocks provide no indication of genetic merit.

INTRODUCTION

It is known that sheep already existed at the Cape when the seafaring pioneers rounded the Cape of Good Hope towards the end of the 15th century. For many years the colonists farmed with the Hottentot or Cape fat-tailed, hairy sheep and it was only during the early part of the 19th century that the first exotic breed, the Merino obtained a foothold in the Colony. This was the result of the first organized breeding improvement programme for sheep in South Africa initiated by the local Government and embracing a planned back-crossing of the best and whitest indigenous sheep to Merino rams as reported by de Geus (1953). The Merino rapidly increased in popularity with subsequent importations from Germany, France, the United States and Australia to become the most important breed of sheep in South Africa.

Subsequent breeding programmes were aimed at improving local or imported breeds and the development of new breeds better suited to local conditions and demands. A system of upgrading to the new or "better" sheep featured prominently in most cases. A great deal of success has been achieved with these efforts—the Merino has, for instance, been improved considerably as a producer of fine wool; the imported German Merino has undergone such a metamorphosis that its name was recently changed to the "South African Mutton Merino" and well adapted local synthetic breeds of increasing popularity, notably the Dorper, Dormer and Döhne Merino have been developed. Most of this was achieved with little conscious use of modern population genetics. Measurement of production, one of the first steps in applying population genetics to animal improvement (Turner, 1964), was very seldom practised. As from 1964, however, the Department of Agricultural Technical Services began providing the necessary facilities for sheep breeding programmes based on measurement through the introduction of two national performance recording schemes for sheep.

AIMS OF RECORDING SCHEMES

The National Mutton Sheep Performance and Progeny Testing Scheme was introduced in 1964 to provide (a) objective measurements for selection and (b) the necessary data for calculating phenotypic and genetic parameters for the more important mutton breeds. The scheme operates as an on-the-farm recording scheme and corrected weight gain up to weaning and ewe production records are the main selection criteria (Campbell, 1974).

The South African Fleece Testing Centre was founded in 1965 to provide central fleece measurement facilities to woolled sheep breeders. The National Performance Testing Scheme for Woolled Sheep was launched in 1973 to provide a more intensive follow-up service to participating breeders. Traits recorded are fleece weight (greasy and clean), clean yield percentage, fibre diameter, crimps per 25 mm, staple length, body weight and birth status. Sheep are ranked for clean fleece weight and body weight. All fleece measurements are done free of charge.

Through both schemes the Department also provides a specialist extension service to the sheep industry. Short courses in the principles of selection are conducted regularly and the attendance at such a course is a prerequisite for participation in the woolled sheep scheme. The educational and advisory service offered through the schemes is perhaps their most important function as indicated by Turner (1973).

APPLICATION OF PERFORMANCE TESTING TO BREEDING PROGRAMMES

The traditional system of stud breeding in South Africa is often not compatible with scientific breeding programmes based on measured performance. In the case of Merinos this is amply borne out by the fact that only 63% of all registered Merino Stud breeders have thus far applied for membership to the performance testing scheme. The overall effect of the scheme on the Merino stud breeding industry cannot, therefore, be of real significance. Furthermore, selection on performance is often so diluted by consideration of other factors, that it becomes ineffective and most of the participating studs regularly purchase rams from the few parent studs where measured performance plays no role in selection. This situation, to a large extent, also applies to most of the other sheep breeds. The most important single group of participants in fleece measurement is the newly established "Breed Society for Performance Tested Merinos". This Society had a membership of only 34 in 1975 but had more than double as many fleece samples analysed as the 798 members of the "Merino Stud Breeders' Association" founded in 1937.

The "Breed Society for Performance Tested Merinos" was formed in 1972 by a group of 12 commercial wool producers who had previously closed their flocks and were selecting on measurement. Members now also include a few registered studs as well as the central flocks of the only two well established group breeding schemes for Merinos in the country. The Society was founded on the belief that:

(a) production improvement in the Merino industry in future will mainly depend on breeding improvement achieved in the studs are not merely on upgrading,

(b) the industry will only maintain a maximum rate of improvement if performance testing is applied and the data properly utilized and

(c) reasonable genetic differences most probably exist between studs but the differences in appearance and performance is largely due to differences in environmental factors and virtually no indication of actual genetic merit (van der Merwe, 1975).

The Society follows the following selection procedure: As a preliminary selection of young replacement material, all available animals are classed subjectively for breed characteristics and serious defects with not more than 10% being culled. For rams the following traits are measured at the age of 18 months: body weight in kilograms (X_1), clean fleece weight in kilograms (X_2), fibre diameter in microns (X_3) and wrinkle score (X_4) using the standard photographs of Turner et al. (1953). The values are incorporated in a selection index proposed by Poggenpoel and van der Merwe (1975) viz. $I = 1X_1 + 7X_2 - 1X_3 - 1X_4$. For ewes the traits body weight in kilograms (X_1), greasy fleece weight in kilograms (X_2) fibre diameter in microns (X_3) and wrinkle score (X_4) are measured. The index for the selection of ewes is: $I = 1X_1 + 4X_2 - 1X_3 - 1X_4$. Rams and ewes with the highest index values are then selected.

Differences in genetic merit between flocks is estimated by making use of a control flock run at the Tygerhoek Experimental Station. The basic flock from which this control was randomly taken can be regarded as having a considerable degree of genetic variability and an inbreeding coefficient of roughly zero (Heydenrych, 1975). It is attempted to keep the control flock genetically stable by randomly selecting 16 ram replacements in such a manner that each father is consistently represented by a son as sire. Rams are individually mated to ten ewes and replaced each year. Ewes are replaced by a second daughter to reach mating age. Where a parent does not, for some or other reason, provide a replacement, a replacement is lotted from the progeny of the other parents but no parent is allowed to contribute more than two members to the following generation of parents. Parent-progeny and brother-sister matings are avoided throughout.

For progeny testing, aimed at estimating inter-flock genetic merit, sets of 12 rams are selected from the control in such a manner that the mean value of each set for each of the four traits mentioned is as close as possible to the mean value of the basic ram population from which the sets are taken. A deviation of not more than two per cent is allowed. Such a set of rams is mated to a random sample of ewes of the breeder who simultaneously mates his own rams to a similar sample of ewes. The two ewe samples must be comparable in all respects. The progeny of the two groups are identified, reared together and evaluated at approximately two-tooth age using the selection index mentioned. It is endeavoured to evaluate a total of approximately 200 or more progeny with a minimum of 70 per group which can be of both sexes.

The difference in breeding value between the breeder's flock and the control flock, expressed as a percentage deviation from the control, is then calculated as follows:

Mean value of progeny from breeder's rams $= \overline{X}_E$

Mean value of progeny from control flock rams $= \overline{X}_K$

Difference between progeny of breeder's rams and control rams:

$$D = \overline{X}_E - \overline{X}_K$$

Difference in breeding value between breeder's flock and control flock:

$$T_D = 2D = 2(\overline{E}_E - \overline{E}_K)$$

Expected mean value of control flock in specific environment:

$$\hat{\overline{X}}_K = 2\overline{X}_K - \overline{X}_E$$

or

$$\hat{\overline{X}}_K = \overline{X}_E - 2D$$

Percentage deviation in breeding value of breeder's flock from control

$$= \frac{D}{\hat{\overline{X}}_K} \times \frac{100}{1}$$

The percentage deviation in breeding value for each specific trait can be calculated thus. It will be used to adjust the within-flock value of each individual trait measured as well as the combined selection index. The performance tests of all rams appearing in the society's catalogue will therefore provide a good indication of true genetic merit for each trait.

The control flock furthermore provides a practically stable gene pool from which samples can be drawn from time to time to measure selection progress in each flock. If a breeder repeats the control test after a number of years, his between-flock performance tests will be adjusted accordingly.

RESULTS OF BETWEEN-FLOCK TESTS

The first round of test matings have already yielded interesting results as illustrated in Table 1 for the property clean fleece weight at 12 months.

TABLE 1

RESULTS OF TEST MATINGS TO CONTROL FLOCK RAMS IN THREE UNRELATED FLOCKS (CLEAN FLEECE WEIGHT AT 12 MONTHS)

Flock	A	B	C
Environment	Winter rainfall pasture	Winter rainfall pasture	Semi-arid scrub
Number of progeny from own rams	88	172	163
Number of progeny from control flock rams	113	211	118
Average clean fleece weight of progeny from own rams (kg)	6.54	4.06	3.08
Difference in breeding value between own flock and control flock (kg) $= T_D$	+0.34	+0.40	+0.27
Expected average production of control flock in specific environment (kg)$= \overline{X}_E - 2D$	6.20	3.66	2.81
Deviation in breeding value from control (%)	5.48	10.93	9.61

These results clearly illustrate that phenotypic differences between flocks provide no indication of genetic merit. Flock A, with a phenotypic value more than twice as large as flock C, in actual fact has a lower breeding value for the trait recorded.

FUTURE DEVELOPMENTS

Although sheep breeding programmes in South Africa are still mainly practised on traditional lines, there is every reason to expect that this will eventually change, albeit not drastically. The provision of free performance testing services for sheep and the consequent formation of a new breed society which is a far cry from what has always been associated with such institutions, has heralded the change to a population approach to breeding.

Constant revision of breeding programmes, supported by research results, is essential. Of higher priority is, however, the bridging of the communication gap between breeder and geneticist. De Lange (1971) points out that, because science does not have all the answers, communication between breeder, animal scientist and geneticist is rather a case of sharing an attitude than sharing ideas. "The latter will, however, follow automatically if the former is achieved," he concludes. Bridging the gap in attitudes should be the immediate object of all persons and institutions concerned with sheep breeding, not only in South Africa, but in all parts of the world where sheep breeding has not yet fully shared in the opportunities offered by modern science.

ACKNOWLEDGEMENTS

I wish to thank the "Breed Society for Performance Tested Merinos" for making available the results of their between-flock progeny tests.

REFERENCES

Campbell, Q.P. (1974). D.Sc. Thesis, University of the Orange Free State.
De Geus, L.H. (1953). In: "South African Stockbreeder and Farmer" (Seal Publishing Co. : Johannesburg).
De Lange, A.O. (1971). South African Journal of Animal Science **1** : 193.
Heydenrych, H.J. (1975). *Ph.D. Thesis, University of Stellenbosch.*
Poggenpoel, D.G. and Van Der Merwe, C.A. (1975). *South African Journal of Animal Science* **5** : 249.
Turner, Helen N. (1964). *Proceedings of the Australian Society of Animal Production* **5** : 21.
Turner, Helen N. (1973). *Zeitschrift für Tierzüchtungh and Züchtungsbiologie* **90** : 278.
Turner, Helen N., Hayman, R.H., Riches, J.H., Roberts, N.F. and Wilson, L.T. (1953). CSIRO Division of Animal Health, Production Division, Report No. 4.
Van Der Merwe, C.A. (1975). Unpublished paper delivered to Merino Sheep and Wool experts' Association of South Africa.

6

SHEEP IN SOUTH-EAST ASIA

I.D. SMITH
University of Sydney, New South Wales, Australia

SUMMARY

There are approximately 4.6 million sheep in south-east Asia and of this population, over 90% are found in Indonesia, the majority in West Java. Burma has 180 thousand sheep and West Malaysia, Thailand and East Timor each have 40-45 thousand.

The sheep populations of Indonesia, West Malaysia and southern Thailand consist very largely of a single basic ecotype, generally described as the Kelantan breed in West Malaysia or the Priangan or native sheep in Java. These animals are of small body size and have short tails; only the rams are horned. They are reasonably well covered with a coarse carpet-wool fleece of variable colour and quality, the face always being bare.

In Indonesia, fat-tailed sheep are also seen, more particularly in East Java, Madura, Sulawesi and East Nusatenggara. They are usually white and larger than the short-tailed sheep, with a more restricted fleece covering.

The native sheep of central Burma are quite different from the sheep of Indonesia and the Malay Peninsula. They are larger, with long pendulous ears and a convex facial profile; both sexes are usually polled.

Although there is little information presently available concerning rates of reproduction, body growth or wool growth of sheep in south-east Asia, plans are in progress for the performance testing in Indonesia of selected indigenous ecotypes or breeds and a limited number of exotic breeds. The major emphasis in this area must be upon meat production, and improvement of fleece production is a much more doubtful proposition.

INTRODUCTION

There is scant information concerning the sheep of south-east Asia and the population is not large, totalling approximately 4.6 million (F.A.O. 1973). The majority of this population is in Java, Indonesia having approximately 3.5 million sheep, compared with 180 thousand in Burma, 40-45 thousand each in West Malaysia, Thailand and East Timor and 29 thousand in the Philippines; none of the remaining countries in the area has a sheep population greater than 13 thousand. Within Indonesia, the greatest concentration of sheep is in West Java, with 65-70% of the total population, followed by Central Java (including Jogjakarta) which has about 20%; other regions where sheep are found are East Java, Nusatenggara (the Lesser Sunda Islands), Sumatra and Sulawesi. Sheep are kept for use in traditional ceremonies, meat production and the value of their skin.

The only breeds from this area that have been described as such are the Priangan and Javanese fat-tailed sheep (kambing kibas), both from Java (Fischer 1955), and the Kelantan sheep from West Malaysia and southern Thailand (Smith and Clarke 1972).

This communication records preliminary observations on several sheep breeds (or ecotypes) from south-east Asia, particularly in relation to their fleece and wool follicle characteristics, and discusses the potential of these sheep and present sheep breeding programmes.

MATERIALS AND METHODS

Flocks of sheep were observed and inspected throughout West Malaysia and Java, and also in central Burma, southern Thailand and west Sumatra. Particular attention was paid to physical conformation and reproductive phenomena; where possible, the reproductive tracts of ewes slaughtered for mutton were examined. On the basis of current oestrous and lambing activity, together with the evidence of ovarian activity or foetal development in slaughtered ewes and the estimated age of lambs at foot, it was possible to make a valid assessment of ovarian activity throughout the year.

Skin and fleece samples were obtained from the mid-side region of adult sheep where conditions permitted. Sampling procedures and laboratory techniques for the skin samples closely followed those described by Clarke (1960), whilst laboratory measurements of the fleece samples were made by the methods described by Chapman (1960).

RESULTS

The Kelantan sheep of West Malaysia and southern Thailand, the Priangan sheep of West Java and the native sheep of Central and West Java, Jogjakarta and Sumatra are of very similar conformation, being of small body size with a straight facial profile and semi-pendulous ears of medium length; many of the sheep (5-35% within individual populations) have only vestigial ears (microtia). Tassels (or wattles) are sometimes present on the neck. The tail is short (usually less than 12 cm in length) and thin. The rams are horned and usually have a mane and ruff of coarse hair whilst the ewes are polled but occasionally have small scurs or horn buds. These sheep are reasonably well covered with a carpet-wool fleece of variable quality and colour, the face always being completely bare. Because of the ubiquitous presence of body lice (*Damalinia ovis*) and the resultant fleece damage, together with the almost total absence of shearing at any time other than immediately prior to slaughter, it was not possible to make any valid estimate of fleece weight—however it would appear not to be greater than about 1 kg. Bodyweight is 30-50 kg for mature rams and 20-40 kg for mature ewes. Ewes are very prolific, often lambing twice within the year and having a high incidence of multiple births. There seems to be no well-defined period of seasonal anoestrus.

In East Java, and, to a lesser extent, Central Java, fat-tailed sheep (kambing kibas) are found: these are usually short-haired and white, usually with a small amount of wool over the back and mid-sides. They are generally larger than the short-tailed sheep described above and have pendulous ears and a straight or S-shaped fat tail. Sheep of this type are also found throughout Sulawesi, Madura and Nusatenggara.

Both the short-tailed and fat-tailed sheep of the Malay Peninsula and Indonesia are usually raised in small flocks of 5-10 animals. During the day they may be tethered or allowed to roam freely, grazing by roadsides, streams and between rice fields. At night, they are usually herded into lightweight bamboo enclosures, with the floor about 60 cm or more above the ground. Food supplements are rarely provided, except in the case of the short-tailed sheep of the Garut region in West Java, which are trained for ram fights.

The mature sheep of central Burma are also larger than the short-tailed sheep of West Malaysia and Indonesia, having long pendulous ears and a definitely convex facial profile: both sexes are usually polled.

The results of the various measurements of wool follicle populations are shown in Table 1.

TABLE 1

WOOL FOLLICLE CHARACTERISTICS (MEAN±S.E.) OF SOME SOUTH-EAST ASIAN SHEEP BREEDS

Breed and location	Native sheep, Jogjakarta, Java			Kelantan sheep, West Malaysia		
Number of sheep sampled	5			53		
Follicles/mm^2	8.4	±	0.4	6.2	±	0.3
Primary follicles/mm^2	3.2		0.1	3.0		0.1
S/P ratio	1.7		0.1	1.1		0.1
Fibre thickness ratio	3.3		0.4	2.6		0.1
Mean fibre diameter (μ)	40.3		1.9	49.6		1.5
% Medullation	32.8		4.8	41.8		2.3

Table 2 shows fleece measurements in sheep from three south-east Asian countries. The fibre diameters and percentage medullation of the Javanese and Kelantan fleece samples were similar to the values obtained from skin sections.

The mean fibre diameter of the Burmese fleece samples was significantly greater than that of the Kelantan (P< 0.01) and both groups of Javanese fleece samples (P< 0.001). Percentage medullation was greatest in the fleece samples from the Burmese sheep and was significantly greater than in the Javanese samples.

TABLE 2

FLEECE MEASUREMENTS (MEAN±S.E.) IN SOME SOUTH-EAST ASIAN SHEEP BREEDS

Breed and Location	Number of specimens	Staple length (cm)	Fibre diameter (μ)	Percentage medullation
Native sheep (Central Java)	20	6.8 ± 0.5	39.6 ± 2.0	42.6 ± 3.4
Priangan (West Java)	5	4.4 0.8	35.0 5.0	32.5 7.0
Kelantan (West Malaysia)	4	4.5 0.7	47.3 3.4	71.9 5.0
Native sheep (Central Burma)	16	3.9 0.1	66.7 5.7	82.1 2.9

The fleece samples of the Priangan and Javanese native sheep did not differ significantly in either mean fibre diameter or percentage medullation.

DISCUSSION

The indigenous sheep population of Java, Sumatra, West Malaysia and southern Thailand appears to consist of a single basic ecotype, described as the Priangan in West Java (Fischer 1955) or the Kelantan in West Malaysia and southern Thailand (Smith and Clarke 1972). The sheep of Java, however, show greater variability than those of the Malay Peninsula or Sumatra. The Priangan is nominally descended from crosses of Australian Merino and Cape fat-tailed sheep with the small, short-tailed native sheep of Java (van der Planck 1949; Fischer 1955), however the present evidence of such infusion is minimal.

Both van der Planck (1949) and Fischer (1955) have described fat-tailed sheep (kambing kibas) as being more prevalent in East Java, Madura, Sulawesi and East Nusatenggara. The larger size and white colour of these sheep was also remarked upon by Fischer (1955) who suggested that they may have been originally introduced by Arab traders: they might just as well owe their origin to the Cape fat-tailed (Africaner) sheep of South Africa, which were introduced into the East Indies by the Dutch.

The fleece of the Javanese sheep (both the Priangan and native sheep) carries less medullated and more fine fibres than that of the Kelantan sheep and this could be interpreted as being due to the infusion of Merino blood. The mean fibre diameter and percentage medullation of the fleece from Kelantan sheep was as previously reported by Smith and Clarke (1972) and Ryder (1974).

The mean fibre diameter of the central Burmese sheep was similar to that previously reported by Lang (1958) and clearly these sheep have little in common with those of southern Thailand, West Malaysia, Sumatra or Java, differing both in fleece type and body conformation. In the past, there have been introductions of Indian sheep into Burma and this has had an obvious effect upon the the conformation on most of the native sheep.

Although the fleece of the Burmese native sheep shows little potential for utilization, the fleece of the Javanese sheep, in particular, appears to be suitable for carpet-wool production, although lacking in staple length. It is, in fact, used for this purpose on a small scale in West Java, where there is a carpet factory at Bogor, however the sheep are rarely shorn except at slaughter.

In recent years, Corriedale sheep have been introduced to central Burma and Suffolk and Dorset Horn sheep to Java. However, critical information concerning the performance of such exotic breeds relative to that of indigenous ecotypes under controlled conditions is lacking.

Since there is a limited domestic and export market for wool of the quality that can be produced and since the fleece may, in fact, be detrimental to the well-being of sheep in this environment there is little current interest in development of wool production. The main features desired in southeast Asian sheep, which provide meat that is acceptable to the majority of the population, with no religious restrictions, and which are ideally suited to smallholder production systems, are prolificacy, high growth rate and disease resistance.

The prolificacy of the short-tailed sheep of the Malay Peninsula has been described by Smith and Clarke (1972) and, in Indonesian breeding programmes, the main concern will be to preserve, and if possible, enhance this prolificacy.

The main purpose in introducing exotic breeds is to increase growth rate and, presumably, mature body size. So far, Suffolk and Dorset Horn sheep have been used, but other breeds considered are the Wiltshire Horn, Dorset Down, Dorper, Awassi and Mandya. Concurrent with the performance testing of any such introduced exotic breeds, and their crosses, selected populations of Priangan, fat-tailed and other indigenous ecotypes will be evaluated. All of the above breeds have extended breeding seasons (9 months or more per annum) and have a significant incidence of multiple births, so that the use of such breeds to improve meat qualities would not be to the detriment of prolificacy.

At present there is considerable interest in Java in sheep breeding programmes, but until there has been a preliminary assessment of the productive potential of indigenous ecotypes under improved management conditions, it would be presumptuous to define or suggest any programme using introduced breeds.

REFERENCES

Chapman, R.E. (1960) CSIRO Animal Research Laboratories Technical Paper No. 3, Appendix 111, p.97.

Clarke, W.H. (1960) CSIRO Animal Research Laboratories Technical Paper No. 3, Appendix 111, p. 92.

F.A.O. (1973) Production Yearbook **26** : 190.

Fischer, H. (1955) *Tierzüchter* **7** : 311.

Lang, W.R. (1958) *Textile Research Journal* **28** : 90.

Planck, G.M. van der (1949) *Tijdschrift voor diergeneeskunde* **74** : 559.

Ryder, M.L. (1974) *Journal of the Textile Institute* 65 : 13.

Smith, I.D. and Clarke, W.H. (1972) *Australian Journal of Experimental Agriculture and Animal Husbandry* **12** : 479.

7

SHEEP BREEDING IN THE U.S.A.

W.J. BOYLAN
University of Minnesota, U.S.A.

SUMMARY

A decline of 45% in the number of breeding ewes has occurred in the last 10 years. This has had a major impact on many aspects of the industry, including breeding. A majority of ewes are Rambouillet (Merino) or related breeds. The Suffolk, Hampshire and other British breeds are also important in numbers, especially in farm flocks. There are few organized breeding programmes and their impact is small. Much of the effort for breed improvement by pure-bred breeders is focused on the show ring.

An organization of sheep producers, called Sheep Industry Development Programme (SID), has an active programme to expand the sheep population and production. SID activities have included organizing educational programmes for producers and developing innovative and pilot production programmes.

The major effort in sheep breeding research is conducted in the North Central regional research project. This is a co-ordinated effort of University and USDA experiment stations involving activities in several disciplines, various environments, geographical areas and management systems. Much of the activity in breeding is focused on breed evaluation and utilization. This research has been motivated by increasing interest in more intensive types of production, increasing efficiency of production and evaluation of the Finnsheep.

INTRODUCTION

The sheep industry of the U.S. has experienced a drastic decline in numbers of breeding ewes since World War II. This decline has been accelerated in recent years. Total number of breeding ewes was 9,334,000 on January 1, 1976 (USDA, 1976). This reflects a decline of 7% since 1975 and 45% in the last 10 years. Some notable changes in the geographical distribution of the sheep population have also occured concomitantly with the general decline. A substantial portion of the decline has taken place in the large flocks in the western range areas and proportionately less in the farm flock states of the corn belt or midwest. A change has also been noted in the relative amount of income from wool *v.* lamb. Total income from wool and lamb has changed from a proportion of 25% from wool and 75% from lamb, 15 years ago, to 17% from wool and 83% from lamb today. The decline in sheep numbers has had a major impact on breeding programmes and many other facets of the industry, including marketing, labour, predator control, management and nutrition.

BREEDS AND BREEDING PROGRAMMES

Accurate and detailed information on the breed composition of commercial flocks and their distribution throughout the U.S. is not known. It has been estimated (Whiteman, 1968) that most of the commercial ewes in the western range country carry some fine wool breeding, namely Rambouillet (or Merino). These flocks may consist of grade Rambouillet or related breeds such as the Columbia, Targhee and Panama. In other areas and also to some extent in the west, many flocks consist of blackface cross-bred ewes produced by mating fine wool type ewes to Hampshire or Suffolk breed rams. Many commercial farm flocks consist of grade Hampshire or Suffolk breed ewes. Other breeds employed to a lesser extent, but in some areas an important segment of the population, are the Cheviot, Corriedale, Dorset, Lincoln, Montadale, Oxford, Romney, Shropshire and Southdown. The largest number of registrations of pedigree animals in 1975 was the Suffolk breed (46,970) followed by the Hampshire (17,157) and Rambouillet (10,041).

There are very few organized breeding programmes and the impact of those that do exist is likely very small. Pure-bred societies have breed improvement programmes, but the extent of breeder participation is variable and often limited. It has been estimated that only 2% of all pure-bred flocks were enrolled in a performance testing programme (Parker, 1968). The principal medium of "breed improvement" is through the show ring and through pedigree selection of related individuals based on show ring standings. Government support for breed improvement of economic traits, such as, lamb growth rate, weaning weight, carcass attributes or ewe productivity has been very limited. Support has usually been in the form of educational programmes. Furthermore, the structure of the marketing system has not usually provided a direct incentive to pure-bred breeders to improve performance or carcass traits within breeds. However, market values have been important for shifts in selection between breeds. Cross-breeding to produce market lambs for slaughter has been well accepted, but much of the effort is haphazard.

SHEEP INDUSTRY DEVELOPMENT PROGRAMME (SID)

Despite the decline in sheep numbers there has developed in recent years a more concerned and concentrated effort on the part of the sheep industry to encourage and accept technological improvements in various aspects of sheep production, including breeding. In some sense a new era of optimism and renewal has developed in the industry. Much of this effort is focused in the activities of the Sheep Industry Development Programme (SID), a production oriented programme of the American Sheep Producers' Council. This organization of sheep producers and lamb feeders has directed its primary attention to marketing and to promoting lamb and wool. The activities of SID have been instrumental in organizing various educational programmes for producers, including symposia on breeding and genetics. SID has also been actively involved in developing innovative and pilot programmes in various phases of sheep production in co-operative efforts with producers. A recent development in the industry, initiated by SID, is the "Blueprint for Expansion Programme". This programme is proposed as a concerted effort to expand sheep production in the U.S. Among the significant goals of this programme is the aim to expand the total amount (weight) of lamb marketed, by 50%, in the next 10 years. Achievement of this goal is proposed by a 10% increase in lamb weight at slaughter (following a consistent trend of 1% annual increase), by a 10% or more increase in number of lambs marketed (also consistent with present trends and use of more prolific breeds such as Finnsheep) and by a 25 to 30% increase in breeding ewe numbers. Other specific programmes related to expansion are concerned with marketing, predator control, environment protection and land use planning, labour and management training programmes, research expansion and co-ordination, animal health and industry organizational structure. The goals for ex-panded production are accepted as realistic by the industry and the general programme has wide acceptance. However, in view of recent declines in the industry much effort will be required for this programme to succeed. Record high prices paid for lambs in 1976 will provide a basic incentive for expansion.

REGIONAL RESEARCH IN BREEDING

Research in sheep breeding is conducted at State (Universities) and Fed-eral (U.S. Department of Agriculture) experimental stations. The research activities of the various state experimental stations are co-ordinated on a regional basis. At the present time there are two regional projects. The prin-cipal breeding effort is in the North Central regional project (NC-111). This effort is composed of contributing projects to the general project title of "Increased efficiency of lamb production". The general objective of this co-ordinated research is to "develop and evaluate systems for more efficient lamb production by combining breeding, nutrition, physiology and man-agement procedures, under varying geographical and environmental condi-tions". Specific objectives are:

1. Develop effective short term methods for utilizing genetic variation in fertility, breeding season, lambing rate, lamb survival, growth and carcass traits under various management systems.

2. Develop and evaluate accelerated lambing systems suited to available genetic stocks.

3. Establish the nutrient needs of sheep under various management condi-tions.

4. Determine land, labour, housing, equipment and health requirements

for lamb production under various management systems. Cost analyses will be included where appropriate.

A brief description of each of the contributing projects of the various state and federal stations to NC-111 is given in tabular form in Table 1. It should be noted that emphasis varies among stations in the kinds of contributing projects. This is noted in Table 1 by reference to the specific objectives listed above (1 to 4) for each station.

A predecessor to the present NC-111 regional project was the North Central regional technical committee on sheep breeding (NC-50). Results of that study have been summarized in part in a regional publication (North Central Regional Publication 198, 1970).

TABLE 1

CONTRIBUTING PROJECTS TO REGIONAL RESEARCH IN SHEEP BREEDING AND PRODUCTION (NC-111)

State	Project — Investigations	Contribution to Objective Number*
California	Long term selection for weaning weight and multiple births in Targhees; evaluate crosses of Targhee and Suffolk with Finnsheep, management systems, drylot *v.* range	(1,3)
Idaho	Evaluate Panama and Panama X Finnsheep, Suffolk terminal sire breed, growth, carcass merit, nutritional requirements, energy levels	(1,3,4)
Indiana	Finnsheep crosses with Columbia, Dorset and Rambouillet, terminal sire breeds, Suffolk, Hampshire, Shropshire, Southdown; intensive production system, accelerated lambing, nutrition, management	(1,2,3,4)
Kansas (Colby)	Hampshire × Rambouillet, Dorset × Rambouillet and Rambouillet ewes, drylot *v.* year-around grazing, accelerated lambing, nutrition, housing, equipment, two management systems	(2,3,4)
Kansas (Manhattan)	Forage utilization, body composition, growth efficiency, environmental effects and performance, intensified production system	(2,4)
Kentucky	Nutrient utilization early weaned lamb, forage use, ewe breed and body size related to productivity	(3,4)
Minnesota	Evaluation of Finnsheep with crosses of Targhee and Suffolk, F_1, F_2 and backcrosses, terminal sire breeds Dorset, Hampshire, Oxford, Suffolk; heterosis and recombination effects, breed development, ewe productivity indices, longevity, lamb growth, efficiency, physiology, out-of-season breeding, management, nutrition, intensive lamb production, forage utilization, buildings, environment	(1,3,4)
Missouri	Evaluate systems of production, accelerated, modified and conventional lambing, Finnsheep, Dorset, Rambouillet cross-bred ewes, nutrition, management, equipment	(3,4)

TABLE 1 *(continued)*

State	Project — Investigations	Contribution to Objective Number*
Oklahoma	Five cross-bred ewe combinations of Finnsheep, Dorset and Rambouillet under intensive production, accelerated lambing, terminal sire breeds are Hampshire, Suffolk and Hampshire × Suffolk cross, management	(1,2,4)
Ohio	Management systems, housing, accelerated lambing, nutrition, use of Finnsheep genetic material, confined *v.* non-confined production, forage utilization, labour requirements	(1,2,3,4)
Oregon	Breed of sire effects on F_1 growth, carcass and maternal performance, eight cross-bred groups (Romney, North Country Cheviot, Dorset, Finnsheep, Suffolk, White face), hill *v.* irrigated pasture, management systems, measure appetite, feed efficiency, growth	(1,3,4)
Texas	Develop and evaluate methods to improve reproduction efficiency, accelerated lambing, endocrine stimulation, selection, management, disease. Finnsheep crosses *v.* Rambouillets	(1,2,4)
Illinois	Management, housing—confinement *v.* pastoral, nutrition, breeds and cross-breds, accelerated lambing, synchronization of oestrus, early weaning	(1,2,3,4)
North Dakota	Confinement systems of production, nutrition, energy levels, maintenance, breed evaluation for confinement management, Hampshire, Suffolk, Cheviot, Columbia	(2,3)
USDA—U.S. Meat Animal Research Center, Clay Center, Nebraska	Identification and development of superior genetic resources, utilization of genetic resources, selection between/within breeds, development of breeding programmes, hormone and nutritional investigations. Develop, evaluate, facilities, equipment, procedures to minimize labour requirements. Develop production strategies, evaluate three production systems with varying management intensities Germ plasm evaluation: reproduction, growth, carcass performance using Suffolk, Hampshire, Dorset, Rambouillet, Targhee, Corriedale and Finnsheep Germ plasm utilization: evaluate individual and maternal heterosis, recombination loss in Suffolk × Hampshire, F_1, F_2, F_3 and backcrosses, develoment of multibreed synthetic combination using (1) Finn, Rambouillet and Dorset, and (2) Finn, Suffolk and Targhee Maternal breed selection criteria: develop and evaluate selection criteria; develop superior germ plasm resources for industry use Maternal breed selection criteria: develop and evaluate selection criteria for heavy market weight lambs; selection in Suffolks Nutrition, ewes and lambs. Out-of-season breeding ewes and rams. Heavy lamb production, carcass attributes, lamb carcass composition and palatability Livestock engineering, shelter, predator control, flock health, disease, management Management systems: models of production, varying management intensity	(1,2,3,4)

TABLE 1 (*continued*)

State	Project — Investigations	Contribution to Objective Number*
USDA Agricultural Research Center Beltsville, Maryland	Nutrition, management, accelerated lambing, utilization of Finnsheep combination with domestic breeds, development of Morlam synthetic for year-around lambing	(1,2,3)
USDA U.S. Sheep Experiment Station Dubois, Idaho	Develop methods to maximize total pounds of market lamb/ewe in the flock. Accelerated lambing, selection, nutrition, management systems for orphan lambs, evaluation of Finnsheep crosses in a range environment, marketing lambs at heavier weights, study death losses, predator control	(1,2,3,4)

* Contribution to specific objective number, 1, 2, 3 or 4 of the regional project (see text).

A second regional project is in the western region and the project title is "Increased efficiency in marketing lamb and mutton" (W-137). The objectives of this project are:

1. To evaluate the marketing potential of heavy lambs (including intact males) as compared to current market lambs.

2. To determine ways of utilizing lamb and mutton to increase consumer acceptance.

A brief description of each of the various state experimental stations contributing projects to the western regional project is presented in Table 2.

Much of the current research in sheep breeding is focused on breed evaluation and utilization. This research has been motivated in part by an increasing interest in the industry in more intensive forms of lamb production, improving efficiency of production and the use of Finnish Landrace breed (Finnsheep). A major stimulus to research in sheep breeding came with the importation of the Finnsheep breed to the U.S. in 1968. The first importation came from Ireland and was a joint effort of the University of Minnesota and the U.S. Department of Agriculture. This breed is widely recognized for its prolificacy and the primary reason for the importation was to assess its genetic potential for increasing productivity of domestic flocks (Oltenacu and Boylan, 1976a).

MINNESOTA RESEARCH IN BREEDING

(a) Finnsheep evaluation and use

Research at Minnesota with the Finnsheep was designed to assess its potential genetic contribution in a series of cross combinations with domestic standard breeds. Finnsheep rams were mated to ewes of the Minnesota 100, Suffolk and Targhee breeds. F_1, F_2 and various back-cross progeny were produced.

The Finnsheep reached sexual maturity at a younger age than domestic standard breeds. About 95% of the Finnsheep ewes produced their first

TABLE 2

CONTRIBUTING PROJECTS TO REGIONAL RESEARCH
IN MARKETING LAMB AND MUTTON (W-137)

State	Project — Investigations	Contribution to Objective Number*
California	Carcass attributes of heavy lambs, efficiency of production, slaughter lambs at various weights from selected and control lines of Targhee, Finnsheep crosses; Suffolk is terminal sire breed	(1)
Idaho	Use of beef and pork fat in processed lamb and mutton products, taste panels, consumer evaluation, major emphasis to develop lamb and mutton products, Evaluate production costs, carcass quality of male lambs, heavy carcasses, Finnsheep cross carcasses	(1,2)
Oregon	Economic study—price response function, marketing lambs	(1)
Utah	Evaluate ram and ewe lamb carcasses, Targhee, Columbia and Rambouillet breeds, quality, composition, taste panel analyses, flavour, texture of light v. heavy carcasses. Develop cured meat products from deboned mutton, mechanical deboning evaluated	(1,2)
Washington	Evaluate heavy v.conventional weight carcasses, costs of production and returns, various breed combinations including Finnsheep. Study use of antioxidants on lamb and mutton flavour	(1,2)
Wyoming	Selection in Columbia and Suffolk based on cutability at 73 kg liveweight. Study physical, chemical and nutritional composition of mechanically deboned meats, emphasis on flavour	(1,2)

* Contribution to specific objective number, 1 or 2 of the regional project (see text).

lambs at 12 months of age v. only 51% of the Targhee ewes. The earlier sexual development of the Finnsheep was also apparent in Finnsheep cross-breds. About 92% of F_1 ewe lambs produced lambs at 12 months of age (Table 3).

Dickerson and Laster (1975) observed that the percent of female lambs reaching puberty at a specified date was far higher for Finn-sired (72%) than for Rambouillet-sired (38%) crosses or pure-breds (34%) of Suffolk, Hampshire, Dorset, Targhee, Corriedale or Coarse Wool breeds.

Mature Finnsheep ewes had markedly larger litters (3.2) at lambing than pure-bred Minnesota 100, Suffolk or Targhee ewes (1.4). F_1 and F_2 ewes also exceeded the pure-breds (Table 3). Meyer and Bradford (1973) reported similar results for litter size for Finn-Targhee cross-bred and Targhee ewes, 2.2 v. 1.5, respectively.

Although Finnsheep sired lambs (F_1) were smaller than standard breeds, 3.3 v 3.6 kg, at birth their vigor and survival to weaning was remarkably good. Finnsheep sired lambs had a survival rate of 84% *v.* 80% for the average of Minnesota 100, Suffolk and Targhee (Oltenacu and Boylan, 1976b). Dickerson, Glimp and Laster (1975) reported that preweaning death loss of Finnsheep sired lambs was only about one half as great as Rambouillet sired lambs (19 *v.* 39%).

Finnsheep F_1 cross-bred lambs had good early growth rates. They compared favourably in body size at weaning at 70 days to the average of the Minn 100, Suffolk and Targhee breeds (17.8 *v.* 17.4 kg). Only the straightbred Suffolk lambs exceeded Finn × Suffolk lambs in weaning weight (21.04 *v.* 19.78 kg). Dickerson, Glimp and Gregory (1975) reported favourable early growth of Finn-cross lambs compared to crosses of several U.S. breeds. When litter size and lamb weaning weights were combined into an index of ewe productivity Finsheep cross ewes, F_1, F_2 and back-crosses, exceeded the performance of the standard domestic breed ewes (Oltenacu and Boylan, 1976c).

Carcass merit of Finnsheep F_1 lambs was slightly less than pure-bred Minnesota 100, Suffolk and Targhee breeds, but still commercially acceptable

TABLE 3

LEAST SQUARES MEANS FOR PERCENT OF EWES LAMBING AT 12 MONTHS OF AGE OF THOSE JOINED TO RAMS AND LITTER SIZE OF EWES LAMBING

Breed of ewe	% lambing at 12 months of age	Litter size 1 year ewe	Litter size Mature ewe
Finnsheep	95± 4	1.7	3.2
Minn 100 (M)	75± 4	0.9	1.2
Suffolk (S)	90± 4	1.2	1.5
Targhee (T)	51± 5	1.1	1.4
Average (Standard breeds)	72± 2	1.1	1.4
F×M	94± 3	1.3	2.0
F×S	87± 4	1.4	2.1
F×T	95± 4	1.4	2.1
Average (F_1)	92± 2	1.4	2.1
F_2 (F×M)	90± 6	1.4	2.2
F_2 (F×S)	82± 7	1.2	2.0
F_2 (F×T)	91± 7	1.2	—
Average (F_2)	88± 4	1.3	2.1
F×(F×M)	95± 6	1.7	2.2
F×(F×S)	97± 6	1.7	2.1
F×(F×T)	85±10	1.3	—
Average (Back-cross to Finn)	92± 4	1.6	2.2
M×(F×M)	75± 7	1.0	—
S×(F×S)	93± 7	1.4	—
T×(F×T)	69± 8	1.2	—
Average (Back-cross to Standard)	79± 4	1.2	—

(Table 4). Finnsheep cross-bred lamb carcasses generally had less external fat than carcasses from domestic breeds. However, significant differences were observed in amount of internal fat. Finnsheep cross-breds had more kidney and pelvic fat than domestic breeds, 4.3 *v.* 3.1% (Table 4). A larger amount of kidney and pelvic fat in ¼ Finn-Targhee cross-bred lamb carcasses compared with Targhees was also noted by Rattray *et al.* (1973). Details of lamb carcass merit of various Finnsheep cross-breds is reported elsewhere (Boylan, Berger and Allen, 1976a). It has also been observed that the palatability of Finn sheep cross-bred lamb carcasses may be enhanced, compared with domestic breed lambs, because of a greater amount of unsaturated to saturated fats, thus reducing the "tallowy" effect sometimes observed in lamb fat (Boylan, Berger and Allen, 1976b).

Results from research suggest that the Finnsheep breed can be effectively used in sheep breeding programmes to achieve a significant increase in productivity Optimum use will depend on the management system employed. Many producers should benefit by using cross-bred ewes of ¼ to ½ Finnsheep breeding (Boylan, 1975). This recommendation is made with the assumption that the Finnsheep cross-bred ewe is mated to a meat type terminal sire breed ram to produce a market lamb for slaughter.

(b) Utilizing heterosis; recombination loss

Utilizing individual and maternal heterotic effects can be distinctly advantageous. Recent studies at Minnesota showed that by using a specific 3-breed crossing system total flock productivity for weight of lamb weaned per ewe could be increased by 20 to 32% over the average of the parental breed performances. Columbia, Suffolk and Targhee breeds were evaluated on straight-bred (pure-bred) performance, single crosses and in a specific 3-breed cross-breeding system (Rastogi, 1972). The specific 3-breed cross-breeding system utilized a cross-bred ewe. The system capitalized on both individual and maternal heterotic effects. The cumulative effects over the life cycle of the lamb was high even though individual heterosis for many lamb performance traits was low; about 4% for daily gain and 3% for age at market weight (Rastogi *et al.*, 1975). The way in which the breeds were combined was important in achieving maximum productivity in the specific 3-breed cross. Heterosis was also observed in this study for feed efficiency during the post-weaning growth phase (from 23 to 50 kg). The average amount of heterosis in feed efficiency was 3.2% (Teehan, 1974). In view of the importance of feed costs in lamb production, significant economic advantages should accrue from utilizing heterosis in feed efficiency.

Recombination loss in heterotic effects was assessed in the cross-breeding study. Generally the recombination loss was small (Teehan, 1974). This suggests that reduction in performance in subsequent generations of a synthetic population would be small. This gives support to the idea of developing superior synthetics (new breeds) from promising cross-bred combina-

TABLE 4

A COMPARISON OF LEAST SQUARES MEANS OF PURE-BRED AND CROSS-BRED (F_1) PERFORMANCE FOR SEVERAL CARCASS TRAITS

Breeding group[a]	N	Hindsaddle wt, kg	Kidney and pelvic fat, %	*Longissimus* area, (cm)²	Fat thickness over *longissimus*, mm	U.S.D.A. yield grade	U.S.D.A. quality grade
Pure-breds ($\bar{P}$)	66	10.66±0.03	3.11±0.12	13.36±0.25	4.19±0.19	2.97±0.05	10.76±0.11
Cross-breds ($\bar{C}$)	169	10.19±0.02	4.27±0.07	13.13±0.16	3.25±0.17	3.08±0.03	10.27±0.07
Difference ($\bar{P}-\bar{C}$)		0.47±0.03**	−1.16±0.12**	0.23±0.26	0.94±0.20**	0.11±0.06	0.49±0.11**

** $P<.01$

[a] Pure-bred values are the mean of least squares means for the Minnesota 100, Suffolk and Targhee breeds. Cross-bred values are the mean of least squares means for the F×M, F×S and F×T breeding groups.

tions. Further research in this area is planned at Minnesota. Development of synthetics has been initiated at the USDA Meat Animal Research Center.

(c)   Foreign breeds

Sheep germ plasm resources available in the U.S. are limited to existing breeds introduced in the early development of the country with the exception of the Finnsheep, introduced in 1968. This limitation restricts the genetic potential for improvement of domestic sheep and effective utilization of natural resources. Imports are limited or restricted because of animal health quarantine regulations. In 1975 additional restrictions were imposed, virtually eliminating introduction of breeds not presently in the U.S. Animal breeders have long advocated the need for access to foreign breeds and have suggested development of procedures for importation (Lush, 1961). This need, with special reference to the sheep industry, was re-emphasized by Boylan (1968). Little has changed and the status quo remains. The present USDA policy lacks imagination and represents a negative approach to utilization of world resources for food and fibre production.

REFERENCES

Boylan, W.J. (1968). Genetics symposium, McGregor, Texas. Sheep Industry Development Program, proceedings.
Boylan, W.J. (1975). Sheep breeding and feeding symposium, Sioux Falls, S.D. Sheep Industry Development Program, proceedings.
Boylan, W.J., Berger, Y.M. and Allen, C.E. (1976a). *Journal of Animal Science* **42**: 1413.
Boylan, W.J., Berger, Y.M. and Allen, C.E. (1976b). *Journal of Animal Science* **42**: 1421.
Dickerson, G.E. and Laster, D.B. (1975). *Journal of Animal Science* **41** : 1.
Dickerson, G.E., Glimp, H.A. and Gregory, K.E. (1975). *Journal of Animal Science* **41** : 43.
Lush, J.L. (1961). "Germ Plasm Resources," R.E. Hogson, editor. (The American Association for the Advancement of Science: Washington, D.C.).
Meyer, H.H. and Bradford, G.E. (1973). *Journal of Animal Science* **36** : 847.
North Central Regional Publication 198. (1970). (Ohio Agricultural Research and Development Center, Wooster, Ohio).
Oltenacu, E.A.B. and Boylan, W.J. (1976a). *Journal of Animal Science* (submitted).
Oltenacu, E.A.B. and Boylan, W.J. (1976b). *Journal of Animal Science* (submitted).
Oltenacu, E.A.B. and Boylan, W.J. (1976c). *Journal of Animal Science* (submitted).
Parker, C.G. (1968). Genetics symposium, McGregor, Texas. Sheep Industry Development Program, proceedings.
Rastogi, R.K. (1972). Ph.D. Thesis, University of Minnesota.
Rastogi, R.K., Boylan, W.J., Rempel, W.E. and Windels, H.F. (1975). *Journal of Animal Science* **41** : 10.
Rattray, P.V., Garrett, W.N., Meyer, H.H., Bradford, G.E., East, N.E. and Hinman, N. (1973). *Journal of Animal Science* **37** : 892.
Teehan, T.J. (1974). Phd. Thesis, University of Minnesota.
Whiteman, J.V. (1968). Genetics symposium, McGregor, Texas. Sheep Industry Development Program, proceedings.
USDA. (1976). "Agricultural Statistics" (Statistical Reporting Service: Washington, D.C.)

8

NATIONAL SHEEP BREEDING PROGRAMMES — GREAT BRITAIN

J.W.B. KING

ARC Animal Breeding Research Organization, Edinburgh, U.K.

SUMMARY

The British sheep industry is relatively small in relation to total livestock production but forms an important enterprise in hill and upland areas. Sheep are reared under widely different conditions using a large number of breeds. Stratified cross-breeding systems are used, many starting with hill breeds and crossing them to give cross-bred ewes which in their turn are mated to Down breeds to produce slaughter lambs. Selection objectives vary between breeds and management systems.

Recording is carried out by the Meat and Livestock Commission and comprises three schemes: i) Commercial—Whole Flock for the collection of physical and financial data on sheep performance ii) Commercial—Individual Ewe for the contemporary comparison of breeds and groups within breeds and iii) Pedigree—Individual Ewe for the selection of breeding stock.

Study of the genetic influence of different breed groups shows the value of concentrating improvement attempts on hill breeds, Longwool crossing breeds and Down breeds. Where the establishment of controlled schemes for selection for commercial characters is difficult, as in hill breeds, special development flocks are being set up.

Much research work relevant to sheep improvement is carried out in Britain. Some of the current work at ABRO is reviewed and the results from comparisons of slaughter lamb sire breeds are presented.

THE BRITISH SHEEP INDUSTRY

In order to understand the problems faced in the improvement of British sheep and the methods being adopted it is necessary to consider the nature of the sheep industry in Britain. This industry is a complex one with different regional patterns and many local variations, but some attempt at summary is nevertheless necessary.

The total sheep population in Britain is about $13\frac{1}{4}$ million breeding ewes. These are kept in around 74,000 flocks and the flock size averaged 174 ewes in 1974. Sheep production tends to be an enterprise of the less fertile areas so that while less than 30% of holdings have sheep in England, about 50% do in both Scotland and Wales. As a proportion of the total output from all livestock, that from sheep is relatively small amounting to only about 6% of the total. Of the returns from sheep only around 10% comes from wool so that the industry is primarily oriented towards the production of meat.

The relatively small size of the sheep industry is not dictated by low consumer demands for the product, unlike the situation in many European countries. The average annual consumption of carcass meat (excluding manufactured products) is estimated at 7.1 kg of mutton and lamb, thus exceeding the consumption of pork (5.5 kg) but falling someway behind that of beef and veal (12.8 kg). The home production of mutton and lamb has never met the demand of the home market and over the years 50-60% of supplies have been imported, mostly from New Zealand. The trading position is changing with decreasing imports from New Zealand and a growing export of fresh lamb carcasses to the EEC.

The average annual lamb crop reared has been estimated by Read 1974 at 105%. This figure is made up of very different levels of performance in different environments. The 44% of breeding ewes kept on hill farms average 0.8 lambs per ewe, while those 17% on upland farms average 1.2 lambs and the remaining 39% of ewes on lowland farms average 1.3 lambs. Despite the low level of performance on the hills, sheep production has remained a major enterprise in these areas. As the only feasible form of livestock production in many remote areas, sheep production in these localities has attracted government support in the form of various direct and indirect subsidies.

It is always a matter of surprise to recall that in Britain there are at least 50 breeds of sheep. Some of these are represented by only small numbers but even excluding breeds with less than 25,000 ewes one is still left with 24 breeds. Most of these have limited distributions and fit in with the main production systems as shown in Table 1.

The hill breeds listed are for the most part confined to particular geographical areas although this localization is not as marked as it used to be. Most of the sheep kept on the hills are pure-breds and usually these breeds stay there for only part of their productive life. Having produced 3, 4 or 5 lamb crops on the hills they are sent for 2 or 3 further lamb crops to less severe upland conditions where a common practice is to mate them to a Longwool ram, usually a Border Leicester or a Bluefaced Leicester. The cross-bred female lambs from this mating are then used for breeding purposes in the lowlands, being mated to Down breeds to produce lambs for slaughter.

The stratification of the sheep industry is a long-established feature and one which has become commercially regularized. Thus particular crosses, such as the Greyface (Blackface x Border Leicester) and the Scottish Halfbred (North Country Cheviot x Border Leicester) are standard articles of commerce. It would be wrong to give the idea that the stratified crossing system is a rigid one because many variations on the established system are

TABLE 1

PRINCIPAL BRITISH BREEDS GROUPED BY BREED CLASS (SOURCE MLC 1972)

Hill breeds	Scottish Blackface	Welsh Mountain
	Swaledale	Improved Welsh
	Dalesbred	South Welsh Mountain
	Derbyshire Gritstone	Black Welsh Mountain
	Lonk	Radnor
	Herdwick	Beulah Speckleface
	Rough Fell	Hardy Speckleface
	North Country Cheviot	Exmoor Horn
	South Country Cheviot	Shetland
Upland breeds	Clun Forest	Greyface Dartmoor
	Kerry Hill	Whiteface Dartmoor
	Devon Closewool	
Romney	Kent or Romney Marsh	
Devon	South Devon	Devon Longwool
Lowland	Dorset Horn	Wiltshire Horn
	Polled Dorset Horn	Lleyn
	Improved Dartmoor	Llanwenog
Longwool	Border Leicester	Lincoln Longwool
	Bluefaced (Hexham) Leicester	Leicester
	Teeswater	Colbred
	Wensleydale	
Down	Suffolk	Southdown
	Dorset Down	Shropshire
	Hampshire	Ryeland
	Oxford	

to be found locally and in response to changes in the market. The latest survey on the breeds kept for lamb production in Britain was carried out by the Meat and Livestock Commission in 1972 and this recorded the existence of over 300 different crosses. Many were present in only small numbers but the survey did reveal that about 20% of Down x Scottish Halfbred ewe lambs were retained for breeding. In addition many draft hill ewes are mated directly to Down breeds for the production of slaughter lambs.

Lambs slaughtered for the national flock are a variable product reflecting the breeds and crosses used and different systems of production. Except for about 2% of lambs produced out of season to obtain high early spring prices, most lambs are produced on grass with a proportion of these being fed rape or root crops as the grazing season ends. The number of lambs slaughtered each month increases from April to October and then declines. The price paid usually changes in inverse ratio to the numbers slaughtered and early lambs are sold at lighter carcass weights in attempts to achieve the highest price.

OBJECTIVES FOR SHEEP IMPROVEMENT

The complexities of the sheep breeding scene in Britain bring with it problems in deciding upon improvement objectives. For sheep which are kept in different environments there may well be different objectives. Thus on the hills, multiple births are by and large regarded as a distinct disadvantage. Those hill farmers with some improved land on their farms may accept a proportion of ewes with twins which can be segregated on the better ground but increasing the proportion of twins would cause them major problems of husbandry. Many lambs from hill flocks are sold in store markets at the end of the season and the lightest animals make very poor prices. To the majority of hill farmers improvements in weight of lambs at weaning would be a reasonable objective but changes in prolificacy is definitely not. In upland production systems increased numbers of twins are an advantage provided they do not place too many additional calls on the limited labour force. Many of the ewes used in these upland areas are, however, older draft ewes of hill breeds so there is seen to be some conflict of interest between the hill farmer who did not require prolificacy and the upland farmer who could use it in limited measure. To some extent the change in nutritional circumstances of the ewe produce the kind of change which is required. Blackface ewes seem to respond well to the improved nutrition and draft ewes may well give lamb crops of around 150%. Other hill breeds such as the South Country Cheviot respond less well (Gunn and Doney 1974) and so are less suitable for use in such a system. For lowground production the situation may differ again. A small proportion of farmers attempt to produce lambs for slaughter as early as possible in the season before the price for lamb falls dramatically and they again prefer singles to twins for their marketing circumstances. For farmers producing the bulk of their lambs in mid to late season additional twins certainly are a major economic advantage.

Other selection objectives also cause difficulty according to environmental circumstances. If we seek to improve growth rate then it is necessary to face up to the problems created by correlations between growth rate and mature body size. If mature body size is increased then it seems inevitable that the maintenance requirements of those ewes will also be increased and that the number that can be kept on the same area of the land must be reduced. Numerical relationships between these factors are, however, largely unknown and changes in grazing pressure on extensive pastures may take many years to manifest themselves. The improvement of growth rate may also run into complications through the demand of the meat trade for carcasses with a certain level of fat cover. The use of a genotype with a higher growth rate may not necessarily be advantageous if the lambs do not reach the required degree of fatness for sale till later in the season, and so enter a period of lower prices, or if such lambs cost more to feed because the natural pastures are becoming exhausted and supplementary feeding becomes necessary. In such circumstances there must be real doubts about the economic value to be attached to growth rate since there is no guarantee that in a particular set of circumstances an increase in growth rate will guarantee an improvement in the overall economics of that system.

No attempt will be made here to examine further desirable selection objectives but it should be noted that in practice a great deal of attention is paid to breed characteristics. In a country where there are so many different breeds and crosses in use, the identification of breeds in the market place is a matter of concern and it is therefore not surprising that breeders should pay considerable attention to the trade marks of their particular product.

DEVELOPMENT OF SHEEP RECORDING IN GREAT BRITAIN

Although many attempts have been made to develop sheep recording and related improvement schemes the majority of early attempts in Britain met with only limited success and none achieved national status. In some localities good coverage was obtained and Mr. H.B. Parry started one scheme in the Oxford area in 1952 and this was recording some 12,000 ewes by 1968. The Eastern region of the National Agricultural Advisory Service also instituted a ewe recording scheme in 1958 and this had grown to some 2,000 ewes by 1968. Allied improvement activities also took place during this decade with the development of co-operative ram performance testing in Wales and the introduction of a scheme by the British Wool Marketing Board offering free assessment of ram fleeces. No nationally available recording scheme was, however, available until the arrival of the Meat and Livestock Commission (MLC) which was created in 1967. Since this body has prime responsibility for sheep improvement in Britain it is necessary to explain briefly the status and purpose of the Commission.

The Meat and Livestock Commission was set up by Act of Parliament in 1967 to promote greater efficiency in the livestock and livestock product industries of Great Britain. It took over the obligations of the Pig Industry Development Authority which had been operating with a similar remit up until that time and extended the work to cover sheep and beef cattle, although still excluding wool, dairy cattle and milk. Although the Commission is concerned with efficient marketing and distribution it is not a marketing board and has no trading powers. The functions to which particular attention is given are livestock improvement, carcass evaluation and classification, the provision of market information and the promotion of sales of home produced meat. Income is obtained from a levy on all animals slaughtered and the current levels are cattle 68p, calves 8p, sheep 4p, pigs 20p, giving an annual total of £3.8 million in 1975.

The Sheep Improvement Services branch of the MLC concentrates its work on the livestock side of the sheep improvement into two main areas.

The first area is in the improvement of management which is seen as presenting an opportunity for very large and rapid returns through the exploitation of the potentialities of existing breeds and crosses. The second area of activity is in genetic improvement and this takes the form of both breed comparisons and within-breed selection. To cope with these various needs recording is organized into three distinct schemes:

(i) commercial, whole flock
(ii) commercial, individual ewe
(iii) pedigree, individual ewe.

The extent of these recording schemes is shown in Table 2.

TABLE 2
NUMBERS OF MLC RECORDED FLOCKS

		1971/72	1972/73	1973/74	1974/75
Commercial, whole flock		548	713	788	827
Commercial, individual ewe		51	58	51	58
Pedigree, individual ewe		264	292	297	294
	Total:	863	1063	1136	1179

The number in the various schemes has not grown greatly over the years due to a policy of restricting entry to those flocks where recording was thought to be of value.

The most widespread scheme is that for commercial whole flock recording. This is designed to obtain physical and financial data on the inputs to and outputs from a flock so that the balance sheets can be drawn up and gross margins calculated. Although the body weight of ewes may be recorded and lambs weighed at 8 weeks of age, no individuals in the flock are identified and this scheme is directed towards providing management information. This form of recording shows up a wide range of results and the factual information collected on the physical and financial performance of the flocks is used extensively for advisory and educational purposes. Ewes are individually recorded in a smaller number of commercial flocks where parentage is known and where it is thought likely that comparative information about different breeds or crosses can be obtained within the same flock. Body weights of ewes are also taken and lambs weighed at 8 weeks of age and if possible at slaughter.

The third category of recorded flocks is that for pedigree flocks which are already recording the sire and dam of every lamb born. This scheme records additional performance data on the individuals in a flock and breeders are given lifetime summaries of the performance of their ewes and a simple index to help in the choice of breeding animals. The Down flocks are well represented in this scheme, and to a lesser extent the lowland breeds but the hill breeds are poorly covered due to the difficulties of recording in extensive conditions.

THE STRATEGY FOR GENETIC IMPROVEMENT

At an early stage of development of the Commission's improvement work with sheep, a Scientific Study Group was set up to review the scientific aspects of the Commission's existing and proposed improvement schemes. That review proved especially difficult for sheep because of the large number of breeds involved, the many production systems employed and the often conflicting nature of objectives for improvement. Some indications of where attention might most profitably be focused was provided by considering the genetic contribution of different breed groups to the estimated national production of lamb carcasses, as shown in Table 3.

TABLE 3

NUMBER OF EWES IN PURE-BRED GROUPS AND
ESTIMATED GENETIC CONTRIBUTION TO
LAMB CARCASS PRODUCTION (AFTER READ 1974)

Breed Group	Million ewes	Genetic contribution (%)
Hill	6.9	36
Upland	0.8	8
Longwool	0.02	13
Down	0.24	37
Lowland	0.6	6

In relation to their number hill ewes contribute relatively little directly to lamb production because of the low level of productivity on the hill and the need to retain a high proportion of ewe lambs as breeding stock. The genetic effect of these breeds is, however, increased by the large numbers of cross-bred breeding stock which derive from them. The Down breeds are seen to make a very important contribution, as would be expected from their widespread use as terminal sires. Although the Longwool breeds have a relatively small effect in total this can be seen as coming from a very small pure-bred population. In national terms, the three preceding groups would account for 86% of the output of lamb carcass meat and it therefore seemed reasonable to concentrate efforts on them.

As a first step the Study Group recommended that breed and cross comparisons should be organized so that commercial flock owners could make better decisions relevant to their own management systems. The MLC has begun the first phase of a comprehensive trial to compare ram breeds as the sires of lambs for slaughter. The breeds being compared are the Border Leicester, Dorset Down, Hampshire Down, Ile de France, North Country Cheviot, Oxford Down, Southdown, Suffolk and Wensleydale.

The creation of improvement schemes for each breed would be an expensive undertaking and the Study Group suggested the establishment of a limited number of 'development' flocks in which the feasibility of selection schemes could be tried. Such flocks were to have defined selection objectives and would be operated under the direction of the MLC with their financial assistance. If these schemes were successful then such flocks could provide a useful source of breeding stock. The MLC has now identified 7 potential development flocks in the Scottish Blackface, North Country Cheviot, Welsh and Swaledale breeds. In the Down breeds, where a number of quite large flocks are individually recorded by breeders, it appears that no special efforts may be necessary to establish selection schemes for commercially important traits.

In addition to the mainstream of MLC activity there are a number of pilot ventures under way which will not be described but do testify to widespread interest in ways and means of improving sheep.

RESEARCH INTO SHEEP IMPROVEMENT

Many Universities, research institutes and advisory bodies in Britain are involved in sheep research, aspects of which are pertinent to sheep improvement. No attempt at an overall summary will be made here but it may be useful to describe some of the current investigations at the Agricultural Research Council's Animal Breeding Research Organization (ABRO) which seem particularly relevant to sheep improvement problems.

Despite the widespread use of cross-breeding in Britain the experimental evidence for the adoption of the practice is often lacking. The use which is made of this breeding method is often designed to utilize complementarity rather than to take advantage of heterosis. Crossing of hill breeds utilizes breeding stock which can be produced cheaply on the hills and the crossing process changes the performance characteristics to make the cross-bred ewe more suitable for use in better conditions. A question which is still largely unanswered is whether there is a useful amount of heterosis in crosses of hill breeds and whether this might be utilized in the exploitation of these severe environments. Under upland conditions some evidence for heterosis was found by Wiener and Hayter 1975 particularly in crosses of the Scottish Blackface and South Country Cheviot breeds. Another experiment is in progress on a Welsh hill farm involving the Welsh and Scottish Blackface breeds and the preliminary results show evidence of a rather low level of heterosis (Purser 1975). More recently a large cross-breeding experiment has been started on a Scottish hill farm and involves crosses of the Scottish Blackface stock with Swaledale, North Country Cheviot, Derbyshire Gritstone and Exmoor Horn rams.

ABRO has also carried out some comparisons of sire breeds, picking out a few breeds which seemed of particular interest at the time the trials were started. Breeds which were comparative newcomers to Britain were used alongside existing Down breeds. The Oxford was used as a control—a small control flock of this breed is maintained—and alongside it the Suffolk and

Dorset Down as common British breeds along with the Oldenburg, Ile de France and Texel as imported breeds. Lambs from these crosses were slaughtered at two weights and subjected to full carcass measurement with a proportion of them being dissected. The results are summarized in Table 4.

TABLE 4

COMPARISON OF SLAUGHTER LAMB SIRE BREEDS —
AVERAGE OF RESULTS AT 35 AND 40 Kg LIVEWEIGHT
SLAUGHTER

	Dorset Down	Ile de France	Olden- burg	Oxford	Suffolk	Texel
No. of lambs	343	343	295	322	339	351
Av. birth wt. (kg)	3.8	3.7	3.8	4.1	3.9	3.8
Av. 8-week wt. (kg)	19.2	18.9	19.4	20.6	19.9	19.2
Age at slaughter (days)	170	173	168	150	158	168
Killing-out %	44.0	44.0	42.1	43.5	43.5	44.3
Carcass grade (points)						
(1 poor, 7 excellent)	3.6	3.3	2.8	3.1	3.2	3.2
No. dissected	40	41	43	46	47	51
Total lean (kg) in left side	4.26	4.33	4.28	4.32	4.29	4.67

Quite marked breed differences were found and illustrate the utility of different breeds for various production systems. The Dorset Down produced lambs that graded well and would be appropriate for early lamb production, although not showing the highest growth rate. From these comparisons the Texel breed stood out as producing a markedly leaner carcass suggesting it might be useful for the production of lambs for slaughter at heavy weights.

Comparative trials have also been carried out with alternative dam breeds including the Finnish Landrace. These will be described in another section of this book and so will not be elaborated on here except to record that extensive experimentation with many different co-operators both from the advisory field and individual breeders does take place as needs arise.

A recent innovation on the British scene is an attempt by ABRO to introduce the concept of group breeding schemes to sheep farmers which is being attempted in two breeds. In Wales discussions are in progress with Welsh Mountain breeders and with the Agricultural Development Advisory Service and similar discussions are taking place in Scotland, this time with the East of Scotland College of Agriculture, with the object of establishing a group of Scottish Blackface breeders. Both approaches will involve a degree of breeder co-operation that has not been attempted before in Britain. The degree of interest which is present in sheep breeding and a growing concern with the financial details of sheep production systems makes the timing opportune.

REFERENCES

Gunn, R.G. and Doney, J.M. (1974) *Proceedings of the British Society of Animal Production* 3: 96.

Meat and Livestock Commission (1972) Sheep Improvement Scientific Study Group Report.

Purser, A.F. (1975) *Proceedings of the British Society of Animal Production* 4: 101.

Read, J.L. (1974) *Proceedings of the Ruakura Farmers' Conference* 26: 42.

Wiener, G. and Hayter, S. (1975) *Animal Production* 20: 19.

9

SHEEP BREEDING IN URUGUAY

R.C. CARDELLINO
Uruguayan Wool Secretariat, Montevideo

SUMMARY

Sheep production is one of the main industries in the country, Uruguay being the fifth largest wool exporter in the world. The present sheep population is estimated at 18 million. Corriedale, Polwarth, Australian Merino and Merilin are the main breeds. The composition of the national flock reflects the emphasis that is placed on wool production and the predominant semi-extensive production systems. The present breeding structure follows the common hierarchical pattern and the difficulties in introducing new breeding systems are almost identical to those of other countries. A performance recording scheme (flock testing service) was started in 1969 with relatively good success and several points concerning the efficiency of the scheme are discussed. More effective extension work at the commercial level and support for the activities of the more enthusiastic breeders are being emphasized. Alternatives to the present system, such as co-operative nucleus breeding schemes will be encouraged.

INTRODUCTION

Uruguay has a total area of 18.5 million ha, of which 85% can be considered as productive. Approximately 13 million ha are under native or improved pastures and are utilized for grazing sheep and beef cattle.

Climatic conditions are quite uniform throughout the country, with an average annual rainfall of 1000 mm and average daily temperatures of 12-13°C in July and 24-25°C in January. There are however, great variations in the average rainfall between years and also within years between seasons.

Most of the farms are mixed and run beef cattle and sheep together. The average ratio of sheep/beef cattle in the country is around two, but ranges from four in the so-called sheep areas to one in the more productive areas, and the average stocking rate is around six dry ewes/ha.

SHEEP PRODUCTION

Sheep production and especially wool production is one of the main industries in the country and represents nearly 30% of the value of total exports, making Uruguay the fifth largest wool exporter in the world. In recent years its importance has declined and there has been substitution of sheep by beef cattle. As a consequence there was a reduction in sheep numbers of 10%. The present sheep population is estimated at 18 million sheep and there are some indications that this is increasing again.

There are 36,363 farms running sheep, distributed according to flock sizes as shown in Table 1.

TABLE 1

DISTRIBUTION OF SHEEP FARMS ACCORDING TO FLOCK SIZES

Flock Size	Number of Farms	%	
1-99	15,376	42.3	
100-299	7,526	20.7	
300-599	4,575	12.6	
600-999	2,917	8.0	
1,000-2,499	3,770	10.4	24.5% of
2,500-4,999	1,441	4.0	the total
5,000-9,999	586	1.6	sheep
>10,000	172	0.5	
Total	36,363		

In general, systems under which sheep production takes place, can be described as semi-extensive, with relatively more use of land and labour than capital, sheep usually being relegated to the poorest pastures. This situation is quite similar throughout the country and there is no specialization of farms to particular productions such as fat lambs, cross-bred ewes, etc.

The main objective of the sheep industry is the production of wool, the home market for fat lambs being very poor. Mutton meat, mainly cast-forage ewes and wethers, is quite popular in the country areas where the consumption reaches 33.5 kg/person/annum, compared with 0.5 kg in the main cities.

There have been marked changes in the proportion of breeds with the years, particularly an increase in the proportion of Corriedales, Polwarth and Merilin * and a decrease in the proportion of Australian Merino, Rambouillet Merino, Romney and crosses. The present breed distribution (Table 2) seems to reflect the characteristic instability of marketing conditions for different types of wool, mutton and lambs.

The present structure of the national flock (Table 3) reflects the emphasis that has been placed on wool production, and the predominant production systems.

Sheep production is very dependent on the variations in pasture production throughout the year, with marked shortages during winter and summer. The growth of lambs after weaning in the summer is usually very slow and as

---

* An Uruguayan breed approximately ¼ Lincoln–¾ Rambouillet merino developed in 1930.

a consequence, on average only 50% of the females are mated for the first time as two-tooths (1½ years). Lambing normally takes place in July-August, except for many Polwarth and Merino flocks that are mated in the spring. As a result of a poor nutritional level in the last stages of pregnancy and very severe climatic conditions during lambing, the average reproductive efficiency is low (70-75% of lambs marked/ewe joined) with very marked variations between years. Lamb mortality ranges from 15% in good seasons up to 35% in bad seasons. It has been shown in Uruguayan conditions (Azzarini and Ponzoni 1971) that the reproduction performance can be considerably increased with spring lambings.

TABLE 2

DISTRIBUTION OF BREEDS

Breed	Sheep Numbers	%
Corriedale	8,500,000	50.1
Polwarth	1,500,000	8.9
Aust. Merino	700,000	4.1
Merilin	1,000,000	5.9
Romney	350,000	2.1
Crosses	4,900,000	29.0

SHEEP BREEDING

(a) The traditional system

There were several stages in the history of sheep breeding in Uruguay characterized by the predominance of different breeds and the utilization of different breeding systems. An initial period of grading up to Merino-type breeds, especially the Negrette from Spain, the Rambouillet from France, the Vermont from U.S.A. and the Australian from Tasmania, was followed by a strong predominance of Lincoln and Romney crosses at the beginning of the century when there was an important demand for mutton meat in Europe. As a consequence of the reduction in the sheep meat market and of good prices for wool, a period of alternated cross-breeding started with the use of Merino rams on cross-bred flocks when the wool became "too strong" or the use of Lincoln or Romney rams when the flocks became "too fine and small". The result was a great variation in wool characteristics within individual clips.

In 1935, a Sheep Improvement Commission was created to orientate sheep breeding in the country and it started a process of grading up to pure breeds such as the Corriedale and Polwarth, which were imported at that time, Merino and Romney. During that process, a system of identifying superior graded-up animals was created, in which both males and females were inspected for the first time before the first shearing. Those without obvious faults, and reaching the breed type standards and a certain level of productivity (subjectively appraised) were given a single tattoo. The single tattooed animals were inspected again the following year and those reaching more strict levels of performance were given the double tattoo. Three different types of ram breeding flocks were differentiated: the registered, the non-registered double tattoo and the non-registered single tattoo, and a clear dependence between the three was originated with the registered flocks on top of the hierarchy.

TABLE 3

COMPOSITION OF THE NATIONAL FLOCK BY CATEGORIES (*)

Categories	Numbers	%
Rams	310,000	2.0
Breeding ewes	7,400,000	50.0
Ewe hoggetts (not mated)	2,150,000	14.6
Male hoggets (castrated)	1,750,000	11.9
Wethers	2,500,000	17.2
Others (†)	600,000	4.3

(*)   Lambs not considered.
(†)   Animals which are slaughtered during the year for consumption on the farm.

This system, which is still operating, was successful at the beginning in orienting the producer and raising the level of production probably as a consequence of a wide genetic variation in traits readily assessed by visual appraisal and the elimination of major faults. Even though there is no evidence of what genetic progress is being made, it is very likely that the present methods are not adequate to ensure high rates of improvement. The registered flocks, whose importance is over-emphasized under the present system, in many circumstances are quite small, and because of the high economic value of the pedigree in itself, culling is usually very light: (Ponzoni, R. 1973). In the non-registered ram breeding flocks, which are larger and more numerous, selective pressure has been stronger but the identification of superior animals (particularly the double tattooed) has been made on a purely subjective basis, including many characteristics not directly related to productivity and on a "between stud" basis.

Because of the presence in most studs of non-registered ram breeding flocks, a study of the pedigree flock books does not give an accurate picture of the breeding structure of the industry. A survey was made last year covering 600 studs with the objective of determining the breeding practices commonly used in the studs and the breeding structure of the sheep industry. Results are still being analysed but so far there are indications that it follows the common hierarchical pattern. However, there appears to be a higher proportion of stud animals in the top level of the hierarchy, while secondary studs are not as tightly associated with a particular top stud, compared with the Australian situation. As in any hierarchical breeding structure, the genetic progress in the whole industry will depend on what genetic progress is being made in the top studs.

The National Sheep Production Survey (Cardellino *et al.* 1972) showed that approximately 10% of all the commercial flocks with more than 600 sheep have a selected nucleus (mainly on the basis of tattooing) in which rams bought from studs are used, to produce rams for their own commercial flocks.

(b) Flock Testing Service, a new approach

A performance recording scheme was started in 1969 with the purpose of introducing more effective breeding methods, through the objective measurement of economically important traits. The service is run by the Uruguayan Wool Secretariat (S.U.L.) and the Sheep Improvement Commission that now belongs to the Secretariat, with representation from all the sheep breed associations. The service is free, but in fact is funded by the whole industry.

Ram hoggets, registered or unregistered, are measured when they are 15-18 months old. At shearing time, records of greasy fleece weight and body weight of the shorn animals are taken and a visual appraisal of the quality number, colour, character and handle of the fleece is made. A midside sample of wool is taken and sent to our wool laboratory where yield %, staple length and fibre diameter by air-flow method are determined. The data are processed in a computer and the animals ranked according to their clean fleece weight. The rest of the traits are expressed as deviations from the average.

The stud breeder is assisted by extension officers at shearing time, in collecting records and wool samples. Once the records are computed, the extension officer returns to the stud in order to advise on the correct interpretation of the results. No selection indices have been used so far. In the studs where pedigree records are kept, correction factors are applied to animals born as twins and being the progeny of maiden ewes.

In order to make the Service more flexible, there are two possible schemes. In Scheme I (Figure 1), ram hoggets prior to shearing are visually classed and only the reserve rams are shorn and recorded. This system is mainly used in those studs which sell rams in full wool before the end of the year.

The effects of this preliminary culling on the selection differentials for traits of economic importance, will depend on the proportion selected as reserve rams, the criteria used in selecting them and its relationship with productive characters.

Figure 1— Flock Testing Service. Scheme I.

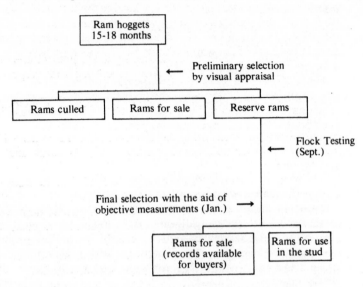

In the Scheme II (Figure 2), the whole group of ram hoggets is recorded, except for a small proportion of animals culled for obvious faults.

Figure 2— Flock Testing Service. Scheme II

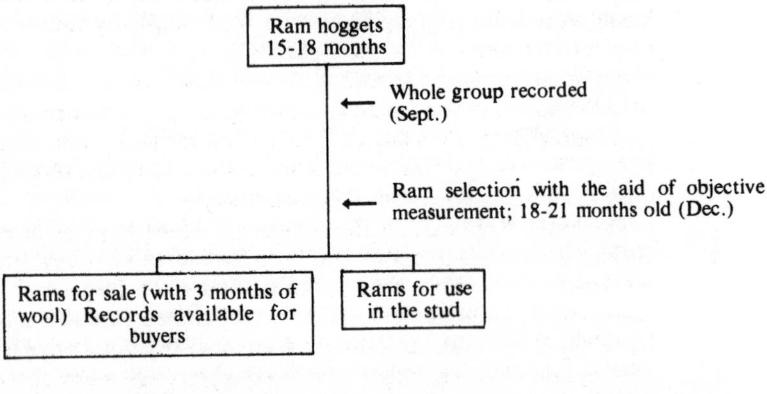

The final selection of the rams with the aid of records is done in Nov.-Dec. at 18-21 months of age, and most of the rams are used for the first time as two-tooth. The rest of the rams are sold at auction in their own studs with 3-4 months of wool and all the records available for buyers. Extension officers assist the ram buyers in the correct interpretation of the results.

Several factors are important in evaluating the efficiency of a performance recording service:

— the number of studs using the service
— their continuity
— their relative importance according to the present breeding structure
— the proportion of animals being recorded with respect to the total available
— the use that is effectively made of the records in their selection programmes.

The number of studs using the service has increased from 12 in 1969 to 70 in 1975 of which 65% are Corriedales, 25% Polwarths and 10% Merino and Merilin.

The level of continuity has been high, with 85% of the studs that used the service in 1970 continuing to use it through 1975.

From the point of view of dissemination of genetic improvement to the whole industry under the present breeding structure, the relative position in the pyramid of the studs using the service is very important in determining the overall efficiency of the service. At present, 40-50% of the top Corriedale and Polwarth studs are using the service regularly.

The number of hogget rams recorded per stud varies from 20 to 300 with an average of 60 and expressed as a proportion of the total number of rams available in each stud, the range is from 25 to 90%, the proportion tending to increase in the big studs.

A performance recording scheme is essential to implement efficient selection programmes but unless an adequate use of the available records is made, very little is going to be achieved simply by measuring the performance of the animals. Some checks have been made to see which rams were actually used from those recorded. The results varied considerably, the best being obtained in some studs measuring the performance of all their hogget rams and effectively using as 2-tooth and for two years, rams from the top 5% in fleece weight; the worst in some studs using the service only to check their traditional selection or for curiosity.

The rate of increase in the number of studs has not been very exciting and the difficulties in convincing stud breeders to use new breeding systems are almost identical to those of other countries. In general, the need for a change in the present methods is not recognized. A stud is a commercial activity whose benefits are closely related to the number of rams sold each year and their prices, rather than genetic progress in any characteristic. This activity is carried out in a system where shows are very important, where much attention is paid to characteristics of very doubtful importance (breed type, constitution, etc.) and where the pressure from the demand (ram buyer) for any change is very small. The adoption of more efficient breeding methods at the stud level is very difficult and slow if there is not a simultaneous change in the attitude of ram buyers, which means that more effective extension work at that level is required. However the attitude of some of the biggest Corriedale studs utilizing the scheme II for some years with economic success and promoting the use of measurements at ram auctions, has had and will probably have a great impact in putting more pressure on other studs to do the same.

In the so-called dual purpose breeds, although the term is not very precise, characteristics related to conformation are normally included in the selection criteria. It has been shown (in preparation) in Uruguayan conditions, working with Corriedale wethers (two and four tooth) and confirming earlier results of Tallis and Turner (1964) with Merinos, that body weight is the best indicator of meat production and that there was a very small variation in the proportion of the hindquarter with respect to the total carcass, thus showing little scope for improvement through selection of the hindquarter to forequarter ratio.

The present programme is far from perfect and was the result of a compromise between the scientific and the breeders' points of view, and adapted to the existing situation. Some modifications will be introduced, mainly the elimination of some subjectively appraised traits such as handle and character and the elaboration of selection indices with different alternatives of economic weights as a complement to the present information.

All the possible avenues of producing significant changes in the current breeding methods will be exploited, especially more effective extension work at the commercial level and continuing support and assessment of the activities of the more enthusiastic breeders.

Alternatives to the present system are being encouraged, such as the formation of co-operative nucleus breeding schemes, and technical assistance is given to those commercial producers running their own ram breeding nucleus flocks.

REFERENCES

Azzarini, M. and Ponzoni, R. (1971) "Aspectos Modernos de la Produccion Ovina". Universidad de la Republica: Montevideo.

Cardellino, R.C., Gonzalez, G., Nicola, D. and O'Brien, G. (1972) "Relevamiento Basico de la Produccion Ovina en el Uruguay" Secretariado Uruguayo de la Lana: Montevideo.

Ponzoni, R. (1973) "Aspectos Modernos de la Produccion Ovina" Universidad de la Republica: Montevideo.

Tallis, G.M., Turner, H.N. and Brown, G.H. (1964) *Australian Journal of Agricultural Research* **15**:446.

10

THE SHEEP INDUSTRY AND SHEEP BREEDING IN RUSSIA

N.A. JHELTOBRUCH

All-Union Research Institute for Sheep Breeding, Stavropol, USSR

Sheep production takes an important place in the national economy of the USSR. The share of sheep production in the total agricultural production is 6.2%, and of the total animal production is 10.1%. In some areas of the country it is considerably greater. For example, in the North Caucasus it comprises 17.8% of the animal production, in Kirgizia 27.5% and in Turkmenia 80.0%.

The development of the sheep industry in the USSR is determined mainly by the requirements of the national economy for meat and wool products. The important basis for developing sheep production in the USSR is the availability of large areas of semi-desert, steppe and hill pastures, which are not suitable for breeding of any other kinds of livestock.

Of the different sheep products wool is the most important, although sheep meat is also important. In terms of the total meat consumption of the country in recent years mutton comprises 8-9%, compared with some Middle Asian countries where it approaches 60%.

In addition to wool and meat production large amounts of Karakul and furcoat pelts are produced which meet the requirements of both the national and foreign markets.

Wide variations in the climatic conditions of the USSR have allowed the evolution of many breeds of sheep with different purposes. Nevertheless, in recent years much attention has been given to the development of fine and semi-fine wool sheep breeds, producing the most valuable wool types.

In spite of the long standing presence of suitable environments for sheep breeding this branch of Animal Husbandry was greatly influenced by the First and Second World Wars and the Russian Revolution. Before the First World War and the revolution Russia had a sheep population of 89 million, including 4.5 million Merinos. By 1924 the total sheep population had been considerably reduced to only 350 thousands of these fine-woolled sheep. In order to correct this reduction the Government took some drastic measures, including the establishment of large specialized sheep farms and the training of sheep specialists, and encouraged further developments of research work. Much was done to convert a low-productive coarse-wool sheep industry into a highly productive one producing fine and semi-fine wool fleeces. This increase was attained by pure-breeding as well as by mass crossing of coarse-wool ewes with fine-fleece rams. Valuable flocks of fine-fleeced sheep and breeding studs producing animals for selling to commercial sheep farms were established in a comparatively short period of time.

Temporary occupation of the sheep breeding regions of the Ukraine, North Caucasus and Volga during the Second World War resulted in another major reduction in the sheep population mainly of fine-fleeced sheep breeds. During the war improvement of sheep production in other regions of Russia was also stopped.

For these reasons it is reasonable to consider that the major improvements in sheep production leading to a raising of its productivity have been achieved during the postwar period, i.e. during the last 25–30 years.

As a result of the widespread use of artificial insemination the conversion of low productive coarse-wool sheep breeds into fine and semi-fine fleeced breeds, alongside the establishment of some new breeds of high wool and meat production, was successfully achieved in a short period of time.

Fine-fleeced sheep breeding in the USSR was moved away from the geographic areas of its evolution—South Ukraine and North Caucasus— and new zones of fine-wool sheep production were established on the Volga and Urals, in West and East Siberia, in Kazakhstan and Kirgizia. With the establishment of new "breeds" the problem of breeding fine and semi-fine fleeced sheep in the hill regions of Caucasus, Kirgizia and Kazakhstan to replace their local coarse-wool animals was overcome.

At present attention is being paid to improving nutrition which should allow further development and perfecting of the fine-fleeced sheep breeds in the country.

The establishment of some semi-fine fleece and dual purpose sheep breeds for the steppe and hill regions of North Caucasus, Volga, Kazakhstan and other zones, was also undertaken.

The increase in sheep population and gross output of wool is shown in Table 1.

TABLE 1

SHEEP POPULATION AND WOOL PRODUCTION IN THE USSR IN RECENT YEARS

	1950	1960	1970	1975
Sheep population (millions)	77.6	136.1	130.7	145.3
Wool production (thousand tons)	172	354	417	504
% fine and semi-fine wool	18.6	54.2	66.6	77.0

As can be seen from Table 1 the gradual increase in the sheep population was accompanied by an increase in wool production. The increase in wool production resulted mainly from an increase in fine and semi-fine wool fleeces.

The most widely spread fine-fleeced breeds of the USSR are the following:

Caucazskaya breed: dual purpose breed found in North Caucasus.
Askanijskaya: dual purpose breed, South Ukraine.
Altajskaya: dual purpose breed, Altai territory.

Soviet Merino: wool and/or wool and meat, breed. It is the major breed and is located in the North Caucasus, the Volga region, Ukraine, Siberia, Kazakhstan.

Groznenskaya: wool breed, semi-desert regions of the North Caucasus.

Stavropolskaya: wool breed, steppe regions of the North Caucasus and Volga.

Kazakhskaya: fine-fleeced dual purpose breed, Kazakhstan.

Kazakhski Archaro: Merino breed, which was bred by hybridization with wild rams (Archaro). They are bred in the hill regions of Kazakhstan.

Kirgizskaya: fine-fleece breed, hill regions of Kirgizia.

Zabaikalskaya: fine-fleece breed well adapted to the severe climatic conditions of East Siberia.

Semi-fine fleece breeds are as follows:

Zigaiskaya breed: South Ukraine, Kazakhstan, Moldavia.

Kuibishevskaya breed: well suited to breeding in areas with high rainfall (Central Russia).

Severo-Caucazskaya: dual purpose breed, steppe zones of the North Caucasus.

Tjan-Shanskaya breed: hill regions of Kirgizia.

Some other important breeds, bred in different regions of the USSR are as follows:

Karakul breed: pelt production, semi-desert regions of Middle Asia and Kazakhstan.

Romanovskaya breed: fur coat pelt production, North of the European part of the country.

Edilbaevskaya (Kazakhstan), Gissarskaya (Tadjikistan and Uzbekistan), Djaidara (Uzbekistan) breeds are of interest as fat-tailed breeds.

The establishment of the large number of highly productive sheep breeds in a short time was the result of continuous and persistent efforts of sheep breeders and research workers.

In his *Manual on the Establishment and Perfection of Fine Fleece Sheep* the scientist M.F. Ivanov, reported that the major objectives of sheep improvement were:

1. The identification of the most productive genotypes from among the best animals (phenotypes) for the different characters of importance to the sheep industry.
2. The fixation of the best genotypes and their rapid reproduction.
3. The creation of the certain highly productive lines from among best genotypes.
4. Replacement of bad genotypes by much better ones.
5. The creation of new, better genotypes by means of combination of different lines.
6. The use of small mutations for the establishment of new valuable genotypes.

For the identification of the valuable genotypes, mainly among rams, progeny testing was used on a large scale.

In the course of establishing a fine-fleeced dual purpose sheep industry with the use of imported English breeds, it became evident that they were not well adapted to the Russian climatic conditions and management. This difficulty led to the application of cross-breeding as the main route for the establishment of semi-fine fleeced sheep of an early maturity. In contrast to other countries, this was achieved by crossing coarse-wool and semi-coarse wool ewes with meat-and-wool rams. New breeds and breeding groups of dual purpose sheep were established followed by careful and intense selection for the most desirable types and individuals.

At present sheep improvement programmes are carried out not only on the best stud farms, but also on commercial farms. Line breeding is the important method of selection in the leading stud farms. It allows the individual traits of the best breeds and animals to be selected and promotes the division of the breeds into different quality groups. These are then combined to produce hybrids which not only allows inbreeding to be avoided under general conditions, but also promotes the phenomenon of heterosis.

One important feature of the sheep stud industry is the fact that each breed has several specific stud farms in different climatic areas, each of which produces specific hybrids for particular regions and are the major reproducers of the young stud animals in the country.

The main method of selection and improvement in commercial flocks is cross-breeding. As the result of intensive research work, the most suitable cross-breeding programmes were developed, taking into account the factors influencing the efficiency of cross-breds, particularly the exploitation of heterosis effects.

Breeding programmes are of great importance in the improvement of sheep production in Russia. There are specific programmes for stud farms, stud stations and individual breeds. The existence of such programmes helps in avoiding errors in selection, gives it purpose and raises the efficiency of the breeding work in general.

Until recently the improvement of breeding and productive traits in the sheep was carried out mainly by the mass conversion of low productive sheep breeds into highly productive ones by the use of the best rams from the best breeds. Along with this type of work, there currently exists the possibility of transmitting large numbers of highly productive ewes to the zones with sheep of low productive potential (breed substitution) which should result in the rapid improvement of sheep breeding in these regions.

The existing methods of sheep production on most farms rely on the extensive use of natural pastures, and in most regions are labour intensive: the level of mechanization is low. The extensive nature of sheep production in many areas of the USSR prevents the widespread application of mechanization, thereby limiting intensification and reducing the costs of the production.

In this connection fundamental changes are taking place at present in some regions. The main direction of the change is in the specialization and concentration of sheep production. One method of intensification is the introduction of new advanced methods of technology on an industrial scale.

Recently a number of small farms have been turned into larger ones with a consequent increase in the level of mechanization. These are found in the regions of intensive crop production, with an integration of pasture within the arable rotation. In summer the sheep are normally grazed on these cultivated pastures and then housed in the winter. The important change in the organization of indoor housing of sheep on large farms relates to changes in the preparation of feeds on a commercial scale. The grinding of hay, straw, silage, mixing them in given proportions and supplying the necessary concentrates, protein and vitamins result in the preparation of complete feeds which are pelleted. Different mixtures for different sex and age groups of animals are prepared to raise their productivity and thereby reduce the cost per unit of production. Such a feeding system provides an increase in nutritive value of rations and allows the process of feed distribution, which is the most laborious part of the operation, to be fully mechanized.

A consequence of intensifying sheep production in the intensive arable areas, even allowing for the provision of fencing and water drinking facilities, is that it allows all feeds to be used efficiently and a reduction in labour costs.

This intensification is taking place on many farms now with 5 to 10 thousand head of sheep. In addition feedlots for sheep fattening are being developed as well as changes in the enterprises providing flock replacements.

Judging from the limited information available at present such mechanized farms are economically viable and efficient. It is still early days yet, and there are many problems still to be solved.

The national programmes for future sheep improvement relate to further increases in the profitability of the sheep industry. The most important of these are:

1. Concentration and specialization of sheep production. This includes the establishment of specialized sheep farms on the basis of new developments; specialization of sheep production within individual farms and regions, as well as within the large-scale industrial regions of the country. Stavropol territory, for example, will be specialized to supply other regions of the country with stud animals. Big capital investments will be supplied for the farm building construction, irrigation and watering to improve natural pastures and to establish new cultivated pastures.
2. Further improvement of sheep breeds and their proper distribution in different areas. The testing of various breeds from an economic point of view in the areas of their localization was initiated to select the most productive and adaptable types for particular regions and climatic conditions.

3. Intensification of the breeding programme. More new breeds and hybrids are being established with the application of modern technology. The basic requirements of these breeds are as folllows: high prolificacy, good meat and wool production, early maturity, disease-resistance at high concentrations of animals (stress-resistance).
4. The wide use of cross-breeding in commercial flocks.
5. The establishment of breeding evaluation units for the rapid and reliable genetic selection of stud animals, mainly sires, with the purpose of using only the most desirable genotypes in sheep breeding programmes.
6. Further perfection of the selection methods, the development of suitable selection indices, the use of computers for the analysis of breeding records, the modification of the methods of sheep tagging, etc.

II

GENETIC SELECTION AND BREED IMPROVEMENT

11

METHODS OF IMPROVING PRODUCTION IN CHARACTERS OF IMPORTANCE

HELEN NEWTON TURNER
CSIRO, Epping, N.S.W., Australia

SUMMARY

The first step is to define the characters of importance, which are given as:

For Wool Production

High clean wool weight per head

Average fibre diameter (medium to fine for apparel wool, coarser for carpet wool)

Percent clean yield (optimal level not clearly defined)

Percent medullation (zero or low for apparel wool, certain percentage for carpet wool)

For all Types of Production (wool, wool+meat, meat)

Reproduction rate (high level)

Easy-care features: Skin wrinkle score (low value)
                                Face cover score (low value)

If Meat is a Product

Body weight (high value)

Early growth rate (high value)

Fleece protection during growth is important, but relevant characteristics are not yet clearly known; there is a considerable amount of current research.

The Objective Measurement Policy Committee of the Australian Wool Corporation indicated clearly that crimp number is not important in processing by omitting it from the list of desired characteristics. The OMPC also considered that between-fibre diameter variation within mobs is unimportant in the Australian clip, but along-fibre variation may sometimes be sufficiently great to cause processing problems. Current research on these two sources of variation is discussed, but it is stressed that more information is needed about the **exact** importance of diameter variation to the processer, and about its tolerance limits, before further selection criteria are added.

The OMPC findings relate to Australian wool sold on world markets; elsewhere there may be a need to study the requirements of local markets or of home consumption.

The classic techniques for genetic improvement are selection and cross-breeding. Australian recommendations for selection are discussed, as well as results of experiments confirming them. A common concept is that cross-breeding will lead to faster improvement than selection; this concept is challenged and some comparisons made of likely gains by each method. In particular, when a large population is to be improved, it should be remembered that selection can be done simultaneously throughout the population, but limitation on the number of males available will restrict improvement through cross-breeding.

The importance of a large genetic difference between a ram-breeding source and its dependent flocks is stressed, as well as a high rate of genetic gain when the ram-breeding source is under selection.

DEFINING IMPORTANT CHARACTERISTICS

Sheep produce mainly wool (apparel or carpet), meat, milk and skins; subsidiary uses (manure and transport) will not be considered here. Besides quantity and quality of commercial products to satisfy his market, the breeder is concerned with the sheep's ability to produce in a given environment at minimum cost. He is therefore interested in many characteristics, but must keep the number low to obtain greatest genetic progress in improving them.

(1) *Wool*

(a) Apparel wool

The Final Report of the Australian Wool Board's Objective Measurement Policy committee (AWB 1972) provides a good starting-point for defining the important characteristics of apparel wool, as required for commercial use on the world market. Other requirements are discussed later.

The OMPC list will be re-arranged under the headings of quantity, quality, and protection during growth.

(i) Wool quantity

Clean wool weight is the commercial product, but would it be per head or per unit of input (feed and other costs)? Most costs are per head, and wool weight/unit of feed ("efficiency of conversion") is correlated with weight per head (Turner and Young 1969, Turner and Dunlop 1974*). Wool weight per head therefore stands as the general definition of quantity-greasy weight for all except final selection of stud rams, since the correlation with clean weight is high ($+0.8$ to $+0.9$).

---

* Reviews are quoted where available, together with subsequent papers.

Efficiency of conversion is important with hand-feeding; is it equally important for pasture-fed sheep with a fluctuating feed supply (Dolling 1960)? Selection for high wool weight increases efficiency, but the gain is estimated as 40-70% of the potential under direct selection for efficiency (Turner and Young 1969). The cost of efficiency tests, and the obligatory reduction in numbers tested, need to be balanced against the value of increased genetic progress in efficiency; this exploration has not been undertaken. Selection on wool weight/body weight is no better for increasing efficiency than selection on wool weight alone, if decreases in body weight are controlled (Turner and Young 1969).

A high *percent clean yield* is important both to producer and processer, but some wax and suint may be needed for fleece protection.

(ii) Wool quality

The OMPC list of important characteristics is:—

Average fibre diameter,

Average fibre length (controlled by breakage),

Distribution of fibre length (also controlled by breakage),

Fibre strength (as far as it is determined by "soundness"),

Coloured fibres (in otherwise white wool).

Pigmentation of whole fleeces was not mentioned; the world market demands mainly white wool, though there is a current fashion for natural colours. "Colour" (degree of whiteness) was also not listed, but the text mentioned the need for scoured wool capable of taking desired shades. A clearer statement about permissible levels of "creaminess' is needed from processers.

Medullation was not mentioned at all; no medullation, or a low percentage, is required for apparel wool. Australian wools meet this requirement, but medullation may be a problem elsewhere.

Two notable omissions from the OMPC list are crimps per unit staple length and fibre diameter variation.

Crimp is unimportant in processing except in a few special circumstances. Further, over the last 3 decades it has been shown to be unreliable in its traditional role as a guide to diameter (Lang 1947, Roberts and Dunlop 1957, Turner, Dolling and Kennedy, 1968, Turner, Brooker and Dolling, 1970, AWC 1973), while, because of its high negative genetic correlation with wool weight, attempts to raise weight while maintaining crimp will greatly hinder selection response.

Variability in fibre diameter can arise between or along fibres. The latter is mentioned by OMPC as important under "fibre strength" and "soundness", but within Australian flocks between-fibre variation was regarded as acceptably low.

The components of between-fibre variation within one mob have been estimated as contributing to total variation (AWB 1972, Dunlop and McMahon 1974)—

	%
Fleeces	19
Body regions	5
Fibres within staples	76
	100%

Diameter variation along the fibre (which may affect soundness) may contribute 20-75% of total variation, depending on seasons.

The OMPC conclusion that between-fibre variation can be ignored is not universally accepted within Australia, and may not be valid for breeds elsewhere. Before yet another selection criterion is added however, further evidence is required from processing trials about the exact effects of diameter variation, with specification of permissible limits.

In the meantime, research on genetic aspects of within-staple diameter variation is in progress, concentrating on the skin, and working with both horizontal and vertical skin sections. On horizontal sections, diameter variation can be split into 4 components:

Difference in mean diameters of fibres from primary and secondary follicles ($\bar{D}_P - \bar{D}_S$),

Variation among fibres from primary (VD_P) and secondary (VD_S) follicles,

Ratio of secondary to primary follicles (S/P).

For the Australian Merino, preliminary investigations indicate that S/P is so large and ($\bar{D}_P - \bar{D}_S$) so small that VD_S becomes the major component. Further, there are more "outlying" fibres of extra high diameter from secondary than primary follicles. More data are needed (Turner unpublished).

The high cost of measuring horizontal sections has led to investigation of a possible genetic correlation between evenness of follicle depth (scored on vertical sections) and between-fibre diameter variation (Nay, in progess). If a correlation is demonstrated, a cheaper score for diameter variation will be available.

Diameter variation along the fibre is largely influenced by nutritional changes, but there is evidence that individuals differ in their ability to maintain an even diameter (Jackson and Downes, in progress).

Fibre length is listed as important, but is expensive to measure; the easily-measured staple length is correlated with it, though not highly. As will be seen later, staple length increases under selection for wool weight, so need not be a separate selection criterion.

Coloured fibres in a white fleece are a disadvantage, but there is no information about their inheritance.

(iii) Fleece protection during growth.

The important features are:—

Weathering (damage from sunlight, dust penetration etc.),
Soundness (already discussed),
Vegetable matter (amount, type, distribution),
Discoloration (due to fly-strike and "fleece rot").

There are many traditions and opinions about characteristics which will protect the fleece during growth, and since the OMPC Report research in the area has intensified. So far there are few established conclusions.

The characteristics under investigation include:—

Grease content (absolute and relative amounts of wax and suint),

Fibre population density,

Staple size,

Crimp (which may be important in holding fibres together, though not in processing).

One characteristic (degree of skin wrinkle) has been shown to be related to fly-strike, plainer sheep being less susceptible.

(iv) Attention to end users.

The OMPC findings relate to Australian wool, most of which is Merino. They can be extrapolated to apparel wool sold on the world market. Local markets, or wool grown for home consumption, may have different requirements. Pigmented wool is demanded in many countries to save dyeing costs in making all-purpose blankets, used also as cloaks, while hand processing may call for a higher average fibre diameter than commercial processing.

The end use of the wool must always be considered in developing selection criteria.

(b) Carpet wool.

No publication similar to the OMPC Report exists for carpet wool. Quantity characteristics are likely to be the same as for apparel wool. Considerable work on desirable quality characteristics of Indian carpet wools has been done at the Wool Research Association, Bombay, under the direction of Messrs. A.D. Sule and G.R. Kulkarni (Wool Research Association 1971-74). Based on this, one definition of desirable characteristics for Indian carpet breeds is (Sule, personal communication):—

Average fibre diameter	36-40 μm
Percent fibres without medullation	At least 40%
Percent fibres with medullation:	
Heterotype (interrupted medulla)	At least 20%
Hair (medulla 60% of diameter)	Not more than 10%
Kemp (shed fibres)	0 or under 2%
Length	3-3½ inches.

TABLE 1

IMPORTANT CHARACTERISTICS FOR WOOL
A=APPAREL WOOL; C=CARPET WOOL
W=WORLD MARKET; L=LOCAL MARKET; H=HOME USE

Characteristic	Type of Wool	Market	Requirements
Clean wool weight per head	A or C	All	High
Average fibre diameter	A	W	Medium to fine
		L or H	Coarser, but no exact specifications available
	C	All	Coarser (within limits)
Percent medullation	A	W	Zero or low per cent
		L or H	Higher levels acceptable (of necessity)
	C	All	Certain percentages desirable
Length	A or C	W	Minimum lengths specified
		L or H	Various lengths used
Pigmented fleeces	A	W	White (except in special cases)
		L or H	Strong demand for pigmented as well as white
	C	W	Generally white, but some demand for pigmented
		L or H	Strong demand for pigmented as well as white
Pigmented fibres in white wool	A or C	W	Undesirable
		L or H	Acceptable
Colour (degree of "whiteness")	A or C	W	Stated to be important but clearer specification required
		L or H	Unimportant
Variability of fibre diameter (between fibres)	A	W	Australian wool: Considered within acceptable limits but clearer specification of "acceptable limits" needed
			Other wool: More information required before Australian findings extrapolated
		L or H	No information
	C	All	No specific information
Protection during growth: Soundness (diameter variation along fibre) Weathering Vegetable content Discoloration	A or C	All	All features important, but more information required about characteristics which might offer protection such as: Wax, suint (content & ratio) Crimp Staple size Fibre population density

The Indian workers stressed that the definition might not apply to breeds outside India. Earlier enquiries (Turner 1973a) led to the conclusion that the main function of medullation was to lower costs, but Mr. Sule (personal communication) considers that heterotype fibres contribute to "springiness".

(2) *Meat*

The main characteristics important for meat production are:—

> Reproduction rate,
> Body weight,
> Early high growth rate.

Reproduction rate (defined as number of lambs born, or weaned, per ewe joined) is important for all aspects of sheep production. The main contributor to a higher rate is the incidence of multiple births.

Body weight or early growth rate are not important to wool production, but become so when meat is considered.

Meat quality is largely affected by pre- and post-slaughter handling; there is little information about related characteristics on the live animal.

(3) *Milk*

Australia has only one sheep-milk farm. The most thorough genetic research on sheep-milk production is done in France, and will not be summarized here.

(4) *Skins*

Skins are an important by-product in Australia, and a main product in some other countries, but little information is available on which selection criteria can be based. Research is in progress in the W.A. Department of Agriculture.

(5) *"Easy-care" features*

Labour costs are so high in Australia that features likely to reduce maintenance costs must be considered. Some which have been investigated are:—

> Skin wrinkle score: Plain-bodied sheep are easier to shear, have fewer shearing cuts, and less fly-strike. In addition, they have a longer staple and higher reproduction rate, and often no less clean wool than sheep with a higher wrinkle score (Turner and Young 1969, Dun and Eastoe 1970).

> Face cover score: open-faced sheep reduce wigging costs. The once-vaunted correlation between face cover and reproduction rate, however, is not high enough to be of value for indirect selection.

(6)   Conclusions

Characteristics important for wool (apparel or carpet) are summarized in Table 1.

End use has been taken into consideration, via three avenues:

W = Sale on world markets, for commercial use,

L = Sale on local markets,(e.g.), from one village to another,

H = Home use, that is, processed by the grower.

In many cases, wool in categories L and H would be processed by hand.

Summarizing from Table 1 and Sections (b-e), the following important characteristics have definite information on which to base selection criteria:—

Main product	Characteristic	Requirement
Wool	Clean wool weight per head	High level
Wool	Average fibre diameter:	
	Apparel wool	Medium to fine
	Carpet wool	Coarser
Wool	Percent medullation:	
	Apparel wool	Zero to low percent
	Carpet wool	Certain percentage
Wool, or Wool+Meat, or Meat	Reproduction rate (number of lambs born or weaned per ewe joined)	High level
Wool+Meat, or Meat	Body weight and Early growth rate	High levels
Wool, or Wool+Meat, or Meat	Easy-care: Skin wrinkle score Face cover score	Low levels

Fleece protection during growth is important for wool, but no definite objective selection criteria are yet available. None are available for skins. Further, body weight is included as a criterion in many programmes, largely because many costs are on a per head basis, but there is no work available to assess its correlation with efficiency, such as exists for wool weight.

GENETIC IMPROVEMENT

The classic techniques for genetic improvement are:—

1. Selection
   (a) Direct
      (i) Mass selection on individual phenotype
      (ii) Using relatives' performance (with or without (i))
   (b) Indirect on correlated characters

2. Cross-breeding
   (a) Changing the genotype
   (b) Exploiting heterosis

In many cases, (2) will be combined with (1).

The usual statement is that heritability level is the main criterion for choosing between (1) and (2), but other points for consideration are:—

Products wanted, and the characteristics defining them,

Levels of these characteristics for the flock in the environment under consideration, as well as for the type considered for crossing (or the cross-bred itself) *in that same environment,*

Genetic correlations between the characteristics, as well as their heritability levels,

Whether or not major genes control any production characteristics (e.g. pigmentation); use of heritability is based on an assumption of additive gene action.

The heritability levels of the characteristics listed on p. 98 are all high enough for successful selection (Turner and Dunlop 1974), there are no antagonistic genetic correlations provided diameter replaces crimp for assessing quality (Turner 1972), and most are controlled by additive genes. But the need for looking at the desired product is illustrated if we consider the impossibility of using selection alone to turn a Merino flock, with zero medullation, into a carpet-wool flock with 30% medullation in the fleece.

Because 75% of Australian sheep are Merino, with apparel wool as the main product, selection has received considerable attention here. The most useful course in this Section is to state the Australian recommendations and what might be achieved by them, then to make a comparison between genetic gains through selection and through cross-breeding. The latter becomes particularly relevant when so many countries, including Australia and New Zealand, are importing, or planning to import, exotic breeds for crossing with existing flocks. Gains by each technique will be considered at two levels:

For individual flocks breeding their own rams, or groups of flocks dependent on a central stud (or studs) or nucleus,

For a large population, such as might be involved in the case of exotic imports.

(1) *Selection*

(a) Wool-producing, or wool-and-meat-producing flocks

(i) Australian recommendations

These are:

- Mass selection (on individual performance) will be used in most cases.

- Performance records on relatives will be needed only:
  - For progeny tests on rams when AI is to be used, or when rams from different sources are to be compared.
  - For selecting rams on reproduction rate, when female relatives are needed. Dam's performance is more profitable than that of any other female relative, but if written records are kept those for different types of relative can be combined.
- The main selection criteria should be based on objective measurement, namely:—
  - Wool weight per head (greasy weight for all except final ram selection, when clean weight should be used).
  - Average fibre diameter (rams only, because of cost).
  - Body weight (not necessary if wool is the only product, but to be included when meat is considered).
  - Number of lambs born (or weaned) (for rams, by the dam; for ewes, combination of dam's and own performance).
- Other characteristics related to maintaining the animal at low cost should be based on definite information, such as:
  - Degree of skin wrinkle.
  - Amount of face cover.

These recommendations are for Australian conditions, under which sheep are mostly run at pasture all the year round in fenced paddocks, without shepherding. Labour costs are high. Written performance records have rarely been kept in the past, except on a few studs, but their use is increasing.

These conditions raise difficulties in identification at birth of multiple-born lambs or multiple-bearing ewes, and the following characters have been investigated for use in indirect selection (parameters from Young, Turner and Dolling 1963):—

Character	Genetic correlation with number of lambs at 1st 3 lambings		Gain under indirect as a percent of gain under direct selection	
	Born ($h^2 = 0.4$)	Weaned ($h^2 = 0.2$)	Born	Weaned
Body weight	+0.2	+0.5	26	65
Skin wrinkle score	−0.2	−0.3	19	29
Face cover score	−0.1	−0.3	10	29

In the case of wrinkle and face score, selection would be for lower values. Selection for plain body has the added advantage of having a direct effect on the ram's fertility (Dun and Eastoe 1970). Clearly no indirect selection is as efficient as direct, but body weight and wrinkle score will confer some advantage if direct selection is impossible.

Hormone levels are under investigation as a means of indirect selection, and are being discussed elsewhere in this book.

The recommended age for selection is prior to first joining (usually at 1 ½ years of age). Preliminary selection techniques are available (Nay 1973), but there is no space here to discuss their effects (Turner and Evans, in preparation).

(ii) Combining characteristics.

Given individual selection criteria, the question arises of how to combine them.

Firstly, average fibre diameter is of great importance, but what diameter should be the aim? Turner (1973b) used results of sales by measured sample to obtain regressions of price on average diameter, and was able to show that selection for increased fleece weight, while keeping diameter from increasing, is in most cases more profitable (in terms of money return per head) than selection for finer diameter. Once the average clean wool weight has reached 4.5 kg, however, reducing diameter (without losing fleece weight) may be more profitable than increasing fleece weight, depending on the existing average fibre diameter.

Diameter can readily be kept from increasing by rejecting rams with a diameter more than one standard deviation (approximately 2μm) above their group mean, even though their wool weights may be high.

The question of combining wool and body weights, when meat production has to be considered, has received some attention. Turner and Evans (in preparation) have developed charts designed to estimate the relative economic values (REV) of wool weight and body weight for a range of prices for wool, lambs and carcase, and from them have suggested the following selection indices:

(6 to 8) × greasy wool weight + body weight

(8 to 10) × clean wool weight + body weight

Selection for reproduction rate can be achieved by selecting on dam's performance, taking twin-born animals if records are not kept, (Young and Turner 1965), or selecting on the dam's lifetime performance if they are (Rae 1963, Turner 1968). Young and Turner (1965) discussed methods of selecting simultaneously for reproduction rate and wool weight, in the absence of written records, but no-one has yet published figures for the optimum combination for wool weight (maintaining diameter), body weight and reproduction rate.

(iii) Confirmation of recommendations.

The estimates of heritability for sheep production characteristics, and of the genetic correlations among them, have been confirmed in selection experiments, which have been listed and discussed by Turner (1976). In only one experiment out of a total of 9 (Pattie and Barlow 1974) has clean wool weight failed to show a continuing response; in all others the response has ranged up to 7% per annum.

There is no experiment in which selection includes all the recommended characteristics, but in one (CSIRO) selection has been based since 1950 on high clean wool weight, with rejection for high fibre diameter and high wrinkle score. The increase in clean wool weight has averaged just over 2% per annum, except for a period of drought years. There are 2 groups, in one of which the selection for quality was changed in 1961 to maintaining crimp instead of diameter. Response in clean wool weight with diameter maintained has been over 2%; with crimp maintained only 1% per annum. Neither body weight nor reproduction rate has changed under selection for high clean wool weight.

In another CSIRO selection experiment, with selection only for reproduction rate, results have been:

Group	Selected for	Begun	Lambs born/100 ewes joined to ewes aged 2-7 years (1972)
B	High No. lambs born	1959	209
T	High No. lambs born	1954	137
O	Low No. lambs born	1954	112

Wool weight has not suffered under selection for high reproduction rate.

Reference should also be made to the several breeds in the world which have exceptionally high reproduction rate—for example, the Finn, Romanov, Chios and Dahman. From what is known of their history, it seems selection in the past has contributed to their high performance.

(iv) Meaning of results in practice.

The CSIRO results are for the actual groups being selected. In practice, there will usually be a central nucleus (or stud) under selection, from which flocks will draw their rams. The relationship between the nucleus and the dependent flock is shown in Table 2.

The ratio of gain in flock and nucleus eventually becomes the same, but the flock will always be 2 generations behind the stud in production level. The time taken to reach the same rate of gain depends on the difference in genetic level at the start. If $d = 2\Delta G$, then flock and nucleus will have the same rate of gain (ΔG) from the start. If $d < 2\Delta G$, the rate in the flock is lower at the start and slowly rises to that of the nucleus. If $d > 2\Delta G$, then the rate of gain in the flock is *higher* initially than in the nucleus, but decreases until it is equal.

There are benefits to a whole system of nucleus and dependent flocks in ensuring that the nucleus has:

The highest possible genetic level,

The highest possible rate of gain.

These points will be discussed elsewhere in this book, but an important point is that the cost of achieving high levels in the nucleus should be balanced against the size of the system which benefits from them.

It is possible to set up a number of model systems and to determine in them the benefits of selection. To do this, a flock composition must be chosen, since this varies at different times of year. The composition at selection time would be:

Breeding ewes mated the previous year,

Young animals aged $1\frac{1}{2}$ years (unselected),

Weaners (both sexes),

Mature rams.

Most of the wool comes from the ewes, so Table 3 gives gains each year under selection for the breeding ewes (before replacements are made) and the $1\frac{1}{2}$-year-old ewes. Columns 2 and 3 show the gain in production level over that in Year 1, while Cols. 3 and 4 show the total *extra* production through selection since Year 1, obtained by cumulating the figures in Cols. 2 and 3.

Table 3 can be used for a nucleus, to estimate gains and balance them against costs. The same can be done for a dependent flock, entering the Table two generation-lengths behind the nucleus. For example, with a generation length of 3.25 years, a dependent flock would reach between Years 8 and 9 the level reached by the nucleus in Year 15.

(b)   Meat-producing flocks

The requirement for meat-producing flocks would be high reproduction rate and high body weight at weaning, rather than at $1\frac{1}{2}$ years. At this age, corrections for birth type (single or multiple) and age of dam and lamb are required for accurate selection. These are built into the performance-recording programmes already in operation in New Zealand and being constructed in Australia.

(c)   Selection in a large population

Selection within flocks, or within nuclei supplying their dependent flocks, can be practised simultaneously throughout a large population. As will be seen later, this is in contrast with the cross-breeding situation.

(2)   *Cross-breeding*

Selection within flocks can lead to steadily increasing gains for the important sheep production characteristics, *provided* there is genetic variation on which to operate. What has cross-breeding to offer?

Selection operates on genes assumed to have mainly an additive effect. Cross-breeding may operate through bringing in new genes with additive effects; either a new genotype is developed with genes from each parent, or crossing is continued till the old genotype is replaced by the new ("grading-up"). This kind of cross-breeding has occurred everywhere in developing new strains or breeds.

TABLE 2

RELATIONSHIP BETWEEN A NUCLEUS AND A FLOCK DRAWING RAMS FROM IT

ΔG = Gain per generation in nucleus, due to selection

d = difference in genetic level of nucleus (N_0)

and flock (F_0) at start

(After Bichard, 1971)

Generation	Production level		Change from preceding generation	
	Nucleus	Flock	Nucleus	Flock
0	N_0	$F_0 = N_0 - d$	—	—
1	$N_0 + \Delta G$	$\frac{1}{2}(N_0 + F_0)$ $= N_0 - \frac{1}{2}d$	ΔG	$\frac{1}{2}d$
2	$N_0 + 2\Delta G$	$N_0 - \frac{1}{4}d + \frac{1}{2}\Delta G$	ΔG	$\frac{1}{4}d + \frac{1}{2}\Delta G$
3	$N_0 + 3\Delta G$	$N_0 - \frac{1}{8}d + \frac{5}{4}\Delta G$	ΔG	$\frac{1}{8}d + \frac{3}{4}\Delta G$
4	$N_0 + 4\Delta G$	$N_0 - \frac{1}{16}d + \frac{17}{8}\Delta G$	ΔG	$\frac{1}{16}d + \frac{7}{8}\Delta G$
5	$N_0 + 5\Delta G$	$N_0 - \frac{1}{32}d + \frac{49}{16}\Delta G$	ΔG	$\frac{1}{32}d + \frac{15}{16}\Delta G$
y	$N_0 + y\Delta G$	$N_0 + (y-2)\Delta G$	ΔG	ΔG

Cross-breeding may also operate by exploiting "hybrid vigour (heterosis)" *if it occurs.* This is a phenomenon which sometimes appears with the crossing of genetically distinct groups (breeds, strains, inbred lines); characteristics in the cross then have levels different from the mid-parent level which would be predicted if genes acted additively. The difference may theoretically be positive or negative, but is only of practical value if positive, and if the cross is superior to the better parent.

The superiority of a cross over the *inferior* parent is sometimes incorrectly called hybrid vigour. Additive genes derived from the superior parent contribute to the cross; hybrid vigour *may* add something more, but is not always present.

Additive gene effects, obtained either through selection or crossing, are permanent, and, in the case of continued selection, cumulative. Hybrid vigour effects present in the first cross later gradually disappear, so that the cross must be continually made to obtain maximum benefit.

The value of hybrid vigour therefore depends on the amount which is present when two groups in question are crossed. The amount of information available for sheep is not great, and no work has been done to estimate the cost of keeping separate populations and continually crossing them, compared with the cost of selection or crossing (without heterosis) + selection. For these reasons the present discussion will be confined to changing the genotype through crossing.

(a) Changing the genotype (without selection).

(i) For one flock.

The procedure is to introduce animals (usually rams) from one (or more) genetic groups into another. For example, a flock owner decides to change the strain of rams he buys, or a country with a low-producing native breed looks to introducing exotic breeds.

Cross-breeding should not be started without an evaluation of the cross. This means not only estimating whether heterosis is present (with sheep it is frequently unimportant), but, more importantly, how the introduced group, or its cross, will compare with the existing group *in the existing environment.*

Let us take one Indian cross-breeding experiment as an example. Results have been variable, but one will serve as illustration (CSWR1 1972). The American Rambouillet has been crossed with various native breeds at the Central Sheep and Wool Research Institute in the semi-arid area of Rajasthan. The Rambouillet at home produces 4-5 kg of greasy wool, with a diameter of 21-23 μm and virtually no medullation. One native Indian breed (Chokla) produces 1-2 kg, with a diameter of 26-30 μm and 20-40% medullation. Predicting from these, a first-cross with no heterosis might produce 2.5-3.5 kg of wool, with a diameter of 23.5-26.5 μm and 10-20% medullation.

Breed	Greasy wool weight (kg)	Fibre Diameter (μm)	Percent medullation
$\frac{1}{2}$R$\frac{1}{2}$C — F1	2.2	21	22
— F2	2.2	22	16
$\frac{5}{8}$R$\frac{3}{8}$C	2.2	20	12
$\frac{3}{4}$R$\frac{1}{4}$C	2.0	17	8
Rambouillet (R)	2.3	17	0
Chokla (C)	2.0	26	45

Fleece weight in the first-cross is below what would be predicted on Rambouillet home performance, but in agreement with its much lower performance in Rajasthan. The diameter of the Rambouillet is finer in Rajasthan; diameter and medullation are both about half-way between the parents in Rajasthan.

An increase in R genes leads to predicted falls in diameter and medullation, but to no further increases in fleece weight—in fact, an actual fall for $\frac{3}{4}$R. The parent fleece weights, however, are themselves not far apart under Rajasthan conditions.

These results make clear the dangers of extrapolating from one environment to another, and emphasize the need for evaluation by contemporary comparisons—not necessarily of the two parent breeds, but of the native breed, the F1, and, preferably, the F2 as well.

An argument advanced in favour of cross-breeding, compared with selection, is that results can be achieved more quickly. Is this true?

To begin with, a cross must be evaluated. This means setting aside part of the existing flock, splitting the part at random into two halves, one to be mated to rams from the existing flock, the other to introduced rams, whether of a different strain or a different breed. This should be done in at least two years, the resulting progeny being run together and compared. Wool can be compared at $1\frac{1}{2}$ years of age which means completion of the comparison 3 years after commencement. It would be preferable to compare ewe lambing percentages as well, for at least the first 2 lambings, which adds 2 years to the evaluation period (up to weaning), making the total 5 years.

The second half of Table 3, which deals with crossing without selection, compares this procedure with selection. Extracts are:

Ewes	At Year	Production level		Total extra production since Year 1	
		Selection	Crossing	Selection	Crossing
Breeding	10	Base+ $4.0\Delta G_a$	Base+0.6c	$12\Delta G_a$	1.2c
flock	15	Base+ $9.0\Delta G_a$	Base+1.0c	$47\Delta G_a$	6.0c
$1\frac{1}{2}$-years-old	10	Base+ $8.0\Delta G_a$	Base+1.0c	$36\Delta G_a$	4.0c
prior to selection	15	Base+ $13.0\Delta G_a$	Base+1.0c	$91\Delta G_a$	9.0c

For equivalent gains, c (the difference between the existing breed and the cross) need to be 8-10 times ΔG_a (the annual genetic gain). If wool weight is under selection and ΔG_a is 2%, then the half-bred must have a fleece weight more than 20% above the existing breed if crossing is to be more benefit than selection, after 10 years. This means that the introduced group (breed or strain) must produce 40% more than the existing group, in the existing environment.

(ii) For a large population.

Selection can be done simultaneously throughout a large population. The problem with crossing is limitation on the number of introduced males. Again various models can be set up, but one example will serve. This was worked out recently in relation to the number of village flocks in India which could be influenced in 20 years by a central stud of 3000 exotic breeding ewes. This stud was assumed to supply rams to 12,000 ewes in centres which would produce cross-bred rams for distribution. Surplus pure exotic rams were assumed supplied direct to village flocks. Using natural service for cross-bred and AI for pure-bred rams, it was estimated that the following numbers of ewes could be brought into the scheme:

In 10 years 1.4 million

In 20 years 2.8 million

The average percent of exotic genes for breeding ewes in the village flocks would be:

	In 10 yrs.	In 20 yrs.
Served by pure-bred then ¾ bred rams	36	53
Served by cross-bred rams (½ and ¾)	20	33

(b) Changing the genotype (combined with selection).

TABLE 3

GAINS UNDER SELECTION AND CROSS-BREEDING (EWE LIFE ASSUMED 5 YEARS)

	Selection				Cross-breeding (without selection)			
	Gain from base year (in terms of ΔG_a*)		Cumulated gain from base year (in terms of ΔG_a)		Gain from base year (in terms of c**)		Cumulated gain from base year (in terms of c)	
Year	Breeding*** ewes	1½-year old ewes	Breeding ewes	1½-year old ewes	Breeding ewes	1½-year old ewes	Breeding ewes	1½-year old ewes
1	0	0	0	0	0	0	0	0
2	0	0	0	0	0	0	0	0
3	0	1.0	0	1.0	0	0	0	0
4	0.2	2.0	0.2	3.0	0	0	0	0
5	0.4	3.0	0.6	6.0	0	0	0	0
6	0.8	4.0	1.4	10.0	0	0	0	0
7	1.4	5.0	2.8	15.0	0	1.0	0	1.0
8	2.2	6.0	5.0	21.0	0.2	1.0	0.2	2.0
9	3.0	7.0	8.0	28.0	0.4	1.0	0.6	3.0
10	4.0	8.0	12.0	36.0	0.6	1.0	1.2	4.0
11	5.0	9.0	17.0	45.0	0.8	1.0	2.0	5.0
12	6.0	10.0	23.0	55.0	1.0	1.0	3.0	6.0
13	7.0	11.0	30.0	66.0	1.0	1.0	4.0	7.0
14	8.0	12.0	38.0	78.0	1.0	1.0	5.0	8.0
15	9.0	13.0	47.0	91.0	1.0	1.0	6.0	9.0
20	14.0	18.0	107.0	171.0	1.0	1.0	11.0	14.0

* ΔG_a = annual genetic gain
** c = difference between existing breed and half-bred with introduced breed
*** Ewes mated previous year, before replacements added.

The figures in Table 3 assumed no selection. Obviously selection can be imposed on crossing, and the situation would then become similar to one in which flocks drew from a superior nucleus (Table 2). There could be central nuclei of pure-bred and cross-bred animals, as described in the Indian example above; in Table 2 the difference between flocks and pure-bred nucleus would be d_1, and between flocks and cross-bred nucleus d_2. Rates of gain in the nuclei would supply the ΔG values.

(c)   Some problems.

Inheritance between breeds is not always the same as inheritance within breeds, and this can cause problems in predicting the results of crossing.

Pigmentation is one example, medullation another. Does the Merino possess genes inhibiting medullation, thus being likely to cause problems in crossing with carpet breeds—or does it not? This question is currently under investigation.

CONCLUSIONS

The scope for this chapter is enormous, and all that has been achieved is to stimulate discussion, rather than to give final answers. We have a lot of information about some important characters and how to select for them. We need more information about others—particularly how to protect the fleece during growth. We should not rush into adding further selection criteria unless we are sure of them—and we should realize the importance of high genetic superiority and rates of gain in central nuclei whether these are pure-bred or cross-bred.

We should also ask questions about the relative merits of selection and crossing before we embark on the latter, and I have tried to indicate what form these questions should take.

REFERENCES

Australian Wool Board (1972). Objective measurement of wool in Australia. Final Report of the Australian Wool Board's Objective Measurement Policy Committee. October 1972.

Australian Wool Corporation (1973). Objective measurement of wool in Australia. Technical Report of the Australian Wool Board's Objective Measurement Policy Committee. October 1973.

Bichard, M. (1971). *Anim. Prod.* **13**: 401-11.

CSWRI (1972). Central Sheep and Wool Research Institute, Aviknagar, Rajasthan, India. Annual Report, 1972.

Dolling, C.H.S. (1960). *Proc. Aust. Soc. Anim. Prod.* **3**: 69-76.

Dun, R.B. and Eastoe, R.D. (1970). Science and Merino Breeding. N.S.W. Govt. Printer.

Dunlop, A.A. and McMahon, P.R. (1974). *Aust. J. Agric. Res.* **25**: 167-81.

Lang, W.R. (1947). *J. Text. Inst.* **38**: T241-56.

Nay, T. (1973). *Wool follicles. A Manual for Breeders.* Prepared by CSIRO. Published by Australian Wool Corporation.

Pattie, W.A. and Barlow, R. (1974). *Aust. J. Agric. Res.* **25**: 643-55.

Rae, A.L. (1963) *Massey University, Sheep-farming Annual*, 1963: 167-82.

Roberts, N.F. and Dunlop, A.A. (1957). *Aust. J. Agric. Res.* **8**: 524-46.

Turner, Helen Newton (1968). Proc. Symposium on Physiology of Reproduction in Sheep, U.S.D.A. Sheep Development Programme, Stillwater, Oklahoma.

Turner, Helen Newton (1972). *Anim. Brdg. Abstr.* **40**: 621-34.

Turner, Helen Newton (1973a). Report on carpet wool to the Expert Panel on Breeds of Sheep and Goats for Import to Australia (appointed by Animal Production Committee). (Unpublished).

Turner, Helen Newton (1973b). Paper 42 in A.W.C. (1973).

Turner, Helen Newton (1976). *Animal Breeding Abstracts* **44**: 1206.

Turner, Helen Newton, Brooker, M.G. and Dolling, C.H.S. (1970) *Aust. J. Agric. Res.* **21** : 955-84.

Turner, Helen Newton, Dolling, C.H.S. and Kennedy, J.F. (1968). *Aust. J. Agric. Res.* **19** : 79-112.

Turner, Helen Newton and Young, S.S.Y. (1969). Quantitative Genetics and Sheep Breeding. Macmillan Co. of Australia, Melbourne.

Wool Research Association, Bombay, (1971-74). Studies in the fine physical mechano-chemical and morphological characteristics of various Indian carpet wools, with a view to evaluate their performance characteristic. Series of Annual and Technical Reports, WRA, Bombay.

Young, S.S.Y. and Turner, Helen Newton (1965). *Aust. J. Agric. Res.* **16** : 863-80.

Young, S.S.Y., Turner, Helen Newton and Dolling, S.S.Y. (1963). *Aust. J. Agric. Res.* **14** : 460-82.

12

RESPONSE TO SELECTION FOR INCREASED FLEECE WEIGHT IN MERINO SHEEP

B.J. McGUIRK* and K.D. ATKINS**

*CSIRO, Division of Animal Production, Epping, N.S.W., Australia
**Agricultural Research Station, Trangie, N.S.W., Australia

SUMMARY

Eight generations (23 years) of selection for increased clean fleece weight at Trangie Agricultural Research Station (Fleece Plus flock) has been successful in increasing hogget wool production by approximately 20%. While response would appear to be continuing, it is at a much slower rate than was achieved in the early years of the experiment.

Correlated responses to selection were as expected, with adult fleece weights, average fibre diameter and yield increasing, while hogget live weight has not. The incidence of dry ewes in the Fleece Plus flock is higher than in a Random control flock, but this difference between the flocks has not increased with time.

INTRODUCTION

Increased clean wool production is one of the major breeding goals for the Australian Merino Industry. In this chapter are described the effects of eight generations of selection for increased fleece weight in the Fleece Plus flock at Trangie Agricultural Research Station. Responses to selection between 1951 to 1965 have been described by Pattie and Barlow (1974) and Barlow (1974). Here the aim will be to bring the results up to date, to relate responses observed to those predicted from estimates of heritability and genetic correlations, and to consider the effects of selection for increased hogget wool production on flock productivity.

MATERIALS AND METHODS

In 1951 ten selection flocks, each of 100 ewes, were formed at Trangie from a base flock of 2100 ewes, the second stud ewes on the Station. One flock (Fleece Plus) was selected for increased clean fleece weight based on performance at hogget shearing. These ewes were joined with five hogget rams, the top rams on clean fleece weight in a flock of 150. These rams selected for the Fleece Plus flock were 1.13 kg (28%) superior to their flock average while the Fleece Plus ewes had an average superiority of 0.23 kg.

Since then the Fleece Plus flock has been closed with hogget replacements being selected from within the Fleece Plus flock. Five hogget rams have been selected each year and on average, 48% of available hogget ewes. The average annual selection differentials achieved by these selections were 0.75 kg and 0.19 kg respectively for the rams and ewes.

Responses to selection in the Fleece Plus flock have been estimated as deviations from an unselected control flock (Random). This flock was also started in 1951, with 100 ewes and 10 rams. Since 1962, 25 hogget rams have been used each year. All hogget replacements in this flock are chosen completely at random, the aim being to maintain the genetic merit of the Random flock at its 1951 level.

RESULTS AND DISCUSSION

(a)   Hogget clean fleece weight

Direct response to selection for increased fleece weight is indicated in Figure 1, which includes both absolute and percentage deviations from the Random flock, averaged over sexes. Selection has been effective, although response has been somewhat erratic. The 1974 drop Fleece Plus hoggets produced 25% more clean wool than their Random counterparts, after 23 years of selection. Nevertheless, initial rates of response have not been maintained, and realized heritability estimates are now below those estimated for the Trangie base population (Morley 1955).

Figure 1 — Response to selection for increased fleece weight in the Fleece Plus flock.

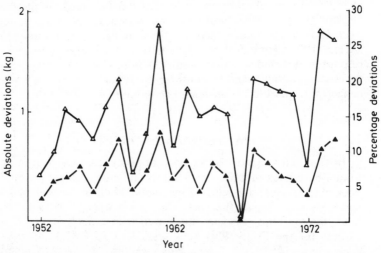

Response calculated as absolute deviations (kg) (▲————▲) and percentage deviations from the Random flock (△————△)

In reviewing the results of the experiment to 1965, Pattie and Barlow (1974) concluded that response had reached a plateau. However, it is not clear if response really has ceased; given the considerable year to year fluctuations in relative performance this would in any case be difficult to assess. In order to remove some of these fluctuations, the percentage differences between the flocks have been averaged for 5 four-yearly intervals (1952-55 drops, 56-59 etc.) and a final three year period (1972-74 drops). The Fleece Plus flock's superiority was then as follows: 11.8, 17.3, 17.7, 16.2, 19.1 and 19.8%. While this break-up of the data is completely arbitrary, it is suggested that response to selection is still being achieved, if at a slow rate.

(b) Adult Wool Production

In selecting for increased wool production, the general aim is to increase flock productivity. Hogget records are recommended as the basis for selections because they are available before animals normally enter the breeding flock, and because hogget performance is thought to be a reliable guide to an animal's breeding value for wool production.

Greasy fleece weights are available for ewes in the Fleece Plus and Random flocks for the period 1959 to 1971. The records were for shearings 2 to 7, when ewe ages ranged from 21 months to 6 years 9 months. The results indicate that selection for increased hogget clean fleece weight has increased greasy fleece weights at all adult shearings, as Brown, Turner and Dolling (1966) also found. For an improvement of 10 per cent in hogget clean fleece weight, adult greasy fleece weight has been increased by 7.7%, or by 0.35 lb greasy wool per head. A proportion of this gain would represent current flock gains, due to positive phenotypic regressions of adult fleece weights on hogget performance.

(c) Crimp and Diameter

Selection for increased fleece weight produced wool with fewer crimps per inch and with a higher average diameter. The most recent information indicates a reduction of 32% in crimp frequency (8.4 v 11.0 crimps per inch) in the Fleece Plus ewes, and an increase of 7% in average fibre diameter (21.2 v 19.9 μm). These changes, when expressed as realized genetic correlations, are in good agreement with expectations. In selecting for increased fleece weight, less selection pressure is lost maintaining average fibre diameter than if crimp frequency was kept constant.

(d) Live weight and feed efficiency

Selection for increased fleece weight has not lead to an increase in hogget live weights, confirming the earlier estimates of the absence of any genetic correlation between these traits. Averaged over the three most recent drops at Trangie (1972-1974), the Fleece Plus ewe hoggets were 3% lighter off shears. Pen-feeding studies indicate that Fleece Plus ewes produce more wool than Random ewes primarily because they are more efficient and not because they eat more (Williams and Winston 1965).

(e)   Wool Yield and Fleece Rot

Selection for increased fleece weight has increased yield, with yields in hogget ewe fleeces (1972-74 drops) averaging 67.7% in the Random flock, and 71.9% in the Fleece Plus flock. An increase in wool yield in the Fleece Plus flock would be expected from estimates of genetic correlations (Morley 1955), and has also been found in CSIRO flocks selected for increased fleece weight (Turner, Dolling and Kennedy 1968).

There is a danger that increasing wool yield will expose the fleece to fleece rot, weathering and dust penetration. For example, fleece rot predisposition has been shown to be positively correlated with yield, both phenotypically within flocks (Hayman 1953) and genetically across strains (Dunlop and Hayman 1958). Barlow (1974) examined data from four drops of ewes and found fleece rot scores (indicating both incidence and severity of fleece rot) to be higher in Fleece Plus ewes than in Random ewes. These data have now been extended to cover 12 drops of both rams and ewes and there is no evidence of any difference between the Fleece Plus and Random flocks. Despite this result, one has to ask if it is wise to continue to increase wool yield.

(f)   Reproductive Performance

Published estimates of the correlation between fleece weight and reproductive performance are in poor agreement, even in sign (Turner 1969). The traits appear to be uncorrelated in the CSIRO Merino flock (Young, Turner and Dolling 1963; Turner, McKay and Guinnane 1972), while Kennedy estimated a strong negative correlation in the Trangie population.

Selection for increased fleece weight at Trangie has reduced the number of lambs born per ewe joined, by increasing the proportion of dry ewes. In the period 1962 to 1969, the average difference in the percentage of ewes lambing in the Fleece Plus and Random flocks was 12%. This difference appears to have existed since the early years of the experiment (Barlow 1974), without increasing in magnitude. Such a result may be peculiar to the Trangie flock, as this is the only population in which the incidence of dry ewes has been found to be heritable (McGuirk, in preparation).

CONCLUSIONS

Selection for increased hogget fleece weight at Trangie has been effective in increasing both hogget and adult wool production. Gains in hogget production have been of the order of 20% in the eight generations of selection since 1951. The initial rates of gain have not been maintained and the realized heritability of fleece weight is currently less than the estimate by

Morley (1955) obtained for the Trangie population. Despite this picture, it is not clear if a plateau has been reached, as Pattie and Barlow (1974) suggested.

Results from the Fleece Plus flock are of considerable relevance to recommended breeding programmes for the Merino industry. However, too much should not be read into the pattern of response in any one selection flock, as considerable variation can be expected between replicate selection lines (Hill 1974). Of much more general interest is the average pattern of response in the various selection lines for increased fleece weight currently maintained in Australia.

Correlated responses to selection for increased fleece weight are generally in good agreement with expectations, at least in direction. The negative genetic correlation with ewe reproductive performance deserves further study; a more detailed description of reproductive performance is clearly needed. The increase observed in wool yield may reduce the fleece's resistance to fleece rot, dust penetration and weathering. If, in selecting for increased fleece weight yield was maintained at a constant level, this would mean markedly reducing the efficacy of fleece weight selection programmes, by the order of 25 to 30%.

REFERENCES

Barlow, R.A. (1974) — *Australian Journal of Agricultural Research* **25** : 973.

Brown, G.H., Turner, Helen Newton and Dolling, C.H.S. (1966) — *Australian Journal of Agricultural Research* **17** : 557.

Dunlop, A.A. and Hayman, R.H. (1958) — *Australian Journal of Agricultural Research* **9** : 260.

Hayman, L.H. (1953) — *Australian Journal of Agricultural Research* **4** : 430.

Hill, W.G. (1974) — *Biometrics* **30** : 363.

Kennedy, J.P. (1967) — *Australian Journal of Agricultural Research* **18** : 515.

Morley, F.H.W. (1955) — *Australian Journal of Agricultural Research* **6** : 77.

Pattie, W.A. and Barlow, R. (1974) — *Australian Journal of Agricultural Research* **25** : 643.

Turner, Helen Newton (1969) — *Animal Breeding Abstracts* **37** : 545.

Turner, Helen Newton, Dolling, C.H.S. and Kennedy, J.F. (1968) — *Australian Journal of Agricultural Research* **19** : 79.

Turner, Helen Newton, McKay, Elaine and Guinnane, Fay (1972) — *Australian Journal of Agricultural Research* **23** : 131.

Williams, A.J. and Winston, R. (1965) — *Australian Journal of Experimental Agriculture and Animal Husbandry* **5** : 390.

Young, S.S.Y., Turner, Helen Newton and Dolling, C.H.S. (1963) — *Australian Journal of Agricultural Research* **14** : 460.

13

SELECTION OF MERINO RAMS FOR EFFICIENCY OF WOOL PRODUCTION

G.J. TOMES, I.J. FAIRNIE, E.R. MARTIN and S.C. ROGERS
Muresk Agricultural College, Western Australian Institute of Technology

SUMMARY

Rams pre-selected for high body weight and high wool production were tested for feed utilization efficiency before final selection for use in a 1975 artificial insemination programme in Western Australia.

Twenty-seven Merino rams from the Australian Merino Society nucleus flock, were individually penned and conditioned on a diet composed of lucerne, formalin-treated rapeseed meal, wheat and mineral supplements, for three weeks. On completion of conditioning, the rams were fed the standard diet (16.7% C.P. 2.64 Mcal. DE/kg) for 100 days. Feed intake in these trials was restricted to maintain constant body weight.

Body weight of rams was correlated with both wool production ($r=0.55*$) and with maintenance requirements ($r=0.93**$). The mean production of clean wool per kg of feed was 16.4 ± 0.71 g; the larger rams being less efficient ($r= -0.463*$). Substantial differences in feed conversion efficiency between animals of similar body weight were also recorded.

During 1975-76, twelve of the previously tested rams were fed at maintenance level for eight weeks while being used in artificial insemination. Their maintenance requirements then were 26% higher than during the period of testing when the rams were not in use.

INTRODUCTION

The rate of wool production is determined not only by the feed intake which reflects the availability and palatability of feed and appetite of individual sheep, but also by the efficiency of conversion of digested nutrients to wool (Ferguson, 1956). Differences in wool production between animals are associated with differences in feed conversion and the higher producers are usually the most efficient (Turner and Young, 1969). Dolling (1970) found no variation in ability of Merino sheep to digest the ration and stated that the efficiency of utilization of digested nutrients determines individual performance. A substantial proportion of the variability in fleece production between animals of similar body weight is caused by differences in the efficiency of food conversion. However, if there is a large difference in body weight, the contribution of efficiency is reduced and the feed intake accounts for a higher proportion of changes in wool production (Pattie, 1973).

The differences between animals are usually more pronounced on high planes of nutrition (McManus *et al.* 1966) with a slight reduction in magnitude under medium levels of nutrition and the possibility of complete elimination under poor nutritional conditions (Piper and Dolling, 1969).

Saville and Robards (1972) recorded that the differences in efficiency between sheep selected for wool production and unselected animals, initially increased and then decreased with increasing intake but the overall ranking was not influenced.

Direct selection for feed conversion efficiency is not recommended in Merino breeding. Selection for wool weight is suggested as the soundest method of indirect selection for efficiency. In a group breeding scheme operated by the Australian Merino Society, rams are tested for feed conversion efficiency as part of the final selection before use in artificial insemination. In this chapter data recorded in the third year of testing for efficiency of conversion of feed to wool are presented. Also the maintenance requirements of rams used for artificial insemination are compared with the requirements of stall-fed rams.

MATERIALS AND METHODS

Twenty-seven 18 months old rams from the Australian Merino Society were used in this trial conducted at Muresk Agricultural College during April-August, 1975.

On arrival, all animals were drenched and vaccinated and placed in an animal house. Rams were tethered in individual slatted-floored pens and offered a standard diet (59% lucerne chaff, 20% formalin-treated rapeseed meal, 20% wheat, 1% mineral supplement) containing 16.7% C.P. and 2.64 Mcal. DE/kg for three weeks. On completion of conditioning, all rams were shorn and weighed then fed the standard diet for 100 days. Feed intake in this trial was restricted to maintain constant body weight. The National Academy of Science (1968) recommendations were used to determine initial feed intake levels. At fortnightly intervals, all animals were starved of food and water for 24 hours, weighed and their daily rations adjusted accordingly. On completion of the trial, all rams were shorn, weighed and their fleeces tested (Shepherd, 1976).

In the second part of this study, 12 selected Merino rams were fed at maintenance level for eight weeks while being used for artificial insemination $(3 \pm 0.52$ collections/day) during December 1975-February 1976 period.

The pelleted diet (19% barley, 21.5% oats, 40% lucerne meal, 17% sweet lupins, 2% molasses, 0.5% mineral and vitamin supplement) contained 16.04% C.P. and 2.86 Mcal. DE/kg. During these trials the moisture content of the feed was monitored and 26 samples were analysed for crude protein and digestible organic matter content.

RESULTS AND DISCUSSION

During the trial, four rams were removed due to low appetite and one with lameness. Three others were older than 18 months when the trial commenced and are not included in the results. Regression analyses were used to examine the relationships between:

(a) Body weight (x, mean = 67.6 kg, SE = 3.46) and feed consumption (y) for maintenance.
$$y = 1.67x + 6.56 \ (r = 0.932**)$$

(b) Body weight (x) and feed consumption (y) for maintenance in A.I.
$$y = 1.94x + 18.0 \ (r = 0.561*)$$

(c) Body weight (x) and clean wool production (y)
$$y = 0.015x + 0.93 \ (r = 0.552*)$$

(d) Body weight (x) and feed conversion efficiency (y) (g of clean wool/kg of feed)
$$y = -0.10x + 23.4 \ (r = -0.463*)$$

$$* \ P < .05$$
$$** \ P < .01$$

Daily production of wool and daily feed requirements are presented in Figure 1.

Figure 1—Effect of liveweight on wool production and feed requirement of rams

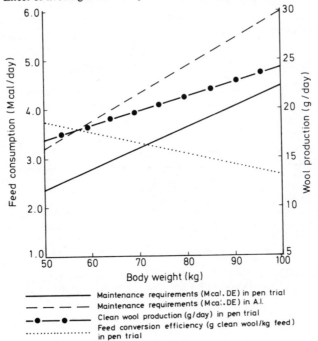

Maintenance requirements (Mcal.DE) in pen trial
— — — — Maintenance requirements (Mcal.DE) in A.I.
Clean wool production (g/day) in pen trial
—●—●— Feed conversion efficiency (g clean wool/kg feed) in pen trial
·············

Maintenance requirements were lower than those indicated by N.A.S. (1968). The expected relationship between food requirements and body weight was

$$D.E. = 2 \times 70 \ W \ kg^{0.75} \qquad N.A.S. \ (1968)$$

but was in fact $D.E. = 2 \times 68 \ W \ kg^{0.75}$

The difference is small and is in part accounted for by increase in wool weight on the ram during the trial period which was accompanied by a corresponding decrease in body tissues to maintain constant overall weight.

Difficulty has been experienced maintaining the body weight of working rams during the A.I. period. The 26% increase in maintenance requirements recorded here (although significant) is insufficient to account fully for the problems encountered. It seems that rams have depressed appetites during the working period as well as a higher maintenance requirement.

The results show an expected increase in maintenance requirement with higher body weight. Wool production increased at a lower rate with the result that feed conversion efficiency was significantly less for large rams than small rams ($P < 0.05$). This confirms trends observed in previous years.

This subject is important because the Australian Merino Society places much emphasis on body weight in its selection index. It seems this may be reducing improvement in the efficiency with which sheep convert feed to wool.

Clearly it is important to have more information about this relationship. Until this is available, it would seem sensible to restrict comparisons of feed conversion efficiency to rams within narrow body weight ranges.

ACKNOWLEDGEMENTS

These trials were financed by the Australian Merino Society and we wish to thank D.J. Curwen, T.A. Beeson and H.R. Wilson for technical assistance.

REFERENCES

Dolling, C.H.S. Breeding Merinos. Rigby, 1970.
Ferguson, K.A. (1956). The Efficiency of Wool Growth. Proceedings A.S.A.P. **1** : 58.
McManus, W.R., Arnold, G.W. and Dudzinski, M.L. (1966). *Australian Journal of Experimental Agriculture and Animal Husbandry* **6**: 96.
National Academy of Sciences (1968). Nutrient Requirements of Sheep (Washington).
Pattie, W.A. (1973). Pastoral Industries of Australia. Sydney University Press.
Piper, L.R. and Dolling, C.H.S. (1969). *Australian Journal of AgriculturalResearch* **20**: 561.
Saville, G.D. and Robards, G.E. (1972). *Australian Journal of Agricultural Research* **23**: 117.
Shepherd, J.H. (1976). Proceedings—International Sheep Breeding Congress, W.A.I.T.
Turner, Helen Newton and Young, S.S.Y. (1969). Quantitative Genetics in Sheep Breeding. Macmillan.

14

SELECTION FOR INCREASED PROFITABILITY IN A FINE WOOL MERINO FLOCK

W.L. WEATHERLY
Streatham, Victoria, Australia

SUMMARY

An integrated selection programme practised on a fine wool Merino flock in western Victoria is described and discussed. The characters selected for, collection and use of data, and integration of the programme is described. Data indicating heritability of susceptibility to tender wool and independently to worm infestation is presented and discussed. Selection for increased Clean Fleece Weight, using a rating derived from the Clean Fleece Weight of each sheep compared to its micron, is explained and compared to selection using Clean Fleece Weight alone.

INTRODUCTION

The shift in stock breeding towards objective assessment and separate selection of individual characters and the division from intuitive holistic selection has been a subject of debate and interest for years. The many attempts to integrate these methods are clearly of interest.

The economic success of a selection programme depends on the total effect of the whole scheme, and the success of any programme is difficult to evaluate, let alone anticipate. The economic value of selection for any individual character depends on the level of that character in the population under selection since the economic value will depend on the rate of progress made, and different rates of progress will be made when the desired character is at different levels. Excessive emphasis on a character can be economically as damaging as neglect—it is unfortunate that characters vary in their ease of measurement, since when measurement is practised those characters measured are often overemphasized. Holistic, intuitive assessment, with objective data presented, is suggested as the only practical way of balancing and integrating a selection programme.

Constitution in stock is considered an important characteristic, though constitution, and its selection, is seldom clearly defined. Heritability of constitution has long been assumed, and in practice demonstrated. Variation between and within populations of sheep exists for the characteristic. The separation of constitution into characters, and the performance of these in independent gene pools, is thus clearly of interest.

The selection programme discussed is part of a commercial enterprise. The sole aim is to increase profitability by:

(a)   Increasing the quantity of production
(b)   Maintaining and improving the quality of production
(c)   Decreasing the cost of production

The programme must be simple, effective and worthwhile; data must be gathered and processed with the minimum of time and effort, to give the maximum of useful material, presented in a usable manner.

MATERIALS

The Blythvale stud and flock is run on a 3000 hectare property in western Victoria at 10 sheep/hectare stocking rate with a 50 cm annual rainfall. The selection programme discussed is practised on the stud which currently holds 500-600 breeding ewes (average of 327 ewes joined p.a. for past 30 years)—this flock breeds rams to service a flock of c. 6,000 ewes.

The sheep are Western District Saxon blood fine wool Merinos of 66-74s wool count. The population is relatively isolated, 8% of stud ewes being mated to rams from other studs over the last 30 years. There is a two way movement of some ewes between the stud and flock—a varying proportion, usually 1-3% of the best flock ewes are classed up to the stud each year. This gives the advantage of a large gene pool to select from, and yet the stud is a manageable size to allow detailed and careful work.

Management of the stud was altered and developed from 1971—before this fleece testing of one year old rams and mothering of lambs and a card system were in practice, but were not being utilized sufficiently to justify the costs.

Shoulder sampling of fleeces, and testing for tender wool, is carried out at shearing, analysis of wool samples being made by the Gordon Institute of Technology at Geelong, Victoria. Data on scouring is gathered when the particular mob are yarded for other operations (dipping, crutching, etc.). Stud lambs are mothered (and presence of primary follicles or "mother-coat", pigmentation and wrinkles are noted) 6-9 weeks after the start of lambing. Mothering is done in a set of yards designed and built specifically to mother and sort out stud sheep in. The sheep are classed by an independent classer (Mr. Geoff Pennefather) at 1 year old and finally at 2 years old when they are categorized for their future breeding life. A card is kept for each stud sheep, and data is accumulated on these. Master cards hold information on sheep that have left the stud (died or lost identification) and have descendants in the stud.

All rams classed for stud sires are now single mated for at least the first year of use in the stud.

METHODS, RESULTS AND DISCUSSION

(a) Increased Production

(i) Wool Production

The quantity of wool production is increased mostly by careful classing using fleece test ratings. All 1 year old rams (to be classed to stud, flock, or sale next year) and all replacement stud ewes are fleece tested.

Fleece test ratings are derived from graphs made independently for each group of ewes or rams that are run together and are of the same age. The graph is made from the Clean Fleece Weight (CFW) of each sheep plotted against its micron (see graph in Appendix), an average line of best fit is constructed to suit the cloud of individual measurements. (The gradient of this line varies from group to group, but a decline in CFW with finer micron is always seen.) Lines are then drawn parallel above and below the line of best fit at 0.23 kg intervals. These lines are used to determine each sheep's CFW/micron rating, all sheep within the same pair of parallel lines have the same rating, which is different to all the other sheep on the graph. The classes immediately above and below the line of best fit are always labelled C and D respectively, and ratings range from A++ to G. There are two advantages of this method. Firstly, as each ram is presented for classing at 2 years old the classer is given a simple, effective and easily comprehensible rating which replaces the confusion of giving two figures that are only of importance in relation to each other. Secondly, as Tables 1 and 2 demonstrate, this method prevents the shift towards the stronger average micron in the next generation that would result from CFW selection independent of micron.

TABLE 1

COMPARISON OF SELECTION OF TOP 50 RAMS USING CFW and CFW/MICRON RATING

Method of Selection	Micron Group				Total
	>20.0	19.9-18.5	18.4-17.0	<17.0	
CFW with micron rating					
Classed in	—	17	26	6	50
Classed out	4	7	20	1	31
CFW alone					
Classed in	2	21	25	3	51
Classed out	2	3	21	4	30

TABLE 2

NUMBER OF SHEEP MISCLASSED USING CFW SELECTION RATHER THAN CFW/MICRON RATING SELECTION

		Micron Group			
	>20.0	19.9-18.5	18.4-17.0	<17.0	Total
No. Rams					
Classed in	2	4	—	—	6
Classed out	—	—	2	3	5

Figure 1—Clean Fleece Weight in relation to fibre diameter (1973 drop rams)

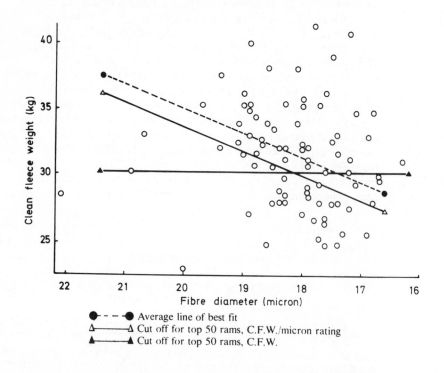

●– – –● Average line of best fit
△———△ Cut off for top 50 rams, C.F.W./micron rating
▲———▲ Cut off for top 50 rams, C.F.W.

Figure 1 shows the line of best fit drawn for the 1973 drop rams. The fleeces were grown between August 1973-74, 4-16 months old. Usually about 50 rams are retained to service the stud and flock; the graph shows the cut-off lines for the top 50 for CFW or for CFW/micron rating. For this group of 81, 13.6% of rams would be misclassed using independent CFW selection, and this would result in a shift towards stronger micron. Sheep with the most different micron from the group ($< 10\%$), particularly at the stronger end of the range, do not usually conform to the trend in the graph.

The replacement stud ewes are given the same fleece test, the top 20% CFW/micron rating ewes are called weight elite ewes. Half of these are mated to the most superior stud sires for CFW/micron rating—to increase the probability of breeding superior wool cutting sheep.

(ii)   Sheep Production

To increase sheep production/ewe mated account of the mother's breeding performance is taken in selecting the stud sires. Rams with mothers that have been both regular and successful (low culling rate of progeny) breeders are preferred. As well, some of the very successful ewes are carried on under special care when their mouths deteriorate (when they would normally be culled) to breed more lambs. At classing some size and fleece weight is allowed for twins, but there is no direct bias for twinning. Twinning should, for a flock run under similar environmental conditions over several generations, already be at the optimal level for the flock. The level of twinning in the flock discussed can be explained by the very high mortality of both twins in a poor season. Selection towards a higher rate of twinning could result in a decrease in lambing success—through increase in the proportion of mortality to births and through increases in culling—unless environmental conditions are altered.

Lambings that require assistance are noted on the cards of the lambs and ewes; passenger and barren ewes are culled.

(b)   Quality of Production

Quality is assessed at classing, and a lot of importance is placed on it. At classing, the sheep that as lambs showed wrinkles and primary follicles are checked for these characters—thus there is little possibility of missing faulty sheep, particularly since these faults are at very low levels (cullings for hair and wrinkles being less than 1%).

(c)   Decreased Cost of Production

Selection towards decreased cost of production has been increased with a shift towards minimal management—with low cost and labour input.

The stud sheep are run under similar or more rigorous conditions compared with flock sheep—classing leading to selection of sheep most suited to commercial conditions. It is a tragedy that the market place forces those whose income comes mainly from the sale of sires to treat their stock to more favourable conditions than the progeny of the sire sold.

Weaner rams are deliberately run on poorer quality pastures during summer to "sort them out". The sheep that continue to produce satisfactorily and do not show tender wool, are of interest. Both young stud ewes and rams are run without supplementary feeding, and under a low drenching regime. Young stud sheep are classed at two years old when 18 months has usually passed since the last drenching.

(i)   Tender Wool

Table 3 indicates difference between stud sires in the susceptibility to tender wool. These differences are repeatable. Thus, no ram showing tender wool as a 1 year old is used as a stud sire.

TABLE 3

TENDER FLEECES, 1973 DROP RAMS

Progeny group	No. tender fleeces	Total in group
A	3 (25%)	12
B	3 (33%)	9
C	—	6
D	1 (10%)	10
E	2 (25%)	8
F Syndicate	1 (7%)	14
G Syndicate	3 (13%)	23

(ii)   Worm Infestation

Groups of the same age run together and are examined for scouring (the staining of wool with faecal material). The amount of scouring is categorized:

Clean—no contamination from scour.

O.K.—some stain, superficial and not sufficient to cause flystrike.

Plug—some dagging on the wool around the anus, restricted to this area but sufficient to allow flystrike.

Very Dirty—extensive scouring covering a considerable area around the anus, and showing accumulation of dags on the back legs and scrotum.

In practice scouring can be attributed to several factors—worm infestation, changes of feed, and illness. Young stud sheep, rams and ewes, have been examined in the last two years when less than 1 year old and at 1.5 years. In each case the whole group, and all individuals within it, had not been drenched for several months, and in each case considerable differences existed between the sires' progenies.

Tables 4 and 5 show the figures for two consecutive drops of rams at a similar age, and the age of the sheep and the conditions at the time allow scouring from causes other than worm infestation to be disregarded. The data clearly indicates heritability of susceptibility to the effects of worm infestation. The ewes that bred these rams were mated randomly with regard

TABLE 4

SCOURING FOR 1973 DROP RAMS* ON 26 NOVEMBER 1974

| Progeny Group | Degree of Scouring % | | | | |
	Clean	O.K.	Plug	Very dirty	No. rams
A		33	50	17	12
B	22		45	33	9
C			50	50	6
D	40	10	40	10	10
E	22	33	45		9
F Syndicate		21	58	21	14
G Syndicate	17	9	61	13	23

* Last drenched 28/9/73, run together since 2 months old.

TABLE 5

SCOURING FOR 1974 DROP RAMS* ON 9 SEPTEMBER 1975

| Progeny Group | Degree of Scouring % | | | | |
	Clean	O.K.	Plug	Very dirty	No. rams
F	50	38	12		8
G	84	8		8	12
J	100				2
L	75	25			8
M	60	20	20		10
R	75		12.5	12.5	8
S	55	33	11		9
W Syndicate	—	50	43	7	14
X	89	11			9
Y	75	8	17		12
Z	80	20			5

* Last drenched 15/8/74, run together since 2 months old.

to this character, and the sires selected to breed these rams were considered the best rams overall.

It is of relevance that the sire of the X1974 drop rams has never scoured and now has not been drenched for 3.5 years, and the sire of this ram was the father of the E1973 drop rams, the most superior group for this character in their drop. The B and C1973 rams made the W1974 syndicate, the C1973 ram siring the majority of lambs. For the 1974 drop rams, the F, S, and W groups were the only ones sired by older rams that were not selected as stud sires after experiencing minimal drenching and care, and these showed the greatest amount of scouring.

Differing resistance to parasites is seen between and within breeds of sheep. The differences observed suggest a value in selecting for increased resistance or tolerance since the cost in production loss, and treatment for worm infestation at high stocking rates is considerable. The differences shown within Tables 4 and 5 suggest that rapid progress can be made.

It can be seen that heritabilities of Clean Fleece Weight, susceptibility to worm infestation and to tender wool appear to function independently. (The B and C1973 drop rams were the relatively heaviest cutting rams, and their progeny were the heaviest cutting group. However, C1973 rams were the most susceptible to the effects of worm infestation, but showed no tender wool, whereas B1973 rams showed a considerable amount of tender wool, though they were not so inclined to scour.)

(d) Integration

The integration and balancing of the selection programme is effected by the sheep classer. His judgment evaluates the characters and assesses their importance in the sheep and in the flock. The success of a breeding programme rests squarely on his shoulders. None the less, it would appear that factors other than those generally selected for may well have a significant influence on the overall efficiency of a sheep enterprise and may be worthy of consideration in the selection of breeding stock.

15

THE IMPACT OF ENVIRONMENTAL FACTORS ON SHEEP BREEDING IN THE SEMI-ARID TROPICS

P.S. HOPKINS, M.S. PRATT and G.I. KNIGHTS
Toorak Research Station, Julia Creek, Queensland, Australia

SUMMARY

Modifications to the existing environmental conditions (e.g. provision of natural shade) can lead to a significant improvement in the production performance of tropical sheep. Alternatively, the selection of "adapted" animals from within the existing gene pool also leads to significantly better production performances. The judicious use of both strategies together with the infusion of genetic material from high producing animals from more temperate regions are discussed in relation to the environmental physiology of tropical sheep.

INTRODUCTION

The harsh semi-arid tropical environment exerts a profound influence on the production characters of sheep. Animals are of small body size, return low wool cuts and exhibit a poor reproductive performance. Any attempts to improve productivity from tropical areas may include an alteration of the natural environment to lessen its impact on the animal or, alternatively, the breeding of sheep which are capable of adequate production under existing environmental conditions. This chapter examines these strategies in relation to current research efforts in semi-arid tropical Queensland.

METHODS AND RESULTS

(a) Alterations to the environment

The provision of shade is an obvious means of alleviating environmental heat stress. This treatment can be included in management strategies at the time of joining to improve the pregnancy rate of Merino sheep (Table 1).

Shade facilities can also be used to improve lamb survival. Heat stress in the last month of pregnancy markedly restricts the growth rate of the foetal lamb in the absence of nutritional intervention (Table 2).

Since the low survival rate ($\approx 20\%$) of small birth weight lambs (< 2 kg) is an important aspect of reproductive wastage in this environment the provision of shade can be of marked benefit to pregnant ewes and newborn lambs.

TABLE 1

THE SIGNIFICANT (p<0.01) EFFECTS OF SHADE AND IODINE SUPPLEMENTATION AT JOINING ON THE PREGNANCY RATE OF SUMMER–JOINED MERINOS

Group	Number of Sheep	
	Joined	Pregnant
Control	196	112 (57%)
Shaded	199	145 (73%)

TABLE 2

THE EFFECTS OF MATERNAL HEAT STRESS ON LAMB BIRTH WEIGHT

Group	Lamb birth weight (kg)	Ewe live weight (kg)
Heated	2.3	39.2
Control	3.6	39.0

Ewes were paired on the basis of foetal size (110 days) and pair-fed during treatment period.

The propagation of natural shade trees is an economical and practical means of altering the environment. A glen of 66 Athel pine (*Tamarix aphylla*) cuttings have attained a height of 2.5 m in the 18 months since planting. This shade glen covers an area of 10,000 m² and as such offers an excellent means of tempering environmental heat stress to maximize fertility.

(b)   Selection for environmental adaptation

Studies in environmental physiology have identified "adapted" Merinos within the existing gene pool. These animals have a significantly lower rectal temperature and respiration rate when exposed to summer conditions (Table 3).

Adapted sheep also exhibit significantly better production performances than their less adapted counterparts in the same flock (Table 4).

The measurement of sweating rates of tropical sheep has shown that this avenue of evaporative water loss (200 ml/h) greatly exceeds the respiratory component (50 ml/h). The significance of this finding in relation to body

TABLE 3

MEAN RECTAL TEMPERATURES AND RESPIRATION RATE OF "ADAPTED" AND "NON-ADAPTED" SHEEP DURING EXPOSURE TO SUMMER CONDITIONS

Group	No. of sheep	Mean rectal temperature (°C)	Mean respiration rate per min.
Non adapted	50	40.1	185
Adapted	48	39.2	110

Differences are significant $p < 0.01$.

TABLE 4

THE RELATIONSHIP BETWEEN EWE BODY TEMPERATURE AND PRODUCTION PERFORMANCE (LIVEWEIGHT, PREGNANCY RATE)

Group	No. of sheep	Ewe liveweight* (kg)	Pregnancy rate*
Adapted	98	39	58%
Non-adapted	103	35	36%

* ($p < 0.01$).

temperature regulation of woolly sheep and the degree to which this index can be used to screen animals for environmental adaptation await elucidation.

Assessments of the respiration rates of sheep exposed to environmental heat stress is currently the most successful means of identifying "adapted" sheep. In screening animals on this basis it is essential to undertake the test at a time when environmental conditions maximize the between sheep differences in thermoregulatory efforts. Significant ($p < 0.01$) rectal temperature differences between "adapted" and "non-adapted" sheep were demonstrated when ambient temperatures exceed 37°C. These differences were not significant when ambient temperatures were below 37°C. This threshold can merely be taken as a guideline, since the efforts of humidity and flock genotype would be necessary considerations for each set of circumstances.

DISCUSSION

When considering the impact of a tropical environment on sheep breeding programmes it is pertinent to first establish the importance of both physiological and behavioural responses to environmental conditions. If an adequate availability of natural shade and pasture allows sheep to adopt grazing patterns which minimize exposure to heat stress then the animal's behavioural responses may be of paramount importance. Alternatively, lesser amounts of shade and available pasture may produce symptoms of heat stress in more temperate regions, particularly when the strain or breed of sheep pastured in that region has poorly developed adaptation mechanisms. It is necessary, therefore, to consider the type of sheep and the prevailing conditions in an integrated fashion in order to make valid judgment of the importance of the type of genetic material which should be propagated in a particular environment.

It is apparent from the results presented herein that it is possible to improve the productivity of tropical Merinos. This can be achieved by either altering the existing environment or selecting adapted sheep from within the existing gene pool. It is not yet clear what further gains can be made by the judicious use of both strategies. The provision of shade is both economically and practically feasible. The selection of "adapted" Merinos is most easily undertaken by an assessment of respiration rates during exposure to natural summer conditions. The physiological mechanisms underlying the expression of this adaptation are however obscure.

It would also be possible to select tropical animals on the basis of production performance. Ram breeding programmes aimed at achieving this goal have been initiated at the Toorak Research Station. These studies assume that the increased production stemming from environmental adaptation can be readily measured and propagated provided that natural environmental conditions prevail at the time of selection. More basic studies of environmental physiology are however indicated so that the gene pool can be prudently extended beyond the bounds of that currently available in the region. The ceiling for production may well be further increased by the infusion of genetic material from more temperate parts of the continent. Some of the likely attributes of this strategy are presented in the next chapter in this book (Stephenson, Tierney and Hopkins).

It is possible that exotic sheep may be of some benefit to the breeding programmes of tropical Merinos if the future wool and mutton markets warrant their inclusion. An evaluation of the environmental adaptation of such animals could be made by preliminary observations of respiratory rates during exposure to conditions which mimic the natural summer conditions in Australia. The integration of this data with more general information pertaining to the physiological, behavioural and genetic implications of their introduction would obviously be indicated. The likely impact of their importation to Australia would need to be substantial if it is to outweigh the long-term disadvantages of changes in wool quality and the establishment of numerous breeds and their associated societies.

16

HUSBANDRY AND GENETIC CONSIDERATIONS AFFECTING SHEEP BREEDING IN THE SEMI-ARID TROPICS

R.G.A. STEPHENSON, M. TIERNEY and P.S. HOPKINS

Toorak Research Station, Julia Creek, Queensland, Australia

SUMMARY

The implementation of breeding programmes and collateral husbandry techniques can increase the productivity of tropical Merinos. Significant increases in lamb survival, lamb growth rate, mature live weight and greasy fleece weight are reported. The impact of these findings in relation to the development of flocks of improved fertility and/or dual-purpose qualities are discussed.

INTRODUCTION

A great disparity exists between the productivity of tropical Merinos and their counterparts depastured in more temperate climates. This disparity highlights the problems associated with sheep breeding in the tropics. Nutritional inadequacies and climatic extremes impose severe restrictions on production throughout much of the year. Survival rates of adult sheep are comparable to those of other environments, however, production and reproductive efficiencies are markedly lower (Moule, 1966; Rose 1972). Average live weight of ewes is 40 kg, fleece-cuts average 3.5 kg and lambs marked average 40%.

This chapter integrates the results of three experiments aimed at improving the productivity of tropical sheep by both genetic and husbandry procedures.

METHODS AND RESULTS

(a) Breeding for lamb survival

Three strains of Merino rams were joined to 1,200 indigenous ewes for one oestrous cycle. Ewes were synchronized with 0.15 mg of stilboestrol and joined in the second cycle after injection. Rams were representatives of indigenous animals (locally bred for longer than 10 generations), transient animals (locally bred for 2-3 generations) and introduced animals (South Australian Merinos).

At an average gestational age of 125 days, ewes were pregnancy tested (Pratt and Hopkins, 1975). No real difference in conception rate occurred between sires (Table 1). Pregnant ewes were divided into the three sire groups and placed in equal sized paddocks with similar amounts of shade and water. Biomass of pastures was in excess of 2,000 kg/ha, this being considered more than adequate at the stocking rate of one sheep to 5 ha. Average maximum and minimum temperatures during the last six weeks of pregnancy and the first four weeks of lambing were 33.6°C and 19.2°C respectively. This temperature range constitutes mild conditions for the tropics. At an average age of four weeks the progeny of the three groups were weighed and survival rates were recorded. Significant differences occurred between the three groups (Table 1).

TABLE 1

REPRODUCTIVE PERFORMANCES OF INDIGENOUS, TRANSIENT AND INTRODUCED SIRE STRAINS

Ram strain	Conception rate (%)	Progeny	
		Survival (%)	Live wt (kg)
Indigenous	58	61† *	7.9†
Transient	59	71†	8.1
Introduced	58	80† *	8.4†

Lamb weights recorded at mean age one month
† $p < 0.05$
* $p < 0.01$

(b) Breeding for diversification

Greasy fleece weights and mature live weights of indigenous and introduced rams and ewes were compared. Introduced sheep were allowed a one year acclimatization period before the commencement of the experiment. The measurements were made under normal paddock conditions when good seasons prevailed. The ram measurements were taken from those animals selected for the sire evaluation study. The introduced ewes were also South Australian Merinos however they were of a different stud origin. The results show significant differences ($p < 0.05$) between indigenous and introduced ewes and rams (Table 2).

TABLE 2

GREASY FLEECE WEIGHTS (GFW) AND MATURE LIVE WEIGHTS FOR TWO MERINO STRAINS

Strain	♀ GFW (kg)	♂ GFW (kg)	♀ Live wt. (kg)	♂ Live wt (kg)
Indigenous	3.50	2.24*	40	51
Introduced	4.27	2.70*	45	60

* GFW for seven months wool growth only
Introduced animals had significantly higher fleece weights and live weights ($p < 0.05$)

(c) Supplementation for lamb survival

Practical aspects of improving lamb survival as a collateral husbandry measure to animal breeding programmes were investigated. Urea was used to improve the nutritional status of lambing ewes. A pen study was designed to compare the growth rate of lambs born to ewes fed an ad lib. basal diet of pelleted pasture hay (5.5% CP) or the basal diet containing 10 g urea per kg ration. Twenty control and 20 treated ewes were introduced to pens two weeks before expected parturition. Measurements were continued until three weeks post-partum.

A significant improvement ($p < 0.01$) was measured in the growth rate of lambs in the treated group. In fact these animals were 48% heavier than the control lambs at three weeks of age.

In a follow up field study 600 pregnant ewes were divided into four groups and run in four half-square-mile paddocks when mild environmental conditions prevailed. Two groups of ewes each received urea at the rate of 1 g/l in the drinking water. Lamb weights and survival rates were recorded three weeks after the last expected birth.

An improvement in the survival and growth rate of lambs (4% and 12% respectively) occurred in the supplemented groups. The improvement in lamb growth rate was significant ($p < 0.05$).

DISCUSSION

The low survival rate of newborn lambs in this environment is *the* major cause of reproductive wastage. If lamb survival could be increased then local producers would be in a position to implement breeding programmes and cull sheep on production performance; a strategy which is currently unavailable to them because of the low net reproductive rate. The results presented in Table 1 indicate that lamb survival may be significantly increased by pertinent genetic manipulations. It is likely that the indigenous lambs have a relatively poor ability to survive under even mild conditions. Conversely, introduced animals of higher fecundity (e.g. South Australian Merinos) may express this trait in terms of improved lamb survival. Evaluation of other strains of introduced animals of high fecundity is a logical progression of this study. The judicious use of such strains may have a desirable impact on the productivity of existing indigenous flocks. However, such use would need to embrace collateral considerations of the environmental physiology of the introduced strains, the ways in which husbandry strategies could temper the rigours of the tropical environment, and the genetic stability and wool quality of the strain evolved.

The future market value of wool and mutton must be regarded as unpredictable. The growing demand for mutton by South-East Asian, Middle East and local mining town markets has prompted an investigation into the feasibility of developing dual-purpose animals in tropical Queensland.

The plane of nutrition and the geographical situation of the area means that it is well placed to provide lean hogget carcasses of approximately 20 kg. The dual-purpose tropical sheep may offer producers in this area a type of diversification necessary to enhance their future viability. The ever spiralling costs of production exert a more profound influence on tropical sheep raising enterprises where the efficiency of present production is lower than that of more temperate areas. The results from Tables 1 and 2 suggest that the development of a dual-purpose animal is possible through further genetic manipulations of Merino strains. The initial studies also indicate that it might be possible to effect a concomitant improvement in lamb survival and wool growth. The perils of propagating introduced species without due regard to studies in environmental physiology may however subsequently reveal themselves in the form of low productivity and even higher mortality rates when harsh conditions prevail in the area. The attributes of the indigenous Merino to survive in these conditions is a pertinent reminder of the hazards which confront the injudicious use of introduced strains in a breeding programme.

The success of a breeding programme depends very largely on the net reproductive rate of the flock. Husbandry techniques aimed at improving lamb survival in this environment are therefore important collateral considerations. The data presented in the results indicates the value of providing urea to pregnant and lactating ewes on a poor plane of nutrition. The implementation of such a strategy could therefore be regarded as an integral part of the development of breeding programmes outlined in this chapter.

The data presented in this chapter are largely preliminary in nature. They do however serve to highlight the possible options which are open to research workers and producers alike in their efforts to improve the efficiency of production of tropical sheep.

REFERENCES

Moule, G.R. (1966) *Australian Veterinary Journal* **42** : 13.
Pratt, M.S. and Hopkins, P.S. (1975) *Australian Veterinary Journal* **51** : 378.
Rose, Mary (1972) *Proceedings of the Australian Society of Animal Production* **9** : 48.

17

THE INFLUENCE OF PARASITES ON SELECTION PARAMETERS IN SHEEP

I.L. JOHNSTONE, F.M. DARVILL, K.E. SMART.
Merck Sharp and Dohme (Australia), South Granville, N.S.W., Australia

SUMMARY

Four years of detailed experimentation has shown that commonly practised levels of parasite control may not enable sheep to express fully, their true characteristics and potential.

Under conditions which favour the parasite, mortalities can seriously reduce the selection potential in a breeding flock. Furthermore, an adverse effect on body size, the ratio of primary to secondary follicles and total greasy wool production resulting from parasitism in the first year of life can be permanent and result in the culling, on an objective measurement basis, of some individuals which may otherwise have found their way into the flock.

The effect of four different levels of parasitism on total wool production, fibre diameter, staple length, yield, tenderness and wool manufacturing qualities indicates the degree of variation from the inherent potential which can result.

This chapter does not attempt to answer the question as to whether it is desirable or not, to select sheep under conditions where the effect of parasites is suppressed, or in areas where parasite damage is minimal. It aims to report the nature and extent of change which can occur in parameters normally taken into account in selection.

INTRODUCTION

It is not unusual for rams to be bred and selected in one environment and used in another environment. The stud can be located in a winter rainfall area and the flock in a dominant summer rainfall location. Similarly, there can be interchange between dry and wet environments.

Selection using various parameters can occur under conditions where there has been minimal exposure or stress from parasites of various species imposed on the population.

This chapter does not attempt to answer the question as to whether it is vironment of the flock or elsewhere but aims to document the nature and extent of the effect of parasites on some of the parameters commonly used in selection. Furthermore, the evidence will suggest the nature of the differences which might appear in a tableland flock based on a stud located in a minimal parasite damage situation when sub-optimum levels of parasite control are practised.

MATERIALS AND METHODS

The data presented in this chapter originated from an experiment conducted at Clover Park, the Applied Rural Research Station of Merck Sharp & Dohme (Australia) Pty. Limited, near Hamilton in the Western District of Victoria. The sheep were fine-woolled Merinos of local origin and were first selected from a flock of 550, discarding 150 which included firstly the wrinkly and atypical-woolled sheep and then similar numbers of the heaviest and lightest sheep in the flock. In January 1972, the 400 five to seven month old wethers were randomized on liveweight rank order to 16 paddocks of 1.56 hectares. Four paddocks of 25 sheep were subjected to each of four parasite control programmes based on the broad spectrum anthelmintic, thiabendazole. After 331 days, the 16 groups were put into one mob and run together under identical management, during 1973 and 1974.

(a) Parasite Control Programmes

Four programes designed to permit four levels of helminthiasis were used. These were:

SUPPRESSIVE—Eleven drenches, given monthly to minimize continuing re-infection.

PREVENTIVE — Three drenches, given on a pre-planned calendar basis in January, (at the start of the trial), and in July and September.

CURATIVE — Treatment when there was a visual indication of parasite damage. Drenches were given in April, July and October.

SALVAGE — Only individual clinically affected sheep were treated to avert death.

(b) Management and Measurements

Except for parasite control, all sheep were managed in the same way and were vaccinated, jetted and injected with testosterone paste to minimize disease other than helminthiasis. A uniform stocking rate was maintained in all paddocks during 1972 by adding replacement sheep when deaths occurred.

At shearing, the fleece and belly-wool were weighed together to give total greasy fleece weight and later the skirted fleece was weighed separately. Individual mid-side samples were taken to determine yield, fibre diameter, staple length, tenderness and for classification of each fleece into Australian Wool Corporation (A.W.C.) types. Skirtings, pieces and belly-wool were aggregated separately on a replicate basis, weighed and sampled for yield and A.W.C. type determinations.

In 1972, live weights and parasitological data were collected at monthly intervals. Later, during 1973 and 1974 when the sheep were running as one flock, these data were collected at less frequent intervals. At the conclusion of the trial, 20 sheep from the original Suppressive and Salvage regimes were selected at random and mid-side skin sections taken. Body size measurements were also made on all the surviving sheep in the same two contrasting groups.

RESULTS

Live weights, greasy wool production and changes in sheep numbers are shown in Table 1 a, b and c.

TABLE 1

LIVE WEIGHTS (kg); GREASY WOOL PRODUCTION (kg)
AND SHEEP NUMBERS

		PARASITE CONTROL PROGRAMME			
		Suppressive	Preventive	Curative	Salvage
(a)	Live weights				
	22. 1.72	21.3	21.4	21.4	21.3
	5.12.72	43.9	40.9	39.7	33.6
	16.11.73	58.0	57.0	57.4	54.3
	29.11.74	59.1	59.6	59.4	57.2
(b)	Greasy wool weight				
	1972	3.59	3.11	2.90	2.46
	1973	5.32	5.27	5.38	4.92
	1974	5.00	4.73	4.95	4.95
(c)	Sheep numbers				
	22. 1.72	100	100	100	100
	5.12.72	99	96	91	82
	16.11.73	96	96	91	82
	29.11.74	95	95	91	82

The effect of the four parasite control programmes on wool quality and characteristics is shown in Table 2.

TABLE 2

WOOL QUALITY AND CHARACTERISTICS IN 1972

	PARASITE CONTROL PROGRAMME			
	Suppressive	Preventive	Curative	Salvage
Clean scoured yield (%)	73.3	71.8	71.7	74.6
Fibre fineness (microns)	17.8	17.2	17.0	16.6
Crimps per inch	16.2	16.7	17.0	18.1
Staple length (cm)	8.8	8.5	8.4	7.8
Sound fleeces (%)	96.0	78.1	70.3	65.4
Actual break (%)	0.0	4.2	3.3	12.3
1976 value (AWC minimum reserve price) ($)	6.92	5.81	5.40	4.86

Wool production is a strongly inherited character in sheep (Jackson *et al*, 1975) and is an important parameter used in selection in Merino flocks. The effect of four programmes of parasite control on the distribution of greasy fleece weights of young sheep has been shown in Figure 1. It is clear that there is not a uniform effect through the flock which decreases fleece weight proportionately, but rather, a sorting out of levels of susceptibility or ability to avoid infection. The histogram changes from the more classic type distribution to a multimodal battleship formation.

The skin sections and body measurements taken after two years of common management showed a trend towards a persisting decreased ratio of secondary to primary follicles from severe helminthiasis in the weaner year, but this was not significant from the 20 sections examined (Table 3). The decreased mean fibre diameter in the salvage programme sheep was significant and medullated fibres were also seen in these sheep, but not with the suppressive programme. All the body measurements indicated that helminthiasis in the weaner year had resulted in some permanent stunting, but none of the differences were found to be significant.

TABLE 3

SKIN HISTOLOGY AND BODY MEASUREMENTS OF THE
SHEEP FROM THE SUPPRESSIVE AND SALVAGE PROGRAMMES
IN FEBRUARY 1975

| | | PARASITE CONTROL PROGRAMME | | |
		Suppressive		Salvage
(a)	*Skin histology*			
	No. of sheep	20		20
	Secondary/primary follicle ratio	24.9	N.S.	23.8
	Mean fibre diameter (microns)	19.4	$P < 0.05$	18.5
	Primary/secondary fibre diameter ratio	0.98	N.S.	0.99
	Medullation (%)	0		0.04
(b)	*Body measurements (mm)*			
	No. of sheep	95		82
	Chest depth	312.60	N.S.	305.90
	Hind cannon bone length	216.34	N.S.	214.11
	Pelvis width	176.49	N.S.	176.38
	Pelvis length	231.49	N.S.	228.94

DISCUSSION

At the end of 1972, substantial differences in live weight had developed between sheep on the different parasite control programmes. Similarly, there was an increase of 1.13 kg or 45.9% in wool production for the year during which the Suppressive programme was used. The other programmes gave less dramatic results but there was still a potential of 0.48 kg greasy wool above the production from a three drench programme commonly practised in the area.

Wool from the four treatment programmes was processed to the stage of combed top and showed significant increases in noil and card loss as the level of parasite damage increased (Lipson and Bacon-Hall, 1976). In the combed top, mean fibre length decreased; "neps" which create problems with the cloth making machinery and in the dyeing processes were also substantially increased.

When the economics of intensive parasite control were examined and the costs of labour, materials and mortalities estimated, the gross margin was still increasing to the highest level of the Suppressive programme (Johnstone *et al*, 1976).

During the second year, when all sheep were given identical treatment, the two intermediate programmes caught up substantially in body weight and wool production, but the Salvage group still lagged behind. By 1974, the gap had substantially closed for both wool production and body weight. In considering the compensatory gain which had occurred, the higher mortality eliminating some of the smaller sheep in the less intensive programmes may have biased the results in their favour. It is of interest to note also, that whereas the wool yields for the Salvage programme were the highest in 1972 (Table 2), in 1974 they were lowest; namely 73.3 for the Suppressive programme and 71.2 for the Salvage groups in 1974. This suggests that some of the compensatory gain in greasy fleece weight may have come from fleece constituents other than wool.

The body measurements of the Suppressive and Salvage groups also indicated a trend towards a permanent stunting from helminthiasis in the weaner year, but these differences were not statistically significant.

These results show that parasites can have an effect on several of the parameters used in the selection of Merino sheep. Selection of breeding stock for studs and for commercial flocks takes place most frequently in sheep between 12 and 18 months of age and the effects of any helminthiasis would still be evident at that time.

In the case of greasy fleece weight, the effect of parasites is not merely to reduce fleece weights proportionally, but differential susceptibility to infection and to the parasite effect results in a skewed and protracted distribution of fleece weights.

Fibre diameters are also important, particularly in the selection of stud rams, as this character is strongly inherited.In this parameter, there can be a decrease by one micrometre in the mean fibre diameter of a group of sheep subjected to helminthiasis. The difference would be greater in individual sheep. Under similar conditions, mean staple length can decrease by one centimetre. The effects on the secondary to primary follicle ratio and on body

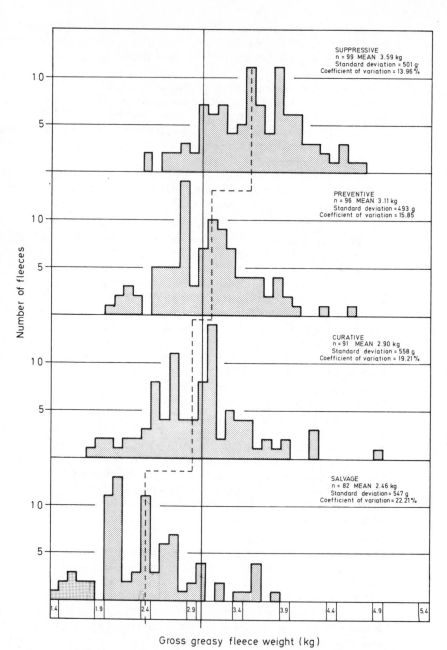

Figure 1—Distribution pattern of greasy fleece weights in Merino wether weaners on four parasite control programmes in 1972. (Growth period 331 days; mean all sheep 3.04 kg).

size were marginal and while they suggested trends in the direction of fewer secondary follicles and smaller sheep following helminthiasis, these trends were not found to be significant.

ACKNOWLEDGEMENTS

Our thanks are due to Merck Sharp & Dohme (Australia) Pty. Limited for permission to publish this data from the Project of Applied Rural Research. Special acknowledgement is made to Chris Banks and Gary Kennedy from Clover Park for the management and supervision of the sheep and the collection of the data. The observations on wool quality were carried out by the School of Wool and Pastoral Sciences of the University of New South Wales and the skin sections were taken and examined by the CSIRO Division of Animal Physiology, Prospect, N.S.W. Mr. Cliff Gray of the CSIRO Division of Mathematical Statistics gave valuable advice and assistance in the handling of the data.

REFERENCES

Jackson, N., Nay, T. and Turner, H.N. (1975) *Australian Journal of Agricultural Research* **26** : 937-957.

Johnstone, I.L., Darvill, F.M., Bowen, F.L., Brown, P.B. and Smart, K.E. (1976) Proc. Aust. Soc. Anim. Prod. **11** : 369-372.

Lipson, M. and Bacon-Hall, R. (1976) Personal communication.

Turner, H.N. and Young, S.Y. (1969) "Quantitative Genetics in Sheep Breeding" (MacMillan Company of Australia: Melbourne).

18

GENETIC IMPROVEMENTS OF FERTILITY AND FLY RESISTANCE

K.D. ATKINS* and B.J. McGUIRK**

*Agricultural Research Station, Trangie; **CSIRO, Division of Animal Production, Epping, N.S.W., Australia*

INTRODUCTION

The traditional procedure in developing sheep breeding plans is to estimate genetic parameters in the population needing improvement and sometimes to provide a check on these estimates by establishing selection flocks. This approach has proved to be completely adequate in developing breeding plans for the improvement of characters such as wool production traits.

A new sheep breeding research programme has recently commenced at Trangie Agricultural Research Station in which this traditional approach has not been followed. Description of this programme is given, attempts are made to explain and justify the procedure adopted and some of the early results which have been achieved are presented.

BACKGROUND TO EXPERIMENT

The aim of the research programme is to develop approaches to improve reproductive performance, and resistance to fleece rot and body strike. Using a traditional approach, the character under selection would be scored or measured on potential breeding stock and then those animals which meet this need (that is, have a high level of fertility, high resistance to fly strike etc.) would be selected. However, with the characters currently under discussion, this approach has limitations.

Selection for reproductive performance is hampered by the sex-limited nature of the trait, and by the limited and discrete range of possible states an animal may have (for example, born as a single or twin, a ewe having 0, 1 or 2 lambs). The discrete nature of the trait is probably partly responsible for the low heritability of fertility characters.

Fleece rot and fly strike tend also to be discrete characters in that an animal either has the conditions or not. Additionally, the incidence of these characters within a flock is very much influenced by climatic factors, especially rainfall. Thus, in many years the incidence of either condition is very low so that selection within a flock is not continuous and relatively ineffective in most environments.

It is therefore hoped to uncover other characters which will enable more rapid improvement to be made in reproductive performance, and fleece rot and fly strike resistance, than with direct selection for these latter characters. These other characters must show considerable variation, have a high heritability and, most importantly, have a high genetic correlation with the traits needing improvement.

EXPERIMENTAL PLAN

Separate flocks of Merinos, sampled from all Merino strains and the principal stud lines within these strains, were purchased in 1974. Fifteen flocks of 100 ewes each were formed in this way and the first complete drop of lambs were born in August-September 1975. It is on these progeny and subsequent drops of lambs that intensive measurement and recording will begin.

By assembling such a wide sample of the Australian Merino it will be possible to apply the results to all strains instead of the traditional approach of working with only one strain and extrapolating to other strains. However, the major reason for the wide sampling is to magnify the variation among sheep for the various characters under consideration, and thus improve the chances of detecting genetic correlations which would be of interest in a breeding programme. This consideration is extremely important in the exploratory stage of a programme where many such possibly correlated traits may have to be looked at, and where some of these measures are expensive and/or difficult to measure.

To overcome these problems, genetic correlations between flocks of sheep will be examined in the hope that these will indicate the likely magnitude of size of the correlation within these flocks. For example in order to estimate genetic correlations within a flock with a reasonable degree of accuracy, one may need to measure all possibly correlated characters on 500 to 1,000 animals, which could be a time-consuming and expensive procedure. However, if such within-flock correlations were mirrored by between-flock correlations then with the 15 flocks, a sample of 15-20 sheep per flock may provide equally reliable information.

The real point of this approach is to use estimates of between-flock genetic correlations as a screen to choose characters which, if correlated highly with reproductive performance of flystrike resistance within flocks, could provide a valuable adjunct or alternative selection criterion in an improvement programme.

The degree of agreement between within-flock and between-flock estimate of genetic correlations is critical. In Table 1 available estimates for the Australian Merino are summarized. It must be appreciated that many of the between-flock correlations shown have large standard errors. The following general points may be made with regard to the agreement between the sets of estimates:

(i) The agreement in sign between the within- and between-flock estimates is good, with the exception of correlations involving body weight and fibre density. Such discrepancies could reflect selection in the development of the different strains and flocks, or the introduction of genes from another breed (for example, a British longwool breed) in the initial development of the medium and strong wool strains of Merino.

(ii) The magnitude of the between-flock correlations is at least equal to, and in most cases greater than the magnitude of the within-flock correlations.

(iii) In general, where sizeable correlations exist within flocks, they are more noticeable as between-flock correlations.

TABLE 1

GENETIC CORRELATIONS ESTIMATED BETWEEN AND WITHIN MERINO FLOCKS

Characters correlated		Between flocks[A]	Within flocks[B]			
			1	2	3	4
GFW*	Yield	0.40(0.38)	−0.09	−0.05		0.06
	CFW	0.97(0.03)	0.80	0.65		0.82
	Staple length	0.98(0.03)	0.29	−0.02		0.70
	CPI	−0.90(0.07)	−0.20	−0.56		−0.87
	Fibre diameter	0.73(0.18)	0.13			0.19
	Fibre density	−0.89(0.12)	0.20			
	Body weight	0.92(0.12)	0.26	−0.11		0.20
Yield	CFW	0.63(0.28)	0.53	0.56		0.64
	Staple length	0.23(0.45)	0.36	0.63		0.54
	CPI	−0.22(0.47)	−0.54	−0.49		−0.47
	Fibre diameter	−0.17(0.49)	0.12			0.03
	Fibre density	−0.21(0.51)	0.15			
	Body weight	0.80(0.28)	0.09	−0.08		0.11
CFW**	Staple length	0.90(0.10)	0.46	0.39	0.37	0.89
	CPI	−0.83(0.12)	−0.53	−0.53	−0.22	−0.96
	Fibre diameter	0.57(0.28)	0.16		0.24	0.16
	Fibre density	−0.83(0.20)	0.30		0.14	
	Body weight	1.01(0.15)	0.27	−0.12		0.33
Staple length	CPI	−0.92(0.06)	−0.54	−0.34	−0.54	−0.75
	Fibre diameter	0.82(0.13)	0.03		0.44	−0.11
	Fibre density	−0.94(0.07)	−0.22		−0.36	
	Body weight	0.81(0.16)	−0.06	−0.26		0.01
CPI*†	Fibre diameter	−0.93(0.07)	−0.10		−0.17	−0.17
	Fibre density	1.01(0.04)	−0.13		0.06	
	Body weight	−0.81(0.16)	0.07	0.05		0.15
Fibre diameter	Fibre density	−0.98(0.03)	−0.63		−0.70	
	Body weight	0.56(0.32)	0.12			0.00
Fibre density	Body weight	−0.74(0.20)	−0.20			

Sources:
A   Jackson and James (1970)
B1   Brown and Turner (1963)
2   Morley (1955)
3   Schinckel (1958)
4   Beattie (1962)

* Greasy fleece weight
** Clean fleece weight
*† Crimps per inch

RESULTS AND DISCUSSION

At this stage, little information has been collected in the programme. The base ewes were joined in 1975 and the proportion of ewes bearing multiples, and the proportion of lambs dying between birth and weaning are presented for each flock in Table 2. Since these base ewes were purchased and not born and reared together, caution is required in interpreting the results.

The 1975 drop progeny experienced heavy rain associated with severe blowfly activity in February 1976. The proportion of animals struck and the proportion of animals affected by fleece rot for each flock is also shown in Table 2. In a comparison of Merino strains in their resistance to fleece rot, Dunlop and Hayman (1958) ranked the strains as fine-wool, medium non-Peppin, medium Peppin and strong-wool in descending order of resistance. A similar pattern can also be discerned from these data, although individual flocks within each strain vary considerably in terms of resistance.

TABLE 2

FLOCK MEANS FOR COMPONENTS OF REPRODUCTIVE PERFORMANCE AND SUSCEPTIBILITY TO BODY STRIKE AND FLEECE ROT

Flock		Per cent* multiple birth	Per cent** lambs dead	Per cent*** lambs struck	Per cent*** lamb fleece rot
Fine-wool	1	48	22	13	23
	2	15	24	4	4
	3	45	33	17	37
	4	33	52	2	8
Medium Non-Peppin	151	19	11	25	
	2	56	30	20	54
Medium Peppin	1	57	25	29	59
	2	29	28	6	30
	3	42	36	3	13
	4	30	41	37	54
	5	3	14	27	64
	6	3	30	14	32
	7	8	19	36	39
	8	71	25	10	39
Strong-wool		21	26	33	67

* percentage of ewes lambing
** percentage of all lambs born
*** percentage of lambs alive

In attempting to correlate, say, litter size with another character, ewes could only be classified into two categories on a within-flock basis, that is bearing singles or multiples. However, on a between-flock basis it is obvious that there is a wide distribution in the proportion of multiple-bearing ewes allowing a correlation to be generated using 15 classifications, that is each flock mean. Similarly, lambs were either struck or not struck by flies on a within-flock basis. But the mean proportion of animals struck in each flock illustrates the wide variation that exists for between-flock correlations.

In conclusion, the approach being used here does appear to magnify among sheep variation and should allow a large number of characters to be screened in terms of their usefulness as alternative selection criteria for the improvement of reproductive performance and resistance to blowfly strike in Merino sheep.

REFERENCES

Beattie, A.W. (1962) *Queensland Journal of Agricultural Science* **19**: 17.
Brown, G.H. and Turner, H.N. (1968) *Australian Journal of Agricultural Research* **19**: 79.
Dunlop, A.A. and Hayman, R.H. (1958) *Australian Journal of Agricultural Research* **9**:260.
Jackson, N. and James, J.W. (1970) *Australian Journal of Agricultural Research* **21**: 837.
Morley, F.H.W. (1955) *Australian Journal of Agricultural Research* **6**: 77.
Schinckel, P.G. (1958) Hawkesbury Conference for Sheep and Wool Officers, Paper 29. CSIRO (Mimeo).

19

PROBLEMS IN THE USE OF EXOTIC GENOTYPES

J.W.B. KING

ARC Animal Breeding Research Organization, Edinburgh, U.K.

SUMMARY

Some of the problems encountered in the importation and evaluation of exotic genotypes are discussed. Available methods of choosing breeds and of sampling them are considered and evaluation procedures after importation are examined.

Although new genotypes may be found to have major advantages they are also likely to display disadvantages of one kind or another. The exploitation of such genotypes presents a new set of challenging problems. There may be a need to evolve appropriate husbandry systems particularly when producing lambs from pasture with marked seasonal constraints.

Consideration needs to be given to the problem of multiplying exotic genotypes and securing investment in its further improvement. The introduction and evaluation of new breeding stocks is likely to be a continuing feature of animal breeding.

INTRODUCTION

The general principles underlying the use of breed resources as a means of genetic improvement have been reviewed by Dickerson 1969 and are discussed by Carter in the present edition. The intention of this chapter is to use some limited experience of experimentation with a few exotic genotypes to consider some of the specific problems which can arise.

CHOICE OF BREEDS

The initial choice of genotypes for importation depends on the acquisition of information from a variety of sources. While there are extensive world lists of breeds in for example Mason 1969, it is rather rare to find well documented accounts of sheep breeds although these may be available for some developed countries. It is possible for breeds represented in small numbers to escape such description and we have, for example, an interesting prolific breed such as the D'man from Morocco only being reported recently (Bouix and Kadiri 1974). This does not happen only in more remote areas of the world as is instanced by the description given by Bekedam (1971) of the Flemish breed found as a relic in the Netherlands. The attention now being paid to genetic conservation is likely to make such examples rather rare in the future and so would make the appearance of hitherto unreported breeds an unlikely event. What seems more probable is that increased knowledge about recognized breeds will reveal possibilities for their exploitation which did not become apparent until their features were better described.

Comparative performance figures are rare and in many instances it is necessary to make inferences from results in different environments without any breeds in common to provide a yardstick for comparisons. In the field of pig breeding a start has been made in Europe with the distribution of one stock of pigs to act as a common control against which other breeds might be assessed (King *et al.* 1976). The use of this technique in sheep is still in the future, but improvements in the freezing of fertilized ova, coupled with ova transplantation would make it feasible.

SAMPLING OF BREEDS

The problems of sampling a breed weigh very heavily on some would-be importers. Often veterinary considerations may restrict the choice of animals so that this question becomes academic but, given freedom of action, should an attempt be made to obtain a random sample or alternatively animals selected by certain criteria? The use of a random sample implies that one wishes to make some pronouncements about the performance of the breed as a whole. This approach is perhaps somewhat naive in supposing that, at the end of the evaluation procedure, breeders will wish to import additional animals when they already have some representatives of that breed, or that veterinary conditions will still allow an import to take place.

Given that there are performance data on which selection can be made, the purchase of animals for importation is a once and for all opportunity for selection which probably should not be bypassed. Since future actions will, however, depend upon conclusions reached upon this particular sample of the breed, precautions should be taken to ensure an adequate sample size (especially of males) so that replacement breeding stock can be produced without inbreeding complications.

EVALUATION PROCEDURES

A great deal of attention needs to be paid to the form of testing used in the evaluation of exotic genotypes. This is not easy if only because the most appropriate form of test will depend upon the way in which the breed is to be exploited and this in turn will depend upon its characteristics which may only be revealed in the testing process itself. The ways in which new genotypes will be used can be classified as: use as pure-breds, use in a cross or as part of a new synthetic population. Because the permutations of possible crosses will be large and the possible ingredients of synthetics also likely to be numerous, there will be a strong inducement to make many assessments on the basis of pure-bred performance. This will undoubtedly be useful but there may be a real danger if the pure-bred is poorly adapted to the new environment in which it finds itself whereas its crosses do show adequate adaptation and therefore perform relatively very much better. Examples of this are the performance of the Border Leicester in Australia (McGuirk 1967) and probably also the East Friesian in Britain both of which show high mortality as pure-breds. On the other hand, cross-breeding tests will be more informative if both pure-bred parents can be included so that a direct measure of heterosis is possible. While many predictions to a first order of approximation can be made neglecting the phenomenon of heterosis there are occasions in which it is very marked. In general the litter sizes of cross-bred ewes of the Finnish Landrace with a variety of breeds could have been predicted from a knowledge of the parental litter sizes. In other reproductive traits the situation has been very different. For example, Land, Russell and Donald (1974) reporting on crosses of the Finnish Landrace with the Tasmanian Merino found that 94 % of cross-bred lambs mated to lamb at one year of age did so, thus resembling the Finnish parent whereas only 5 % of the Merinos lambed.

FURTHER PROBLEMS IN EXPLOITATION

The testing process, whatever form it takes, will provide additional characterization of the imported breed and confirmation or otherwise of those particular merits which prompted its importation. In the case of the Finnish Landrace the prolificacy of the breed was rapidly confirmed. What was not anticipated was that the combination of low milk yield (Crowley and McGloughlin 1972) and a relatively small body size would combine with the size reduction inherent in multiple births to give very low growth rates. This disadvantage coupled with a poor carcass quality by conventional standards has resulted in the Finnish Landrace being rejected by the sheep farming community in Britain. It is to be hoped that this judgment is premature in that the extra prolificacy is a real asset, not readily achieved by within-breed selection. Further attempts have been made to exploit the breed in a variety of different ways. Crosses of the Finn and Dorset Horn do find favour for intensive lamb production but this is only a very small and specialized part of the industry in Britain. For use in grassland systems, encouraging results have been obtained using the Finnish Landrace x Border Leicester as a crossing sire in place of the Border Leicester, and an extract of the report by Deeble and Barker (1976) is given in Table 1. The records derive from a field trial involving about 1,200 ewes.

The Finn/Border Leicester crosses produced a bigger lamb crop, particularly when lambing at one year of age and although the individual weights of their lambs were smaller the disadvantage was not great when measured at 10 weeks of age.

Since the maternal contribution to the early growth of the lamb is known to be important, pilot experiments were also carried out with crosses of the Finnish Landrace and the East Friesian milk breed. These trials evolved

TABLE 1

PERFORMANCE OF FINN × BORDER LEICESTER (F/BL) AND BORDER LEICESTER (BL) CROSS EWES

Lambing at	1 year		2 years		3 and 4 years	
	F/BL	BL	F/BL	BL	F/BL	BL
Ewes lambed per 100 ewes mated	72.6	60.0	92.0	90.9	91.2	90.1
Lambs born per 100 ewes mated	95.9	69.8	178.5	160.8	192.0	175.0
Lambs reared per 100 ewes mated	72.2	51.2	143.0	127.7	149.5	131.3
Lamb wt. at birth (kg):						
Singles	3.7	4.2	4.6	5.0	4.6	5.1
Multiples	2.8	3.3	3.3	3.8	3.7	4.2
Lamb wt. at 10 weeks (kg):						
Singles	22.5	23.7	23.7	25.2	23.3	24.7
Multiples	17.4	19.5	19.4	20.9	20.7	22.0

TABLE 2

COMPARISON OF ABRO DAM LINE AND BORDER LEICESTER CROSSES

	Ewe Type	
	Dam Line × Blackface	Border Leicester × Blackface
All ewes		
wt. (kg)	53	60
fertility 1 yr old %	69	48
2 and 3 yr old %	95	95
Ewes lambing		
% assisted	15	28
No. born alive	1.83	1.63
No. at 8 weeks	1.72	1.58
Lamb performance		
Birth wt. (kg)	3.3	3.8
8-week wt. (kg)	19	20
age at slaughter* (days)	170	165
killing-out %	44.0	43.5
carcass score (1-7 points)	3.2	3.2

* Adjusted to same litter size.

into the creation of a new synthetic Dam line, also intended for use in crossing. The genetic contributions to this line were Finnish Landrace 47%, East Friesian 24%, Border Leicester 17% and Dorset Horn 12%. The results are now available from crosses of this ABRO Dam Line in comparison with a Border Leicester cross and are summarized in Table 2. The performance records have been obtained in an experimental flock of 600 ewes lambing at 1, 2 and 3 years of age.

The cross-bred ewes from the Dam line were smaller than the conventional Border Leicester cross-bred but produced and weaned more lambs particularly in their first lamb crop. A disadvantage in growth rate remained but by existing standards there was no difference in carcass grading of the lambs.

Despite an attempt to produce a crossing sire for use in a conventional manner, it turns out that changes in husbandry practices would be necessary to take full advantage of the new synthetic genotype. Because these sheep are smaller a higher stocking rate seems appropriate and to take advantage of early sexual precocity mating to lamb at one year of age also seems desirable. Consideration of the detailed economics consequent on changes in growth genotype and husbandry practices reveals a rather complex situation. No doubt linear programming methods could be employed to arrive at an optimum situation but reasons for the apparent conservatism of breeders soon become apparent. The moral of this is that with a farming system which attempts to utilize the seasonal growth of grass, no genetic change is likely to be simple and will therefore bear detailed examination of all the factors involved in the sheep production system.

MULTIPLICATION AND IMPROVEMENT OF EXOTIC GENOTYPES

Consideration also needs to be given to the ways in which introduced stock will be multiplied and further improved for the benefit of the industry as a whole. For species with a low reproductive rate, such as the sheep, the existing investment of breeders in their present stock is considerable. Suggestions for change may involve considerable redeployment of capital which may not be easily achieved in the breeding industry where margins are traditionally low. The problem of financing further improvement is a continuing one and experience in Britain would suggest that companies find it difficult to recoup the costs of a genetic programme in sheep. Because importation costs are also high there will often be commercial pressures to realize the investment which has already been made without further provision for the future.

CONCLUDING REMARKS

Experience of the recent boom in exotic cattle breeds may suggest that the importation of exotic genotypes is a passing phase which perhaps recurs at intervals. I would like to suggest instead that the evaluation of exotic genotypes should be a continuous process. The emergence of new markets for different products will provide new opportunities in the future. This factor coupled with improved methods of controlling animal disease will, in the long term, make the international movement of stock that much easier and developments in the ability to store both semen and fertilized ova in the frozen state should reduce transport costs. With progress in the characterization of individual breeds there is likely to be increased emphasis on evaluation of strains within the breed and the emergence of breeding groups will in turn call for measurement of differences between their products. Although adaptations of particular breeds to local environments will slow down the rate of breed substitution, further identification of the critical features in such adaptation should in turn lead to further rationalization.

REFERENCES

Bekedam, M. (1971) 10th International Congress of Animal Production. Paris-Versailles.
Bouix, J. and Kadiri, M. (1974) Options Méditerranéennes, No. 26, 87.
Crowley, J.P. and McGloughlin, P. (1972) *Irish Journal of Agricultural Research* **117** : 119.
Deeble, F.K. and Barker, J.D. (1976) European Association of Animal Production, Sheep and Goat Commission. Zurich.
Dickerson, G.E. (1969) *Animal Breeding Abstracts* **37** : 191.
King, J.W.B., Curran, M.K., Standal, N., Power, P., Heaney, I.H., Kallweit, E., Schroder, J., Maijala, K., Kangasniemi, R. and Walstra, P. (1975) *Livestock Production Science* **2** : 367.
Land, R.B., Russell, W.S. and Donald, H.P. (1974) *Animal Production* **18** : 265.
Mason, I.L. (1969) A Dictionary of Livestock Breeds.
McGuirk, B.J. (1967) *Wool Technology and Sheep Breeding* **14**(2) : 73.

20

EXPLOITATION OF EXOTIC GENOTYPES

A.H. CARTER
Ruakura Agricultural Research Centre, Hamilton, New Zealand

SUMMARY

New breeds could be used to replace existing breeds, to contribute desired qualities in synthesizing new breeds or strains, or to cross with local breeds in stratified or rotational cross-breeding systems. Information on general and specific combining abilities for both offspring and maternal traits is needed to decide optimal utilization strategies.

Sound choice of breeds for importation demands clear definition of breeding goals and assessment of the genetic merit for important production traits of overseas relative to local breeds. The limitations of overseas performance information and the desirability of international co-operation in breed comparison studies are pointed out.

Local testing of new breeds is essential to determine their improvement potential and to guide effective utilization. Progeny test comparison of exotic versus local sires over native ewes provides a satisfactory initial screening, subject to adequacy of numbers and of genetic sampling.

Brief mention is made of alternative importation procedures and of the importance of animal health safeguards.

Investment appraisal of livestock importation involves balancing costs, including testing resources and disease risks, against present and future benefits through genetic improvement, increased flexibility and indirect stimuli to higher production. Rapid initial evaluation under experimental conditions should be followed by controlled on-farm testing and demonstration before general release to the industry.

Present New Zealand experience is discussed in terms of improvement needs for sheep production, nomination of promising overseas breeds, animal health and quarantine considerations and importation procedures. An experimental importation in 1972 comprised Finn, Oxford Down, East Friesian and German White Headed Mutton sheep from Britain and Ireland. Early performance of the pure-breds is summarized and the initial evaluation programme described. Likely future outcomes and developments are indicated.

It is concluded that the controlled introduction and testing of new breeds chosen to meet specific improvement needs should constitute a sound national investment.

INTRODUCTION

Traditional acceptance of and implicit faith in established breeds have in the past two decades been shaken by the "exotic revolution". Likewise cross-breeding is almost attaining respectability in the pure-breeding citadels of yesterday. Yet far from being novel, the introduction and crossing of breeds have been basic ingredients in the evolution of animal production in most regions. How many present sheep industries rely on truly indigenous races? And how many improved modern breeds are strictly "pure"? For present purposes exotic will be defined as a breed or strain for which commercial stock are not, or have only recently become, locally available. "Genotype" will refer to any specific breed or cross.

The resurgence of interest in new breeds has been stimulated by the need for more efficient production, particularly in the face of changing markets and conditions and increased farm costs; by greater awareness of the merits of other races through performance recording and better communication; and by technical advances in disease control and semen storage permitting readier interchange of genetic material. Whereas formerly choice of the breeds imported was largely fortuitous, confined to those familiar and available to the human migrants, the possibility now exists for more rational selection among a far wider range of breeds. The success of the plant breeder in exploiting exotic genotype is surely a challenge to the animal improver.

The evaluation, conservation and utilization of breeds have been under close scrutiny by the United Nations Food and Agriculture Organization, with particular reference to the needs of developing countries. General principles and procedures were discussed by an expert panel (FAO 1967). Subsequent studies have concerned cattle, pigs, poultry and breed conservation, but not specifically sheep. Aspects of exotic livestock importation as an improvement strategy were considered by Carter (1970, 1975). Importation and potential uses of new sheep breeds in relation to local industry needs have been discussed by Boylan (1968) in USA, Rae and Wickham (1970) and Carter (1972) in New Zealand, Turner (1971) and Australia Animal Production Committee (1976) in Australia.

PURPOSES

Introduction of new sheep breeds can serve the following purposes:—

- produce a new product, e.g. lambskin fur or milk for cheese
- exploit a new environment such as desert, tropical or mountain areas, or for environmental conservation itself, e.g. heather sheep
- meet a changing market e.g. heavier weight lamb, coarser or finer wool
- better exploit a changing environment or management system e.g. improved hill country or intensive husbandry with more frequent lambing

- improve production under existing marketing and farming conditions
- provide a wider range of genetic material for research study.

Non-genetic benefits could include profitable re-export and psychological stimulus to betterment of local breeds.

Although emphasis in this paper will be on genetic improvement of present production, it should be remembered that genetic diversity confers powerful flexibility to meet changing future market requirements and production systems.

The basis of genetic improvement in a population is increased frequency of favourable genes. The primary agent in such change is selection, which can operate both within and between breeds. As pointed out by Dickerson (1973) breed differences can be exploited very rapidly and as accurately as the differences are evaluated. The expression of genetic potential is influenced also by the way in which genes are combined, which can to some extent be controlled by the mating system. The contribution of exotic breeds to livestock improvement is thus seen to be dependent firstly upon their genetic superiority over existing stock for desirable productive traits, secondly upon the extent to which they may "nick" with local breeds and thirdly upon the breeding system adopted.

METHODS OF UTILIZATION

New breeds can be utilized to improve productive efficiency in four main ways:—

1. To replace existing breeds, as in the early change from Merino to Romney flocks in New Zealand.

2. To contribute desirable qualities in a cross-bred foundation stock for subsequent interbreeding to form new "synthetic" breeds or strains, past examples being development of the Corriedale, Poll Dorset and Drysdale.

3. As crossing sires to produce superior commercial progeny (not used to breed replacements) in a stratified or "tiered" cross-breeding system, exemplified by mating prolific sire breeds with hardy local ewes to produce fat lamb dams.

4. As component breeds in a rotational cross-breeding programme designed to exploit hybrid vigour, a procedure successfully applied in pig and poultry breeding.

In practice all these methods involve some degree of cross-breeding, breed substitution itself being most rapidly effected by "grading-up". The theoretical basis of cross-breeding has been greatly elucidated in recent years, applications to livestock improvement being discussed by Dickerson (1973). Dickerson (1969) defined the parameters needed to determine optimal breed utilization strategies and outlined appropriate experimental designs. Operational and economic implications of alternative breeding systems were considered by Robertson (1971) and Hill (1971), who stressed the importance of considering the potential for future selection improvement as well as short-term gains. Statistical and sampling aspects of breed evaluation have been studied by Hill (1974a) and Jansen (1974). Timon (1974) discussed sheep breed evaluation and breeding strategies, citing experimental comparisons in Ireland as examples.

In general terms, choice among alternative breeds and breeding systems requires information on additive or directly transmissible genetic breed differences (general combining ability) and on heterosis in breed crosses (specific combining ability), for both individual (offspring) and maternal traits. As pointed out by Moav (1966) continued crossing to combine desired but different qualities of two breeds may be more profitable than interbreeding the cross, even in the absence of hybrid vigour for economically important component traits. This applies particularly in combining superior maternal and individual performance traits of different breeds in situations, common in the sheep industry, where dam overhead costs are high and where genetic antagonisms militate against effective joint selection for maternal and individual attributes. High levels of heterosis are the only justification for recurrent cyclical crossing among breeds of similar productive merit. Heterosis can also contribute to successful stratified crossbreeding and to breed synthesis.

CHOICE OF BREEDS

Sound decision on importation itself and on choice of breeds demands clear definition of breeding goals, adequate description of the performance potential of existing stock, and prediction of the genetic merit for important productive traits of overseas relative to local breeds. The appropriate breeding goal should be to maximize product value relative to total input costs, but allowing for likely future production and marketing trends. Obviously the emphasis on different sheep production traits will vary according to the nature of the enterprise and the farming conditions. In the harshest environments, ability of the ewe to survive and reproduce with a minimum of shepherding and to clip an acceptable weight and quality of wool is of prime importance. At the opposite extreme, intensive feeding systems demand high fertility, growth rate and voluntary feed intakes and perhaps a long breeding period, with wool production of secondary importance.

Choice among candidate breeds for importation is necessarily guided by available overseas performance information. It is unfortunately true however that few countries have yet fully quantified the performance attributes of their own established breeds. A cardinal requirement here is comparative assessment of breeds treated alike. Replication of the comparisons over a range of farming conditions will widen the generality of the conclusions and provide useful information on the importance of genotype x environment interactions. Comparative evaluation is vastly more difficult among breeds farmed in different countries under quite different management systems, particularly in the absence of commercial flock recording information. A strong plea is made for international co-ordination of research aimed at comparing sheep breeds and crosses. This could be achieved on the one hand by including common breeds and if possible comparable production systems in trials in different countries, and on the other by standardisation of recording procedures and criteria, for example including fleece information even when this is not locally important.

LOCAL TESTING

Because of possible genotype x environment interactions, breed performance rankings overseas, even if accurately known, might not reflect productive merit under different conditions. Likewise combining ability established in crosses with overseas breeds may not accurately predict advantages in crossing with different local breeds because of specific genotype x genotype interactions. Local testing of new breeds is therefore essential to determine their improvement potential and to guide effective utilization.

The consensus of present evidence suggests that for the main productive traits in sheep and considering the wide spectrum of breeds potentially available, breed differences in general combining ability are likely to be of greater practical importance than in specific combining ability or heterosis A noteworthy exception to this general statement is environmental adaptability. This is particularly relevant in comparisons of introduced pure breeds with local stock. The ability of animals to adapt, over time, to diverse conditions is well testified by the ubiquity of the Merino, the lowland lamb-producing prowess of the hardy Scottish Blackface and the New Zealand hill country performance of the erstwhile Romney Marsh. But in the short term, change to an alien environment can seriously prejudice performance comparison of a new breed relative to established stock. Poor overall performance of the Border Leicester in comparative trials both in Australia and France can be cited as an example; in both cases however crosses with local stock performed creditably, resulting in high apparent levels of heterosis in productivity due simply to better adaptation of the crosses than of the "exotic" purebred. Perhaps the salient lesson here is that, in practice, the performance of crosses should be compared with that of the "better" parent rather than the parental average (in assessing overall productive merit).

The test evaluation of exotic breeds should clearly relate to the purpose for which they are imported. In seeking improved terminal sire breeds for lamb production, assessment of first-cross progeny out of local ewes for survival, growth and carcass traits would be appropriate. Improvement of the breeding ewe on the other hand demands appraisal of a much broader range of traits, some of which are expressed only in the female and are subject to large maternal and possibly heterotic effects. Initial evaluation based on comparative performance for all economically important traits of crosses with native ewes, relative to contemporary local pure-breds and crosses, should serve as a screening process to identify the more promising breeds. If performance of an exotic cross is not materially superior to that of locally derived stock, the breed in question may be dismissed as unlikely to contribute to productive improvement. If it is superior, subsequent effective utilization of the breed may then require further testing to determine additive genetic and hybrid vigour components of individual and maternal performance.

Estimation of heterosis and maternal effects is facilitated when all breeds are represented by pure-bred females, permitting derivation of reciprocal crosses. The needed information can alternatively be obtained from the generation and performance comparison of first crosses, inter-bred crosses and back-crosses, based on local ewes and requiring use only of males of the exotic breeds. Although this procedure is more time-consuming and demands greater experimental resources, it may be preferable in overcoming the problem of poor adaptation of imported pure breeds and in yielding results more directly applicable to breed utilization in the industry. Whichever testing method is adopted, the importance of adequate numbers and adequate genetic sampling needs emphasis.

IMPORTATION PROCEDURES

Carter (1975) discussed alternative methods of importing new genetic material, numbers of animals needed for satisfactory establishment of breed and animal health implications. He concluded that introduction of about 30 pregnant ewes per breed would permit most efficient use of limited importation and quarantine resources in establishing new sheep breeds. Perfection of embryo storage techniques may however favour future transport of fertilized ova *in vitro* rather than *in vivo*. Disease prevention is of obvious importance not only in protecting existing livestock industries but also in some cases in safeguarding market outlets.

COSTS AND BENEFITS

The value of importation and use of new breeds must be assessed in relation to alternative methods of improvement, in particular selection within or among, and cross-breeding with, existing breeds. For sound decision making such assessment must cover costs as well as benefits. Much study has recently been given (Moav 1973; Hill 1974b; Cunningham 1974) to the cost-effectiveness of alternative breeding policies, taking into account the different times at which expenditure is incurred and benefits are realized by commercial discounting procedures.

Benefits from importation will depend primarily on the actual superiority of the exotic over local breeds and this can be accurately determined only on the basis of local testing. The main cost components relate firstly to direct purchase and importation; secondly to necessary testing and development; and thirdly to quarantine requirements and potential animal health hazards. The risk of introduction of disease cannot readily be quantified nor can it be dismissed, but it should surely be minimal with modern advances in veterinary science and technology.

INDUSTRY APPLICATION

The advantages from introduction of superior breeds will accrue only as they are effectively utilized in the industry. The desirability of rapid dissemination and use of new breeds must however be offset against the need for adequate testing to ensure that they are in fact superior and to guide their most efficient use.

Some degree of control by Government or appropriate national organization over importation, testing and distribution of new breeds is considered necessary in the interests of national livestock improvement. It is important that breeds and animals imported be soundly chosen and that numbers are sufficient to permit effective subsequent multiplication and selection. In most situations experimental evaluation will provide more rapid and accurate information on relative breed or cross-bred performance than will on-farm testing, and will thus expedite initial screening of new breeds. Experimental findings should however be checked by field tests covering a wide range of commercial environments. Convincing demonstration of the productive superiority of new breeds or derived crosses will undoubtedly promote their acceptance and use by the industry.

An appropriate procedure would be conditional release, following initial experimental testing, of rams of "promising" breeds or crosses to co-operating breeders, the conditions including an undertaking to provide valid progeny test information on the rams in comparison with local sires. The outcome of such field testing, together with the further experimental information available, would then determine appropriate policies for dissemination and use of the new breeds in the industry.

NEW ZEALAND EXPERIENCE

The principles and problems of controlled livestock importation are illustrated by recent experience in New Zealand, whose economic well-being is closely dependent on the production and export of animal products. Recognizing the potential benefits from exotic breeds, yet mindful of the paramount need to avoid introducing animal diseases, Government established a small but high-security quarantine station on Somes Island, in Wellington Harbour. The specific purpose was to permit experimental importation of promising breeds in order to assess their potential contribution to improving productivity.

The predominant sheep breed (70%) is the Romney, unexcelled for weight of fleece produced per kilogram live weight, well adapted to a very wide range of farming environments and a satisfactory fat lamb dam in many conditions. General improvement in pasture and animal management, increasing value of meat relative to wool and change in market demand towards heavier carcasses have however highlighted the urgent need for greater prolificacy and the desirability of increased lamb growth and ewe milk production. The Cheviot and the Border Leicester, crossed with the Romney to form the Perendale and the Coopworth, have improved lambing performance in many environments. Selection within these breeds as well as the Romney itself is contributing further gains, albeit slow.

In considering choice of overseas sheep breeds, primary emphasis was for the above reasons given to importation of "high fertility" genes. Attention has been paid also to growth rate, milk yield, wool production and avoidance of pigmented fleeces. The breeds nominated as the most likely, on available evidence, to contribute to improvement of productivity in New Zealand flocks were the Finnish Landrace or Finn (Finland), East Friesian (Germany), Texel (Holland), Bleu de Maine (France), White Headed Mutton (Germany) and Oxford Down (United Kingdom). For each breed it was planned to introduce 30-40 unrelated ewes in lamb to a wide genetic range of sires.

The geneticist proposes, the veterinarian disposes! Practical choice of species and of sources, and consequently to some extent of breeds, for importation must necessarily be subject to animal health considerations. For veterinary reasons, experimental importation of cattle took precedence over sheep despite greater estimated potential advantages from introducing new sheep breeds. Likewise animal health factors dictated Britain and Ireland as the source of the only experimental sheep introduction to date, and required that all animals be at least four years of age. The capacity of the Quarantine Station limited the consignment to 110 sheep.

Of the breeds nominated, adequate samples could be obtained only of the Finn and the Oxford. A small number of German White Headed Mutton and East Friesian was however included to permit preliminary local testing. Animals were very carefully selected on the basis of pedigree and performance background and of course animal health clearance. The ewes were mated to rams which had also been carefully screened.

A total of 99 ewes and 10 rams finally reached Somes Island in December 1972. The ewes lambed in February-March and were mated with the imported rams to lamb again in September-October 1973. They and their progeny were than transferred to a second quarantine location, Mana Island, off Wellington's west coast. Of 200 hectare, this island permits the grazing of sheep but is an exposed and difficult environment. Provision of shelter and supplementary concentrate feeding have been necessary for the imported pure-breds, whose performance cannot therefore be properly compared with the large experimental flock on Mana of Romney ewes and derived crosses.

TABLE 1

NEW ZEALAND SHEEP IMPORTATION: PURE-BRED PERFORMANCE

Breed	Finn	East Friesian	White Headed Mutton	Oxford	Romney
Ewes (rams) imported	46(4)	5(1)	8(2)	40(3)	
Litter size—Somes	2.7	2.5	1.9	1.8	
Weaning %—Somes	240	230	190	130	
—Mana	175	175	80	115	90
—Progeny*	140	130	70	80	75
Live weight ranking	100	120	140	150	120
Fleece weight ranking	100	140	210	190	250
Fibre diameter (μ)	25	30	35	33	32

* New Zealand-born pure-bred ewes, lambing at Mana at $1\frac{1}{2}$-$2\frac{1}{2}$ years of age.

The performance to date of the exotic pure-breds, summarized in Table 1, must be interpreted with caution, firstly because of the small numbers, particularly of the East Friesian and White Headed Mutton breeds, and secondly because of the very atypical conditions under which they have been managed. Results for the Romneys are shown merely as a rough guide. Under the largely indoor management at Somes, litter size and weaning percentage of the four imported breeds were broadly in line with reported overseas figures. Transfer to Mana resulted in a sharp drop in weaning percentage due mainly to lower fertility (particularly the White Headed Mutton) and higher lamb losses, but with some reduction also in litter size. Relative rankings for live weight, greasy fleece weight and fibre diameter (at hogget shearing) are also as expected, except perhaps for the comparatively high wool production of the Oxford. The apparent superior wool production of the Romney over the much heavier White Headed Mutton is of interest.

Collaborative research on these imported pure-breds is involving many scientific disciplines—genetics, physiology, behaviour, wool metrology and biochemistry. An initial aim in the programme is rapid multiplication of the pure-breds to expedite their future research and commercial use. By mating in 1976 the numbers of breeding ewes (2-tooth and older) had increased four-fold for the Finn, two-fold for the Oxford and at intermediate levels for the other two breeds. Superfoetation of the imported females and embryo transplantation to Romney host ewes is being explored as a means to augment natural increase in numbers. Animal health requirements stipulate that the imported animals themselves may never leave Mana and their progeny may not be released for industry use before 1979.

Because of small numbers of exotic pure-breds and their poor adaptation, already apparent, to New Zealand pastoral farming conditions, initial evaluation of these imported breeds will be based on performance assessment of their crosses with Romney ewes, relative to crosses sired by the local Border Leicester, Cheviot and Dorset breeds and to the straight Romney. The recent availability from Australia* of rams of the reputedly highly prolific Booroola Merino strain has permitted inclusion of this breed in the comparisons.

The design of the first phase in the evaluation trials is illustrated in Table 2. The aim is to generate and compare progeny out of Romney ewes by the nine sire breeds together with interbred Finn, East Friesian and White Headed Mutton crosses and Finn and East Friesian backcrosses to the Romney.

TABLE 2

NEW ZEALAND SHEEP IMPORTATION:
MATING DESIGN FOR EVALUATION PROGRAMME

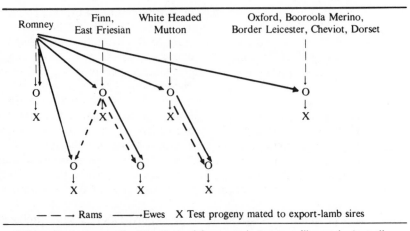

— — → Rams ———→Ewes   X Test progeny mated to export-lamb sires

*Sheep can be imported to New Zealand from certain "approved" areas in Australia.

All progeny will be assessed for survival, growth and wool production; major emphasis will however attach to the reproductive and weaning performance of the females over four or more lambings. Each of the 14 breed types compared will be represented by 150-200 ewes, mated to Down rams when not required to generate further test progeny. The second stage of the programme will develop and test breed combinations and breeding systems designed to exploit the merits of those breeds shown to produce superior cross-bred progeny.

The generation of exotic and local crosses was initiated at Mana in 1974 but obviously limited by available stock carrying resources. An unexpected early finding has been the good lamb growth of the Finn x Romney cross, superior not only to the Romney itself but also to the Border Leicester and Cheviot crosses. Of interest also is the strong mating preference of Finn, East Friesian and White Headed Mutton rams for ewes of their own breed versus Romney ewes. The resulting lower apparent fertility of Romney ewes mated to exotic relative to Romney sires was partially offset by better survival of the cross-bred lambs.

In 1976 the Romney ewes, in lamb to exotic sires, and the available cross-bred progeny were transferred to a third "sub-quarantine" mainland property to join a comparable group of Romney ewes in lamb to local or Booroola sires. This property, with a carrying capacity of about 5,000 ewes plus replacement stock, will henceforward be the main test location until general release of the stock is permitted in 1979 or 1980.

One, perhaps fortunate, consequence of the long overall quarantine period is that useful performance information on the exotic breeds and crosses will be available before they can be released to the industry. Although the testing will still not be complete, and will relate to only one effective environment, it should at least indicate the likely merits of the imported breeds and how they might best be utilized. Continued research and development will clearly be necessary—and rewarding—to ensure most effective utilization of both exotic and established breeds.

The prospects for real productive improvement are indeed exciting, particularly in terms of more efficient and profitable lamb production. An infusion of 25 percent Finnish or East Friesian "blood" should lift lambing percentage of the Romney by 35%, corresponding to about 40 years intensive selection for fertility. This could be rapidly achieved by wide use of half-bred Finn or East Friesian rams. Through judicious choice of the other component in this half-bred ram, for example Border Leicester or Oxford Down, the derived quarter-bred Finn or East Friesian sheep could substantially surpass the present lamb production performance of the Romney while yet retaining its versatility and wool production efficiency.

CONCLUSION

In the quest for more efficient sheep production we cannot afford to ignore the potential contribution of new breeds. Wise selection in relation to national improvement needs is the key to success, both in choice of imported breeds and in promoting subsequent genetic improvement. The introduction of exotic breeds may be expected to provide a challenge and a stimulus to existing breeders. New breeds, to succeed, must adapt to local conditions; for full exploitation of their potential, farmers also may need to adapt and to modify their husbandry practices. The costs of importation and testing may be great but the potential benefits, both direct and indirect, should amply justify the investment.

REFERENCES

Australia Animal Production Committee (1976). Report of Expert Panel on Exotic Breeds of Sheep and Goats for Import to Australia. May 1975.

Boylan, W.J. (1968). *Sheep Breeder and Sheepman Magazine* **87** (5).

Carter, A.H. (1970). *Proceedings of 20th Lincoln College Farmers Conference* : **108**.

Carter, A.H. (1972). *New Zealand Journal of Agriculture* **125** : 25.

Carter, A.H. (1975). Proceedings of the III World Conference on Animal Production, Melbourne, 1973 (ed. R.L. Reid, Sydney University Press, Australia) : 608.

Cunningham, E.P. (1974). Proceedings of the Working Symposium on Breed Evaluation and Crossing Experiments with Farm Animals (IVO, Zeist, Netherlands) : 107.

Dickerson, G.E. (1969). *Animal Breeding Abstracts* **37** : 191.

Dickerson, G.E. (1973). Proceedings of the Animal Breeding and Genetics Symposium in Honour of Dr Jay L. Lush (American Society of Animal Science, Champaign, Ill., USA) : 54.

FAO (1967). Report of the FAO Study Group on the Evaluation, Utilization and Conservation of Animal Genetic Resources, Rome, 1966 (FAO, Rome, Italy).

Hill, W.G. (1971). *Annales de Genetiques et de Selection animale* **3** : 23.

Hill, W.G. (1974a). Proceedings of the Working Symposium on Breed Evaluation and Crossing Experiments with Farm Animals (IVO, Zeist, Netherlands) : 43.

Hill, W.G. (1974b). 1st World Congress on Genetics Applied to Livestock Production (Editorial Garsi, Madrid-28, Spain) II : 291.

Jansen, A.A.M. (1974). Proceedings of the Working Symposium on Breed Evaluation and Crossing Experiments with Farm Animals (IVO, Zeist, Netherlands) : 67.

Moav, Rom (1966). *Animal Production* **8** : 193, 203 and 365.

Moav, Rom (1973). Agricultural Genetics: Selected Topics (John Wiley & Sons, New York) : 319.

Rae, A.L. and Wickham, G.A. (1970). *Sheepfarming Annual 1970* (Massey University, New Zealand) : 87.

Robertson, Alan (1971). Report 10th International Congress on Animal Production, Paris-Versailles 1971 : 57.

Timon, V.M. (1974). Proceedings of the Working Symposium on Breed Evaluation and Crossing Experiments with Farm Animals (IVO, Zeist, Netherlands) : 367.

Turner, Helen Newton (1971). *Wool Technology and Sheep Breeding* **18** : 42.

21

A CROSS-BREEDING EXPERIMENT WITH FINNISH LANDRACE, ILE DE FRANCE AND TEXEL. SOME PRELIMINARY RESULTS

A.H. VISSCHER
Research Institute for Animal Husbandry "Schoonoord". Zeist, Holland

SUMMARY

A cross-breeding experiment with Finnish Landrace (FF), Ile de France (IF) and Texel (TS and TC) is outlined. The Finnish Landrace is chosen for prolificacy and length of breeding season, the Ile de France for length of breeding season and carcass quality and the Texel (TS) for growth and carcass quality. The Texel Control line is the reference for the traditional system.

The aim of the whole project is to investigate under grassland farm conditions the improvement in both biological and economical terms of the efficiency of the three-way cross (TS×IF×FF) submitted to a system of 3 lambings in 2 years, compared with the traditional system of one lambing per year (TC). In this chapter only the preliminary results of some characteristics of the crosses are briefly mentioned and discussed.

INTRODUCTION

The profitability of the sheep industry in Holland as in most West-European countries depends largely on the number of lambs and the weight and the quality of their carcasses produced per ewe per year. The Texel breed is almost the only breed in the Netherlands. This originally Dutch breed is mostly kept in small flocks of about 50 breeding ewes on dairy farms with permanent grassland. They lamb once a year in spring. Under these conditions the breed produces a lamb with a rather unique slaughter quality. A breed comparison has indicated that the Texel breed can be ranked among the best breeds in this respect (More O'Ferrall, 1974). Observations on commercial farms have revealed that the lamb production is 1.4-1.6 lambs weaned per ewe lambed (Sturkenboom, 1971).

The quickest lamb production improvement can be obtained by the use of prolific breeds in cross-breeding programmes (Moav, 1966; Nitter and Jacubec, 1970 and Dickerson, 1973).

These programmes demonstrate that in a three-way cross the breed differences in both maternal and individual performance, the maternal heterosis in the cross-bred ewe and the individual heterosis in the terminal slaughter lamb are utilized completely. Furthermore, the lamb production can be improved when the "dam" breeds or cross-bred ewes have an extended breeding season, so that lambing is possible at 8 month invervals (Dickerson, 1973). So a highly profitable three-way cross could very likely be obtained by crossing first a very prolific, light breed with a meat type breed. Both breeds should have an extended breeding season. The F_1 ewe, which should combine good prolificacy and frequent lambings (at 8 month invervals) with moderate mature weight and slaughter quality, should be mated to a ram of a meat-type breed with excellent slaughter characteristics. The aim of this project is to investigate if such a three-way cross could be promising for the Dutch sheep industry. Results of the comparison of the pure-breds are described by Visscher (1974). In this chapter some preliminary results of cross-breeding will be presented and discussed.

MATERIAL AND METHODS

(a)   Breeds and mating types

The project was started in 1971 with the acquisition of Finnish Landrace as a very prolific light breed and the Ile de France as a meat type breed. Both breeds have an extended breeding season. For the reason mentioned before a Texel Selection line (TS) is involved as sire breed. A Texel Control line (TC) is used as a reference line for the traditional lamb production system. In respect to prolificacy the foundation stock of both Finnish Landrace (FF) and Ile de France (IF) comes from the top half of the original populations (Visscher, 1974). Each line is bred as a self-contained flock of about 60 ewes and 10-12 rams. The mating system is based on avoidance of full-sib and half-sib matings. The choice of sires for the next generation is done within sires. A short generation interval is maintained by using young rams (6 months of age) for breeding as soon as possible and by a ewe replacement of about 30%.

After three years of adaptation, multiplication and evaluation of the pure-bred lines the reciprocal crosses between FF and IF line were made during two consecutive years with a total of 6 mating periods (Figure 1 phase 1) in order to estimate heterosis, differences between breeds and reciprocal crosses. In order to get as many F_1-females as soon as possible the ewes were submitted to an intensive lambing system with 8 month intervals. So most of the ewes in the mating periods 1,3 and 5 were the same ewes. The same was true for the ewes in the mating periods 2, 4 and 6. Before the first and second mating period they were randomly divided in two groups: pure breeding vs. cross-breeding. If a ewe produced a pure-bred litter the next litter was a cross-bred one and the reverse. This is done in order to get as much genetic variation out of the available animals as possible. Before each mating period each group was randomly divided in 10-12 subgroups.

In phase II (Figure 1) the cross-bred females and pure-bred contemporaries will be submitted to an intensive lambing system with 8 month intervals with a maximum of 4 litters per ewe. The TS and TC ewes will produce one lambing per year. The numbers of females involved are approximate numbers.

Figure 1—Experimental design for "Schoonoord" cross-breeding project.

Phase 1	Females					
	Finnish Landrace (FF)			Ile de France (IF)		
Mating period	1	2	3	4	5	6
Males	Aug/Sep	Dec/Jan	Apr/May	Aug/Sep	Dec/Jan	Apr/May
12 Finish Landrace '74/'75	19	17*	13	18	12*	18
12 Finnish Landrace '75/'76	17	13	12	17	13	18
10 Ile de France '74/'75	15	15*	12	19	13*	17
10 Ile de France '75/'76	21	15	14	17	11	17
		* ½ year old females				
Phase II	Females					
Males	FF×FF	FF×IF	IF×IF	IF×FF	TS	TC
10 Texel Selection	90	55	60	80	60	
10 Texel Control						60

(b) Expected composition of the mean of the mating types

Under the assumption that the environmental contribution to each mean is the same in each mating period the means of the four different mating types of phase 1 have the following composition:

breed of sire		breed of dam	: composition of mean
FF	×	FF	$: a_{FF} + p_{FF} + m_{FF} + e$
FF	×	IF	$: \frac{1}{2}a_{FF} + \frac{1}{2}a_{IF} + p_{FF} + m_{IF} + h_{1,2} + e$
IF	×	FF	$: \frac{1}{2}a_{IF} + \frac{1}{2}a_{FF} + p_{IF} + m_{FF} + h_{2,1} + e$
IF	×	IF	$: a_{IF} + p_{IF} + m_{IF} + e$

where a = average additive breed contribution

p = paternal contribution

m = maternal contribution

$h_{1,2}$ and $h_{2,1}$ = heterosis contribution

e = environmental contribution

For each trait it is possible to estimate from these means

(1)   the heterosis $\dfrac{h_{1,2}+h_{2,1}}{2} = \dfrac{(FF\times IF + IF\times FF) - (FF\times FF + IF\times IF)}{2}$

(2)   the difference between reciprocal crosses, if $h_{1,2}=h_{2,1}$ and $p_{IF}=p_{FF}$, is namely: $FF\times IF - IF\times FF = m_{IF} - m_{FF}$

(3)   the difference between sire breeds is namely:
$(IF\times IF + IF\times FF) - (FF\times FF + FF\times IF) = a_{IF} - a_{FF} + p_{IF} - p_{FF}$

The means of the first four mating types of phase 2 (Fig. 1) have the following composition under the assumption that the environmental contribution to each mean is the same in each mating period:

breed   breed
of   ×   of   : composition of mean
sire   dam

TS   ×   FF   : $\frac{1}{2}a_{TS} + \frac{1}{2}a_{FF} + p_{TS} + m_{FF} + h_{3,1} + e$

TS   × FF×IF : $\frac{1}{2}a_{TS} + \frac{1}{4}a_{FF} + \frac{1}{4}a_{IF} + p_{TS} + \frac{1}{2}m_{FF} + \frac{1}{2}m_{IF}$
           $+ h_m + \frac{1}{2}h_{3,1} + \frac{1}{2}h_{3,2} + e$

TS   × IF×FF : $\frac{1}{2}a_{TS} + \frac{1}{4}a_{IF} + \frac{1}{4}a_{FF} + p_{TS} + \frac{1}{2}m_{IF} + \frac{1}{2}m_{FF}$
           $+ h_m + \frac{1}{2}h_{3,1} + \frac{1}{2}h_{3,2} + e$

TS   ×   IF   : $\frac{1}{2}a_{TS} + \frac{1}{2}a_{IF} + p_{TS} + m_{IF} + h_{3,2} + e$

where a = average additive breed contribution
       p = paternal contribution
       m = maternal contribution
       h_m = heterosis contribution of dam
$h_{3,1}$ and $h_{3,2}$ = heterosis contribution of individual
       e = environmental contribution

From these means it is possible to estimate for each trait the maternal heterosis

$$h_m = \frac{(TS\times(FF\times IF) + TS\times(IF\times FF) - (TS\times FF + TS\times IF)}{2}$$

With this four mating types it is not possible to make an estimate for $h_{3,1}$ and $h_{3,2}$. Therefore it is necessary to make reciprocal crosses between TS and FF and between TS and IF.

(c)   Experimental procedure

All animals are kept together in a single flock on the experimental farm "'t Gen" of the Research Institute of Animal Husbandry "Schoonoord". The mating periods are 34 days except the springmating period which is 51 days. Vasectomised FF-rams run continuously with the ewes from 17 days before the beginning of the mating period till the end of the mating period.

The heat control of the ewes is carried out twice daily. Ewes in heat are handmated. The animals are kept on pasture for most time of the year. Fig. 2 gives some details about the management system of both ewes and lambs. Ewes housed indoors are fed with ad lib. hay and concentrates according to the requirements for weight, stage of pregnancy and expected litter size or lactation. Lambs fattened indoors are fed with ad lib. hay and concentrates. The lambs are weighed at regular intervals till slaughter weight. The aim is to slaughter the lambs at a constant carcass weight of 22 kg. Therefore lambs from the pasture are slaughtered at a weight of about 47 kg. Lambs fattened indoors are slaughtered at a live weight of about 45 kg. Slaughtering occurs after a twelve hours period of fasting. Slaughter quality is determined by applying a method of classification of fleshiness and fat covering. This classification system is adapted from the beef classification system used in this institute (de Boer and Nijeboer, 1973). It has a scale from 1 to 6 and each class is divided in three subclasses. For fat covering the optimum class is about 2.67. For fleshiness the highest score is the best.

Figure 2—Management system

| | LAMBING PERIOD | | | |
	JAN/FEB	MARCH/APR	MAY/JUNE	SEP/OCT
PERIOD PREGNANT EWES HOUSED BEFORE LAMBING	about 2 months	about 2 months	about 1 week	about 1 week
PERIOD LACTATING EWES HOUSED AFTER LAMBING	2 months	1 week	1 week	2 months
LAMBS FATTENED AFTER WEANING (PASTURE/ INDOORS)	pasture	pasture	pasture	indoors
PERIOD OF ADDITIONAL CONCENTRATE FEEDING AFTER WEANING	first 50 days	first 50 days	till slaughter weight	till slaughter weight (ad lib.)

RESULTS AND DISCUSSION

(a) Reproductive performance of pure-bred dams (phase 1)

Table 1 shows the original data from which the heterosis effects as well as the differences between the reciprocal crosses and between the sire breeds are calculated (Table 2). As Table 2 shows the heterosis contribution to the percentage ewes lambed per 100 ewes mated is only positive in two out of

TABLE 1

REPRODUCTIVE PERFORMANCE (PHASE 1)

Sire breed		FF	FF	FF	FF	FF	FF	IF	IF	IF	IF	IF	IF
Dam breed		FF	FF	FF	IF	IF	IF	IF	IF	IF	FF	FF	FF
Lambing period		Jan/Feb	May/June	Sep/Oct	Jan/Feb	May/June	Sep/Oct	Jan/Feb	May/June	Sep/Oct	Jan/Feb	May/June	Sep/Oct
Ewes exposed	'75	19	17*	13	18	12*	18	19	13*	17	15	15*	12
	'76	17	13	12	17	13	18	17	12	17	21	15	14
Percentage mated	'75	84	94*	85	100	100*	17	100	85*	47	80	100*	92
	'76	82	100	—	100	100	—	100	92	—	81	100	—
Percentage lambed (of those mated)	'75	94	100*	72	78	78*	67	100	91*	75	83	87*	100
	'76	86	92	—	100	69	—	100	73	—	100	80	—
Ave. litter size	'75	3.40	2.25*	3.13	2.00	1.20*	2.00	2.32	1.10*	1.33	3.20	2.46*	3.18
	'76	3.08	3.67	—	2.18	2.00	—	2.12	2.13	—	3.12	3.42	—
Percentage dead (within 24 hours)	'75	3.9	8.3*	16.0	3.6	0.0*	0.0	6.8	9.0*	0.0	18.8	15.6*	8.5
	'76	10.8	4.6	—	2.7	5.6	—	2.8	23.5	—	1.9	12.2	—

* ewes about 1 year at lambing.

TABLE 2

HETEROSIS, DIFFERENCE BETWEEN SIRE BREEDS AND RECIPROCAL CROSSES (PHASE 1)

Lambing period		Heterosis			Difference between * reciprocal crosses			Difference between * sire breeds		
		Jan Feb	May June	Sep Oct	Jan Feb	May June	Sep Oct	Jan Feb	May June	Sep Oct
Percentage lambed	'75	−16.5	−13	+10	−5	−11	−33	+11	0	+36
	'76	+7.0	−0	—	0	−11	—	+14	−8	—
Ave. litter size	'75	−0.26	+0.16	+0.36	−1.20	−1.26	−1.18	+0.12	+0.11	−0.62
	'76	+0.10	−0.19	—	−0.94	−1.42	—	−0.02	−0.12	—
Percentage born dead (within 24 hrs)	'75	+5.9	−0.9	−3.8	−15.2	−15.6	−8.5	+18.1	+16.3	−7.5
	'76	−4.5	−5.2	—	+0.8	−6.6	—	−8.8	+25.5	—

* + means IF > FF; − means FF > IF.

five lambing periods, the average litter size in three out of five. The percentage lambs born dead is less in dams carrying cross-bred lambs. The differences between reciprocal crosses indicate that the FF-dams conceive better in almost every mating period. As could be expected the difference in the average litter size is consistently in favour of the FF-line. In four out of five lambing periods the difference in the percentage of lambs born dead is in favour of the IF-dams.

The data concerning the difference between sire breeds suggest that IF sired dams conceive better than FF sired dams. In three out of five lambing periods the viability of FF sired lambs is clearly better, in two lambing periods it is worse than IF sired lambs. As expected the difference in litter size between the two sire breeds is about zero in most lambing periods.

(6) Growth and carcass traits (phase 1)

In Table 3 the growth and carcass traits of the first three lambing periods are summarized. The trait "growth from 10-50 days" is chosen as an estimate for the milk production of the ewe. In order to estimate the milk production for the different mating types the dams reared not more than two lambs. When more than two lambs per litter were born these extra lambs were artificially reared. After weaning from the bucket the treatment was the same as for dam-reared lambs. This partly explains the difference in

TABLE 3

GROWTH AND CARCASS TRAITS (PHASE 1)

Sire & Dam breeds Lambing period		FF×FF			FF×IF			IF×IF			IF ×FF	
		Jan/ Feb	May/ June	Sep/ Oct	Jan/ Feb	May/ June	Sep/ Oct	Jan/ Feb	May/ June	Sep/ Oct	Jan/ Feb	May/ June	Sep/ Oct
Traits (males only)													
Growth from	n	16	14	10	13	6	1	16	5	2	10	9	10
10-50 days	x̄	228	260	261	300	373	315	294	422	302	252	263	313
(g/day)	Sx	65	32	32	82	38	—	79	27	26	29	28	71
Slaughter age	n	13	18	13	14	7	1	9	4	2	12	12	18
(days)	x̄	197.5	216.6	161.8	155.5	151.3	127	177.9	165.8	127	165.5	202.4	143.3
	Sx	32.2	23.0	14.6	10.7	5.7	—	33.0	37.1	1.4	10.7	30.0	12.9
Warm carcass	x̄	21.2	22.5	22.7	22.8	23.6	23.5	21.7	22.6	23.3	21.4	23.5	23.8
weight (kg)	Sx	1.0	1.9	1.1	1.6	1.0	—	1.1	1.8	3.3	1.1	1.1	1.5
Net gain (g/day)		107.3	103.9	140.3	146.6	156.0	185	122	136.3	183.5	129.3	116.1	166.1
Carcass yield	x̄	48.9	51.9	51.4	48.9	51.7	51.1	48.1	49.9	51.0	48.7	52.3	53.2
%	Sx	1.4	2.1	1.6	1.4	1.5	—	2.1	2.1	0.9	1.6	1.6	1.6
Fleshiness	x̄	1.87	1.64	2.23	2.88	3.05	3.00	3.22	3.24	3.66	2.69	2.72	3.02
(score 1-6)	Sx	0.28	0.24	0.28	0.25	0.23	—	0.33	0.17	0.47	0.36	0.30	0.24
Fatcovering	x̄	2.82	2.14	2.89	2.47	2.52	3.00	2.63	2.75	3.00	2.63	2.83	3.25
(score 1-6)	Sx	0.57	0.28	0.39	0.43	0.18	—	0.42	0.32	0.0	0.66	0.44	0.24
Percentage of lambs (male and female) that reached slaughter weight													
Actual number		93.9	80.6	80.0	96.4	100	100	86.4	90.9	87.5	81.3	81.3	85.7
in brackets		(46)	(29)	(25)	(27)	(11)	(4)	(38)	(10)	(7)	(26)	(26)	(30)

TABLE 4

HETEROSIS, MATERNAL AND SIRE BREED EFFECTS (PHASE 1)

Lambing period	Heterosis in F_1			Difference between * reciprocal crosses			Difference between * sire breeds		
	Jan Feb	May June	Sep Oct	Jan Feb	May June	Sep Oct	Jan Feb	May June	Sep Oct
Traits									
Males only									
Growth 10-50 days (g/day)	+30	−46	+65	+48	+110	+2	+18	+52	+39
Net gain (g/day)	+46.6	+31.9	+27.3	+17.3	+39.3	+18.9	−2.6	−7.5	+24.3
Carcass yield (%)	+0.6	+2.2	+1.9	+0.2	−0.6	−2.1	−1.0	−1.4	+1.7
Fleshiness (score 1-6)	+0.48	+0.89	+0.13	+0.19	+0.33	−0.02	+1.16	+1.27	+1.45
Fatcovering (score 1-6)	−0.35	+0.46	+0.36	−0.16	−0.31	−0.25	−0.03	+0.92	+0.36
Percentage of lambs that reached slaughter weight									
(male and female)	−2.6	+9.8	+18.2	+15.1	+18.7	+14.3	−22.6	−8.4	−6.8

* + means IF > FF; − means FF > IF.

number of lambs involved in the growth and carcass traits. The other part of the explanation comes from the fact that pure-bred ram lambs were used as sires to produce the next generation. Net gain is defined as the ratio between the mean carcass weight in grams and the mean slaughter age.

From Table 4 it can be seen that the heterosis in the cross-bred is consistently positive for net gain, carcass yield and fleshiness. For growth from 10-50 days of age there are quite large differences in heterosis between the lambing periods. The same could hold true for fat covering because under the more intensive systems the cross-bred seems to produce more fat than the pure-bred. The heterosis in viability expressed as the percentage of lambs that reach slaughter weight is only positive in the last two lambing periods.

For growth from 10-50 days, net gain, fat covering and percentage of lambs that reach slaughter weight the difference between the two reciprocal crosses is consistently in favour of the IF dams. This is probably due to their smaller litter size (higher birth weight) and their higher milk production so that their progeny grow faster and have less fat at slaughter weight. As expected the difference between the two sire breeds for fleshiness is in favour of the meaty IF sire. The viability expressed as the percentage of lambs reaching slaughter weight is in favour of the FF sired lambs.

(c) *Reproductive performance of cross-bred ewes (phase 2)*

The reproductive performance of the first two lambing periods of the two types of cross-bred ewes and their pure-bred contemporaries are shown in Table 5. Of two mating types only very few ewes were available in the second lambing period. This makes the estimate of the maternal heterosis from this lambing period less reliable. Despite this, Table 6 shows that the maternal heterosis for all traits is similar in both lambing periods and positive. Except for the percentage of lambs born dead which has a negative value. The maternal heterosis in the percentage ewes lambed of those mated is quite substantial. This holds to a much lesser extent for the percentage of ewes mated and for the litter size.

TABLE 5

REPRODUCTIVE PERFORMANCE (PHASE 2)

Sire breed Dam breed	TS FF×FF		TS IF×FF		TS IF×IF		TS FF×IF	
Lambing period	March April	May June	March April	May June	March April	May June	March April	May June
Traits (age at lambing)	1	1	1	1	1	1	1	1
Ewes exposed to the ram	22	23	13	15	18	3	13	4
Percentage mated	95	96	100	93	94	67	100	75
Percentage lambed of those mated	81	86	100	87	71	50	92	100
Ave. litter size (dead + alive)	2.41	2.74	2.15	1.69	1.58	1.00	2.08	2.33
Percentage dead (within 24 hrs)	9.8	21.2	7.1	4.5	5.3	0.0	12.0	28.6

TABLE 6

MATERNAL HETEROSIS

Lambing period	March/April	May/June
Percentage of ewes mated	+5.5	+5.0
Percentage lambed (of those mated)	+20	+25.5
Ave. littersize	+0.12	+0.14
Percentage born dead (within 24 hours)	+2.0 *	+6.0 *

* + sign indicates increased % born dead (see text).

Calculated from the percentage ewes lambed of those exposed to the ram in each lambing period of pure breeding of phase 1, the mean lambing frequency of Finnish Landrace and Ile de France ewes is respectively 1.14 and 1.03 per year. If in the cross-bred the maternal heterosis for this trait is consistently about 20% as the first results suggest, a lambing frequency of 1.3 should be obtainable. With a lambing interval of 8 months the theoretical maximum lambing frequency is 1.5 per year. The lambing frequency in the Texel breed is about 0.9. If there is no substantial maternal heterosis in litter size the average litter size of the mature cross-bred will be about 2.6 lambs, which is the mean of the average litter size of FF-line (3.2) and IF-line (2.0). With estimated losses of about 20% the number of weaned lambs per litter will be about 2.1 lambs. So a cross-bred ewe could produce about (1.3 × 2.1 =) 2.73 lamb carcasses a year. In the traditional system of the Texel breed this is about (0.9 × 1.6 =) 1.44 lamb carcasses. This means about a doubling of the biological efficiency of lamb meat production with breeding measures only.

REFERENCES

Boer, H. de and Nijeboer, H., 1973. Stereo diapositives as an aid in carcass assessment. *Wrld. Rev. An. Prod.* 9: 3, p. 50-57.

Dickerson, G.E., 1973. Inbreeding and heterosis in animals. Proceedings Animal Breeding and Genetics Symposium in honour of Dr. Jay L. Lush. Amer. Soc. Anim. Sci. and Amer. Dairy Sci. Assoc., p. 54-77.

Moav. R. 1966. Specialised sire and dam lines. I. Economic evaluation of crossbreds. *Anim. Prod.* 8: 193-202.

More O'Ferrall, G.J., 1974. Which is the best breed of ram for fat lamb production? *Farm and Food Research* 5(3): 69-71.

Nitter, G. and Jacubec, V., 1970. *Suchtplanung für Gebrauchskreuzungen in der Schafzucht. Züchtungskunde* 42: 436-446.

Sturkenboom, H.J., 1971. Financiele Resultaten van de schapenhouderij. *Het Schaap* 15.2.

Visscher, A.H., 1974. A cross-breeding and selection experiment with three breeds of sheep. Proc. Working Symposium Breed Evaluation and Crossing Experiments, Zeist.

22

THE NEED FOR DUAL-PURPOSE SHEEP

A.H. BISHOP, L.J. CUMMINS, J.K. EGAN and J.W. McLAUGHLIN

Department of Agriculture, Victoria Pastoral Research Station, Hamilton, Australia

INTRODUCTION

In Australia the total number of sheep is more than ten times the human population and the export of surplus production of the grazing industry has always made a major contribution to the national economy. Variability in price in response to economic factors, political and technological changes, and seasonal conditions are a feature of export markets and, in the case of meat which is perishable and has high storage costs, these problems are accentuated.

Despite this it has been possible to recognize long term trends influencing the production of meat and also in respect of the demand for meat in importing countries. A number of organizations have studied these trends and have published projections of future demand and supply indicating likely surpluses or shortfalls for various categories of meat.

These reports are reviewed to indicate the likely future markets for sheep meats and the consequences and opportunities for the Australian grazing industry. The implications for priorities of objectives in genetic improvement of sheep and appropriate genetic means to achieve these are discussed.

PROJECTED SUPPLY AND DEMAND

The Food and Agriculture Organization of the United Nations has a continuing programme of projecting future trends and checking the resulting estimates against the outcome, and a well established methodology has been developed.

In 1971 F.A.O. released projections for the world supply and demand for all meats up to 1980. With respect to red meats they concluded that the combined exportable supplies of the surplus producing countries would probably still be below likely effective import requirements. The export surplus and import demand and consequent balance are set out in Table 1.

In 1974 F.A.O. undertook a major revision of the 1971 projection in which more up to date production statistics and improved techniques were used (Table 1). This supported the conclusions of the earlier projection. It is significant that surpluses in Australia and New Zealand constitute 85% of the projected supply of exports in this study.

In 1975 F.A.O. published a summary of selected national meat production and demand outlook studies. These estimates of production and con-

TABLE 1

PROJECTIONS OF TOTAL WORLD EXPORT AND IMPORT REQUIREMENTS FOR RED MEATS AND BALANCE* FOR 1980 ('000 TONS)

Projection	Beef and veal				Mutton and lamb			
	Export Surplus	Import Demand	Balance ±	% of export	Export Surplus	Import Demand	Balance ±	% of export
FAO 1971 [1]	4,179	5,823	−1,650	−40	1,086	1,687	−601	−58
FAO 1974 [2]	4,453	5,536	−1,083	−24	2,110	2,566	−456	−22
FAO 1975 [3]	5,343	4,986	+420	+8	1,307	2,332	−1,025	−73

* Balance: − denotes shortfall; + denotes surplus
[1] "F.A.O. Meat Production and Demand Projections to 1980"
[2] "Review of the F.A.O. Meat Production and Demand Projections to 1980"
[3] "Summary of selected national meat production and demand outlook studies"

sumption prepared by the major beef and sheep meat production and consuming countries were used to revise the balance of world trade as projected by F.A.O. in 1974 (Table 1). The difference between these estimates and those projected by F.A.O. in 1974 was largely due to the decline in sheep numbers in exporting countries in the period 1970 to 1973 in response to very depressed wool prices which was not taken into account in the F.A.O. study.

In this series of studies succeeding estimates indicate greater shortfalls for sheep meats and smaller shortfalls for beef. However there is little change in the estimated shortfall for red meats so that the original conclusions continue to be supported in general.

The above considerations do not take account of the 1973-74 changes in world prices for petroleum and consequent effects on the affluence and capacity to pay for imports of the oil rich countries in the Middle East.

A study by the West Australian Department of Agriculture (Neil personal communication) estimated that the demand for sheep meats in Middle East countries would reach 500,000 tons (including live sheep) by 1980, but the basis of the estimate is not stated. This is about 30% higher than the most recent estimate by F.A.O. for these countries. If it proved correct it would further increase the projected deficit in 1980.

In the event, by 1980 the strong requirement for imports of sheepmeats relative to the projected export supplies could be expected to increase prices and encourage production until the export surplus matches the import demand. Thus, prices for sheep meats can be expected to rise and may approximate to those for beef as they have done in the past when supplies of sheep meats have been limited.

FACTORS AFFECTING TURN-OFF

New price relations amongst red meats are likely to influence the ratio of returns from wool and meat in sheep enterprises and hence the economic basis of existing priorities for genetic change in sheep.

In particular a higher priority is likely to apply to factors affecting the turn-off of animals for slaughter from the flock. Marshall (1973,1974) concluded that an increase in the proportion of breeding ewes was necessary in order for turn-off to increase whilst maintaining a steady population. However true this may be, it is likely that the sex composition of the flock will vary as a result of culling policies with respect to aged ewes, surplus young ewes and prime wethers, rather than as the result of a deliberate decision to change the proportion of breeding ewes in the flock.

We have used a simple mathematical model of flock structure—involving essentially similar assumptions to those described by Egan *et al.* (1972) but including a weighting of 1 ewe = 1.25 wethers = 1.25 yearling sheep to maintain comparable stocking rates—to investigate the effects of increased fertility and growth rate upon turn-off and sex composition. Table 2 presents some generalized results.

TABLE 2

CALCULATED EFFECT OF LAMBING % AND SLAUGHTER AGE OF WETHERS, UPON TURN-OFF AND SEX COMPOSITION

Lambing %	Age of wethers at slaughter years	*Turn-off	Ewes in flock %
Normal	3½	†100	38
Normal	2½	119	44
+15%	3½	107	34
+15%	2½	128	40
+30%	1½	173	47

† Base = 100   *Turn-off = $\dfrac{\text{total sales}}{\text{flock size}}$

The calculations suggest that the response of turn-off to increased fertility is limited by the need to reduce the proportion of breeding ewes to enable the additional young sheep to be maintained on the property until slaughter age. However, the full benefit of a higher reproduction rate can be realized if a concurrent improvement in growth rate permits the sale of young wethers for slaughter earlier than was previously possible. It may also be seen that the flock structure is essentially established by these factors. Calculations by Egan *et al.* (1972) suggest that wool production will remain fairly constant over the range of structures described in the table.

OBJECTIVES AND MEANS FOR GENETIC CHANGE

Acceptance of the probability of changes in the future demand for sheep meats indicated by the projections in Table 1 may justify changes of emphasis in the goals for genetic change. When objectives for genetic improvement are expanded to embrace factors which influence turn-off throughtout the industry, and carcass quality in major sectors, a larger number of characteristics are involved than the few which serve when increased wool production is the major objective. Specifically these may include fecundity, survival of lambs from multiple births, milk production and post-weaning growth rate. Other characteristics related to carcass quality such as fat depth in relation to carcass weight may assume a higher priority of specific markets are to be satisfied (Suiter 1976).

Avenues for improvement in meat production include changes in the base wool producing breed, the Merino, to increase the turn-off of wether sheep, changes in cross-breeding systems for lamb production, and the development of dual purpose breeds or crosses for producing apparel quality wool plus lamb or young wether mutton in the wheat/sheep or high rainfall zones.

(i)   *Meat production from Merino woolgrowing areas*

There is sufficient evidence both in the literature and in the sheep industries of overseas countries, to suggest that it is possible to develop mutton qualities without prejudice to the range of qualities of wool produced within Australian apparel wool breeds. From a review of the literature Turner (1972) concluded that insofar as fleeceweight, meat and milk are concerned, there are no strong negative genetic correlations to prevent concurrent selection for all three. Furthermore, the Merino, which by virtue of its quality wool will probably continue to be the dominant breed type in Australia, has been shown to be responsive to selection for weaning weight (Pattie 1966), bodyweight (Turner, Brooker and Dolling 1970) and fecundity (Turner 1969).

The hierarchical structure of the Merino stud industry provides valuable resource of genetic material which might be better utilized to increase meat production. While within strain selection for a large number of characters can only lead to slow rates of improvement in individual characters, commercial producers have the opportunities for both between strain selection and for developing systematic cross-breeding systems if worthwhile amounts of heterosis can be shown to exist. This in turn could lead to the identification of a small number of appropriate genetic goals capable of being achieved by selection within ram breeding flocks.

(ii)   *Lamb production systems*

The Merino is also the base for a highly efficient stratified cross-breeding system which has evolved for prime lamb production based on the Border

Leicester × Merino ewe. Whilst the efficiency of this system has been confirmed (Matthews 1920; Miller and McHugh 1955; McGuirk 1967) it has several deficiencies in that it is not self replacing and the supply of suitable lines is variable, the wool has a high fibre diameter and hence lower value, and the breeding season is limited.

There is a need to examine alternate systems for lamb production, particularly as future export markets may require less fat and heavier carcass weights than those produced at present. Terril and Sidwell (1975) demonstrated that multiple cross ewes in a continuous breeding system which could be self replacing would lead to substantial improvement in efficiency. Likewise a programme at the Animal Breeding Research Organization at Edinburgh which is showing considerable promise involves the development of a new breed from a gene pool including Finnish Landrace, East Friesland, Border Leicester and Dorset Horn.

(iii)   Dual-purpose breeds or cross-bred sheep

A major contribution to meat production can come from producers of apparel quality wool in the more favoured environments by the development of true dual-purpose types of sheep capable of producing quality wool and lamb or young wether mutton with satisfactory carcass characteristics for potential markets.

The only well established Australian efforts at developing dual-purpose sheep are represented by the Corriedale and Polwarth, both breeds evolving from a Merino and Lincoln ancestry. Most comparisons of the meat production of these breeds have been based solely on the weight and grade of prime lamb produced. Both the Corriedale (McGuirk and Scarlett 1966) and the Corriedale × Merino (Miller and McHugh 1955; Simms and Webb 1945) were shown to be superior to the Polwarth and the South Australian Merino but all were inferior to the Border Leicester × Merino.

In the Peppin Merino and Corriedale crosses, Iwan *et al.* (1971) have reported moderate levels of heterosis for reproduction rate and lamb growth rate. However, although heterosis for weight of lamb weaned per ewe joined was 10-11 %, production of the cross-breds did not exceed that of the pure Corriedales. The final evaluation of this work must await the analysis of wool production data.

By contrast in South Africa recent attempts at developing dual-purpose breeds have been more successful (Hofmeyr 1976). The Dohne Merino, derived from crosses between the German Mutton Merino and the South African Merino appears to have increased reproduction by 20%, improved growth, and maintained wool weight and quality.

Since in Australia mating plans of a dual purpose nature (Comeback, Corriedale and Merino crossed with British breed rams) in the sheep/wheat and high rainfall zones involve more than 25% of the ewes mated, there is an urgent need for specific work in this area.

THE IMPLICATIONS FOR RESEARCH

Research programmes require a substantial time between inception and completion so that aims should be related to future rather than present needs of the industry. After considering the effects of recognizable trends in supply and demand for sheep meats and in the absence of comparable projections for wool, it appears that research programmes should relate to an industry in which more exportable sheep meats are required, together with the continuing production of apparel quality wool.

Because of the wide range of objectives which would need to be sought if meat production from sheep is to be substantially increased, selection within breeds can be expected to achieve only very slow rates of progress. A change in emphasis to other means of genetic improvement is imperative if Australia is to accept the challenge of future markets for sheep meats.

There is an urgent need for much more work in the comparison of breeds, strains and crosses, the examination of the variation between breeds and strains and the relations between productive characters across a range of breeds. Taylor (1972) has discussed the potential use of between breed variation for genetic improvement with respect to cattle. Results need to be interpreted in terms of their effects on flock structure for the important environments to provide a basis for new production systems.

It is clear from the foregoing that the importation of new genotypes could play a vital role, particularly with respect to breeds with valuable fleeces and high fertility and growth or with superior survival of lambs. Because so many characteristics are involved the need and role of new breeds should be established before they are widely distributed to the industry.

Research in these terms will not only produce knowledge of existing and imported breeds and the means of combining them to best advantage, it will produce prototypes of new synthetic breeds. The projects could become the basis for the dissemination of new breeds to the industry by using A.I. The Coopworth and Drysdale in New Zealand and the Dohne Merino in South Africa are successful examples of this approach.

REFERENCES

Egan, J.K., Bishop, A.H. and McLaughlin, J.W. (1972). *Proceedings of the Australian Society of Animal Production* **9** : 71.

Hofmeyr, J.H. (1976). *Proceedings of the Australian Society of Animal Production* **11** : 517.

Iwan, L.G., Jefferies, B.C. and Turner, Helen Newton (1971). *Australian Journal of Agricultural Research* **22** : 521.

McGuirk, B.J. (1967). *Wool Technology and Sheep Breeding* **14**(2) : 73.

McGuirk, B.J. and Scarlett, E.C. (1966). *Proceedings of the Australian Society of Animal Production* **6** : 210.

Marshall, T. (1973). *Journal of Agriculture Western Australia* **14** : 166.

Marshall, T. (1974). *Journal of Agriculture Western Australia* **15** : 58.

Matthews, J.W. (1920). Agricultural Gazette of New South Wales **31** : 761,846.

Miller, W.B. and McHugh, J.F. (1955). *Journal of Agriculture, Victoria* **53** : 385.

Pattie, W.A. (1966). Wool Technology and Sheep Breeding **13**(2) : 71.

Simms, H.J. and Webb, C.G. (1945). *Journal of Agriculture, Victoria* **43** : 277.

Suiter, R.J. (1976). Personal communication.

Taylor, St. C.S. (1972). A.R.C. Animal Breeding Research Organization. Report—January 1972 pp.15.

Terrill, C.E. and Sidwell, G.M. (1975). *Proceedings of the Third World Conference on Animal Production* (Sydney University Press : Sydney) pp.620.

Turner, Helen Newton (1969). *Animal Breeding Abstracts* **37** : 545.

Turner, Helen Newton, Brooker, M.B. and Dolling, C.H.S. (1970). *Australian Journal of Agricultural Research* **21** : 955.

Turner, Helen Newton (1972). *Animal Breeding Abstracts* **40** : 621.

23

DEVELOPMENT OF THE GROMARK — A NEW HIGH PERFORMANCE DUAL-PURPOSE SHEEP BREED

A.C. GODLEE

Tamworth, N.S.W., Australia

SUMMARY

A fixed half-bred of the Border Leicester and Corriedale breeds has been selected for live weight, twinning, wool and to a limited extent reduced fat since 1965 on fertilized natural pasture in northern New South Wales.

The average live weight of rams is: weaning (adjusted) 35 kg, yearling 61 kg, 2 year old 90 kg. Yearling rams grow 450 g of 30μm diameter average wool per month with a low level of hairy birth coat and hairy britch. Ewes rear approximately 100% lambs. Sheep are hardy with little body strike, a minimum of lambing trouble and have a quiet temperament. The use of objective measurement of production characters coupled with final selection on appearance has produced a breed that should compete with cross-breds.

INTRODUCTION

During recent years in New South Wales prime lamb production has become more firmly established on the cross-breeding system with a significant reduction in the number of sheep in self-replacing dual-purpose flocks (A.B.S. 1976.)

In the 1950s attempts were made to form a dual-purpose breed from the popular first-cross Border Leicester-Merino but these projects appear to have lapsed. Since then the trend has been to develop meat types growing wool of a spinning count of 60^S (23μm diameter) and finer. This approach was endorsed by the report of an expert panel (Animal Production Committee 1970).

The Gromark project was commenced in 1965 with the object of developing a dual-purpose breed with the emphasis firmly on meat production. Another aim was to strongly apply selection pressure for production characters in a commercial situation.

The genetic principles employed in this development have been documented by Turner and Young (1969) and Dolling (1970) and have been summarized by Victoria Department of Agriculture (1974).

The project has been conducted on aerially fertilized natural pasture hill country adjoining the city of Tamworth on the north-west slopes of New South Wales. Rainfall averages 650 mm per annum with a summer incidence. Pastures are predominantly summer growing native perennial grasses in conjunction with winter growing naturalized annual legumes of *Trifolium spp.* and *Medicago spp.* Grass seed infestation from *Aristida sp.* and *Stipa sp.* is a feature of the environment.

BREEDING PLAN

430 Corriedale ewes from an old established flock on the northern tablelands of N.S.W. were joined to 12 stud quality Border Leicester rams. The rams which were mainly twin or triplet born represented many of the leading blood lines in the eastern states and were selected from sub-strains that expressed large size and prolificacy. Following the first cross the sheep have been inter-bred to fix them at the half-bred level.

The breeding flock consists of 400 ewes with three sire lines each using two or three rams in a syndicate mating for no more than two seasons. This should have kept inbreeding to a reasonable level and may have allowed some natural selection for libido in rams.

Characters selected for improvement in order of importance have been:

> Live weight
> Twinning (prolificacy)
> Wool (weight and quality)

Little attention has been given to carcase quality except for the use of the scanogram to measure fat depth on the 1973 ram drop as yearlings. Rams showing the higher fat measurement (>4 mm) were rejected. Body shape has been allowed to emerge naturally and has moved towards the longer, leggier and presumably leaner type.

Live weight was chosen as the top priority character for improvement because:—

> It did not require detailed recording of ewes.
>
> It has a moderate heritability at weaning and high heritability at the yearling stage (Turner and Young (1969)).
>
> It has a positive genetic correlation with reproductive rate (Turner and Young (1969)).
>
> Large size of breed means that a lamb carries less fat at a given weight than smaller breeds (Searle and Griffiths (1976)).

Hardiness and easy care features have been built into the selection programme.

Deliberate selection for finer wool has not been considered a worthwhile use of available selection pressure in a breed of this type. Judged by price differentials per μm diameter in recent years it would not lead to a worthwhile increase in monetary return. It would also tend to move the wool type into the zone of fleece rot susceptibility which would go against the easy care aspects of the breed (Henderson 1968).

SELECTION METHOD

Direct selection methods have been used mainly on rams to apply pressure to the chosen characters. Marketing principles dictate that the breed must maintain an attractive appearance and this has been achieved by selecting acceptable types from the highest production grades for use as sires.

Rams are individually ear tagged at birth, and the date, type of birth and age of dam recorded. Tag colour is used to indicate sire line. Actual birth weights are not recorded but estimates of 5 kg for singles and 4 kg for twins are used in calculating liveweight gains per day.

Rams are weighed unshorn at weaning (approximately 4½ months) and adjusted for age, type of birth and age of dam. They are again weighed at yearling stage (approximately 14 months) immediately after shearing and greasy fleece weight at this second shearing is recorded. The rams are then graded for type of birth (S=single, T=twin, Tr=triplet), each of the live weights and greasy fleece weight.

Independent gradings of the production characters are given as follows:—

Top 20%   A grade=potential stud
next 30%   B grade=potential selected flock
remainder C grade=flock

Rams are downgraded for visual faults but are never upgraded for visual excellence. The majority of rams used have graded T.A.A.B. (twin, A grade adjusted live weight at weaning, A grade live weight as yearling and at least B grade greasy fleece weight).

RESULTS AND DISCUSSION

The following production levels have been achieved on fertilized natural pasture without the aid of improved pastures or supplementary feeding.

Ram lambs (weaning, average age 4½ months)

Live wt. Singles av. 35 kg=liveweight gain per day 225 g
         Twins av. 28 kg=liveweight gain per day 180 g

Yearling rams (14 months old) most of drop retained.

Liveweight (off-shears) average 61 kg, range 54-73 kg.
Greasy fleece production averages 0.45 kg per month.
Fibre diameter average 30μm, range 28-33μm.
Scanogram fat depth 12th rib adjusted to 68 kg liveweight, average 3.2 mm, range 1-7 mm.

2 year old rams (sires and reserves)

Live weight (immediately after shearing) 80-100 kg.
Greasy fleece weight 6 to 7 kg, length 12-15 cm.

Ewes

> With unsupervised lambings ewes rear approximately 100% lambs, although with supervised lambings 64% of ram lambs tagged at birth ex mature ewes are twins. Few triplets are produced.

Wool averages about 50^S spinning count (30 μm diameter) and may have fined up slightly due to a preference for a more compact staple tip. There is a very low level of hairy birth coat and hairy britch.

Some indication of the limitations of the environment are illustrated by the large difference in the live weight of singles and twins at weaning and the rarity of triplets although many twins are produced. Even under sub-optimal conditions the live weight gain for this breed compares reasonably well with the general average performance of 230 g live weight gain per day for prime lambs in Australasia (Tribe and Coles 1966: Kirton 1974). The overall performance of the sheep under the conditions imposed indicates an excellent result under more favourable circumstances.

The good growth rate of the breed is a desirable feature to complement higher lambing percentages, marketing at heavier weights and mating of ewe lambs. The large mature size should also enable the production of leaner carcases and attract buyers of rams and breeding ewes as it reduces the cash costs of running a given biomass per hectare.

The general fitness and resilience to adverse conditions and the low level of body strike and lambing trouble should reduce husbandry costs. Being very plain bodied the sheep lends itself to the use of mechanized shearing devices. The quiet and tractable temperament means that the sheep are easy to work and handle.

By rigorous selection for production characters a breed has been produced that should increase production and at the same time reduce costs. As a self-contained breed it should compare favourably with cross-breeding at least for the production of prime lamb dams. With changes in markets and methods of marketing the pure-bred Gromark lamb may compete with the specialized terminal sire breeds.

The Gromark is similar in concept to the Coopworth breed in New Zealand which is a fixed half-bred of Border Leicester and Romney breeds. The Coopworth is slightly stronger in the wool having a spinning count of 46^S-50^S (35-30 μm diameter) and the main selection pressure has been applied to prolificacy and considerable attention is also given to fleece weight and live weight at weaning (Coopworth Flock Book 1972).

ACKNOWLEDGEMENTS

The efforts of innumerable people who have contributed to the theory of population genetics in sheep breeding is gratefully acknowledged.

The encouragement of my colleagues in the N.S.W. Department of Agriculture has been appreciated.

REFERENCES

Animal Production Committee (1970). *Journal of Australian Institute of Agricultural Science* **35** : 30.

Australian Bureau of Statistics (1976). New South Wales Statistics of Livestock 1975.

Coopworth Flock Book (1972). Coopworth Sheep Society of New Zealand.

Dolling, C.H.S. (1970). "Breeding Merinos" (Rigby : Adelaide).

Henderson, A.E. (1968). "Growing Better Wool" (Reed : Wellington).

Kirton, A.H. (1974). *New Zealand Journal of Agriculture.* **129** : No. 3. 23.

Searle, T.W. and Griffiths, D.A. (1976). *Proceedings of the Australian Society of Animal Production* **XI** : 57.

Tribe, D.E. and Coles, G.J.R. (1966). "Prime Lamb Production" (Cheshire : Melbourne).

Turner, Helen N. and Young, S.S.Y. (1969). "Quantitative Genetics in Sheep Breeding" (Macmillan : Australia).

Victoria Department of Agriculture (1974). Operations Manual; Flock Performance Recording Scheme.

24

GROWTH OF DISSECTIBLE FAT IN THE MERINO

G.H. WARREN

University of Sydney, N.S.W., Australia

SUMMARY

A study of the growth and partition of dissectible fat was carried out on pasture reared Peppin Merino sheep. The weight of dissectible fat in the internal, intermuscular, and subcutaneous fat depots was recorded for sheep slaughtered at intervals from birth to almost three years of age.

The relative percentage growth of the major fat depots, internal, intermuscular, and subcutaneous depots, and the partition of fat between these three depots is presented.

INTRODUCTION

The important role of fat in its effects on carcass composition, evaluation by the meat trade and on profitability has been recognized and well documented. Butterfield (1963) showed that selection had no effect on the proportional distribution of muscle in a beef carcass, and Seebeck (1968) found only minor difference in muscle distribution between sheep breeds. It seemed logical then, that for further advances in the production of a carcass, the growth and partition of fat should be studied, as fat would appear to be the tissue to be most amenable to alteration by selection.

MATERIALS AND METHOD

Animals of the same age were selected randomly, weighed and slaughtered without prior fasting. Twenty-six ewes, 26 wethers, i.e. males castrated at birth, and 31 rams were killed at specific ages from birth to almost three years of age.

The pelt and gut contents were weighed and subtracted from the live weight to obtain a pelt-free empty body weight (PFEBWt).

The kidney fat and the fat depots of the abdominal (omental, mesenteric and pelvic) and thoracic cavities (mediastinal fat and coronary fat) were weighed during the dressing procedure. The carcasses were then deep frozen, transported to the dissecting laboratory and kept frozen until required. After thawing the carcasses were halved by soft siding (Walker, 1961), and the left side dissected within 36 hours.

Subcutaneous fat was defined as fat superficial to the plane of the muscles but included the fat beneath the *cutaneous trunci m. (panniculus adiposus m.)*. Muscles were removed individually, the fat removed from their surfaces was recorded as intermuscular fat. Individual muscle weights were also recorded.

The growth rate of the fat depots is expressed as the proportion of the final weight reached at the specified intervals, and is called the relative percentage growth. It is a modification of the method used by Wilson (1954). This method of presenting the data allows the growth between any two consecutive stages to be examined but has the disadvantage that statistical analyses or predictions cannot be applied. An allometric examination and statistical comparisons have been recorded elsewhere (Warren, 1974).

The mean weight of fat in each major depot at each slaughter interval is expressed as a percentage of the mean total dissectible fat weight. These percentage figures indicate the partitioning of the total fat between the major depots.

RESULTS

Results are presented in the tables below, Tables 1 to 6.

The results bearing an asterisk are from sheep from the same flock, killed at the same time but derived from a study by Lohse (1971). They were used in order to gain more information on the changes that appeared to occur during early post-natal life and for which there is little corroborative evidence in the literature for Merino sheep.

TABLE 1

RELATIVE PERCENTAGE GROWTH OF SUBCUTANEOUS FAT

		Ewes			Wethers			Rams	
Age (Days)	N =	Mean PFEB Wt.	Relative %age growth	N =	Mean PFEB Wt.	Relative %age growth	N =	Mean PFEB Wt.	Relative %age growth
0	6	3.179	0.1	8	3.44	0.2	6	3.398	0.2
7	2	3.941	0.7	5	4.63	0.8	3	4.412	0.5
18	3	6.604	7.1	3	5.42	2.3	3	6.629	4.5
28*	1	9.730	2.6	2	8.361	3.5	2	7.967	2.6
84	2	17.65	26.5						
110	3	16.69	17.6	3	15.02	23.1	3	15.80	24.4
180	4	18.00	24.2	4	18.62	26.6	5	18.24	22.2
365	3	15.05	7.4	4	18.18	13.1	5	20.03	17.6
730	2	25.05	32.1	2	22.71	22.6	2	22.43	10.6
1010	2	28.09	100.0	3	33.57	100.0	4	39.44	100.0

Final SC fat wt. (g) 2619　　Final SC fat wt. (g) 2698　　Final SC fat wt. (g) 2485

* from Lohse (1971)

TABLE 2

RELATIVE PERCENTAGE GROWTH OF INTERMUSCULAR FAT

		Ewes				Wethers				Rams	
Age (Days)	N =	Mean PFEB Wt.	Relative %age growth	N =	Mean PFEB Wt.	Relative %age growth	N =	Mean PFEB Wt.	Relative %age growth		
0	6	3.179	1.7	8	3.44	1.5	6	3.398	1.7		
7	2	3.941	5.2	5	4.63	3.7	3	4.412	3.0		
18	3	6.604	8.8	3	5.42	3.2	3	6.629	7.7		
28*	1	9.730	6.0	2	8.361	5.1	2	7.967	5.3		
84	2	17.65	34.8								
110	3	16.69	28.4	3	15.02	31.0	3	15.80	32.2		
180	4	18.00	42.2	4	18.62	38.3	5	18.24	34.7		
365	3	15.05	19.8	4	18.18	19.9	5	20.03	22.8		
730	2	25.05	46.4	2	22.71	33.2	2	22.43	18.2		
1010	2	28.09	100.0	3	33.57	100.0	4	39.44	100.0		

Final IEM fat wt. (g) 2189　Final IEM fat wt. (g) 2698　Final IEM fat wt. (g) 2465

* from Lohse (1971)

TABLE 3

RELATIVE PERCENTAGE GROWTH OF INTERNAL FAT

		Ewes				Wethers				Rams	
Age (Days)	N =	Mean PFEB Wt.	Relative %age growth	N =	Mean PFEB Wt.	Relative %age growth	N =	Mean PFEB Wt.	Relative %age growth		
0	6	3.179	1.0	6	3.44	1.2	6	3.398	1.3		
7	2	3.941	2.5	2	4.63	2.1	3	4.412	1.6		
18	3	6.604	4.4	3	5.42	2.1	3	6.629	4.15		
28*	1	9.730	4.4	1	8.361	5.3	2	7.967	4.5		
84	2	17.65	20.2								
110	3	16.69	15.2	3	15.02	21.9	3	15.80	18.2		
180	4	18.00	24.8	4	18.62	24.3	5	18.24	25.3		
365	3	15.05	11.3	3	18.18	17.1	5	20.03	15.0		
730	2	25.05	33.1	2	22.71	22.4	2	22.43	17.9		
1010	2	28.09	100.0	2	33.57	100.0	4	39.44	100.0		

Final int. fat wt. (g) 3367　Final int. fat wt. (g) 3319　Final int. fat wt. (g) 2743

* from Lohse (1971)

TABLE 4

THE PARTITION OF DISSECTIBLE FAT IN MERINO EWES

							Percentage of total fat		
Age (Days)	N =	Mean PFEB Wt.	Mean SC*	Mean IEM**	Mean Int.* +	Total Fat	% SC	% IEM	% Int.
		(kg)	(g)	(g)	(g)	(g)			
0	6	3.179	3.0	36.6	34.7	74	4.0	49.2	46.7
7	2	3.941	18.5	114	84.2	216	8.5	52.5	38.9
18	3	6.604	187	193	149	529	35.3	36.5	28.2
28+	1	9.730	69.4	131	149	350	19.8	37.5	42.7
84	2	17.65	694	761	680	2135	32.4	35.6	32.0
110	3	16.69	462	622	513	1598	28.9	39.0	32.1
180	4	18.00	635	923	835	2393	26.5	38.6	34.9
365	3	15.05	193	433	381	1007	19.1	43.0	37.9
730	2	25.05	842	1015	1115	3072	27.0	36.3	36.3
1010	2	28.09	2619	2189	3367	8176	32.0	26.8	41.2

   * SC = subcutaneous fat
  ** IEM = intermuscular fat
 * + Int. = internal depot fat
   + from Lohse (1971)

TABLE 5

THE PARTITION OF DISSECTIBLE FAT IN MERINO WETHERS

							Percentage of total fat		
Age (Days)	N =	Mean PFEB Wt.	Mean SC*	Mean IEM**	Mean Int.* +	Total Fat	% SC	% IEM	% Int.
		(kg)	(g)	(g)	(g)	(g)			
0	8	3.44	5.0	41.0	39.5	85	5.8	47.9	46.2
7	5	4.63	21.8	100	70.5	192	11.3	52.0	36.7
18	3	5.42	62.6	85.4	70.6	223	28.1	40.2	31.7
28+	2	8.361	95.3	136	177	409	23.3	33.4	43.3
110	3	15.02	625	836	728	2189	28.5	38.2	33.3
180	4	18.62	717	1034	806	2558	28.4	40.4	33.2
365	4	18.18	354	537	568	1460	24.3	36.8	38.9
730	2	22.71	609	896	742	2247	27.1	39.8	33.0
1010	3	33.57	2758	2698	3319	8875	31.2	31.5	37.3

   * SC = subcutaneous fat
  ** IEM = intermuscular fat
 * + Int. = internal depot fat
   + from Lohse (1971)

TABLE 6

THE PARTITION OF DISSECTIBLE FAT IN MERINO RAMʒ

							Percentage of total fat		
Age (Days)	N =	Mean PFEB Wt.	Mean SC*	Mean IEM**	Mean Int.* +	Total Fat	% SC	% IEM	% Int.
		(kg)	(g)	(g)	(g)	(g)			
0	6	3.398	5.2	43.0	38.0	86	6.0	49.9	44.1
7	3	4.412	12.0	74.0	64.8	150	8.0	48.9	43.1
18	3	6.629	112	189	123	424	26.5	44.6	28.9
28+ 84	2	7.967	63.6	130	124	318	20.0	40.9	39.1
110	3	15.80	607	793	500	1903	32.0	41.7	26.3
180	5	18.24	551	857	695	2203	25.0	43.4	31.6
365	5	20.03	438	561	411	1411	31.1	39.8	29.1
730	2	22.43	263	448	490	1201	21.9	37.3	40.8
1010	4	39.4	2485	2465	2743	7693	32.3	32.0	35.7

  * SC = subcutaneous fat
 ** IEM = intermuscular fat
 * + Int. dep. = internal depot fat
 + from Lohse (1971)

DISCUSSION

(a)   Relative percentage growth of dissectible fat

At birth, in the lambs examined there was a very small amount of subcutaneous fat. Of those lambs dissected at birth, 50% had less than 2 g, 40% less than 6 g, and the remainder between 6 g and 13 g of subcutaneous fat. When fat occurred it occurred most commonly on the flank and shoulder regions.

Between birth and 18 days the subcutaneous fat showed a greater relative increase in weight in relation to its weight at birth than did the intermuscular or internal fat. However, the absolute increase in fat weight in all three depots was similar.

This greater relative increase in subcutaneous fat in the first 18 days is reflected in the figures for the partition of fat, with the proportion of the total fat occurring in the subcutaneous depot increasing rapidly. The biological impetus for the redistribution of the total dissectible fat could be that the subcutaneous fat acts as an insulating mechanism before the fleece has grown sufficiently to take over this role, and thus aids in maintaining a thermal homeostasis.

The seasonal fluctuations in available pasture and other factors affecting growth are reflected in the data by the fluctuations in body weight. The subcutaneous and intermuscular fat were more severely affected by a body weight loss than the internal fat.

At the same age the relative percentage growth values for intermuscular fat were higher than those for subcutaneous fat within the same sex except for the final slaughter weight. This indicates that the subcutaneous fat is later maturing than the intermuscular fat. However, to use the term later maturing may be misleading in the light of its overall growth pattern.

By plotting the relative percentage growth values obtained from the data in this experiment against the PFEBWt it appeared that the point at which the rapid deposition of subcutaneous fat began was at approximately 16 kg PFEBWt for ewes, 18.6 kg for wethers and 22 kg for rams.

For the intermuscular fat, the stage at which rapid fat deposition began was at a slightly higher body weight. However at these weights the intermuscular fat had reached a much greater proportion of the weight recorded at the final slaughter stage.

The individual fat depots composing the total internal fat fell into three distinct groups with regard to their growth. However as the omental and kidney fat depots contained most of the internal fat only these depots shall be considered. It is pertinent to state that, before the stage of accelerated growth of these depots they contained 51% of total internal fat in the rams and wethers, and 63% in the ewes. At the final slaughter stage they contained 71%, 85% and 78% of the total internal fat for the rams, wethers and ewes respectively.

With those internal fat depots under consideration in this paper the PFEBWt at which rapid fattening occurred appeared to vary between the sexes. It occurred earlier in the ewes than in the wethers and later again in the rams. Although the difference between the wethers and rams was not as great as between the wethers and the ewes.

When all three fat depots are considered it was estimated that the rapid fattening stage began when the ewes had reached 58%, the wethers 55% and the rams 68% of the final muscle weight measured. The PFEBWt at which this occurred were between 16.7 and 18 kg for ewes, 18-22 kg for wethers and at or in excess of 22.5 kg for rams. The sexes show the same relative order of magnitude in their body weights in this respect as Berg and Fukuhara (pers. comm.) found in cattle. The weight at which rapid fattening began in the wethers concurs well with the weight for transition from a prefattening to a fattening stage of growth demonstrated by Searle and Griffiths (1975) in a similar strain of Merino.

(b)   Partition of dissectible fat

Subcutaneous fat was much the smallest depot at birth, however, by 18 days it had achieved the proportion of the total depot about which it fluctuated until the last stage of the study. In the period from 180 days to 365 days, when the ewes lost 12% of their body weight and the wethers body weight remained static, the subcutaneous fat was more severely affected than the intermuscular or internal depot fat and fell as a proportion of the total dissectible fat.

At birth, and seven days for all "sexes", the intermuscular fat contained the highest proportion of the dissectible fat. From this point it showed a fluctuating decline with increasing age and body weight.

The internal depot fat was at its maximum as a proportion of the total dissectible fat at birth, then fell rapidly to 18 days of age, when it formed 28%, 32% and 29% of the total dissectible fat for ewes, wethers and rams respectively. From this point it showed a slow but fluctuating increase through to the oldest animals, where it formed on the average over the three sexes, 38% of the total dissectible fat.

As the subcutaneous fat is more sensitive to an inadequate nutritional level than the intermuscular fat or internal fat, the partition of fat between the major fat depots is altered by weight loss. Thus it would be expected in animals which have not suffered a weight loss, the weight of subcutaneous fat would exceed that of the intermuscular fat at a lighter body weight than in those animals which had not suffered a weight loss.

The data suggests that the partition of fat between the subcutaneous and intermuscular regions occurred in three phases. The early phases during which the subcutaneous fat underwent a rapid growth relative to the intermuscular fat was concluded by about the eighteenth day after birth. From about 18 days until rapid fattening commenced, the ratio remained stable at about .75 : 1, except for the period of weight loss. This, the second phase, it is postulated, is the period when muscle growth is taking precedence over the growth of fat. The third stage is concurrent when the rapid fattening and the proportion of dissectible fat in the subcutaneous fat depot increases, that in the intermuscular depot decreases, while the internal fat shows an increase over the proportion at 180 days.

During the triphasic changes in the relative percentage of subcutaneous and intermuscular fat of the total dissectible fat the percentage of internal fat also changed. During the first phase up to 18 days the internal fat fell from about 46% to 30%, with a concomitant increase of subcutaneous fat (Tables 3-6). The internal fat then slowly increased its proportional contribution to the total fat, apparently unaffected by the seasonal weight loss of the sheep.

CONCLUSION

The results obtained are in agreement with Searle and Graham (1972) in defining a phasic growth pattern for the growth of fat. However with the techniques used in this study a triphasic growth pattern was postulated as opposed to the four phases of Searle and Graham. There is close concurrence with the results of Searle and Griffiths (1975) in the live weight at which rapid fattening began.

The results of this study suggest that for a lean carcass not only should an animal have a high mature body weight but it should also have a high mature muscle weight.

In the Merino strain under study a large proportion of the dissectible fat occurs in commercially valueless regions and an area for improvement would be to attempt to shift more of the internal fat into or onto the carcass. In view of the high and independent heritability of the various fat depots demonstrated in other species (Duniec *et al.,* 1961), this should be possible as techniques are available for estimating fat cover.

REFERENCES

Butterfield, R.M. (1965) Thesis PhD University of Queensland.

Duniec, H., Kielanowski, J. and Osinka, 2. (1961) *Animal Production* **3** : 195.

Lohse, C.L. (1971) PhD Thesis, University of Sydney.

Searle, T.W. and Graham, N. McC. (1972) *Australian Journal of Agricultural Resources* **23** : 339.

Searle, T.W. and Griffiths, D.A. (1975) *Proceedings of the Australian Society of Animal Production* **XI** : 57.

Warren, G.H. (1974) Thesis PhD, University of Sydney.

Wilson, P.M. (1954) *Journal of Agricultural Science* **44** : 67.

III

GROUP BREEDING SCHEMES AND STUD BREEDING

25

THE THEORY BEHIND BREEDING SCHEMES

J.W. JAMES

University of New South Wales, Kensington, N.S.W., Australia

SUMMARY

In the traditional stud hierarchy, gene flow is essentially from elite studs through multiplier studs to the general population. Long term responses are thus governed by the rate of progress in the elite studs, selection at lower levels of the hierarchy affecting only the genetic gap between levels. If the sire breeding nucleus is opened to gene inflow from lower levels, there are two main effects. First, the genetic gains from selection at lower levels are introduced to the nucleus, and can be passed from the nucleus to the whole population. Second, by the regular introduction of less closely related individuals to the nucleus, the rate of inbreeding is reduced.

The increased rate of genetic gain through opening the nucleus is only appreciable when about 50% or more of females are needed for replacements, and may then be 10% to 15% greater than in a closed nucleus. The best structure then is to have about 10% of the population in the nucleus, about half of the nucleus female replacements introduced from outside, and all surplus females born in the nucleus used for breeding in the other flocks. However, these values may be varied considerably without much effect on genetic gains. The rate of inbreeding in an open nucleus will usually be about half of that in a closed nucleus of the same size. This may be of value in small nucleus systems, but is of negligible importance in large scale schemes.

INTRODUCTION

Genetic responses to selection within populations of domestic animals are usually not very rapid, and in the slower breeding species such as sheep may take some time to spread through the population. For these reasons, the costs of running a selection programme may not be justified if they have to be recouped by the increased productivity of the flock in which selection is practised. However, this problem may be overcome if selection is carried out in special flocks (studs) so that the costs of obtaining genetic improvement are spread over the increased production of all animals using the studs as the source of superior genes. In fact, not only is this the common pattern in the livestock industries, but there is usually a further structure, with elite studs at the apex of a pyramid, acting as a source of breeding stock to a layer of multiplier studs, which in turn pass on genetic improvements to the commercial producers. This pattern acts to multiply many fold the increased production resulting from a given genetic gain at the top of the pyramid. The structure of the Australian Merino breed was described in these terms by Short and Carter (1955).

In such a system, the direction of gene flow is essentially from the apex of the pyramid to the base, and the result of this is that the long term rate of genetic change in the whole breed is that of the genetic sources at the top of the pyramid. The effects of selection carried out at lower levels of the hierarchy are in the long run only to determine how far apart genetically the different levels of the hierarchy are. The rate at which this long term situation is approached depends on the rates of gene flow between the levels. The effects of such a hierarchy on breed improvement were recently summarized by Bichard (1971).

OPEN NUCLEUS SYSTEMS

In recent years there has been a great expansion of interest in a different structure, usually practised in the context of co-operative group breeding schemes. However, the structure could in principle be applied in other contexts, such as on a single property, or indeed it could be adopted by the traditional stud hierarchy. Therefore the structure will be referred to here by the more general name, open nucleus breeding system. The characteristic of this system is that the sire breeding nucleus is not closed, so that gene flow occurs not only from apex to base of the pyramid, but also in the reverse direction, from the base to the nucleus. Normally, in group breeding schemes, only females are transferred from the base to the nucleus, but this is not a necessary restriction, at least in theory. If open nucleus systems are used in the traditional hierarchy, there may well be transfer of males from daughter studs to parent studs. However, only transfer of females from base to nucleus will be given detailed consideration in this chapter, since in group schemes sire selection in the base will be impracticable for managerial reasons in most cases. The rates of genetic gain which can be obtained by opening the nucleus to females from the base have been discussed by Jackson and Turner (1972) for an Australian Merino system, and by Rae (1974) for New Zealand Romneys, and in both cases optimal structures for such a system have been sought. A somewhat different approach has been taken by James (in press) who attempted a theoretical analysis in more general terms.

The essential effect of opening the nucleus on the rate of genetic gain is that because genes now flow from base to nucleus as well as from nucleus to base, any genetic gains made in the base by selection of base replacements will contribute also to genetic gains made in the nucleus, the extent to which this happens being dependent on the rate of gene flow from base to nucleus. So compared with a closed nucleus, an open nucleus has two sources of long term gain, of unequal magnitude.

If we compare the two possibilities of either closing or opening the nucleus after it is established, we see that the difference in mean breeding value of nucleus replacements in the two systems arises because in the open nucleus female replacements come partly from the base, and therefore can be very highly selected compared with the replacements which would otherwise have to be chosen from within the closed nucleus. But since the mean breeding value of the base is lower than that of the nucleus, this genetic difference offsets to some extent the expected genetic superiority of the females from the base due to their being highly selected. Making optimum use of base females in the nucleus thus involves balancing these two factors in the best way possible. Looked at in this way, the problem becomes one of choosing nucleus female replacements whose expected breeding values are as high as possible, and the cut-off points in nucleus and base should be chosen so as to have equal expected breeding values. For selection on a character with heritability h^2, this is equivalent to adding A/h^2 to the phenotypic value of each nucleus born female, where A is the mean breeding value difference between nucleus and base, and then selecting the best females on these adjusted values as nucleus replacements. (To allow for any environmental differences the phenotypic values should be taken as deviations from the corresponding means). The only problem is in knowing A, which may vary in the early stages of the programme before settling down to its long term value. If it is assumed that A will be relatively constant then it is possible to work out theoretically what the selection cut-off points should be in nucleus and base, and from tables of the normal distribution convert this to proportions selected from nucleus and base, which is a more convenient method, since comparisons need only be made on a within flock basis.

The optimum fraction of the population to be used in the nucleus can be considered in a similar way. The smaller the fraction of the population in the nucleus, the more intensely can females be selected from the base as nucleus replacements, and so the greater will be their genetic superiority over the mean of the base. On the other hand, as the nucleus size decreases the sires for use in the base must be selected from among a smaller number of males born in the nucleus, and so their genetic superiority over the mean of the nucleus decreases. The result will be that this tends to increase the genetic gap between nucleus and base, and thus negate the beneficial effect of the more intense selection of females. The average genetic value of female replacements for the nucleus will be greatest when the nucleus size is chosen so as to give the best balance between these opposing forces. The choice of nucleus size may then be regarded as a problem of balancing some loss of genetic gain on the sires-of-dams path against an increase of genetic gain along the dams-of-sires path.

RATES OF GENETIC GAIN

Equations which can be used to predict the consequences of open nucleus systems have been worked out (James, unpublished data). For instance, the steady rate of gain when the system settles down, G*, is given by the following equation.

$$G^* = \frac{(1+y)C_N + xC_B}{(1+y)l_N + xl_B} \qquad (1)$$

In this equation x is the fraction of breeding females in the nucleus which were born in the base, y is the fraction of base breeding females which were born in the nucleus, and l_N and l_B are the average ages of parents in nucleus and base respectively. C_N and C_B are the genetic selection differentials in nucleus and base, measured from the means of the groups in which selected animals were born. In selecting for a trait with heritability h^2, the genetic selection differentials are h^2 times the phenotypic ones. If A* denotes the stable difference in mean breeding value between nucleus and base, then

$$A^* = \frac{2(l_B C_N - l_N C_B)}{(1+y)l_N + xl_B} \qquad (2)$$

The genetic selection differentials are averages and can be calculated as follows. Let D_{MN} be the genetic superiority of males selected as nucleus sires over the mean of all males born in the nucleus, D_{FN} be the genetic superiority of nucleus-born females selected as nucleus dams over the mean of all nucleus-born females, and d_{FN} be the genetic superiority of base-born females selected as nucleus dams over the mean of all base-born females. Then

$$C_N = \frac{1}{2}(D_{MN} + (1-x) D_{FN} + xd_{FN}) \qquad (3)$$

Let D_{MB} be the genetic superiority of rams selected as base sires over the mean of all nucleus-born rams, D_{FB} be the genetic superiority of nucleus-born ewes selected as base dams over the mean of all nucleus-born ewes, and d_{FB} be the genetic superiority of base-born ewes selected as base dams over the mean of all base-born ewes. Then

$$C_B = \frac{1}{2}(D_{MB} + yD_{FB} + (1-y)d_{FB}) \qquad (4)$$

Thus C_N and C_B can be calculated, as can l_N and l_B from the vital statistics, so that for any proposed nucleus breeding plan, we can work out its consequences for ultimate rate of genetic gain and the genetic gap between nucleus and base.

It may be helpful to show how equations (1) to (4) are applied, and this is done using the example shown in Figure 1 of Jackson and Turner (1972). In this system there is a nucleus of 20 rams and 1000 ewes with 2 ram and 4 ewe age groups. The base has 400 rams of 3 age groups and 20,000 ewes of 5 age groups. The lambing rates are 70% in the nucleus and 74% in the base. Half the nucleus ewe replacements come from the base and all surplus nucleus-born ewes are used as base dams, so x=0.5. Since 125 nucleus ewe replacements are selected from the 350 nucleus-born ewes, the other 225 are used as base ewe replacements, so $y = 225/4,000 = 0.05625$.

The best 10 rams are selected as nucleus replacements from 350 born each year, giving $D_{MN} = 2.289$ (in units of $h^2\sigma$). Also 133 base rams are replaced yearly, so in all 143 of 350 are selected, giving a selection differential 0.949. Then since $\frac{10}{143} D_{MN} + \frac{133}{143} D_{MB} = 0.949$ we find $D_{MB} = 0.848$.

The best 125 out of 350 nucleus-born ewes are selected as nucleus dams, giving $D_{FN} = 1.045$. Since the other 225 are used as base dams 125 $D_{FN} + 225 D_{FB} = 0$, since the mean deviation for the whole group is zero. Thus $D_{FB} = -0.580$. The best 125 out of 7400 base-born ewes are selected as nucleus dams, giving $d_{FN} = 2.482$. Since 225 base ewe replacements are nucleus-born, the other 3775 are base-born, and in all $3775 + 125 = 3900$ base-born ewes are selected, with a differential of 0.755, from the 7400. Then since $125 d_{FN} + 3775 d_{FB} = 3900(0.755)$, we find $d_{FB} = 0.698$. Then

$$C_N = \frac{1}{2}(2.289 + (0.5)(1.045) + (0.5)(2.482)) = 2.026$$

$$C_B = \frac{1}{2}(0.848 + (0.056)(-0.580) + (0.944)(0.698)) = 0.737$$

The average ages of sires are 2.5 years in the nucleus and 3 years in the base. From age-specific fecundities in Table 2 of Jackson and Turner (1972) the average ages of dams are 3.72 years in the nucleus and 4.28 years in the base. Then $l_N = 3.11$ years and $l_B = 3.64$ years. Thus

$$G^* = \frac{(1.056)(2.026) + (0.5)(0.737)}{(1.056)(3.11) + (0.5)(3.64)} = 0.49$$

and

$$A^* = \frac{2((3.64)(2.026) - (3.11)(0.737))}{(1.056)(3.11) + (0.5)(3.64)} = 1.99$$

To find an optimum breeding plan we have to repeat the calculations for a range of values of the different variables and find that combination which gives the best results. Clearly, given the number of variables involved, this may be a rather tedious operation.

For this reason, a fairly general survey was carried out, using a slightly simpler model (namely, assuming generations do not overlap) in order to get ideas on when an open nucleus system gives appreciably greater gains than does a comparable closed nucleus system, and to what extent the optimum structure of the scheme depends on variations in the basic parameters. The detailed results are to be published elsewhere, but the general conclusions are as follows.

First, by looking at optimum schemes for different combinations of selection intensities in males and females, it was found that the advantage of opening the nucleus was not very great unless the fraction of females born needed for breeding was about 50% or more. If the selection intensity in females is 20% then the best that can be done by opening the nucleus is to increase the rate of gain by about 5% of that attainable in a closed nucleus. On the other hand, if the selection intensity in females is 50% to 80%, then an open nucleus may give genetic gains 10% to 15% greater than can be obtained in a closed nucleus. The selection intensity in males does not greatly affect these results, though the gains are slightly greater, for a given female selection intensity, when a smaller fraction of males is selected. The selection intensities which favour opening the nucleus are of course those common in sheep and beef cattle breeding.

Under these conditions (1% to 2% of males needed for breeding, 40% to 80% of females needed for breeding) the ultimate rate of genetic gain was greatest when about 10% of the whole population was in the nucleus, when about 40% to 50% of the nucleus female replacements were introduced from the base, and when all nucleus born females not needed as replacements in the nucleus were used as replacements in the base. However, the rate of gain was not very sensitive to changes in these parameters. For example, if 1% of males born and 80% of females born are needed as replacements, less than 10% of the *extra* gain obtained by opening the nucleus is lost by varying the nucleus size between 4% and 20% of the whole population, by varying the fraction of nucleus female replacements introduced from the base between 30% and 70%, or by using all or none of the surplus nucleus-born females as replacements in the base. Since the extra gain is only 15% of the total, such variations would reduce the total rate of gain by only 1% to 2%. There is therefore plenty of scope for varying these parameters for practical reasons without losing much potential for genetic gain. This is perhaps not surprising, since the effect of opening the nucleus is not great, though undeniably valuable.

RATES OF INBREEDING

In the same paper a method of calculating the effect of opening the nucleus on the rate of inbreeding was developed. It is not proposed to discuss the method here, since it involves matrix algebra and is a little complicated. However, a brief summary of the main conclusions can be given. The introduction of breeding animals from outside into the nucleus reduces the average degree of relationship between nucleus sires and dams below the value it would have in a closed nucleus of the same size, and therefore results in a lower rate of inbreeding. The parameter with the greatest effect is x, the rate of introduction of genes into the nucleus. When x = 1, and all nucleus dams are introduced, the effective size of the nucleus is about twice that of a closed nucleus of the same size. Most of the increase in effective size of the population is achieved by raising x from zero (i.e. a closed nucleus) to 0.5, rather than by raising x from 0.5 to 1. Thus, very roughly, we can say that an open nucleus in which about half the female replacements are introduced from the base would have about twice the effective size of a closed nucleus of the same actual size. Since it would have about twice the effective size, it would have about half the rate of inbreeding. This is so over a wide range of values of the relative size of nucleus and base. For instance, a closed nucleus of 25 sires and 400 dams used each generation would have an effective size of 95, and thus an inbreeding rate of about 0.5% per generation. An open nucleus with $x = 0.5$, $y = 0$, when the base used 100 sires and 1600 dams per generation, would have an effective size of 208, while if the base were five times as large (500 sires, 8000 dams) the effective size of the population would be 219. In these cases the rate of inbreeding would be about 1% every 4 generations. In systems as large as this, inbreeding is a very minor consideration. Only in much smaller systems would the inbreeding effect be worth taking into account in designing the system. It should be noted that the size of the nucleus rather than the size of the whole system is the important feature in relation to inbreeding.

OTHER CONSIDERATIONS

In planning a nucleus system, the long term features are determined only by the procedures adopted for selection and transfer of breeding stock, as can be seen from equations (1) to (4). The initial conditions, namely the source of foundation stock and the way in which foundation stock are distributed between nucleus and base, have no influence on either the final rate of genetic gain, or on the genetic gap between nucleus and base. But this does not mean that they are unimportant. The initial level of breeding value will determine the base line above which gains are made, and a system which starts x units ahead in breeding value can expect to stay x units ahead in breeding value. So the average genetic value of the foundation stock should be as high as practicable. It is intuitively reasonable, and can be proved mathematically, that the best available animals in the foundation stock should be placed in the nucleus, because this leads to the most rapid gains in the early years of operation. This is because the passage of genes from nucleus to base is used to lift the base towards the level of the nucleus, and the higher the level of breeding value in the nucleus, the greater the lift given to the base. However, it must be remembered that the overall level of the whole population at the beginning is not altered by allocation of animals to nucleus or base, and that although by very intense initial selection a highly superior nucleus may be established, this genetic differential in the nucleus will be only partly passed on to the next generation of the base.

Nevertheless, it will be worth seeking the best possible genotypes for the nucleus, and in a co-operative group scheme there may be some advantage in looking for superior foundation stock in other flocks. The question of whether to use highly selected individuals from many flocks or to choose foundation animals from one genetically superior flock cannot be answered in general. An answer depends on how many flocks are available and on the relative magnitude of between-flock and within-flock genetic variation. This question has been discussed by James (1966) and Jackson and James (1970). Usually it would seem best to select from several flocks, especially as this would be expected to increase the genetic variability available for selection. However, the decision must be made in the light of individual circumstances.

A variety of questions about organization and its genetic effects can be answered quickly using equations (1) to (4). In the example, we can ask if it is worth using the cast-for-age rams from the nucleus in the base, replacing some of the rams 2 years younger which would be used otherwise. These rams have a genetic differential 2.286 over the mean of their own drop, but compared with the drop 2 years later this is reduced by (2)(0.49) to 1.306. These would replace the worst 10 of 143 selected under the scheme described above. These 10 would have an expected genetic differential of 0.269, so if they were replaced by the older nucleus rams the genetic mean of the base rams would be increased by $\frac{10}{143}$ (1.306−0.269)=0.078 and since half of this would be passed on to their progeny, the gap between nucleus and base would be reduced by about 0.04 $h^2\sigma$. The optimum operating conditions would be slightly altered, as would the rate of genetic gain. If this were the only change made, the rate of gain would be increased to 0.495$h^2\sigma$ while A* would be reduced to 1.95$h^2\sigma$, as can be checked by substitution in the equations. These are of course comparatively small changes.

DISCUSSION

In the preceding sections attempts have been made to show the important features of nucleus breeding schemes and how their genetic effects may be calculated. There have also been discussions of operating conditions which give greatest genetic gains in the steady phase. However, these are not necessarily the economically best conditions. In particular, since the rate of genetic gains is not very sensitive to variation in the parameters, these may well be varied in order to produce a significant cost reduction with only a negligible change in genetic gains. For instance the size of the nucleus may be halved with very little effect on genetic gains, but costs of the nucleus would be greatly reduced. However, the economic optimization of a scheme will depend on many factors specific to the particular situation, and cannot be profitably discussed here.

Though the equations given here are quite general, the numerical results discussed have all been based on the assumption of individual selection for a continuously distributed trait. This has made it possible to ignore heritability, since all responses are proportional to heritability. If such methods as progeny testing are considered, further parameters such as numbers of sires tested, and relative usage of proven and untested sires need to be considered. Again, it is probably best that these be considered in the context of specific schemes.

It is possible that open nucleus systems may be somewhat more valuable for all-or-none traits which are rare. It is well known that the effective heritability of such traits is low, but increases as the trait becomes more common, and then falls again. If a large scale screening is carried out to initiate the nucleus, the level of incidence may be raised enough to make a significant improvement in effective heritability. However, there might then be little advantage to keeping the nucleus open.

Perhaps it is worth reiterating the basic principles on which a nucleus breeding scheme should be based. Considering all replacements, male and female, regardless of their place of use, it is clearly best that selection should be on estimated breeding value, taking into account any genetic differences between their places of origin. This will maximize the mean breeding value of the whole of the next generation, which will not be affected by the mating pattern adopted. Setting aside a sire-breeding nucleus enables the best sires and dams to be mated, so that in the progeny the sires will be selected from a group with an above average mean, and the problem to be solved is to find the best balance between this effect and the reduced selection intensity involved in selecting sires from a small fraction of all males born. There will be some further complications if we allow for genetic differences between age groups, but the general principles will not be affected.

REFERENCES

Bichard, M. (1971). *Animal Production 13*: 401.
Jackson, N., and James, J.W. (1970). *Australian Journal of Agricultural Research* **21** : 837.
Jackson, N., and Turner, H.N. (1972). *Proceedings of the Australian Society of Animal Production* **9**: 55.
James, J.W. (1966). *Australian Journal of Agricultural Research* **17** : 583.
Rae, A.L. (1974). *Massey University Sheepfarming Annual 1974*: 117.
Short, B.F., and Carter, H.B. (1955) C.S.I.R.O. Bulletin No. 276.

26

THE PROMOTION OF BREEDING SCHEMES AMONG STUD BREEDERS

B.C. JEFFRIES
Department of Agriculture and Fisheries, Adelaide, South Australia

SUMMARY

The entrenched hierarchical structure of the stud sheep industry has closed the parent studs to new genetic material. This structure and traditional selection based on visual appraisal has slowed or even halted genetic progress over the last 30 to 40 years. Renewed progress in stud and commercial flocks depends on the wider adoption of appropriate breeding schemes based on objective measurement in flocks large enough for adequate genetic progress.

To break the hold of traditional methods, we need firstly to identify those social and economic forces in the industry which tend to maintain the *status quo*, and to recognize the influence on producers' decision-making of groups such as stock and station agents, sheep and wool classers, bankers and consultants. At the same time, it will be necessary to create an awareness of the need for change in all sections of the industry. This is best done by demonstrating the superiority of commercial sheep bred on objective criteria over those bred by traditional methods—by demonstrating the economic benefits of breeding for production rather than for blood.

Changes in breeding methods will almost certainly produce social problems for some individuals and sections of the industry. These possibilities must be foreseen and extension must attempt to minimize the problems of adjustment.

To date too few people have been charged with the task of promoting the adoption of recommended breeding programmes and these have not always had adequate training and experience in sheep breeding or in extension. We need to expand the training of appropriate personnel.

INTRODUCTION

The hierarchical structure of the stud Merino sheep industry has been described by Short and Carter (1955); this structure provides an efficient method for disseminating changes, both good and bad, that may be made in the parent studs (Pattie, 1973). Bichard (1971) has examined the rate of dissemination of genetic improvement in such a structure and shown there is a lag of approximately two generations between tiers of the hierarchy when average males are sold to the next layer. The lag will be more or less if below or above average males are transferred.

McGuirk (1976) has stated that permanent genetic improvement in commercial flocks of Merinos is entirely dependent on selection policies adopted in the stud; as pointed out by Pattie (1973), however, the parent studs are closed to introduction of new genetic material and so breeders have restricted genetic variation with which to work.

The importance of promoting breeding schemes amongst stud breeders has been stressed by Pattie (1973) who states "A search for ways in which the present system, so rigidly closed by tradition, could be modified to allow introduction of new genotypes and the use of scientific breeding methods would seem to be a most urgent project at present facing the Merino and other breed societies".

In this chapter the extension methods which have been used in promoting breeding schemes will be discussed; a model communicating innovations described by Rogers and Shoemaker (1971) is used to categorize these extension efforts and to comment on the effectiveness of these approaches.

METHODS AND MATERIALS

Rogers and Shoemaker (1971) describe four basic stages in the Innovation-Decision Process:—

Knowledge — the individual is exposed to the innovation's existence and gains some understanding of how it functions.

Persuasion — the individual forms a favourable or unfavourable attitude towards the innovation.

Decision — the individual engages in activities which lead to a choice to adopt or reject the innovation.

Confirmation — the individual seeks reinforcement for the innovation-decision he has made, but he may reverse his previous decision if exposed to conflicting messages about the innovation.

This model has been used to study thirteen cases, in Australia and overseas, in which the author has been involved in the promotion of breeding schemes. Key factors in these case studies are shown in Table 1 under the stages of the Innovation-Decision Process. The factors listed are those considered important in the extension of breeding schemes and are not the results of in-depth sociological surveys.

RESULTS AND DISCUSSION

While recognizing that evaluation of these case studies is by no means a detailed sociological study of the way in which the innovation-decision process works in the sheep industry, it does provide a basis on which the promotion of breeding plans can be considered in a general way and allow a discussion of solutions to the problem described by Pattie (1973).

It can be seen from Table 1 that the means by which individuals gain knowledge of breeding schemes are diverse. The traditional extension methods of breeding schools, bulletins, press and radio have not always been as effective as might be expected. Other means such as hogget competitions used as teaching aids, satisfaction from breeding schemes with cattle, and the influence of farm consultants, woolbuyers and successful breeders using these recommendations have also influenced the spread of knowledge about breeding schemes. The prestige of organizations in developing countries like FAO, has hastened acceptance of new information promoted by their experts.

TABLE 1

KEY FACTORS IN THE PROMOTION OF BREEDING SCHEMES

Stage of innovation decision process	Key factors	
Knowledge	Hogget competitions	Mass media-radio
	Breeding schools	Beef recording scheme
	Wool buyers	Australian Merino Society
	Farm consultants	An Accountant
	Extension bulletins	Visit existing schemes
	FAO sheep raising project	
Persuasion	Visits to schemes in operation	Exercises in selecting rams on measurement
	Visits to studs and satisfied clients using measurement	Contact with author
Decision	Compared progeny of rams from 3 sources	
	Compared crosses of ewes and progeny.	
	Decided to establish "control" flock	
	Compared nucleus progeny with remainder	
Confirmation	Full adoption of schemes (7)	
	Recently commenced (2)	
	Interest waning (2)	
	Opposition by sheep classer (2)	

Experience has shown that encouraging interested breeders to visit schemes in operation and discuss them on a producer level is a worthwhile extension activity. Sheep breeders seem to react more favourably when breeding schemes are on their own or local flocks. They are less receptive to demonstrations on government research centres. This clearly indicates the need for establishing demonstration flocks with co-operators. Breeders using these schemes are some of the most effective extension agents.

In two of the four cases studied in which interest is either waning or has ceased, professional sheepclassers have played an influential role in presenting conflicting information and attitudes. Many sheepclassers see these breeding schemes as a threat to their livelihood and extension officers must recognize this fact. Indeed officers should try to work with and through professional sheepclassers who are widely accepted in the industry.

Breeders in these case studies were motivated by low returns, the unavailability of rams of the quality required at a price they could afford to pay, and high labour inputs. They might have been interested if sound economic assessments had been available in an understandable form. Indeed there is a need for articles such as those by Thatcher and Napier (1976), Napier and Jones (1976), and Spriggs (1975) to be published as extension material for use by extension officers in the field.

(a)   Extension activities

The key factors listed in Table 1 suggest areas in which extension activity should be expanded.

(1)   Demonstration flocks should be established in the local area. Utilize co-operators using objective measurement to promote breeding schemes.

(2)   At hogget competitions and sheep classing demonstrations used as teaching aids get people to estimate fleece and/or body weight. Then shear and weigh the wool and body as a check on accuracy or lack of it.

(3)   Write bulletins on breeding plans and budgets with step by step procedures, e.g. Clark and Bennett (1973), Jefferies (1976), and Turner (1973).

(4)   Economic articles must be widely promoted, but at present they exist in research publications such as Hill (1974) little understood by laymen.

(5)   Schools should be organized to train breeders and sheepclassers in principles of breeding and schemes which could be promoted Australia-wide.

(6)   Alternative roles will be required for people whose professional activities would be curtailed by the introduction of scientific breeding schemes. For example, sheepclassers could become salesmen for studs using objective measurement.

(7) Schools for stock agents have been held at least in Tasmania and Western Australia. These could be expanded to reach all those engaged in selling stud stock and in ram selection.

(8) Liaison with wool testing laboratories to help interpret objective measurement results to the breeder as an aid to ram selection.

(9) Co-operation with breed societies, such as the Poll Dorset production competitions (Stafford and Dolling 1975), helps to promote measurement among breeders.

There would be a greater chance of acceptance by sheep breeders if they could see less of a threat and more of the economic, labour and management advantages of the proposed changes in breeding plans for themselves and the industry as a whole. For example, there is less work in managing a nucleus flock than a complex family system.

(b) Extension personnel

It appears that the most effective extension officers in sheep breeding are those with a good academic training in genetics, and wide experience in breeding projects under the guidance of recognized geneticists on properties locally, interstate and overseas. Schools and workshops on genetics supported by work with co-operating breeders to train the officer to handle the practical problems associated with sheep breeding. This sort of experience helps officers to keep their feet on the ground.

More officers need to be trained to work in this field. There are only one or two specialist extension officers working on sheep breeding in each State.

(c) Sociological Surveys

There is an urgent need for sociological data on why producers have not accepted recommended breeding plans. This data needs to be collected from those for and against change as well as other members of the strong social structure in the stud industry. This information could then be used to help design more effective Extension Programmes.

ACKNOWLEDGEMENTS

The co-operation of many breeders with whom I have worked is gratefully acknowledged. My thanks are also due to colleagues who have assisted over the years and to Mrs. Ellen Bennett and Mr. J.R.W. Walkley, who have helped prepare this paper.

REFERENCES

Bichard, M. (1971). *Anim. Prod.* **13** : 401.

Clark, A.R. and Bennett, N.W. (1973). *Agric. Gazette of N.S.W.* **84** : 172.

Hill, W.G. (1974). Proceedings 1st World Congress on Genetics applied to Livestock Production, Madrid.

Jefferies, B.C. (1976). Extension Bulletin No. 3 : 76. S.A. Department of Agriculture and Fisheries.

McGuirk, B.J. (1976). *Proc. Aust. Soc. Anim. Prod.* **XI** : 13.

Napier, K.M. and Jones, L.P. (1976). *Proc. Aust. Soc. Anim. Prod.* **XI** : 17.

Pattie, W.A. (1973). In 'The Pastoral Industries of Australia', Edited by G. Alexander and O.B. Williams. Sydney University Press.

Rogers, E.M. and Shoemaker, F.F. (1971). The Communication of Innovations 2nd edition. The Free Press, New York.

Short, B.F. and Carter, H.B. (1955). *C.S.I.R.O. Bulletin No. A189*, Melbourne.

Spriggs, J.S. (1975). *Agricultural Record* **2** : 69.

Stafford, J.E. and Dolling, C.H.S. (1975). *Wool Tech. Sheep Breed.* **XXII** : 27.

Thatcher, L.P. and Napier, K.M. (1976). *Animal Production* **22: 261.**

Turner, Helen N. (1973). Keeping up with the Times in Sheepbreeding. Published by Economic Wool Producers Ltd.

27

SOCIOLOGICAL, ECONOMIC, BUSINESS AND GENETIC ASPECTS OF SHEEP GROUP BREEDING SCHEMES

G.R. PEART

Gulargambone, New South Wales, Australia

SUMMARY

The birth and development of sheep group breeding schemes represents a major departure from the established selection, business and sociological patterns long accepted within the industry. This is the first large scale adoption of research results announced 20 years ago and expensively extended in the interim. Geneticists had not previously considered group breeding and currently a research gap exists on the potential gains between starting and the steady state. Close ties between farmers and researchers, a management oriented extension service, and open Government support have given New Zealand a sound basis for group breeding. In Australia group breeding schemes have grown largely neglected by the huge organization designed to assist the wool industry. Group breeding has caught the imagination of farmers and is already a major force for genetic and general change. Groups are likely to continue to increase and producer inspired vertical integration within the industry has already started. The traditional sheep industry, the advice of scientists and the development of group breeding are reviewed in the light of sociological, economic business and genetic factors of importance. The need for a sound legal and financial basis for a long term investment in animal breeding is covered and the likely future development of group breeding schemes discussed.

INTRODUCTION

In 1976 the author was granted a three-month fellowship by the Australian Agricultural Council to study livestock group breeding schemes in New Zealand and Australia. This involved visiting many of the groups and research workers, accountants and extension officers involved in various aspects of group breeding schemes.

In 1967 the first schemes began in New Zealand and Australia and have expanded rapidly. New Zealand now has 27 sheep and 7 cattle groups while the major Australian sheep group has 420 members and 1.25 million ewes representing 10% of the Western Australian and 3% of the Australian Merino ewes joined to Merino rams. A description of the findings of the fellowship follows with specific reference to sociological, economic, business and genetic aspects.

THE TRADITIONAL SYSTEM

(a)   The Business of Ram Breeding

The rapid increase in the number of farmers joining group breeding schemes (GBS) during the last six years represents a major departure from established business and sociological patterns long accepted within the sheep industry. Within Australia most Merino sheep men rigidly adhere to a single blood line within the three tier system of parent studs, daughter studs and commercial flocks. Flock rams are bought from a stud during an annual visit on a fee scale set by the parent studs. In many cases the stud classer assists in choosing the rams and annually culls the maiden flock ewes and thus plays a key role in maintaining the close business association between flock and stud.

(b)   The Notion of Quality

Preservation of and progress in sheep quality have been in the hands of the studs, studmasters being well paid and respected for their acknowledged skills in this area. It has been calculated that in terms of genetic gain a two-generation advantage is held by parent studs over daughter studs and by daughter studs over commercial flocks, (assuming that genetic progress is being made). As only less desirable rams are sold down through the ranks a hierarchy of quality exists. Quality assessment is a visual-tactile skill taught superficially through such systems as the annual show and jackeroo training but ultimately inherited by studmasters and some classers. Studs are the authority on and repository of sheep quality.

(c)   The Social Organization

From the pinnacle of the 23-parent studs unfolds a well defined social organization. The functions are divided among ram breeders, multipliers and users who co-operate to produce wool and improve its quantity and quality. From the top level flow not only the genes but also the concept of quality. This value transmission takes place in many ways but forms the pivot on which the present sheep industry depends. Many of the present sheep breeders and their fathers were trained as jackeroos on the parent studs so both by education and family example the studs notion of quality is reinforced. Assent is annually renewed at the visit to collect the rams, a major social outing and one in which many wives participate, enjoying the hospitality of the studmaster and his wife. The commercial sheep breeder and his peer group are socially tied to the stud and all trust, accept and practise the stud's notion of quality. The country and State shows and the media coverage of sheep champions chosen by the visual assessment of studmasters completes the national acceptance of the traditional system.

While many facets of agriculture have been recently moving towards a more "urban" organization of specialization, stricter differentiation and professionalisation of jobs, these forms have long existed in the sheep industry. Assessment of quality is a professional skill held tightly by the sheep classers who through classing the maiden ewes and selecting the rams for the flock owner completely control genetic progress. Such a stable social organization controlled by the parent studs and the professional sheep classers is well able to resist change and generally has done so.

In recognition of the apparent contribution being made to improving the quality of the Merino sheep the government decided not to sub-divide the large stud properties for soldier settlement. This decision helped keep the stud owners in a position of wealth, power and prestige which reached its peak following the 1950s wool boom. This was the time (1955) chosen by scientists to announce to the Merino sheep industry that the very basis of selection used in the studs was wrong. Twenty years of repetition has done little to improve the acceptability of the message which was unacceptable within the sociological organization and threatened the economic position of the rulers. Economic recession has reduced the wealth of the studs but until the advent of GBS their power and prestige in the sheep industry remained unchallenged.

SCIENTIFIC RESULTS

Following research on sheep selection methods scientists claimed that much faster improvement could be made in characters of economic importance by the use of objective measurement. Fleece weight and quality were the first characters examined but the potentials for progress in many others, particularly reproductive rate, have since been defined. For major reviews see Turner 1969, Dun and Eastoe 1970. From 1955 onwards (CSIRO 1955) an extension programme was undertaken to bring about the adoption of selection on fleece measurement. Turner (1973) stated that the campaign to influence stud breeders was generally unsuccessful and today only two major daughter studs in N.S.W. are said to be using fleece measurement effectively.

Scientific recommendations conflicted with traditional selection procedures and the accepted notion of quality. In the face of these new findings sheepmen turned to their professional advisers on matters of quality, the classers, who dismissed the scientists' claims. The classers maintained that they could pick by eye heavy cutting sheep; scientists had ignored the many other factors essential in a "quality" sheep and anyhow non-sheepmen couldn't possibly give sound sheep advice. Most sheepmen accepted this as it agreed with their own feelings.

Selection on fleece weight was not a completely new concept and the work of Mr C. Euston Young in the 1930s in Queensland stimulated scientists to research the benefits of this method (Turner 1973). Austin (1904) states that the "Wanganeila" rams were selected on clean fleece weight from

1878 onwards. The breakthrough made by both men was lost when traditional classing methods took over with their passing.

ADOPTION OF MODERN SELECTION METHODS

The major extension effort for twenty years to sell selection on fleece weight has had some success but mostly at the flock level. Department of Agriculture sheep officers were able to explain the potential benefits of selection on fleece weight and some flock owners agreed to select maiden ewes on greasy fleece weight. Following the work of Morley (1955) larger commercial flock owners were encouraged to form their own ram breeding nuclei within their flocks. Management difficulties at a time of low sheep profitability have led many innovators to revert to the traditional system. A few nucleus breeders have survived and now sell rams. Sheep extension officers have usually only been accepted as advisers on sheep quality by fringe members of the traditional social organization. Their comparative youth, lack of acceptable social background and regular replacement makes them poor competition for the stud classer. Farmers who took up fleece weighing met opposition from leaders and peers, their sheep showed no visual response and rams selected on fleece weight were unavailable. Most returned to the fold.

DEVELOPMENT OF GROUP BREEDING SCHEMES

For GBSs to be as widely accepted as they have been there must have existed a number of ram buyers dissatisfied with the quality of rams available. This feeling is partly generated by the widely publicized results of scientific research and partly by a lack of any visual or measured progress in sheep quality over many years. Organizing this feeling required the presence of an acceptable catalyst person to spark the idea and an organized social unit within which to speak and from which could come the first GBS members. The first two men (Parker 1970 and Shepherd 1975) who thought up the GBS concept were stud masters of medium sized studs, convinced by scientists about selection on measurement, acceptable to sheepmen, and able to speak and convince some of their group of ram buying clients. Such a break with tradition is helped by a degree of economic independence from possible repercussions. A climate of innovation in the farming community is also important because if farmers have made one break with tradition they more easily make the next. This is particularly evident in Western Australia where many new-land farmers left their fathers in the eastern states and adopted a whole new farming technology. In New Zealand GBSs have been particularly successful within the new sheep breeds where the first innovation was the decision to drop Romneys.

Successful catalysts may have a range of backgrounds but must be acceptable, have an audience and be able to convince a proportion of members of the genetic and economic benefits of selection on measurement and the promotion of superior ewes. Farmer catalysts must be established leaders capable of convincing members of their peer group to join the scheme. The other common catalysts are farm management consultants who convince a group of their existing clients to form a group. Once started and organized the forces within the group are self sustaining and non-farmer members are unnecessary and undesirable in any position of dominance.

Major departure from an accepted "norm" within a group usually brings alienation because it is a challenge to those in power; in this case both economically and socially. For the ordinary farmer, success of the new system demonstrates that he made the wrong decision in not joining the scheme and so he would prefer to see it fail. Sons have been strongly opposed by their fathers who see their own prestige and life's work in terms of the ongoing success and standing of the family farm. Such pressure is hard to resist in a close farming family where knowledge, values and assets are all inherited. Stud opposition to GBSs has been strong with expulsion from breed societies of early innovators. This excommunication was followed by social isolation of both the farmer and his wife from the former social organization of stud breeders and ram buyers. Opposition made starting a GBS difficult but has helped keep members together, firmed their resolve to succeed and helped publicise their ideas.

The promotion of top flock ewes is a very appealing concept for ordinary members who feel they are making a personal contribution to progress in the sheep industry. This satisfaction is often demonstrated in a renewed interest in the sheep enterprise and an improvement in management standards and general innovative drive.

Within the founding group not only must the members be acceptable to one another but the quality of the sheep in the individual flocks must also be acceptable. Members who consider their sheep are superior or different are reluctant to pool them. Many groups insist on assessing the acceptability of any new applicant and inspecting his flock to maintain harmonious relations within the group.

FACTORS IMPORTANT IN THE ESTABLISHMENT OF A GBS

Economic and business aspects have tended to be neglected in the enthusiasm of forming a GBS, but a sound economic and business base is essential for a long term venture such as animal breeding. GBSs start with the accepted aim of improving the economic productivity of the sheep. Thus, drawing up the selection programme is a natural first undertaking. The mechanics of the selection programme need to be clear so that both members and nucleus manager understand and efficiently carry out their work. In New Zealand members started by contributing all twinning maidens into the nucleus and entry requirements have slowly expanded. Geneticists have given some suggestions on optimal flock structure in the steady state (Jackson and Turner 1972; Rae 1974) but no precise guidance exists for the build-up years. It is known that the initial screening of a large number of ewes gives a valuable leap forward in genetic progress. Progress in the early years is then heavily dependent on the genetic merit of the foundation sires chosen for the first two joinings.

For farmers considering forming a group breeding scheme a benefit-cost analysis is necessary to weigh the increased costs of measurement and transport plus the formation costs against the value of the extra wool produced in unknown quantities until the steady state is reached. From initial screening, it is four years before the first ram is available and ten years before the flock contains at least all FIs, so at the flock level benefits only come slowly.

Some work has been published on the economic benefits of selection (Spriggs 1975; McGuirk 1976) but sufficient data is not yet available to examine the economic benefits of GBSs. Few of the present schemes have attempted a feasibility study before commencement. And almost all owe their success to ram sales rather than major gains in productivity although this is now being demonstrated in some. While precise genetic gains during the build-up phase are unknown, reasonable data exist for the steady state situation and economic estimates can at least be based on this. Benefits and costs must be related to both the member and the nucleus manager. Calculations indicate that assuming no sale of rams to non-members, the saving in ram costs covers the costs of measurement and gives some extra return to the nucleus manager. Establishment costs can only be covered in the long term by ram sales or rises in productivity. It is worth noting that with a potential annual gain in fleece weight of 2% and an 18% inflation rate and a static wool price one has to be a believer in survival economics to start a GBS. The nucleus manager is the key to the success of most GBSs so his appointment is vitally important. Group recognition of him as a top sheep manager and good visual judge of sheep is important. Many managers previously ran studs and usually have the desired qualities. If they are also community leaders then this personal acceptability lends acceptability to the sheep and since early flock progress is slow the nucleus ewes and particularly the rams for members must be fed so that they look like stud sheep. Members need visual proof that they have made the right decision. A sound legal structure is essential in such a long term undertaking to protect the members investment and set out acceptable financial and working arrangements. Three major decisions govern the legal structure of any GBS. These involve the ownership of the nucleus sheep, control within the group and

distribution of rewards. Sheep ownership may rest with the original contributor, nucleus manager or a corporate body such as proprietary limited
company or co-operative.Rules must be formulated to cover distribution of
the assets if the group disbands. Control usually starts with an elected committee which provides the officers of any corporate body that may be
formed. When nucleus managers own the sheep the committee has only an
advisory role with reduced powers. Distribution of rewards depends on the
ownership structure, the simplest being for the nucleus manager to retain
the profits after distributing 1 ram per 5 ewes contributed. Where individuals retain ownership the nucleus manager is paid an agistment fee
designed to give him a profit approximately 50% greater than his commercial sheep return as a reward for his extra time, and managerial skill. Corporate bodies must devise a payment system which generously rewards the
nucleus manager, who takes the major responsibility in the GBS, and profits
are then distributed on a share basis which is usually equal.

While many groups enthusiastically sell the co-operation by gentlemen's
agreement concept, many prefer to ensure their place and investment in the
group with a documented legal and business structure. This helps stabilize
and organize the original idea and cements the commitment of all members.

FACTORS IMPORTANT IN THE ONGOING SUCCESS OF GBSs

The long term viability of GBSs once the enthusiasm and newness have
worn off, requires rather different factors from those which start them. Support and encouragement no longer come from advisers or scientists who,
having contributed their ideas, leave the farmers to get on with the job. Support must come from within the group itself. Meetings to work out legal,
business and selection programs, at which each member needs to feel he is
making a contribution, help keep the group together while the slow business
of breeding gets underway. Newsletters and perhaps visits to other GBSs
help fill in the 3-4 year lag period before the first rams are available. Com-
radeship and peer group support from within the GBS have overcome the
high drop out rate among farmers taking up fleece measurements. This support is important in living with the added management problems, the opposition of neighbours and friends and the long wait while genetic gains
materialise at the flock level.

A new enthusiasm arises when visual and economic improvement can be
positively demonstrated and this phase has been reached in groups in New
Zealand and Western Australia. At this stage the fence-sitting neighbours
join or buy rams and the members go out and publicise the proven wisdom
of their early stand and thereby enlist new members. The two founding
groups have risen from a start of about 10 members to 100 and 420 ram
buying clients respectively. The need to believe that progress is being made

is important if the GBS is to survive. Methods to demonstrate progress need devising; the progeny test every 10 years using frozen semen from the top rams is the most accurate. Financial rewards from genetic progress sufficient to cover formation and annual costs have been made in many New Zealand groups where lambing percentage has risen 40% and selection for easy care sheep has reduced labour requirements. A more obvious reward has been the profit from the sale of surplus rams to new members and commercial buyers. Increased profits for the manager and members are the ultimate proof of the original idea and that genetic progress is being made.

THE FUTURE OF GBSs

A new stable social organization with a new value system is developing around the present GBSs. Assistance in value transmission in terms of sheep quality is given through the agricultural colleges and universities who teach future farmers. The acceptability and social position of members tends to change once a scheme proves its success and this is acknowledged by outsiders wishing to buy rams or by an increasing membership with original members holding the dominant position. The GBS mirrors the stud system in terms of social strata but with democratic rather than dictatorial rule. GBS leaders with their reinforced confidence and wider contacts among rural, research, and marketing people, tend to rise as leaders in other spheres. The growth of GBS into major commercial ram sellers in direct opposition to the studs will continue. This success has led in New Zealand to the formation of stud GBSs and a partial change to selection on measured characters by many individual studs. Changes in the Australian stud system will be very slow because increases in wool production are less obvious than improvements in reproduction rate and because sheep quality assessment is in the hands of a small professional class. The Western Australian group with 1.25 million ewes and 420 members, convinced of their progress and improved returns, is likely to increase membership rapidly. This group has sufficient production to completely supply a major woollen mill and will probably vertically integrate its whole operation. In New Zealand one GBS bought a small wool store to interlot wool and now owns a wool scour and are 25% shareholders in a spinning mill. An exciting future for GBS seems assured. They have overcome the opposition of the traditional study system and demonstrated a co-operative and innovative skill not thought to exist in the farming community. If the wool production industry is to survive it will probably need to follow the trends set by such other agricultural products as pigs and poultry. This will involve a revolution in wool handling and vertical integration of the industry. In New Zealand a close association exists between Universities, Agricultural Colleges, extension workers, and farmers, all aided by a Government heavily dependent on rural export earnings. Government support of livestock improvement is obvious in the early development of a computerized sheep selection plan and encouragement of GBS by people at all levels. The Government has given further aid by taking a 25% holding in the farmer initiated spinning mill. In Australia attitudes in the extension service range from interest to opposition. No effective association exists in the chain between researchers and farmers. In Australia, agricultural research appears to be done for the sake of Science whereas in New Zealand it appears to be done for the sake of farmers. Active encouragement of vertical integration is essential if commercial opposition is to be neutralized. GBSs have made the breakthrough advocated by scientists twenty years ago and have made a major contribution to improving the earning capacity of the Australian sheep industry, however assistance is apparently withheld at all levels in case the stud industry objects. GBS now have a sound social and financial base, contain considerable innovative and leadership skill and are market oriented. Vertical integration is now possible and with this they will assume a position of dominance in the wool industry and survive where others will not.

ACKNOWLEDGEMENTS

The granting of an Australian Agricultural Council fellowship allowed the study tour, the findings of which are reported in this chapter.
Support from the Australian Wool Corporation was given to attend the conference.

REFERENCES

Austin, A. (1904). Letters to the editor. *The Australian* 19 July.

C.S.I.R.O. Australia (1955). Selecting Merino sheep: Proposals for increasing wool production by accurate measurement. C.S.I.R.O. Leaflet Series 13.

Dun, R.B., and Eastoe, R.D. (1970). Science and the Merino breeder. N.S.W. Department of Agriculture. Sydney: N.S.W. Government Printer.

Jackson, N. and Turner, Helen Newton (1972). *Proceedings of the Australian Society of Animal Production* **9** : 55.

McGuirk, B.J. (1976). *Proceedings of the Australian Society of Animal Production.* **11** : 93.

Morley, F.H.W. (1955). *Agriculture Gazette of N.S.W.* **66** : 400.

Parker, A.G.H. (1970). *Journal of Wool Technology and Sheep Breeding* **17** : 19.

Rae, A.L. (1974). *Proceedings of the first conference of the New Zealand Federation of Livestock Breeding Groups 5.*

Shepherd, J.H. (1975). *Proceedings of the second conference of the New Zealand Federation of Livestock Breeding Groups 35.*

Spriggs, J. (1975). *Agricultural Record* **2** : 69.

Turner, Helen Newton and Young, S.S.Y. (1969). *Quantitative Genetics in Sheep Breeding.* Macmillan: Melbourne.

Turner, Helen Newton (1973). *Zeitschrist Für Tierzüchtung und Züchtungsbiologie.* **9** : 278.

28

ADVANTAGES OF GROUP BREEDING SCHEMES

A.G.H. PARKER
Wairunga, Havelock North, New Zealand

Livestock breeding groups were first established in New Zealand in 1967. They were formed by people who realized that in every large flock or herd, there is a small number of very high-performing animals. The basis of a group breeding scheme is to be able to identify these animals on several properties, and bring them into a central nucleus where they are mated to animals of similar high performance. The male offspring are used in the contributing flocks or herds, and hence improve them genetically.

Traditional livestock breeding has a structure which can be represented by a triangle. At the apex are the stud breeders. Below these are the flock ram or run bull breeders, or propagators, and at the base are the commercial farmers. The stud animals at the top are registered and no unregistered animals can enter the registered flocks or herds. The only movement of stock is downward, involving sires. Because of this, high-performing females are scattered throughout the triangle. Development of the breed, however, and any increase in performance throughout the industry is limited to the progress of the group of breeders at the apex.

The more modern breed societies allow for the identification of high-performing females, regardless of their position in the triangle, and encourage their use in the apex for the breeding of sires. Thus there is a movement of stock in both directions; high-performance females moving up, and high-performance-tested sires moving up and down.

The establishment of group breeding schemes allows for the screening of very large numbers of ewes at the base of the triangle for the purpose of identifying high-performing females, and so the progress that can be made is further maximized.

It should be pointed out at this stage, that animal improvement in the future is unlikely to be confined only to group breeding, and that there will continue to be a place for individual breeders, provided they operate reasonably large flocks or herds, and use modern recording techniques to the fullest extent. Furthermore, the industry as a whole, would be best served if a balance is maintained between individual breeders, and group breeders. There are, however, several advantages to group breeding which should be mentioned.

Genetic Advantages

Because of the large numbers of animals being screened to find the highest producers, only the very best need enter the central nucleus. In many cases each member of the group runs a nucleus within his own flock or herd, where females coming in from the screened commercial section are bred and recorded. It is only the best of these females which enter the central nucleus each year, and then only after careful selection and scrutiny for structural soundness and freedom from important faults.

Many of the advantages arising from group breeding are as follows:

(1)   Animals coming into the central nucleus by this means have been tested under practical farm conditions, where their offspring will later perform.

(2)   Records are taken for traits of commercial and economic value, and selections are based on these records, subject to health and established standards of soundness.

(3)   Replacement sires for the co-operating farms come mostly from the nucleus unit, so the improvements are quickly spread throughout the group.

(4)   A rapid generation turnover of breeding stock can be maintained.

(5)   There can be an avoidance of inbreeding which may arise with small breeding units.

(6)   There is likely to be a continuity in breeding objectives over the years, giving improvement in traits of commercial importance.

(7)   By joining together with other breeders, a breeder can benefit from a co-ordinated policy, from pooled experience, from shared facilities, and from the improvements achieved.

Other Advantages

Because of its size and potential, a breeding group has an identity greater than that of an individual breeder. This helps in getting the breeding group, and the commercial goals it stands for better known among farmers. In the past it has often been difficult for a breeder who has improved his stock to obtain a financial reward for his efforts, because he had little impact on the market. With a co-operative venture, more can be spent on promotion and advertising, and the improved stock can be better identified and marketed. Farmers can be more confident of the standards attained, continuity of supply, and the performance of the stock they buy.

The economies of scale of operation can be applied to large-scale group schemes without loss of accuracy or standards of overall excellence of the sires produced.

The group will have a common set of breeding goals, of recording systems, and breeding methods. These will be agreed upon, only after much discussion, argument, and negotiation among group members. A common agreed policy is likely to be more sound, convincing and effective than a series of different plans of independent breeders.

Because of their size and co-operative nature, group breeding schemes can act as a focus for advisory, technical and other aids to breed improvement, and obtaining the advice of a geneticist on breeding methods is more feasible with a group. Groups may also be able to exercise more influence than the individual breeder, in political and commercial circles.

Groups are in a good position to give technical advice and services to farmers using their stock, just as large commercial companies do, and in practise, such follow-up services and liaison with farmers, and the feedback from their reactions and experience with the improved stock, may be crucial to the success and future of breeding groups.

Results from Breeding Groups

Figures giving reliable information on progress being obtained in breeding groups are not readily available because of the short time that groups have been in existence, and because scientifically-controlled tests have not been undertaken.

The New Zealand Romney Development Group is one of the oldest groups and some comparative figures are available on two-tooth ewe performance.

On the property at Wairunga where the central nucleus of the group is located, the central nucleus sheep are mixed in, for management reasons, with the Wairunga flock.

The results presented in Table 1 concern the lambing performance of two-tooth ewes in the central nucleus which are the progeny of contributed ewes screened from members' flocks. The performance of these ewes, which in most cases have been recorded for one generation only, is compared in six successive years with that of two-tooth ewes born in the Wairunga flock in each of the same years. It should be mentioned that the ewes in the Wairunga flock have been on an intensive, within-flock selection programme for lamb production for 20 years.

TABLE 1

LAMBING PERFORMANCE OF TWO-TOOTH EWES IN THE CENTRAL NUCLEUS

	1970	1971	1972	1973	1974	1975	Av.
No. nucleus two-tooths							
mated	6	43	58	86	87	113	65
% docked	116.7	111.6	134.5	124.7	116.1	129.4	122.2
Wairunga two-tooths							
mated	376	305	251	227	216	212	264
% docked	93.5	114.1	115.5	116.3	106.5	112.8	109.8
Margin for central nucleus %	+23.2	−2.5	+19.0	+8.6	+9.6	+16.6	+12.4

Table 2

FULL BREAKDOWN OF LAMBING PERFORMANCE IN TWO-TOOTH EWES

	Nucleus Flock 2-tooths						Wairunga 2-tooths					
YEAR	1970	1971	1972	1973	1974	1975	1970	1971	1972	1973	1974	1975
NO EWES MATED	6	43	58	86	87	113	367	305	251	227	216	212
EWE PERFORMANCE												
% reared lambs	83.3	83.7	88.0	91.3	81.6	90.3	73.0	87.2	85.7	85.9	82.4	85.4
% lost all lambs	16.7	11.6	8.6	6.2	14.9	5.3	14.7	7.2	10.7	10.1	13.0	7.1
% barren	0.0	0.0	1.7	2.5	2.3	4.4	10.6	4.9	3.2	3.1	4.6	7.5
% deaths, tupping/lambing	0.0	4.7	1.7	0.0	1.2	0.0	1.7	0.7	0.4	0.9	0.0	0.0
% Ewes assisted at lambing	0.0	0.0	0.0	1.2	0.0	1.8	0.0	1.0	0.4	0.0	0.5	2.5
Lambing												
% lambs born/Ewes mated	150.0	130.2	153.4	140.7	140.2	139.4	118.3	140.7	138.2	137.0	133.8	128.1
Proportion singles/total lambs born	33.3	39.0	27.0	39.5	37.7	38.2	48.5	37.3	41.2	40.2	43.3	43.8
Proportion twins/total lambs born	66.7	61.0	69.6	57.9	62.3	57.9	49.8	54.1	54.1	59.8	54.7	58.2
Proportion triplets/total lambs born	0.0	0.0	3.4	2.6	0.0	3.9	1.4	7.7	3.5	0.0	2.0	0.0
Proportion quads/total lambs born	0.0	0.0	0.0	0.0	0.0	0.0	0.0	0.9	1.2	0.0	0.0	0.0
% Lambs docked/ewes mated	116.7	111.6	134.5	124.7	116.1	129.4	93.2	114.1	115.5	116.3	106.5	112.8

Table 1 shows that the two-tooth progeny of the contributed ewes have docked an average of 12.4% more lambs over the six-year period than the two-tooth progeny of the intensively-selected Wairunga ewes. In one year the Wairunga two-tooths performed better than the two-tooths from the screened ewes, but by only 2.5%. The ewes of the central nucleus and those of the Wairunga flock have at all times been run as one mob and mated randomly to the same rams. The full details of the performance of the two groups are given in Table 2.

Structure of Breeding Groups

To operate effectively, groups must have a formalized structure. The form and organization of group-breeding schemes varies widely between groups, reflecting their different origins, and the needs and facilities of each group. Several types of organization are in use at present, such as the private company, Industrial and Provident Society, Partnership, Incorporated Society, or simply a gentleman's agreement. In some cases ownership of the stock in the nucleus remains in the hands of the contributors, in others the company or Society, and in some cases, the nucleus is owned and operated by one member of the group.

The long-term success of breeding groups, however, will depend largely on the effectiveness of their organization, and the more formal structures will give the best guarantee of long-term success.

Conclusion

It is confidently believed that livestock breeeding groups will continue to evolve and develop, as circumstances dictate, and that, by being able to maximise the advantages of modern animal breeding methods, they will play a major role in the development of the livestock industries in most sheep and cattle raising countries.

29

THE AUSTRALIAN MERINO SOCIETY NUCLEUS BREEDING SCHEME

J.H. SHEPHERD

The Australian Merino Society, Kwolyin, Western Australia

INTRODUCTION

The Australian Merino Society, as it is currently structured, was not the result of a predetermined Breeding Plan. Quite the opposite. It actually evolved to meet a need—and once one need was satisfied, quite often another developed, requiring further adaptations and innovations.

As more and more people became involved, decision making was, of necessity, spread to make use of many people's talents. Today over 500 sheep breeders and several associated, interested, non-rural people, have a viable, functional and democratic process of combining several disciplines, not the least of which include Wool Testing Laboratories, Agricultural Colleges, Research Institutions, Farm Management Consultants, etc., in an attempt to maximize returns from genetics, nutrition, management and marketing.

The relatively simple process of breeding rams for sale—under a system whereby a sheep, once it had a piece of Blue Paper saying who bred it, immediately inherited some mystical superiority, which alone qualified it to be a superior animal—namely a stud animal, did not encourage change.

Once measurement for production became the hallmark of qualification, certain problems appeared in the "ram breeding" business. The first and foremost being the relatively small differences between the supplier and the supplied. It soon became all too clear that there were equally good ewes in the clients' flocks as there were in the Stud. It also became all too apparent that it was almost a national sin not to be using this superior genetic material. As it stood, the vast number of superior ewes running in ordinary farmers' flocks around the countryside, breeding wethers and flock ewes, were contributing nothing to the nation's genetic progress in sheep production.

Once sheep men were made aware of this situation they readily joined together and isolated their superior ewes into a common Central Nucleus to breed their ram requirements. As a prelude to this rather dramatic change in sheep breeding, certain background criteria need to be understood in order to evaluate the reasonably successful establishment of such an active group of people.

HISTORY

The Parent Flock, which was eventually to be the basis of the Central Nucleus, had been a Registered Merino Stud, established in 1943 and supplying by 1960, up to 1,000 rams annually to the industry. This Stud had been classed along traditional lines, until 1953, when fleece weighing commenced on the reserve 1½ year old rams. This policy continued until 1957, when all 1½ year old ewes were fleece weighed. By 1960, with Stud ewes at 4,000 head, artificial insemination was commenced, using superior sires selected on fleece weight criteria alone.

The use of AI was continued for seven years. Body weight and average fibre diameter as additional factors were used from 1962 onwards. During the early and mid-60s, distribution curves of fleece weight ranking and body weight ranking were drawn annually, and progress measured for increasing productivity at the top end of the scale. By 1966, through co-operation with daughter flocks, it was found, by simple progeny testing, that a few elite sires were emerging in the daughter flocks which were virtually equal to sires available from the Parent Stud. It was then realized by the Studmaster that it would be necessary to

1. increase the size of the gene pool,
2. make use of the selection differential of the few elite females in daughter flocks,

if the past rate of progress was to be maintained.

From this thinking emerged the co-operative breeding plan now employed.

BREEDING PLAN STRUCTURE

Once motivated, wool growers have to be given a concrete and concise as well as a practical, simple and functional breeding plan, which all can follow and take part in.

The actual breeding plan became more sophisticated as the actual size of the enterprise grew. The dramatic growth of the Society must be appreciated in order to understand some of the basics behind the criteria used in selection.

The table below portrays the annual progression of selection pressure applied through the Central Nucleus Flock which was operated on a two-tier system until 1971: i.e. daughter flocks returned the top ½% of maiden ewe progeny direct to Central Nucleus, which used these superior ewes for breeding rams which were supplied direct to contributors, after culling on objective as well as subjective factors.

Year	Number of ewes mated from which ewe progeny were selected	Number of maiden ewes contributed to C.N.
1967	: 20,000	20
1968	100,000	100
1969	350,000	400
1970	500,000	700
1971	700,000	800
1972	800,000	900
1973	900,000	720
1974	1,000,000	750
1975	1,100,000	800
1976	1,200,000	800 (anticipated)

From 1971 onwards, a three-tier system was developed with the establishment of regional ram breeding co-operatives which were used as screening sub-flocks to the Central Nucleus. Currently there are 30 such co-operatives operating throughout Western Australia and 2 in South Australia. A further 3-4 are planned for 1977.

Initially, via the ram breeding co-operative flocks, which desirably service 30-40 farmer contributors and up to 50,000 ewes (one has over 60,000), the top ½% of maiden ewe progeny from *each contributor's* flock was sent to Central Nucleus, and the next 2% in ranking order were screened off to the ram breeding flock. This operated for 1971 and 1972.

The Three-Tier Breeding System 1971-77

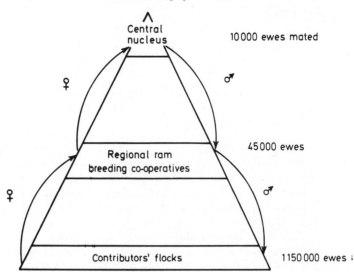

From 1973 onwards, ewes to Central Nucleus came only from the ram breeding flocks, the number of maiden ewes contributed being equal in number to 2% of the total ram-breeding ewes mated at each co-operative.

Farmer contributors gave maiden ewes equal in number to 1% of their total ewes mated, to their respective ram breeding co-operatives. The co-operatives receive top ranking sires from Central Nucleus and in turn supply top ranking sires bred in each co-operative flock to original contributors.

Ram breeding co-operative managers are usually chosen for their knowledge of genetics, husbandry and managerial skills, as well as acceptability by group members.

SELECTION CRITERIA

Prior to any selection, a decision on what to select for, is a necessary prerequisite.

The changing emphasis of selection criteria has been largely influenced by industry demand. Development of Persian Gulf markets for live shipping wethers of over 60 kilograms liveweight from Western Australia, led to increased emphasis on bodyweight in the late 1960s and early 1970s. Wool industry emphasis shifting from subjective to objective appraisal in the last few years, has led to a closer look at trade requirements and what is obtainable from competitive fibres.

Thus, from 1963/70 selection criteria were based solely on a fleece weight/body weight index, using a pundit square grid method of isolation of desirable sires and dams, with an upper limit on average fibre diameter on rams only.

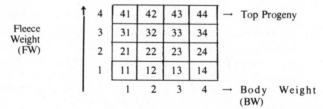

This was really only an arithmetical expression of superimposed distribution curves for fleece weight and body weight. It should be pointed out that culling for normal undesirable visual traits was part of normal practice.

Because individual ranking of ewes became necessary, from 1970 onwards, the selection criteria were based on the Berryman formula of F.W. × 10 + B.W.

The reasoning behind this formula was to give equal economic loading to wool and meat in the index, i.e. historically wool has normally been worth ten times in value per unit more than mutton in Western Australia.

With the advent of more sophisticated methods of measuring the physical characters of wool it became necessary to quantify the more important of these, and to try to assess their relevance in a breeding plan.

Before deciding on which characters should or should not be subjected to selection, it is necessary to ascertain

1. whether the character is measurable,
2. whether the character is heritable.

This required further refinement into characters which are controlled by the environment and characters which are influenced by genetics.

FACTORS OF GENETIC ORIGIN

What effect do such characteristics as:

  (i) clean fleece weight,
  (ii) tensile fibre strength,
 (iii) average fibre diameter,
 (iv) range in fibre diameter (i.e. coefficient of variation),
  (v) ratio of number of secondary follicles to number of primary follicles,
 (vi) ratio of diameter of secondary follicles to diameter of primary follicles,
(vii) average fibre length,

have on

(a)   processing—e.g. topmaking and spinning?
(b)   fabric quality?

(i)   Clean Fleece Weight

"Research in the CSIRO and elsewhere has shown that it is possible to increase wool production by selecting sheep for fleece weight rather than the conventional procedure which is influenced by staple crimp. We have found such wool to give higher top and noil yields, better combing tear and greater mean fibre length of top, leading to stronger yarn, than the conventionally selected control." (Dr. M. Lipson, former Chief CSIRO Division of Textile Industry.)

(ii)   Tensile Fibre Strength

Tensile strength has obvious importance and although environment plays a major role, Dr. Lipson has shown that there is "increased yarn strength with wool from sheep selected on fleeceweight".

(iii)   Average Fibre Diameter

(a)   B.A.E. reports show a high correlation between price paid for greasy wool and average fibre diameter (other things being equal).

(b)   Dr. Lipson has shown "for fabrics of identical structure, fibre diameter is the major factor affecting handle of wool"

(c)   Fibre diameter is recognized as the most important physical attribute of raw apparel wool in terms of the fabric into which it may be processed (Dunlop and McMahon, 1974).

(iv)   Range in Fibre Diameter

This is a function of variation from four different sources, each contributing differently to the total.

Source of variation	Percentage of total variance
Between sheep	20
Between sites over the sheep	10
Between fibres within sites	66
Between points along the fibre	4
	100

Measurements have shown that for a single staple of wool, with an average diameter of say 23 microns, the actual range can be as low as 10 microns, or as high as 40 microns and higher.

Most evidence to date has shown that variations in fibre diameter have little or no effect on either wool processing or the final product. It is the average figure that is of importance.

The fibres at the coarse end of the range (e.g. fibres exceeding 28 microns) are known to the trade as "coarse edge". Professor K.J. Whitely of the University of N.S.W. has as recently as June 1976, stated "the available technical evidence shows there is no good reason why the serious breeder should concern himself with the coarse fringe at this time".

In order to readily understand the full significance of these larger diameter fibres in a normal staple of wool, let us look at a few specific examples. In tests on samples of wool from six different rams, the following results were obtained from A.W.T.A. Melbourne.

Ram No.	Mean Fibre Diameter	% Fibres exceeding 28 microns	Range in Diameter from finest to strongest
1	23.0	4.6	18 microns
2	23.0	10.6	30 microns
3	23.0	13.3	32 microns
4	23.0	18.5	40 microns
5	21.5	13.0	30 microns
6	20.0	6.0	28 microns

It is quite difficult to comprehend how, if trade says 21 micron wool is better than 22 micron wool, and 22 micron wool is better than 23, and so on, meaning of course that one or two microns make all the difference to the price they pay, as per the following price schedule, that someone can say

AUSTRALIAN AVERAGE WOOL PRICES
19 Micron to 30 Micron Cents Per kg Clean
and Differential Per Micron

Selling Season	19M	21M	23M	25M	27M	30M
1964/65	224	207	192	178	171	226
1965/66	234	213	201	191	181	176
1966/67	229	209	199	180	168	150
1967/68	235	208	179	150	136	103
1968/69	244	218	184	150	132	118
1969/70	228	187	162	139	119	99
1970/71	174	140	123	115	106	99
1971/72	161	153	145	130	126	119
1972/73	419	371	360	338	317	292
1973/74	405	356	336	308	279	237
Average Price	255.3	226.2	208.1	187.9	173.5	161.9
Differential						
C per kg	29.1	18.1	20.2	14.4	11.6	
C per Micron	14.55	9.05	10.1	7.2	3.9	

that ram sample 1 (above), with 4.6% fibres greater than 28 microns and a range in diameter of 18 microns, performs exactly the same as sample 4 with 18.5% fibres over 28 microns and a range of 40 microns. Similarly, for ram No. 5, this wool (other things being equal) should bring more money than ram No. 1, yet it has 8.4% more fibres over 28 microns than No. 1. Ram No. 6 would be classed as a fine wool even though it had 1.4% more fibres over 28 microns than No. 1 and a range of 28 microns compared to 18 microns.

What do some processors have to say about this?

On a recent visit to wool processing complexes in five different countries, the opportunity was taken to seek information from as wide a spectrum of the industry as possible. Some of the information obtained is summarized.

(a) Woollen mill No. 1 in the U.K.

This mill processes only top quality fine wools from Australia. The managing director stated that average fibre diameter contributed only some 70% towards the variable known as "handle" and that 30% was due to other factors. Upon testing a number of "tops" taken at random from the best hand-picked top quality mill-run, no fibres exceeding 28 microns could be found.

(b)   Woollen mill No. 2 in the U.K.

This mill stated they did not agree with the Australian statement concerning variation in fibre diameter and that they supplied "tops" conforming to a specification which was as follows:

No more than   3½% fibres exceeding 30 microns in a 21 micron top
No more than   7 % fibres exceeding 30 microns in a 23 micron top
No more than 15 % fibres exceeding 30 microns in a 25 micron top

They also stated that if they exceeded these limits, their customers, the spinners, would be on the phone in half an hour asking "What is wrong with this top? We are having trouble with "ends down".

Before we rush to conclusions and say either our advice in Australia is wrong or that the processors don't know what they are doing, let us have a rational look at the situation.

What sort of information has been available to the spinner over the years?

1.   Average fibre diameter.
2.   Average top fibre length.

He has never had access to information (or only at a high cost) regarding variation in fibre diameter. As new and more rapid methods of recording individual fibre diameter become available, we find topmakers are now starting to produce tops which conform to more stringent limits and are actually starting to measure fibre diameter range as a normal practice. It was found that at least one topmaker was not actually quoting this information to spinner clients (only average fibre diameter). Because they are now starting to control this coarse edge, they are able to supply their clients with a better product as far as speed of processing is concerned, and they know full well that until their competitors do likewise, their top will perform better. Once this starts to happen it becomes

1.   highly competitive, and then
2.   eventually accepted as the norm.

(c)   Spinning mill No. 3

Spinners try to produce as much yarn (that is, spin down to as fine a yarn as possible) in as short a time as possible. In the past, spinners had a highly labour-intensive industry, but in keeping with most other industries, capital is slowly replacing labour.

The spinners have covered themselves by not spinning down to limits, but stopping short with the yarn having a greater number of fibres in cross section than the theoretical limit to which it could have been reduced. Each time they tried to speed up the spinning process, they had trouble with breakages. Woollen mills were visited where wool was being processed at only 80% of the speed of synthetic yarn. Over the centuries, the spinners have learnt to live with wool and its imperfections and have devised machinery to do just this. Average Merino wool has certain characteristics, e.g. "coarse edge", and no one really worried about it. In fact, the final report on objective measurement of wool in Australia by the specially appointed Objective Measurment Committee of the Australian Wool Board in October 1972 came out with the profound statement that "variations in fibre diameter of sound fleeces grown within Merino flocks have been found to fall well within acceptable commercial limits".

Of course they do, because the acceptable commercial limits were set to suit the Merino clip, and not vice versa. Why? Because if the limits were set any tighter, there would not be enough wool to process!

In a recent paper delivered by a director of a large woollen mill it was pointed out how very significant a small excessive percentage of coarse fibre becomes, and that it would seem, therefore, that the traditional habit of spinning well above (in terms of numbers of fibres in cross section) the theoretical limit spinners have been protecting themselves against imperfections or shortcomings in wool fibre and not maximizing productivity.

It would appear from overseas evidence that we should at least be paying some attention to the "coarse edge" and giving it some consideration in our selection criteria.

(v)   Ratio of number of secondary follicles to number of primary follicles

Research has shown that fleeces from flocks selected for high Clean Fleece Weight, process better. Also, as CFW increases, so does the S/P ratio. Is there a correlation?

Quoting from Dr. I. Fairnie's address to the W.A. branch of A.S.A.P. in 1973 "High S/P ratios are also correlated with fibre uniformity, i.e., low fibre diameter variation".

(vi)   Ratio of diameter of secondary follicles to diameter of primary follicles

Selection of sheep with little differences in diameter between secondary and primary follicles would also reduce fibre variation (Ryder and Stephenson 1968). This also confirms the work of H.B. Carter who has previously shown a direct relationship between coefficient of fibre diameter variation and ratio of diameter of primary to secondary fibres.

(vii)   Average fibre length

"In general, increased fibre length tends to improve spinnability and strength of the resultant yarn, although there are limits beyond which problems can occur in drawing" and "we have observed increased yarn strength with wool from sheep selected on fleece weight, which is due to the greater mean fibre length of such wool, compared with that of conventionally selected controls". (Dr. M. Lipson).

So much for factors of genetic origin.

FACTORS OF ENVIRONMENTAL ORIGIN

Is there anything we can do in our selection programme which may be of benefit in the areas where the environment has influence on our wool?

1.   Fibre diameter variation along the fibre.

Which animals show the greatest degree of along-fibre diameter variation? Are they the ones with a wide variation in follicle size—not only within the primary group and within the secondary group, but between these groups?

2.   Dust penetration and the subsequent yield loss as excess noil in top-making.

Is this also due to wide variation in fibre diameter; is it a function of the amount of wax in the wool, and could it be related to S/P ratio?

3.   Fleece rot—is this also related to S/P ratio, yield, suint, diameter, etc.

There are quite a number of factors which have not been really related to variables like the environment and processing losses. Future research may show up new areas where animal breeders, in this case the Merino ram breeder, may be able to select to help the processor more profitably handle his fibre.

With these thoughts in mind, it should be possible to devise a breeding plan based on trade requirements and breeders' profitability.

The Breeding plan incorporates

    1.   rams selected by an index.
    2.   artificial insemination.

The current ram selection index for body weight and fleece characteristics is:

$$\frac{W}{D^2} + B\,W + S/P - 2\ \text{S.D. fibre diameter}$$
$$- 2\ \text{S.D. follicle diameter} - 10\frac{dp}{ds} - 2 \times \%\ \text{suint}$$

where: W   =   clean wool weight for 12 months × 1,000.

               D   =   average fibre diameter.

            S/P   =   secondary follicle/primary follicle ratio.

S.D. fibre diam. = standard deviation of individual fibre diameters.

S.D. follicle diam. = standard deviation of individual follicle diameters

$$\frac{dp}{ds} = \frac{\text{diameter of primary follicles}}{\text{diameter of secondary follicles}}$$

       B W = body weight.

The only other selection factor is efficiency of food conversion. Once rams have been selected according to the above index, the 30 top ranking ones are placed in individual feed pens and maintained at zero body weight change, i.e. kept on a maintenance diet for 90 to 100 days. At the completion of this period, the amount of clean wool produced is divided by the amount of food eaten, and animals are thus ranked for efficiency of conversion of food to wool.

This value is then added to the index to give a final ranking. The highest performing rams then go into the A.I. programme.

PROGRESS OF ARTIFICIAL INSEMINATION

Initially, Central Nucleus and a few ram breeding co-operatives used A.I. It was later extended to include the flock members at the third level of the co-operatives.

Year	Ewes Inseminated
1973	19,300
1974	39,200
1975	63,000
1976	?

MARKETING

Once an organized, controlled and performance-tested breeding plan is implemented, the way is then open to produce large lines of a predictable product, e.g. sufficient to supply the total requirements of a mill with, for example, 1,000 bales of wool per week.

Spinners have one basic requirement—a continuous supply of an extremely uniform and predictable top. The whole design of the topmaking process is based around one basic function—to present as nearly as possible, a perfectly uniform top.

Under a large scale Group Breeding Scheme, strictly but democratically controlled, the possibilities for producing large scale uniform products become much more a reality.

Conversely, the demands of a Breeding Plan designed to cover these ambitious goals, may be beyond the resources, the abilities and the ambitions of individuals operating alone, and much organized, professional and joint effort may be required. Whatever the emotional and self-interested screams and yells of traditional Merino stud breeders, the cold hard facts of life are such, that small, isolated and intensely jealous ram breeding enterprises may just not be able to keep up with industry demands.

FUTURE

The Australian Merino Society looks to the future with confidence and a faith built on logic, measurement, assurance and last but not least— trust. Members believe they have a destiny—one which is controllable, constructive and highly creative. They know where they are heading and it will take a lot to stop them. They are indeed a dedicated group of animal breeders.

REFERENCES

Dunlop, A.A. and McMahon, P.R. (1974). *Australian Journal of Agricultural Research*, **25**: 167.

Ryder, M.L. and Stephenson, S.K. "Wool Growth", Academic Press, 1968.

30

STUD MERINO BREEDING

F.M. COOKE
Western Australian Stud Merino Breeders' Association

If one takes a look at the Australian Stud Merino breeding industry, one will find that it is made up of hundreds of individuals operating under an accepted set of breeding principles, but at the same time doing very much "their own thing".

Some have stayed strictly within a major bloodline, others have fused one or more bloodlines, some have scrapped one bloodline and changed to another.

Some breed stud sheep to sell to other studs, some breed to sell only to flock owners and others—about 40%—breed to provide their own flock ram requirements.

On a different theme, some breeders go in for measurement, others use some, and yet others, none. One must add that measurement is not new to the stud industry. Austin has described studs using objective measurement principles well before the turn of the century.

At the Deniliquin Show in 1874 washed wool weights were used as a judging criterion.

ATTITUDES TO CHANGE

Stud men have rarely been hasty however to adopt major changes in contradiction to practical experience.

The great depth of experience going back to John McArthur, Samuel Marsden and others in the late 1860s, underlies much of this caution.

The word Merino in Spanish signifies "fugitive without a regular home".

The fine woolled flocks that had the run of the Spanish country-side for centuries were known as the "Ovegas Merino" travelling sheep.

The Merino was unique to Spain and the much sought after fine wool produced by the Spaniards brought great wealth to the Nobility and the Church that owned the various cabanas or flocks. These flocks developed individual characteristics through differing environmental and management factors that became relatively fixed over the centuries by random breeding from within.

The most famous of the Cabanas were the Escurial or royal flocks, the Negretti, the Infantata, the Paula and the Guadalupe and collectively they founded the great Merino flocks of the world.

The sequence of events leading to this began in 1765 with the Elector of Hanover receiving a gift of 300 Escurials from the King of Spain. Later in 1786 the King of France received 400 Infantatas thus founding the famous flock that exists to this day at the Royal Farm, Rambouillet. The English King George III received Negrettis to establish the Royal flock at Kew in 1791.

Captain Waterhouse later brought the first Merinos into Australia in 1798 which were Escurials from South Africa.

In the USA, Colonel Humphries initiated what was to become the famous Vermont strain with the importation in 1808 of 100 Infantata Merinos.

The Napoleonic war in 1809 destroyed the Merino industry in Spain. The old Cabanas were either sold or slaughtered. Many went to the USA, others mostly Paulas were shipped to England and the balance driven overland by Napoleon to France and Central Europe.

From 1810 onwards the Merino scene shifted to Europe, the USA, and Australia.

In England such great interest was shown in Merino breeding during this period that in 1811 the world's first Merino Society was formed with Sir Joseph Banks as president. Demonstrating that there is little that is new under the sun.

By the early eighteen hundreds the centre for stud Merino breeding was Saxony. The Merino sheep industry had developed to such an extent that in 1802, of the fourteen million sheep on the German side of the Rhine, four million were Merino.

The methodical Germans, had by this time much improved the yields and bodyweights from those of the original Spanish sheep, and the famous Electoral, Steiger and Gadegast flocks of Saxony and Lichnowski of Silesia were to become the foundation blood of the Australian stud industry.

Our industry in Australia began as mentioned earlier, with the introduction by Captain Waterhouse of Escurials from the South African flock of Colonel Gordon.

These sheep led to the development of Australia's two original studs, Camden owned by Colonel McArthur and Burrundulla by the Cox family, Captain W.A. Cox having purchased the Merinos from Waterhouse on his leaving the colony.

McArthur subsequently purchased Negrettis from George III's flock at Kew in 1804. He also pioneered the introduction of Merinos from Saxony. Some of his first importations from the Electoral flock in 1812, were sent on to James Cox of "Clarendon", Tasmania, who was a son of W.E. Cox, of "Burrundulla".

"Clarendon" thus became the first of the famous studs to evolve in that State.

By 1830 there were 10 Merino studs in N.S.W. and 7 in Tasmania, all were based on Merinos imported direct from Saxony, such as was William Riley's famous Raby Stud in 1824.

Others were founded on Camden or Burrundulla ewes and rams, imported from Saxony.

The Saxons reigned supreme in Australia until 1860. The sheepmen concentrated on fineness and purity of fibre. Tasmanian, Victorian and Mudgee studs dominated the ram breeding scene.

The studs provided the flock rams and most stud rams were imported from Saxony with a few known as "Anglo Merinos" from England.

Towards the end of that period the French Rambouillet made its first appearance.

The climatic restrictions applying to fine wool Merinos, mainly in the newly developed plain country of N.S.W. led the Peppin Brothers George and Fred to seek to produce a long stapled, broader woolled sheep that would thrive under much wider environmental conditions and cut more wool. The Peppins founded their stud in 1861 with the purchase of Rambouillet rams which included the sire "Emporer". The premier family that resulted from crossing this ram with Cannally and selected station bred ewes; plus the later establishment of and cross mating with the "Warrior" family founded by the imported Vermont ram "Grimes", led to the development by these remarkable breeders of the now famous Peppin blood.

While the above-mentioned long wool Rambouillet/Vermont cross sheep were under development at Wanganella, significant developments had already begun in South Australia. As early as 1841 John Murray at Mt. Crawford took steps to establish what subsequently became known as the Murray blood. From the original flock consisting of Camden blood ewes and another 100 ewes mated to Tasmanian rams, the famous Murray blood was evolved from continuous selection and breeding from within for over 130 years.

The Mount Crawford sheep were the foundation of many South Australian studs that evolved with the emphasis on strong wool.

It was from such flocks that the Australian strong and medium wool Merino as we know it today evolved.

What type of sheep that went into the original mix, will never be known, except that there is understood to have been an infusion of English Leicester cross from the ewe side to broaden the wool and give more size to the Saxon.

The breeding methods used have changed little to this day and go back to Robert Bakewell. They may be summarized as linebreeding, heavy culling, firm conviction on type and consistency in selection.

Such disciplines, plus an element of luck in the quality of the genetic material that came their way, rewarded a number of breeders with a medium strong wool Merino suited to the arid Australian pastoral conditions. It is from these flocks (but a relative handful of the hundreds that never made the distance) that the great parent studs were evolved as we know them today.

These new Peppin and South Australian strains had barely begun to filter through the existing Australian flock, when the Vermont invasion swept through the established Saxon studs.

THE VERMONT "INVASION"

It began in 1884 when Patrick McFarland brought in "Matchless" from the USA. Weighing over 200 lbs and carrying a fleece of 28 lbs (albeit a yield of only 8 lbs scoured). His purchase was emulated by Samuel Mc-Caughey, then Australia's largest flock owner. These men felt the New Vermont sheep would be the answer to their search for larger sheep, with heavier fleeces than on the Saxons at the time.

The first crosses were a marked improvement. However, instead of judicious infusion of the new blood, they went for broke. "The more wrinkles and grease the better," became the standard and most of Australia's major Merino studs of the period rushed to jump on the bandwagon of the foolish new fashion.

The New Vermonts were most dissimilar to the "Grimes" of the 20 years before. They were a product of American expediency in that they had been developed with excessive wrinkles and excessive grease to fill bales.

The American woolgrower sold his wool on the farm to private buyers at a district price. Quality therefore did not matter, quantity did and the New Vermont filled the bill.

The effect of the New Vermont on Australia's Merino industry was disastrous.

The Vermont breeders introduced extreme corrective mating, grease, wrinkles, and accentuated the growing problem of primary fibres. Wool quality and constitution suffered and the great drought of 1902 marked the end of the Vermont era in Australia.

Australian woolgrowers from then on looked to the Peppins, the Murrays and a few other studs that had resisted the Vermont craze for their breeding base.

These studs continued to line breed their sheep and by 1936 had largely "fixed" the Australian Merino type that still stands today.

Australia's Vermont era served as a lesson to our studs not to hasten into major change.

THE SCIENTIFIC ERA

The next major changes to stud breeding practice were suggested in the late 1950s by two scientists, Dr. Dun from the N.S.W. Department of Agriculture and Dr. Newton Turner from CSIRO.

Having successfully contributed to the elimination of Australia's greatest rural problem, the rabbit, with the discovery of myxomatosis, the CSIRO was naturally an organization held in great esteem by the man on the land.

Thus when Drs. Dun and Newton Turner expressed doubts on the ability of the stud industry to service effectively the needs of the flock owner, their views had to be taken very seriously.

What upset the Merino stud men was the proposal by these scientists that studs should forget what the sheep looked like and go by measurement alone.

No doubt the intention of these scientists was to achieve an improvement in the rate of gain of wool production.

But to the stud breeder it appeared that they had totally disregarded his ability as a Merino breeder and as a business man who had to work within a fiercely competitive marketing framework.

Whereas the scientific disclosures of the late 1950s and early 1960s got a cool reception from the stud industry, they were seized upon by various opportunists who saw the chance to "knock" the stud system.

CO-OPERATIVE BREEDING

That leads me of course to the Australian Merino Society whose founder, Mr. Jim Shepherd is a contributor to this book.

Mr. Shepherd has indicated that the rams that have been used over the Central Nucleus ewes are from his Bungaree stud.

He also indicated that the progeny of the nucleus ewes are inferior to the stud Bungaree progeny born on the same farm and run under the same conditions.

I could suggest that the whole system would work better if the Central Nucleus was scrapped and Bungaree ram hoggetts were used in place of these apparently inferior Central Nucleus-born rams.

But I would be guilty of the same lack of insight that scientists have been displaying. I would be forgetting that ram breeding is a business.

The Australian Merino Society ram breeding business functions in a way that is quite different from the stud breeding business.

Their ewe contribution system is an integral part of their ram marketing because the contribution of ewes leads to an involved and committed clientele.

The A.M.S. must be seen for what it is—an extremely effective marketing system based on co-operative emotionalism.

BUSINESS BASIS

Since 1960 the Australian Stud Merino industry has learnt to live with scientists—and they more and more so with the professional sheep breeders.

With the passing of every year, through their own work with the sheep, and their contact with commercial flocks, scientists' increasing experience draws them to the desirable situation of having more and more in common with the Merino studmaster. They have come to realize the stud Merino industry is complex and that there is more to it than breeding methods alone.

It is a highly competitive, laissez-faire industry where marketing and public relations play a major role. Like any business, it can only exist if it makes a profit.

It is just as valid for a sheepbreeder to market his offering in show condition as it is the used car dealer to polish his cars or the supermarket to place the milk at the back of the shop!

They do these things for sound business reasons.

Science has the great advantage of objectivity. Herein lies its usefulness to our industry: to pry and probe, experiment, to disseminate information and be the fire engine for emergency research. We see that as their role in the ball game, but not that of calling the shots.

Self-appointed experts have been wrong in the past no matter what the field, and too much is at stake for breeders to blindly follow any "Pied Piper" of science.

However, breeders would be similarly foolish to ignore science.

It is acknowledged that science has largely been behind all the progress in agriculture in response to needs that have arisen with the passage of time. In Australia, where would we be today without myxomatosis, trace elements, vaccines, organophosphates, fertilizers, the mules operation, new pasture species and the broad spectrum drench—most of which are great contributions by science.

Objective measurement likewise, fills a need that has arisen in the sheepbreeding industry.

There are simply not enough subjective experts to get around a national flock of 150 million sheep. They are also expensive—yet a discipline must be imposed on flock management if the national flock standard is to come close to that of the Merino studs.

The time is fast approaching where sheep that cut less than the national average of about 4 kg will not be worth shearing.

Science's guidelines of culling with the use of scales in the shearing shed in conjunction with the woolclasser, is a very effective answer to this problem. A great side benefit also is that the method *involves* the flock owner, which makes him more conscious of how his flock performs and results in greater interest, leading to better management.

With the encouragement of the various State departments of agriculture, many stud breeders use objective measurement for sire selection—and as a marketing aid. Experience has shown that though micron indicators are prone to variation due to environmental changes and age of the sheep, they do indicate the position of the animal in relation to the average. Measured clean fleece weight is of significance with sire selection and has great influence on what sires a breeder decides to keep and what to sell.

Most important of all, such information serves the studmaster as his record of progress (or otherwise) over the years.

It is important to enumerate the objectives and current breeding practice as it applies in the stud Merino breeding industry of Australia today.

In 1975 there were over 2.5 million stud Merino sheep in Australia belonging to 1,623 registered studs, which in turn delivered to flock owners over a quarter of a million rams in that year.

These stud flocks may be generally grouped into four distinctive bloodlines.

The Saxon (17-20 micron) the Peppin (20-22 micron) Collinsville (21-23 micron) and the Bungaree (23-25).

AUSTRALIAN BLOODLINES

The bloodlines occupy sections of the Merino wool micron spectrum and the latter ones were evolved from the Saxon to meet the specific needs of various environments, roughly in the order of their severity.

Accordingly, the Saxon is found in areas of high rainfall and mountain conditions, mainly in Victoria, Tasmania and the "Divide" country of N.S.W.

The Peppin and Collinsville serve the more favoured pastoral and agricultural regions and the Bungaree in areas of greater stress from low rainfall and where the annual summer drought is longest.

The Australian Merino flock register exists as a record of the background of the various Merino bloodlines and strains as well as a history of the studs involved. It also functions as security for the investment and asset value of pedigree sheep and as a patent record of the performance of those that are successful.

Generally studs develop to become strains within bloodlines from which families are formed through line breeding and then outcrossing with parallel families developed either from within the stud or with purchases from other studs.

The performance of studs varies considerably depending on the quality of genetic material with which they started and from subsequent purchases. The skills involved with general management and selection procedures also contribute to variation between studs.

The responsibility of the studs is to set a standard of excellence that directly relates to economic importance and to aim for bloodlines that breed true to type.

This is an extremely tall order, but constant perseverance in this direction has enabled a great number of studs to market rams that their clients know from experience will breed according to their like.

For a stud the key to the exercise is to develop or acquire sires of desirable phenotype that are relatively homozygous for desirable attributes. This takes time, effort, perseverance, talent and money, plus more than a grain or two of luck.

Having on progeny test found such an animal, a family within the blood can be developed and concentrated by line breeding (herein lies great scope for A.I.). This together with heavy culling and perseverance with type leads to the goal of all stud breeders—to reduce ancestry and genetic scatter to a minimum and produce a strain of Merino that breeds uniformly for the characteristics they are seeking and that will in turn pass such characteristics on to the flock owner.

It is to this area that genetics gives a commonsense interpretation of some of the more difficult problems of animal breeding. For example Mendelism gives the explanation why animals of similar appearance do not necessarily breed true. Only subsequent breeding performance will decide what genes the creature is carrying. Similarly it explains why the performance of similar strains may vary so greatly, also, why crossed strains tend to breed back to ancestral strains.

It certainly explains why uniformity can only be achieved by line breeding and the use of tested sires which in turn requires so much consistency and patience. It underlines the importance of a sire that performs genetically. They are comparatively rare, and extremely valuable animals and when they occur the objective must be their widest possible use. Here again science can render great assistance with the practicality of semen storage.

REALIZING ON POTENTIAL

At this stage it may be appropriate to inject a little controversy with a declaration that closed parent studs have gone about as far as they can go with Merino sheep breeding. After all, many of the British breeds of sheep have changed very little since 1830 and like the internal combustion engine the situation at the top end of our stud flocks has largely matured. On that plane the breeder's major job is to maintain the levels that have been achieved. Stud and flock situations that are trailing must catch up to the potential of the leaders. This is possible and what a giant stride it would be in itself.

It is all very well to work for better production figures for dairy cattle, laying hens etc., because any environmental deficiencies can be artificially remedied. They are not expected to have to survive in a drought. We do not want to develop a Merino that needs assistance at the first sign of adversity. It is this area that can give rise for concern with current practice in our industry.

The tremendous competition that exists is driving more and more of our studmasters into artificial environments which could lead to strains being developed that will not perform in the natural environment.

STUD CRITERIA

To conclude one must explain what are believed to be the desirable characteristics that stud breeders aim for in breeding.

One such characteristic still used by the studmaster but for which he is often roundly criticized is constitution.

Constitution, of course means carriage and outlook—the subjective assessment of the animal's ability to thrive and produce thrifty stock. Body weight measurement is a good guide to constitution, as unthrifty animals weigh less than the average. However, it remains for the subjective assessment to weed out off-type sheep that would pass on bodyweight, but fail in skeletal shape, hocks or dropped pasterns to name but a few.

Having passed the constitution, the conformation hurdle, the sheep unlike most other animals is only half way there. It must pass in covering as well.

Desirable wool characteristics are fleece weight, length of staple, style, staple formation, yield, evenness of covering, and absence of any coloured fibre.

It is the strict adherence of the studmasters of Australia over the years to standards set for covering on Merino sheep that has won Australia its premier place in the world as a producer of quality wool. Australian wool stands out in the world's market place for its handle, spinnability, yield, and its absence of coloured fibre.

The first thing the stud breeder has instilled in him during his apprenticeship is the importance of how to "read wool that is warm and living on the sheep's back". The remarkable coincidence is that what the studmaster looks for is also what the wool user wants despite the average breeder's lack of knowledge of the end user's needs.

What the studmaster seeks is a bulky covering of wool that experience has taught will resist the ravages of the environment. This is a wool that has to be "stiff" to resist weathering and dirt penetration.

To be "stiff" it has to have a well defined crimp. The greater the development of this crimp definition the greater the apparent bulk of the fleece. This assists the surface of the fleece to remain sealed.

Hand in glove with definition comes handle which among other things indicates that sufficient wax is present to protect the fibre all the way up to the tip and also contributes to the sealing of the surface of the fleece.

Handle is also the assessment of the degree of softness and elasticity. Any deviation towards poor handle is rejected by the studmaster as a breeding fault.

Lock, the other important stud criterion is the subjective indicator of the number of fibres per unit area of the skin. Experience has shown that better lock or staple formation leads to improved fleece weight and structure.

With such a fleece structure as has been indicated, the minimum amount of yolk is needed to service the fleece. Experience again has shown that excessive yolk attracts the blowfly and lowers the thriftiness of the animal.

The result is that well bred wool in Australia is a gutsy high yielding product of great style, handle and colour throughout the micron range, and generally spins better than crimp indicates.

To many scientists these criteria are outmoded, but should still be the subject of further research.

Scientific evidence still contradicts actual wool buying practice as spinners in both the U.K. and Japan have indicated.

The question that remains is: "Can a stud breeder afford to deviate from these traditional principles without irrefutable evidence that these principles are wrong?"

Or is it a safer alternative to maintain the status quo and at the same time, use objective measurement as an aid to selecting higher performing animals?

31

THE USE OF FLEECE MEASUREMENT IN SIRE SELECTION BY WESTERN AUSTRALIAN STUD BREEDERS

B.R. BEETSON

Western Australian Department of Agriculture, Australia

SUMMARY

In 1974, 107 registered Merino studs using the Western Australian Department of Agriculture's Flock Testing Service were contacted in a mail survey to assess the use being made of fleece measurement in sire selection.

There were 57 respondents of whom 30 (53%) were achieving selection differentials for clean wool weight which were at least 60% as high as those which could have been achieved if clean wool weight was the only selection criterion.

Evidence is presented which indicates that 18% of rams sold by registered studs in Western Australia are produced by breeders who apparently base their selection programmes on objective measurement.

INTRODUCTION

In the selection of sheep for wool production visual appraisal varies from 30% (Riches and Turner, 1957) to 50% (Robertson, 1973) as effective as objective measurement.

Despite many Australia-wide extension programmes designed to encourage sheep breeders to adopt programmes based on objective measurement, a recent Australian Workshop on Improvement of Genetic Progress in Sheep Production (Anon., 1975) resolved that,

> ". . . the recommended breeding programmes have achieved very limited acceptance by Merino breeders."

For many years the Western Australian Department of Agriculture has been involved with such extension programmes. The Department began its Flock Testing Service in 1958 to provide measurements of clean wool weight and (later) fibre diameter for inclusion in Merino studmasters' selection programmes.

Only 20% of registered Western Australian Merino stud breeders were using the service in 1973. However these breeders sell 40% of rams sold by Western Australian registered studs (Beetson, 1974).

While these statistics show that the Service is being used by a considerable portion of the industry, they do not indicate how effectively it is being used in sire selection. The study reported in this paper was designed to provide information on the extent to which fleece measurement results are being used in sire selection.

257

MATERIALS AND METHOD

In 1974 a questionnaire (Appendix 1) was mailed to all (107) studs which requested to use the Flock Testing Service that year. The questionnaire form was accompanied by a stamped envelope addressed to the Flock Testing Service. The studmasters were asked to list the identification numbers of each sire they selected for breeding (within the stud) from among the rams fleece measured each year from 1969 to 1973, inclusive.

For each stud the selection differential for clean wool weight (average clean wool weight of selected sires – average clean wool weight of group from which sires were selected) was calculated for each group of selected sires. Sires selected in any one year were treated as a group. If it was indicated that sires were selected from within families, then they were treated as separate groups each year.

The maximum selection differential which could have been achieved by selecting an equivalent number of sires with high clean wool weight as the only criterion was also calculated. Then, by expressing the achieved selection differential as a percentage of the maximum possible selection differential, a measure of "efficiency of selection" (Riches and Turner, 1957) was obtained for each group of sires.

Finally, an "average efficiency of selection" (AES) was calculated for each stud using the following formula:

$$AES = \frac{\sum\limits_{i=1}^{r} \dfrac{N_i \, D_i}{M_i}}{\sum\limits_{i=1}^{r} N_i}$$

where AES = Average efficiency of selection

N_i = Number of sires selected from the ith group ($i = 1, r$)

r = Number of groups

D_i = Selection differential achieved in selection of sires from the ith group

M_i = Maximum selection differential which could have been achieved in the ith group.

The percentage of rams measured which were retained a sires was also calculated for each stud.

The number of rams sold by each stud in 1972 and 1973 was recorded from the Australian Stud Merino Flock Registers (Anon. 1972, 1973).

No attempt has been made to sample studs which failed to reply.

RESULTS

Of 107 questionnaires sent, 61 (57%) were returned. Three of those returned were discarded when measurement data could not be found for the sires listed and one when the respondent stated that the information requested was not at hand.

The remaining 57 replies were listed in Table 1 in descending order of AES. Other data presented in Table 1 are the percentage of rams measured which were retained as sires, the number of rams selected for which measurement data was found and the number of rams sold in 1972 and 1973.

TABLE 1

SIRE SELECTION AND RAM SALE DATA FOR INDIVIDUAL STUDS

AES ranking	AES range	Sires selected		Rams sold	
		Number	Per cent	1972	1973
1		8	40	50	48
2		4	14	231	246
3		4	9	169	163
4		6	2	80	60
5	80-100	5	18	—	—
6		19	21	—	—
7		21	14	54	161
8		4	8	58	89
9		3	6	340	390
10		10	5	24	—
11		66	21	96	109
12		7	11	127	187
13		29	4	184	136
14		3	43	38	25
15		41	7	53	20
16		76	17	56	99
17		16	16	661	555
18		18	4	463	610
19		2	4	70	57
20	60-79	3	6	486	473
21		10	8	142	161
22		5	8	10	29
23		11	3	96	83
24		20	15	110	110
25		14	11	152	231
26		88	9	222	436
27		4	17	90	110
28		27	14	90	84
29		12	23	180	172
30		2	3	160	125
31		37	9	35	42
32		28	23	20	22
33		12	12	320	348
34		261	5	—	66

TABLE 1 *(Continued)*

AES ranking	AES range	Sires selected		Rams sold	
		Number	Per cent	1972	1973
35		4	3	40	56
36		77	21	1,548	1,120
37		34	11	285	400
38	40-59	22	64	196	222
39		4	4	134	135
40		8	4	107	135
41		16	5	332	339
42		4	2	105	120
43		14	18	30	37
44		227	14	535	480
45		4	8	42	46
46		10	9	30	42
47		6	29	—	—
48		80	7	227	369
49		11	10	25	20
50		4	9	165	235
51	<40	10	20	584	827
52		0*	0	59	73
53		0*	50	32	33
54		6	12	105	118
55		11	11	58	132
56		0*	100	142	164
57		13	100	124	115

* No data were available but each respondent stated that selection was not based on objective measurement.

The frequency distribution of studs in relation to broad AES categories is shown in the upper histogram in Figure 1. The lower histogram in Figure 1 shows the relation between AES category and ram sales in 1972 and 1973.

A little over half (54%) of respondents had AES greater than 60% but between them they sold less than half of the total rams sold by all respondents in 1972 and 1973 (46%).

DISCUSSION

In this study, breeders have been assessed in terms of the efficiency of selection of sires after fleece measurement. While this measure of AES can stand as an objective assessment of a breeder's use of fleece measurement results, it cannot stand on its own as a measure of genetic progress in his stud.

It is possible for a breeder to attain a very high AES yet be retaining as sires a large proportion of the rams measured and be achieving a very small selection differential for clean wool weight (see Table 1, Studs 1 and 14).

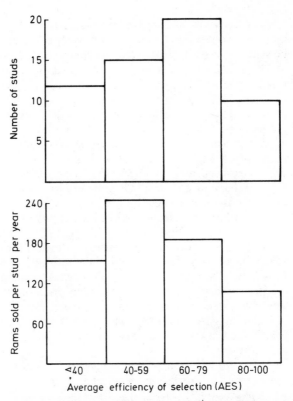

Figure 1—Frequency distribution of studs and ram sales [†] per stud in each AES group

† Australian Stud Merino Flock Registers 1972 and 1973.

On the other hand it is possible for a breeder to attain a very low AES and yet be achieving large selection differentials for greasy wool weight before fleeces are sampled for measurement (Stud 36 in Table 1—during conversation with the author the studmaster has stated that he places heavy selection emphasis on greasy wool weight when choosing potential sires which are then fleece measured).

A third example might be the breeder who achieves a high AES for a group of rams which only account for a small proportion of the total number of replacement sires (see Table 1, Stud 9).

Because of such anomalies, it is probably safer to view AES assessment in groups of studs as an overall measure of the use of fleece measurement.

It is encouraging that 30 breeders (53%) among those who replied to the questionnaire are achieving AES greater than 60%. Even the 15 breeders in group 3 who achieve 40 to 59% AES are probably making worthwhile genetic gains at this final stage of sire selection.

However, without knowing how much selection pressure is being applied to greasy wool weight prior to fleece sampling and measurement little can be said about likely genetic advancement in these studs.

There seemed to be a tendency for those studs with the highest AES to be smaller (Figure 1). However the highest AES group still averaged 108 rams sold per year in 1972 and 1973 which is nine more than the average number of rams sold by each registered Merino stud in Western Australia per year in the same period.

Mention should be made of breeders who did not reply to the questionnaire. Flock Testing Service records show that 22 of these breeders have been measuring more than 50 samples regularly for more than 10 years. Eight of these have indicated in conversation with Sheep and Wool Officers of the Department of Agriculture that they place heavy emphasis on clean wool weight in sire selection. The total number of rams sold per year by this group of eight breeders in the 1972 and 1973 season was 4,848.

When this subjective information is added to the data from respondents achieving an AES of 60% or more, the resulting evidence suggests that approximately 18% of rams sold by Western Australian Merino stud breeders are produced by studs whose breeding programmes are based to a considerable extent on objective measurement.

There may be other studs who did not reply to the survey or use other testing facilities who also base their breeding programmes on objective measurement.

These results are surprising in view of the following statement made in a report from the workshop mentioned earlier (Anon., 1975).

"The Workshop recognized that recommended breeding programmes have largely been ignored by the sheep industry . . ."

ACKNOWLEDGEMENTS

The author wishes to thank Dr D.E. Robertson for his advice and criticism. Thanks are also due to B. Growden who helped collate the data.

The Australian Wool Corporation is gratefully acknowledged for its financial support during the early years of the Flock Testing Service.

REFERENCES

Anonymous (1973) *Australian Stud Merino Flock Register* (The Australian Association of Stud Merino Breeders: Sydney).
Anonymous (1974) *Australian Stud Merino Flock Register* (The Australian Association of Stud Merino Breeders: Sydney).
Anonymous (1975) *Animal Production Committee:* Workshop on Improvement of Genetic Progress in Sheep Production. Sydney, December, 1974.
Beetson, B.R. (1973) *Journal of Agriculture, Western Australia* **14** : 220.
Riches, J.H. and Turner, Helen N. (1955) *Australian Journal of Agricultural Research* **6** : 99.
Robertson, D.E. (1975) *Journal of Agriculture, Western Australia* **14** : 191.

APPENDIX 1

Dear Sir,

FLOCK TESTING SERVICE

To help us assess the use of fleece measurement in ram selection would you kindly list at the bottom of this page each ram you selected as a sire from the group which were fleece measured in 1973. If you have records of the rams selected in previous years that information would also be helpful.

Please use the identification number corresponding to the one in the fleece measurement records. This sheet can then be posted in the enclosed stamped and addressed envelope.

Your assistance will be appreciated.

Yours faithfully.

(H.G. Neil)
Officer-in-Charge,
Sheep and Wool Branch.

Identification Number of Each Sire Selected

Flock	1969	1970	1971	1972	1973

32

FACTORS INFLUENCING THE EFFICIENCY WITH WHICH MERINO STUDS SELECT FOR INCREASED WOOL PRODUCTION

P.H. SAVAGE and B.J. McGUIRK
Department of Agriculture and CSIRO, N.S.W., Australia

SUMMARY

Surveys have been made of selection practices in New South Wales studs to obtain information on specific factors which influence the effectiveness of selection programmes for increased fleece weight.

In a sample of 53 Daughter and General studs, an average of 84% of sires were home-bred. 35% of studs bred all sires used in 1975. In this sample of studs, plus an additional 8 Parent studs, reserve rams constituted an average 16% of all rams classed as reserve or flock rams. In only 8% of studs was the selection of reserve rams made with the aid of production records. In the majority (78%) of studs surveyed, rams were first used as four-tooths, and in 94% of studs, rams were not cast at a set age.

Fleece measurement services have been used by only a minority of studs in the State in any one year, with a maximum of 15% in 1964. Most studs currently using the services would appear to test at least all of their reserve rams.

These findings are discussed with reference to extension programmes designed to foster the use of objective measurement as an aid to sheep selection.

INTRODUCTION

Merino studs currently provide the majority of commercial flocks rams in the industry. While precise estimates are not available as to the percentage of flock rams that the studs provide (Short and Carter, 1955), a figure as high as 80% is not unrealistic.

It is widely appreciated that in a breed so structured, with a separate ram-breeding stud sector, the long term rate of genetic improvement is determined by the selection and breeding practices of the studs. However, apart from the studies of Short (1962) and Dolling (1970), little has been done to describe these practices, and the present study was initiated to rectify this deficiency. The study forms part of an extension programme of the New South Wales Department of Agriculture, designed to foster the adoption of recommended breeding programmes and the use of objective measurement as an aid in sheep selection.

MATERIALS AND METHODS

(a)   The Merino Stud Industry

The Australian Stud Merino Register for 1973 (Volume 50) lists 1,649 studs in Australia, with a total of 1.25 million ewes. This figure represents an estimated 2.5% of all Merino ewes annually mated to Merino rams (Roberts, Jackson and Phillips, 1975). In 1972 there were 515 studs in New South Wales including nine of the 16 parent studs of the breed. The total Merino stud ewe population in the State was 400,170, with an average of 777 ewes per stud.

(b)   Scope of the Study

This study attempted to identify factors influencing likely rates of genetic gain for wool production within the stud section of the Merino industry.

Currently recommended breeding programmes for Merino studs stress the importance of using information on clean fleece weight and average fibre diameter when choosing replacement sires. To ensure a rapid rate of progress, superior stock should be identified and used at an early age (generally as hoggets) and generation lengths should be short. In addition, within the hierarchy of Merino studs (Short and Carter, 1955), progress in Daughter and General studs is also affected by the proportion of sires they purchase (McGuirk and Savage, in preparation).

With these points in mind surveys have been conducted among Merino studs in New South Wales to assess:

   (i) The proportion of rams used in studs which are home-bred.
  (ii) The proportion of young rams which are considered as potential sires and the extent to which fleece measurement information is used in the initial screening or final selection of these rams, and
 (iii) The age at which breeding stock are first used and the average age of rams and ewes in the breeding flock.

Answers to most of these questions were obtained in face-to-face interviews with stud breeders. To date information has been obtained on 61 studs; ultimately 200 will be surveyed. The present sample comprises 8 Parent, 28 Daughter and 25 General studs, and all data obtained relate to the 1975 joinings. More detailed information on age structure of breeding flocks has been obtained for a small number of studs.

Fleece measurement laboratories provided information on the extent to which their services were used by New South Wales studs in the period 1954-1971, with some laboratories providing more recent results. Data were obtained on the average number of rams tested and their age at testing.

RESULTS AND DISCUSSION

In the Daughter and General studs surveyed, the average proportion of home-bred sires being used was 84%. However, the proportion of home-bred sires increased with the size of the stud, so that of the 2,527 sires used in

1975, 95% were home-breds. All sires were home-bred in 35% of the Daughter and General studs, a very similar result to the 32% reported by Connors & Reid (1976) in an Australia-wide survey in 1972. In the present sample of studs, those breeding all of their sires are larger than average with a mean of 3,900 relative to an overall mean of 2,400 ewes for all Daughter and General studs surveyed.

Traditionally, young rams are classed into three grades; culls, flocks rams and reserves. This last category includes potential sires as well as rams considered acceptable for showing and sale as stud or special flock rams. Classing of young rams is usually done when they are in mid- or full-wool. Of the 61 studs surveyed, only 8 had production information available to assist in the identification of reserves. On average, 16% of saleable rams were classed as reserves, with a range of from 6 to 31% for individual studs.

Fleece measurement information on stud rams has been obtained by at most 15% of New South Wales studs in any one year, in 1964. In 1971, the most recent year for which complete information is available, the figure was 11% of studs. Only 63 studs in New South Wales, 12% of all studs in the State in 1971, sent samples in at least 5 of the 10 years in the period 1962-1971, so that the number of consistent users of fleece measurement services was small.

Of the 61 studs surveyed, 54 had used fleece measurement at some time and 36 were current users. Of these, two-thirds tested as a minimum all of their reserve rams, as hoggets.

The majority of studs first used their rams as four-tooths rather than as two-tooths (72% versus 28%). However in the former group of studs rams of exceptional quality are sometimes used as two-tooths.

On average 39% of sires in a stud were being used for the first time. Only 6% of studs cast rams at a set age, most studs rejecting rams when their physical condition or wool quality/production had deteriorated to an unacceptable level. This situation is similar to that described by Short (1962) for a Parent stud.

Detailed information obtained on the age distribution of rams used in six studs followed the same policy of not casting rams at a set age. The mean average age of rams in these studs, at lambing in 1975, was 4.3 years.

On average, hogget ewes made up 22% of the breeding flock in the surveyed studs. In the same six studs for which ram age information was obtained, the mean average age of ewes in the breeding flocks was 4.4 years.

CONCLUSION

It is generally accepted by research and extension workers that recommended breeding programmes for improving wool production in the Merino have largely been ignored by the Merino industry. This assessment is supported by the present results, which indicate that only a small proportion of studs can be regarded as consistent users of fleece measurement. In addition, generation lengths in the studs are much greater than optimal in a fleece weight selection programme, where hogget wool production is a reliable guide to an animal's breeding value. There is little evidence that the industry as a whole uses progeny test performance for wool production, for example, to indicate which rams should be culled.

In promoting the adoption of recommended breeding programmes by the Merino industry it is important that extension and research workers appreciate that the studs already have well-defined selection procedures. Some of these practices hinder the use of objective measurement as a selection aid, such as the practice of selling flock rams in full-wool. This prevents any check on the reserve rams' superiority on greasy fleece weight and also prevents any ranking of flock rams on other than a visual assessment of merit. In other words it will often be necessary for studs to alter their shearing times if objective measurement is to be used in the screening and ranking of rams.

With the current costs of fleece testing it is unrealistic to expect studs to test all rams both flock and reserves, for yield and average fibre diameter. It would be difficult to justify such extensive testing, given that flock ram ranking or selection policies can only influence short-term gains in commercial flocks (McGuirk, 1976). At the same time it is important that rams classed as reserves are clearly superior to the flock average on fleece weight. If this is true for greasy fleece weight then the subsequent fleece testing of only reserve rams would seem to offer a sensible compromise between maximal genetic gains and economic commonsense.

ACKNOWLEDGEMENTS

We wish to thank Professor E.M. Roberts for permission to publish information from the Flock Testing Service, University of New South Wales and Professor Roberts and Dr. N. Jackson for their help in the collation of this material. The New South Wales Department of Agriculture's project on "The Extension of Objective Measurement to the Merino Stud Breeding Industry" is supported by the Australian Wool Corporation.

REFERENCES

Connors, R.W. & Reid, R.N.D. (1976) *Proceedings of the Australian Society of Animal Production* **11** : 9.
Dolling, C.H.S. (1970) "Breeding Merinos" (Rigby : Adelaide).
McGuirk, B.J. (1976) *Proceedings of the Australian Society of Animal Production* **11** : 13.
Roberts, E.M., Jackson, N. and Phillips, Janice M. (1975) *Wool Technology and Sheep Breeding* **12** : 6.
Short, B.F. (1962) *Australian Journal of Experimental Agriculture and Animal Husbandry* **2** : 170.
Short, B.F. and Carter, H.B. (1955) C.S.I.R.O. Bulletin No. 276.

33

THE COOLAH HIGH FERTILITY MERINO FLOCK — A CO-OPERATIVE GROUP BREEDING SCHEME

B. EVANS
Cobbora, New South Wales, Australia

SUMMARY

The Coolah Merino Flock is a commercial ram co-operative breeding scheme selecting for high fertility and fleece weight while maintaining wool quality and body weight.

The success of this venture as a commercial enterprise is shown by the willingness of the members to continue to support it.

INTRODUCTION

This Group evolved from work being undertaken at The University of New South Wales, Roberts and Scott (1975), where Merino flocks had been selected for high fertility for a number of years with encouraging results. Having reached this stage in the experimental programme it was considered necessary to extend the programme to commercial conditions.

METHOD

(a) Ewes donated to Central Flock

In February, 1971 a group of animal producers in the Coolah area each donated a number of ewes to a Central Flock which was to be managed by a member of the group. The initial donation totalled 98 ewes while additional ewes have been donated to the Central Flock each year since.

Before any ewe was accepted into the Central Flock it had to meet certain standards. Firstly a ewe must have had twins at its maiden lambing and have wool measuring 23 microns or less. Secondly each ewe satisfying the initial requirements was classed for obvious wool or anatomical defects. Except for such obvious physical defects, no subjective classing was performed.

Each ewe was tagged in both ears with a metal tag showing year of drop, identification of donor and an individual number.

(b) Ewes born in Central Flock

As with the donated ewes, each ewe born in the Central Flock had to meet the set standards before she was allowed into the Central Flock. Ear tags for Central Flock born ewes showed year of drop, an individual number and identification that they were born in the Central Flock.

(c)　Sires for Central Flock

In 1971 and 1972 two rams from the University of N.S.W. flocks at Hay, were used as sires. In 1973 the first Central Flock-bred rams were used in conjunction with the Hay rams, while since the 1975 joining only Central Flock-bred rams have been used.

(d)　Storage and manipulation of data

All lambing and individual data records for the Coolah Group have been collated and stored in a computer.

Reports produced by the computer have been used each February/March in the culling of ewes and selection of sires for the Central Flock and members' flocks.

(e)　Culling of ewes on lambing percentage

After its third and subsequent lambings each ewe was evaluated in respect to its lambing performance. If a ewe did not maintain a lambing percentage of 160% or greater it was culled (lambing percentage is lambs born divided by number of lambings expressed as a percentage).

(f)　Evaluation of rams

Each year the eighteen month old rams were evaluated for wool and body characteristics as well as for their dams' lambing ability.

Recorded data are used to calculate an index on which each ram is ranked.

$$INDEX = a\ CFW + b\ FD + c\ BW + d\ LP$$

where CFW = clean fleece weight

FD = fibre diameter

BW = live weight

LP = dam's lambing performance

a,b,c, and d are the respective weightings for each of the four characteristics.

The weightings were calculated originally from figures published by Turner and Young (1969) but as knowledge of the flock increased it was decided to alter the emphasis on certain characteristics. The emphasis on clean fleece weight and dam's lambing performance was increased while that upon live weight and fibre diameter was greatly reduced.

(g)　Selection of sires

The top four rams on index were used in a syndicate joining for the Central Flock. The principle was to replace half of the rams each year with hogget rams.

At the rate of one ram for four donated ewes in the Central Flock, each member selected his entitlement from the remaining rams.

Any rams not selected were available for sale to the members or other interested persons.

RESULTS

The lambing results for the Central Flock are presented in Tables 1 and 2.

TABLE 1

LAMBING RESULTS FOR NON-MAIDEN EWES IN THE CENTRAL FLOCK

| | Year of Drop | | | | |
	'71	'72	'73	'74	'75
Number of ewes joined	92	93	71	116	113
Lambs born/ewes joined (%)	114	145	156	136	142
Lambs marked/ewes joined (%)	98	131	117	128	119
Ewes twinning/ewes joined (%)	17	49	62	38	44

TABLE 2

LAMBING RESULTS FOR MAIDEN EWES BORN IN THE CENTRAL FLOCK

| | Year of Drop | | |
	'73	'74	'75
Number of ewes joined	41	37	51
Lambs born/ewes joined (%)	117	105	127
Lambs marked/ewes joined (%)	71	103	122
Ewes twinning/ewes joined (%)	34	19	35

Where Number of ewes joined = number of ewes joined and alive at lambing.
       Lambs born   = lambs born alive and dead.
       Lambs marked   = lambs alive at 3 weeks.

DISCUSSION

As this venture is a ram breeding co-operative, the following describes the comparative performance of the rams being produced.

Five rams were chosen as sires for the Central Flock in 1976. The method of description will entail considering each of the characteristics of the five rams as shown in Table 3 and explaining why they were selected.

TABLE 3

SELECTION DATA FOR 1976 SIRES

Ram Tag	Clean Fleece Weight (Rank)	Fibre Diameter (Microns) (Deviation from Mean)		Live Weight (Rank)	Lambing Performance (Rank)	Index
74-038C	2	22.7	-0.1	3	2	95.4
74-041C	3	23.9	1.1	6	2	93.7
74-089C	6	22.5	-0.3	1	3	94.4
BP-24	1	23.9	1.7	5	2	79.1
RP-41	3	20.0	0.6	1	1	81.2

The 4 year average fibre diameter of the ram lambs measured at approximately eighteen months, was 21.6 microns. The range was 19.4 to 22.8 microns.

(a) Ram 74-038C

Ram 74-038C was chosen as the top ram of the 1974 drop because of the following factors.

Its fibre diameter of 22.7 microns (0.1 microns below the average) was acceptable while its dam's lambing performance of 200% was equal second on rank. The dam of 74.038C had produced 12 lambs in 6 years. It was ranked second and third for clean fleece weight and live weight respectively.

This ram was not ranked first for any characteristic but its very high placing for all characteristics made it the best overall ram on index.

(b) Ram 74-041C

Ram 74-041C had a fibre diameter of 23.9 microns (1.1 microns above average) and a dam's lambing performance of 200% giving it a ranking of equal second for fertility.

On fibre diameter and dam's lambing performance this ram is of equal quality to the top ram 74-038C, but its ranking of third and sixth for clean fleece weight and live weight respectively put it slightly behind on index.

(c) Ram 74-089C

Ram 74-089C had a fibre diameter of 22.5 microns (0.3 microns below average) but was inferior to the above rams with a dam's lambing performance ranking of third. This ram's dam produced eleven lambs in six years. Even though it had top ranking for live weight its ranking of sixth for clean fleece weight put it behind the other two rams.

(d)   Ram BP-24

Ram BP-24 was a 1973 drop ram which was used as a sire in 1975.

At the time of its 2-tooth selection in 1975 this ram had a fibre diameter of 23.9 microns (1.7 microns above the average) while it ranked second on dam's lambing performance. It was ranked first and fifth on clean fleece weight and live weight respectively.

BP-24 was considered the top ram of the 1973 drop rams and so was held over as a sire with the 3 younger rams.

(e)   Ram RP-41

In 1973 RP-41 at eighteen months of age was chosen as the top ram of the 1971 drop ram hoggets.

A fibre diameter of 20.0 microns (0.6 microns above average) ranked first for both dam's lambing performance and live weight and ranking third on clean fleece weight placed this ram clearly as the superior ram of the drop.

RP-41 had been used as a sire for several years and even though a superior ram, it was considered that there was enough of this ram's blood in the flock to exclude it from use again. For this reason it was kept as a reserve only.

CONCLUSION

The whole operation of the Group is a combination of objective measurement and practical sheep judgment as exercised by the members. This can be seen from Table 3 where index was used as a guide to a ram's worth but the members used their practical experience in deciding that an animal of slightly lower index was the better ram because of an outstanding value for a certain characteristic e.g. clean fleece weight.

REFERENCES

Roberts, E.M. and Scott, R.F. (1975). *The Journal of the Australian Institute of Agricultural Science* **41** 56.

Turner, H.N. and Young, S.S.Y. (1969). "Quantitative Genetics in Sheep Breeding" (Macmillan: Melbourne).

34

THE DEVELOPMENT OF CO-OPERATIVE BREEDING SCHEMES IN NEW ZEALAND

A.L. RAE

Massey University, Palmerston North, New Zealand

SUMMARY

The development of co-operative breeding schemes in New Zealand is described briefly. At present, most of the schemes are in the Romney, Coopworth and Perendale breeds where the important objective is to improve reproductive rate. The common element in the operation of these schemes is the identification of high-producing ewes from different flocks and bringing these animals together to form a central flock which is usually maintained under an "easy care" system of management and supplies rams to the contributing flocks.

An examination is made of the elements in the New Zealand sheep industry which have assisted in the development of these schemes. These include the fact that lamb production is the most important trait to be improved, that the farming systems adopted make it relatively easy to identify highly prolific ewes in commercial flocks and that an interest in breeding for increased productivity had been generated among breeders and commercial producers.

Some discussion is devoted to the theoretical and practical aspects of implementing selection plans in co-operative breeding schemes.

The appliction of the co-operative breeding approach to improvement in meat-sire breeds is discussed and some examples are described. Developments in this area are likely to be important for the use of co-operative breeding in other countries, a topic which is also considered.

Finally, consideration is given to the future of co-operative breeding ·schemes. It is clear that they are now a widely-accepted and well-established part of the sheep breeding industry in New Zealand. From the viewpoint of long-term breed improvement, it appears that competition between co-operative schemes and the more traditional ram-breeding flocks will continue and will be important in keeping both at peak efficiency.

INTRODUCTION

Co-operation among breeders to achieve genetic progress in livestock production is by no means new. Bull circles or bull associations for the co-operative ownership of bulls and community co-operation in breeding have been discussed by Lush (1943) and were not uncommon ways of organizing improvement in small herds and flocks. In fact, one may regard the activities of the Herd Improvement Council of the New Zealand Dairy Board as being the result of dairy farmers who are recording butterfat production co-operating to organize a progeny testing and artificial insemination service for the dairy industry. Indeed, the writer's own initial experience in co-operative breeding started in 1956 in amalgamating the breeding operations of a group of small New Zealand poultry breeders to compete with the very much larger Australian breeding companies then entering the New Zealand market. The co-operative breeding schemes to be discussed in this chapter are of more recent origin and refer specifically to sheep.

CO-OPERATIVE SHEEP BREEDING SCHEMES IN NEW ZEALAND

The first co-operative sheep breeding scheme, the New Zealand Romney Development Group, was established in 1967. This was followed by a further two private schemes and the scheme established by the Department of Lands and Survey at Waihora in 1968 (Hight *et al.*, 1975), three in 1969, eight in 1970 and six in 1971. By 1975, 25 sheep schemes involving 322 breeders and seven breeds of sheep were in operation (Parker, 1975). Of these, eight are Romney, seven Coopworth, six Perendale and one each of the Border Leicester, Corriedale, Polled Dorset and South Dorset Downs breed.

A common element in the working of these group breeding schemes as they are more commonly called in New Zealand is the identification of high-producing ewes from different flocks and bringing them together to form a central flock which in turn acts as the main source of sires for the contributing flocks. It is this process, usually called "screening", which has the main genetic influence in the initial establishment of a co-operative breeding scheme. The 25 schemes covered by Parker (1974) are reported to be screening a total of 646,000 ewes and 65,000 of these are being recorded through the National Flock Recording Scheme.

A distinction between these group breeding schemes can be made on the basis of the type of population which is being screened. Some groups are screening from commercial flocks only, others from registered flocks only and still others from both registered and commercial flocks. The distinction between the first two needs to be made because the genetic consequences, at least to the breed as a whole, are different. In screening from commercial sheep, advantage is taken of the opportunity to have high producing sheep from the whole population contributing to ram production and to genetic change. The structure usually consists only of one central flock and its associated contributing flocks, i.e., it is an open nucleus breeding scheme with a two-tier structure which allows the genetic gain to be passed on to the commercial population with the minimum time lag. Where screening is from registered sheep only, the central flock functions in very much the same way as does a nucleus flock in the typical pyramidal or hierarchical structure so commonly found among registered flocks. Thus, apart from the fact that there is no contribution from commercial population, the structure involved is three-tiered. Consequently, the time lag to the commercial flock is more or less doubled compared with the two-tier system.

(a) The basis for screening

In the co-operative schemes operating in the breeding-ewe breeds (such as the Romney, Coopworth and Perendales), the trait upon which initial screening is based is number of lambs reared by the ewe. The standard set for this trait varies from scheme to scheme. A common requirement is that of rearing twins after an unassisted lambing as a two-tooth although in more intensively recorded groups, ewes are required to have lambed as hoggets and have produced twins in their two subsequent lambings. In some cases, hoggets with high body weight and showing oestrous activity have been taken into the central flock. Further selection is done among the available sheep for freedom from structural defects, for fleece weight and quality and for any other requirements laid down by the group members.

It should be noted that in most schemes, the selection and management in the central flock is designed to discriminate against ewes which require assistance at lambing.

(b) Group organization

The central flock is operated by a flockmaster who is responsible for the management and recording of the flock and for supplying information about the flock performance to contributors. The financial organization of the groups differ markedly. Most groups work on an exchange system whereby for each 4 or 5 ewes supplied, a contributor receives one ram from the central flock. The contributor also has the opportunity to buy further rams at reasonable prices. Most groups sell rams surplus to their needs to other farmers.

The form of business organization adopted will depend greatly upon local legislation and income tax procedures and cannot be covered in this discussion. While many of the groups formed in New Zealand started off with merely "gentlemen's agreements" between the participants, increasingly they have become organized as private companies, industrial and provident societies, or incorporated societies (McConnell, 1974). It must be emphasized however that, because the central flock rapidly becomes an asset of very substantial value, it is important that the business aspects of group organization be established firmly at the outset.

(c) Genetic gain in co-operative breeding schemes

The genetic advantages in terms of increased selection differentials, reduced time lag in spreading improvement through the group, selection within a commercial environment and some avoidance of inbreeding problems have been outlined by Rae and Hight (1969), Hight and Rae (1970), Hight and Dalton (1974) and Parker (in this edition). Indications have also been given of the comparative rates of expected genetic gain by Jackson and Turner (1972), Rae (1974) and Clarke (1975).

Actual field or experimental evidence as to the amount of genetic gain being achieved is however scarce. Since they are commercially-oriented, none of the groups has maintained a control population to evaluate progress. The most useful information available to the present comes from the study carried out by the Department of Lands and Survey to evaluate the performance of the progeny of rams bred in the Waihora scheme compared with the progeny of rams purchased from private breeders (as reported by Hight *et al.*, 1975). In summary, two-year-old ewes sired by Waihora-bred rams weaned 22% more lambs than those sired by purchased rams in 1973. In 1974, a less favourable season, the difference for two-year-old ewes was 10.4% and 7.0% for three-year-old ewes. In 1975, the two-year-old ewes sired by Waihora rams weaned 10.9% more lambs while the combined three- and four-year-old ewes weaned 13.8% more lambs. The increased number of lambs weaned was due partly to a higher multiple birth rate and to better lamb survival. These figures represent almost entirely the gain achieved by the initial screening and would suggest that genetically the nucleus is averaging at least 20% above the population of flocks which it is serving. These figures would be approximated in theory if the screening rate had been about one ewe out of a total of 20. Since the population screened in setting up the flock of about 8,000 ewes was about 300,000 ewes, it would seem that the expectation is being achieved.

(d) Some reasons for the success of co-operative breeding schemes

There are a number of elements in the New Zealand sheep industry which are believed to have contributed to the successful establishment of co-operative breeding schemes. The first reason is that number of lambs reared or reproductive rate is the most economically important trait requiring improvement in the basic breeding-ewe breeds in New Zealand. It is a particularly convenient trait because (i) it is relatively easy to set standards of performance (such as twinning at two-year-old lambing) which do not require detailed recording in the contributors' flocks and (ii) the farming systems are such that identification of the ewes can be undertaken with sufficient accuracy. It should be noted that screening of ewes from commercial flocks is particularly relevant to the improvement of maternal traits because these are measured directly on the ewe.

A second element of importance has been an interest among breeders and commercial producers in breeding for increased productivity. In fact, the members of co-operative breeding schemes have been the most consistent group in the industry in emphasizing the importance of performance recording and selecting on the basis of performance records and in seeking improvements in the recording services to meet their special needs. It is noteworthy that about a third of the total ewes being recorded on the National Flock Recording Scheme are from co-operative schemes.

The availability of advice and assistance in the technicalities of setting up co-operative schemes has also been important in their development. This assistance has been supplied by research and university personnel, the Advisory Officers (Animal Husbandry) and Sheep and Beef Officers of the Ministry of Agriculture and Fisheries and, above all, by members of already established groups. The early formation of the New Zealand Federation of Livestock Breeding Groups on the initiative of Mr. A.G.H. Parker has also been of substantial value in encouraging widespread discussion of problems peculiar to co-operative breeding as well as facilitating interchange of information between groups and promoting the use of improved techniques in livestock breeding.

The reaction of breed societies to co-operative breeding has been mixed. The long-established breed societies have generally been antagonistic to the idea of co-operative breeding (more especially when it is involved with commercial sheep). In some of the more-recently formed societies, notably the Coopworth and Perendale Sheep Breeders Association, there has been a willingness to accept group breeding and to make it relatively easy for high-producing sheep to be accepted.

Finally, evidence of the financial success of many of the co-operative schemes, even under circumstances where the traditional breeders were finding it hard to compete has been a potent factor.

(e) Some problems of co-operative breeding schemes

While the members of co-operative breeding schemes have been remarkably successful in solving many difficult organizational and genetic problems, there are still several areas of concern where further effort is required. The first problem is posed by the presence, at least in some groups, of substantial differences in the performance of sheep screened at the same level of performance from different flocks when they are brought together into the central flock. This may be caused simply by inaccurate recording of the performance of the ewes in some flocks, or may be the result of real genetic differences between flocks. It is also possible that it is the result of carry-over effects of the environment on the farm before entry to the central flock. It is difficult to sort out these possibilities but it may well be that the comparisons between ewes screened directly into the central flock are not as satisfactory as one would wish.

A second source of contention in co-operative schemes is the question of the level of environment adopted in the central flock. In most cases, the decision has been to run the flock at normal commercial stocking rates and under normal management consistent with the need to obtain the records required. This seems to be a sensible approach, but experience has shown that this has been a common source of dissatisfaction by contributors with the performance of their sheep in the central flock.

A question which will need to be answered in the near future concerns increasing the standard of performance in lamb production required for acceptance into the central flock. As the level of lamb production improves in the contributing flocks, there is need for new standards of performance to be set for screening ewes out of these flocks. As noted earlier, the standards which have been used to date have been simple and easily recorded in the contributing flocks and have been successful in selecting out a high-producing group. As the level of twinning in the young ewes increases, however, more rigorous standards which may become progressively more time-consuming and expensive to operate are required.

Some genetic problems still remain. The question has been posed by the results of co-operative breeding schemes whether the heritability of reproductive rate is greater at high levels of lamb production than it is at lower levels. To what extent do particular epistatic combinations of genes, which are likely to be broken up at the next segregation contribute to genetic variation at high reproductive rates? Some schemes have drawn their contributors from a wide area covering different environmental conditions whereas others have concentrated on screening stock from a limited area. Is genotype-environment interaction likely to be of importance in this situation?

SELECTION PROGRAMMES WITHIN THE CENTRAL FLOCK

The breeding problems which arise once the central flock has been established have been examined by Jackson and Turner (1972), Rae (1974) and Clarke (1975). In the New Zealand scene, the co-operative breeding schemes have relied on essentially the same selection procedures in the central flocks as are available to other ram-breeding flocks. Because almost all co-operative schemes are in the National Flock Recording Scheme (and its successor Sheeplan), they have been selecting young rams and ewes on the basis of their dam's performance for weight of lamb weaned or reproductive rate, and on their own records for hogget fleece weight and fleece merit. These traits have also been combined into a selection index. So far only a few co-operative schemes have considered more detailed programmes involving half-sib selection, progeny testing and use of artificial insemination.

Rae (1974) has considered the effect of using records of the paternal and maternal grandams and of the half-sisters of the young ram in selection for weight of lamb weaned. Use of the records of the paternal and maternal grandams increases the accuracy of selection by about 3-4% with little extra effort. The use of the lambing records of the half-sisters of the young rams where these records are available gives an increase in accuracy of 10-11%.

The production of sire summaries in Sheeplan (Rae, 1976) supplies the key information for the use of this method of selection.

The co-operative breeding scheme structure with its contributing flocks has always been regarded by the writer as offering perhaps the best opportunities for the combined use of progeny testing and artificial insemination for improvement in reproductive rate. Rae (1974) has discussed three methods of progeny testing within a co-operative scheme: (i) Testing within the central flock. This approach is the least effective way of using progeny testing but may be the only way if recording is not possible in the contributing flocks. (ii) Using a separate testing flock and mating all the central flock ewes to proven sires. This gives about 3-4% greater gain than testing within the central flock and is being operated by at least one co-operative scheme in New Zealand (Williams, 1974). (iii) Testing ram lambs in contributing flocks. This system linked with the use of artificial insemination has been examined in detail by Clarke (1975) who showed that the increase in genetic progress was of the order of 17 to 21% over that achieved by selection for reproductive rate on dam's performance alone. It is noted that progeny testing has its greatest potential advantage when it is associated with the more intense ram selection that is possible through artificial insemination. The cost of this system is likely to be high and there is need for ingenuity in devising labour-saving methods of recording to make it feasible to undertake progeny testing in the contributing flocks. The discussion of artificial insemination by Shepherd (1975) is relevant to this problem. Clarke (1975) has also stressed the effects of flock size on inbreeding under the intense ram selection made available by artificial insemination and has noted the need for care in this regard.

The costs of these more detailed programmes would seem to justify the examination of a system in which only little effort is devoted to improvement within the central flock but reliance is placed on successive screenings from the population for genetic progress.

Studies of the optimum structure of co-operative schemes (Jackson and Turner, 1972) suggest that about one-half of the ewe replacements for the central flock should come from contributing flocks and one-half from the ewes born in the central flock. In practice, this poses a problem, not yet satisfactorily resolved, of comparing the performance of the young ewes bred in the central flock with those screened from contributing flocks under a less intensive level of recording. Further study of this question is needed to achieve a satisfactory method of selection.

CO-OPERATIVE SCHEMES FOR MEAT BREEDS OF SHEEP

It is noteworthy that at this stage only two relatively small co-operative schemes have been started with the specialized prime lamb sire breeds in New Zealand. This is probably the result of the fact that screening of ewes is not really relevant to the objectives of improvement in these breeds. The importance of these breeds in the New Zealand sheep industry nevertheless justifies an examination of ways in which co-operative breeding schemes could be applied. The Southdown breed has been chosen as an example.

Within the Southdown breed, increasing growth rate up to weaning, maintaining low levels of lamb mortality in crosses and producing a carcass with a maximum amount of muscle and no excess fat would be objectives acceptable to many breeders. It should be noted that flocks of the prime-lamb sire breeds are maintained only to breed rams; there are no commercial flocks from which to screen ewes.

An essential requirement for planning a programme of improvement of a breed is information about the structure of existing ram-breeding flocks. A study of the flock records in Vol. 47 of the Flock Book published by the Southdown Sheep Society of New Zealand in 1973 showed that the average number of ewes in the 770 flocks for which returns were given was 106 ewes. The distribution of flock size was such that 61% had less than 100 ewes, 25% between 100 and 199 ewes, 8% between 200 and 299 ewes, and under 6% greater than 300 ewes. Thus on the one hand, there is a large number of small flocks which can only function as multiplying flocks in the present structure and a small number of larger flocks which are likely to be the main contributors to genetic change within the breed. This conclusion was reinforced by a study of the source of the registered (or single-entered) rams in the 1973 Flock Book. In all, 187 flocks (24.3% of the total) single-entered one or more rams. There were however only 22 flocks which entered six or more rams and 11 which entered 10 or more rams. These flocks are in effect the nucleus of the breed and the analysis suggests that major consideration has to be given to methods of selection for these flocks which contribute so substantially to the registered rams used in other flocks.

Estimates of heritability for weaning weight in Southdown sheep indicate a figure of 0.10 (Rae, 1975). This suggests that progeny testing may well be worth using at least in the larger nucleus flocks of the breed. The situation examined is that of a 400 ewe flock with an effective lambing percentage of 100% and an annual loss through death and culling of 10%. For progeny testing, it is assumed that there is available an associated testing flock of 500 Romney 3-year and older ewes. The progeny test procedure is as follows: (i) 25 ram lambs are chosen from the 200 available on the basis of their adjusted weaning weight; (ii) Each ram lamb is mated to 20 ewes in the testing flock in March. When the lambs are born, they are identified to sire and sex, birth and rearing rank are recorded. All lambs are weighed prior to first drafting or at weaning and this weight is used after correction for calculating the sire averages and breeding values. This programme is expected to give an increase in genetic gain over individual selection on weaning weight of about 40%. In part, this arises because it does not lead to an increase in the generation interval since in practice breeders who are using individual selection do not use their rams till they are 18 months of age. This system also makes available to other flocks a number of progeny tested sires ranking below those chosen by the breeder. Thirty flocks operating this system would be able to produce a substantial proportion of the total number of sires required each year by the breed.

The second possibility is the use of a co-operative breeding approach, in this case to operate a central testing flock to progeny test ram lambs contributed by breeders. The procedure involved would be the same as that mentioned above. The ram lambs to be tested would be contributed by breeders, both big and small, from the surrounding area. The ram lambs could well remain the property of the breeder but there may need to be some regulation requiring services to be available to other breeders in the scheme. Assessment of this system from the viewpoint of a breeder of a 120 ewe flock requiring one ram per year for replacement and contributing either three or five ram lambs for testing has been made. The rate of genetic gain when five rams are tested is a little less than achieved in the earlier scheme but this difference is reduced if the rule is adopted that the best rams are used in the contributing flocks irrespective of the flock from which they were derived.

The main advantage of the central testing flock is that it allows breeders with small flocks, but with great enthusiasm for breed improvement, to make their contribution. There is no reason why this co-operative scheme should not operate side-by-side with the method suggested earlier. In fact, as with the screening of registered ewes in dual-purpose breeds, it is contributing to the enlargement of the size of the nucleus of the breed. In addition, the competition between the two methods will be likely to keep both at high efficiency.

DEVELOPMENTS IN CO-OPERATIVE BREEDING IN OTHER COUNTRIES

Parker, in this book, has described the development of co-operative breeding schemes which have taken place in Australia, South Africa, Peru and the United Kingdom. The details will not be presented here.

The following principles would seem to apply to the development of co-operative breeding schemes. Where there is a well established stratification of breeds in the sheep industry, then the main approach in the basic breeds will be that of screening ewes out of commercial flocks into a central flock and the establishment of a two-tier open nucleus structure. The basis for screening may be, as in New Zealand, the reproductive rate of the ewe. In other situations, as for example the Scottish Blackface in Scotland where the genetic potential of the breed for reproductive rate is adequate, the criterion may be the milking and mothering ability of the ewe as expressed in the weaning weights of her lambs. In general, however, the procedures discussed for the dual-purpose breeds in New Zealand would apply.

A further component in a typical stratification is the meat sire breeds. In this case, a plan along the lines of that suggested for the Southdown breed may be worth examining, particularly where the size of the ram breeding flocks is small.

In the case where there is an intermediate stage in the stratification in which a breed with high reproductive rate and mothering ability is crossed with hill breeds to produce first cross ewes for prime lamb production (e.g., the Border Leicester in the United Kingdom), no example of co-operative breeding is available. While the objective in such a breed is to improve maternal traits, there are no commercial flocks from which to screen ewes. There may however be a case for screening ewes into a central flock from the ram-breeding flocks in order to form a nucleus for the breed.

Where no stratification exists but there are distinct breeds which are well adapted to the local conditions, then screening the most productive sheep from the population so that they can contribute to breeding rams would seem to be worth trying. This approach may in some circumstances be more successful than the introduction of poorly-adapted exotic breeds for crossing purposes.

In assessing whether co-operative breeding has a part to play, New Zealand experience would suggest that:

1.   There is need to examine the structure of the breed and the present source of rams for the breed.

2.   It is necessary to define clearly the objectives of improvement in the breed and these objectives have to be accepted by any breeders who wish to form a group.

3.   There is need to establish a technique and standard for screening ewes from the contributing flocks.

4.   An adequate business organization has to be set up to control the group and to manage the central flock.

CONCLUSION

There is little doubt that the co-operative sheep breeding schemes are well established in the basic breeding-ewe breeds such as the Romney, Perendale and Coopworth in New Zealand and will continue to be a feature of these breeds. An indication of the stability of many of these schemes is that they have purchased farms to provide grazing for the central flocks. It is to be expected that the number of schemes in these breeds will increase and there are a number of indications that they will spread to other breeds such as the Merino and Drysdale. It seems likely that there will be development of the co-operative approach in some of the meat breeds. But it is no expected that group breeding schemes will take over the whole of the ram-breeding function for these breeds. Indeed, it seems likely that both co-operative breeding schemes and the traditional stud breeder will exist together, the competition between the two approaches encouraging efficiency in genetic improvement in both segments of the industry.

REFERENCES

Clarke, J.N. (1975). *Sheepfarming Annual* 1975 : 157.

Hight, G.K. and Rae, A.L. (1970). *Sheepfarming Annual* 1970 : 73.

Hight, G.K. and Dalton, D.C. (1974). *Sheepfarming Annual* 1974 : 128.

Hight, G.K., Gibson, A.E., Wilson, D.A. and Guy, P.L. (1975). *Sheepfarming Annual* 1975 : 67.

Jackson, N. and Turner, H.N. (1972). *Proc. Aust. Soc. Anim. Prod.* **9** : 55.

Lush, J.L. (1943). *"Animal Breeding Plans"* 2nd ed. (Iowa State University Press : Ames).

McConnell, G.R. (1974). *Sheepfarming Annual* 1974 : 160.

Parker, A.G.H. (1975). *Sheepfarming Annual* 1975 : 57.

Rae, A.L. (1974). *Sheepfarming Annual* 1974 : 121.

Rae, A.L. (1975). Unpublished data.

Rae, A.L. (1976). Proceedings International Sheep Breeding Congress, WAIT, 1976.

Rae, A.L. and Hight, G.K. (1969). *New Zealand Journal of Agriculture* **118**(5) : 16.

Shepherd, J.H. (1975). Second Conference of the N.Z. Federation of Livestock Breeding Groups : 95.

35

PRODUCTIVE LEVELS OF SHEEP IN NEW ZEALAND AND PROGRAMMES AIMED AT IMPROVEMENT

T.D. QUINLIVAN

Animal Production Services Ltd., Waipukurau, New Zealand

SUMMARY

Productive data derived from national surveys conducted on New Zealand's national flock indicate substantial drops in reproductive performance, wool weight per head and weight of lambs slaughtered. This serious depression in sheep performance is considered due to high stocking rates, low levels of short and long term financial input into the industry, adverse seasonal conditions and the dominance of a sheep breed known to have low reproductive rates.

Clearly, in order to maximize production, emphasis must be given to the development of programmes designed to improve per head performance for meat and wool. This can be achieved by judicious cross-breeding or within-breed improvement.

The essential ingredients within these programmes relate to the annual identification and removal of poor producing females, a clear understanding of the principles of nutrition of the ewe flock, disease control and the establishment of ram breeding systems which ensure the availability of sires with the potential for high production.

INTRODUCTION

The New Zealand economy is largely dependent on the export income derived from its farm livestock population of 56 million sheep and 9 million beef and dairy cattle. Approximately 75% of export earnings are derived from the pastoral industry. The efficiency of this industry is the direct concern of many people both in the private and State sectors.

This chapter deals with aspects of the present performance levels of the sheep industry and particularly that of the Romney—the country's dominant sheep breed. From the information presented it is clear that higher levels of per head performance are required and emphasis is given to breeding programmes which have shown that effective gains can be made.

TRENDS WITHIN THE SHEEP INDUSTRY

(a) Sheep and cattle numbers

In efforts to increase farm production, farmers were enticed to increase stock carrying capacity in the early 1960s. The national flock increased in size from 50 to 61 million sheep to 1967/68 and this level was maintained until 1972/73. Since then, sheep numbers have declined to 55 million in 1975/76 and projected estimates place this figure at 57 million for 1976/77.

Over the same period, beef cattle numbers increased from 4.5 million to a peak of 6.5 million in 1974/75. Total stock units (both sheep and cattle) have been relatively steady at 100 million since 1968/69, although there have been significant changes in the overall composition of sheep and cattle.

(b)   Per head performance

Although the main factor controlling overall farm production is livestock numbers, the key parameter in the production equation is per head performance. Table 1, compiled from national statistical sources, shows the lambing percentages (lambs tailed from ewes to ram) and mean fleece weights from 1964/65 to 1975/76 (estimated).

TABLE 1

MEAN LAMBING PERCENTAGES AND WOOL WEIGHTS

Year	Sheep numbers (million)	Lambing %	Wool Clip per head (kg)
1964/65	51.292	97.4	5.52
1965/66	53.748	99.5	5.86
1966/67	57.343	99.2	5.62
1967/69	60.030	98.8	5.50
1968/69	60.474	94.8	5.49
1969/70	59.937	97.0	5.47
1970/71	60.276	93.1	5.54
1971/72	58.912	94.4	5.47
1972/73	60.883	92.0	5.08
1973/74	56.684	90.9	5.03
1974/75	55.880	95.1	5.27
1975/76	55.300	96.2	5.42

It is evident from the data that both lamb and wool production fell from the mid 1960s to an all time low in 1973/74, this coinciding with the peak of sheep numbers in the national flock. In addition to these changes, the mean weight of lambs slaughtered has fallen by 2 kg. These trends were not confined to any one or group of farming systems or to any particular geographical location. The reasons for this depression in productin can be summarized as follows:

1.   Clearly, the marked increase in stock numbers influenced per head production.

2. Although climatic factors were partially responsible, of considerable significance was the lower volume of farm input in terms of short and long term finance. Thus, pasture improvement, farm subdivision, fertilizer application and improved feed supplementation did not match the increased stock carrying capacity of farms.

3. Although this period witnessed a dramatic change in the breed structure of the sheep industry, cross-breeding failed to halt the slide. However, the Romney still retained its dominance and should bear the responsibility.

Economic sources consider that following higher levels of input during 1972/73 and 1973/74, coupled with lower sheep numbers and perhaps more favourable seasons, per head production increased in 1974/75 and is expected to improve still further in the current season.

(c)   Breed structure of the industry

From a complete dominance during the 1950s the Romney breed has declined in popularity. Currently, 66% of all sheep in New Zealand are of this breed. It has been replaced on second class hill country by the hardier Perendale and in the higher fertility areas by firstcross Border Leicester × Romney and Coopworth. The classic breeding structure of the Romney industry is still centred on the easier hill country of the North Island and Southland-Otago.

The main reason for this shift in popularity away from the Romney lay in its poor reproductive performance in commercial breeding flocks, particularly under high stocking rates. The emphasis on wool production in the mid 1950s led to the evolution of a small framed, high fleece weight animal which was a reproductive disaster.

(d)   Breed performance

Within this climate of declining breed popularity, increased stock numbers and lower returns, a national survey was conducted to examine the reproductive performance of the Romney breed, to define the reasons for deficiencies and to develop improvement programmes. This study was conducted during 1962-72 and the findings have been published (Quinlivan *et al.* 1966; Quinlivan and Martin 1971 a,b,c,d).

The investigation, which defined performance levels in both commercial and stud flocks throughout New Zealand, set basic reproductive parameters for the breed and highlighted the modifying effect of the environment and management on sheep reproductive rates. An example of these effects is shown in Table 2 which demonstrates the differences in lambing percentages between stud flocks and commercial flocks throughout New Zealand. For analysis, the country was divided into 6 geographical areas based on latitude and the commercial flocks further grouped on the basis of farming type.

1 — Unimproved hill country
2 — Improved hill country
3 — Rolling hill country
4 — Intensive flat land

TABLE 2

COMPARISONS BETWEEN ROMNEY STUD AND COMMERCIAL
FLOCKS LAMBING PERCENTAGES

Island	Location	Stud flock means	Commercial flocks				
			Property type				Mean
			1	2	3	4	
North	Northern	102.8	97.2	92.7	98.9	—	94.5
	Central	110.0	85.1	92.6	95.1	—	92.9
	Southern	112.4	87.8	92.3	99.7	105.1	93.7
South	Central	120.8	—	90.7	110.1	112.2	108.2
	Southern	121.7	89.5	108.9	112.3	118.6	111.9
	Mean	115.4	88.2	94.1	102.2	115.5	98.3

As the commercial flock owners were ram clients of the stud breeders in this study, the data highlighted the effects of farming type on sheep production. These effects are clearly evident from the table. Clearly, the environmental effects far outweighed any possible superiority in stud flocks. It was concluded that in general, Romney stud flocks were genetically stagnant. However, it was also concluded that because of its numerical strength and the large variations in observed reproductive levels, that this dual-purpose breed possessed the ability for higher levels of production. Emphasis in future development must be centred on improved total lamb production, lower assistance rates at lambing and improved lamb survivability and thrift.

It is unfortunate that no comparable data is available on the national performance of other sheep breeds in New Zealand, as estimates of their genetic variability under various farming systems should provide information on their productive potential.

SHEEP IMPROVEMENT PROGRAMMES

Within the context of the above study it was logical to develop programmes aimed at improving levels of production in sheep flocks. Although these programmes were aimed at within-breed improvement it was well recognized that breed changes and particularly cross-breeding, could well be used as an alternative approach. Although some of this work commenced during the tenure of the Romney Survey considerable effort has been devoted to this cause since the survey's termination in 1972.

The entire approach is based on the premise that within any animal population the full range of productive merit exists. By removal of animals at the left hand side of the production distribution curve and with the use of sires with the potential for high production, subsequent generations should show higher levels of productivity than the base flock at commencement.

(a) Commercial flock improvement

Following a request for assistance interviews are conducted and basic data collected. The following procedures are adopted.

1. Definition of problem

By careful initial data collection relative to the history of the property and the flock, deficiencies affecting production can be identified. Information is requested on the following. Where applicable, data from the previous 10 years is required.

(a)　Location and property classification
(b)　Area of property
(c)　Numbers of stock wintered—sheep and cattle and classes
(d)　Fertilizer programme
(e)　Stock performance
　　　(i) Sheep — lambing percentages, breeding policies, source of rams, wool weights, mortality rates.
　　　(ii) Cattle — pregnancy rates, calving percentages, cattle weights, breeding policies.
(f)　Economic analyses of income derived from meat and wool
(g)　Disease problems and control

The lambing percentage data are further analysed to determine patterns over the previous 10 years.

(a)　Numbers of ewes joined
(b)　Date rams joined and length of mating
(c)　If mating crayons used—the number of ewes covered and conceiving during each 17 day cycle of mating
(d)　Patterns of lambing
(e)　Number of no-lamb ewes
(f)　Number of lost-lamb ewes
(g)　Number of ewes requiring compulsory assistance at lambing
(h)　Ewe deaths — mating to pre-lambing
　　　　　　　　　 — pre-lambing to lamb marking
(i)　Lambs born dead
(j)　Lambs born alive
(k)　Lambs tailed
(l)　Lambs weaned

These data are compared with prepared base tables to compare the reproductive performance of the flock with mean values. A detailed analysis of the property is possible from these data. Should it be necessary to define further aspects of performance, or to examine the effects of various treatments (e.g. trace elements on hogget thrift) these are conducted on the property.

The client's flock is then placed on a basic improvement programme, the main points of which are as follows:

2.   Sire replacement policies

All rams used must conform to certain basic standards.

(a)   Free from ovine Brucellosis by blood test and vaccinated.
(b)   Genitally sound
(c)   Semen examination pre-mating, only if ram fertility considered to be a problem.
(d)   Of particular importance is that all rams used are derived from recorded stud flocks who themselves are exhibiting high levels of performance. Above average sheep are sought and assistance in ram selection given.

3.   Mating policies

(a)   Date of joining is determined to coincide with the known period of maximum reproductive efficiency within the breeding season for the breed and geographical location. Final decision depends on likely feed availability during pregnancy, anticipated weather patterns at lambing and time of lamb drop relative to patterns of spring growth.
(b)   The recommended ratio of ewes to disease-free selected rams is 1 : 75. This is modified depending on paddock sizes, terrain etc.
(c)   Restricted mating period—usually 34-40 days. Follow up harnessed rams used.

Under this mating system it is anticipated that at least 95% of ewes should be pregnant at the end of a 34-40 day mating.

4.   Feeding policies

The principles outlined by Coop (1972) and those presented in a British Meat and Livestock Commission publication "Feeding the ewe", form the basis of all recommendations to clients. Space precludes detailed description of the principles employed.

In order to monitor the effects of nutrition in various classes of sheep, widespread use is made of weighing scales and condition scoring at strategic times of the year. Both techniques are employed to check hogget growth rates, suitability of premating nutrition and systems are now being developed to check on ewe body weight and condition during pregnancy as it relates to lowered assistance rates, improved lamb survival and better milking ability and hence improved weaning weights. From this information it is now possible to set target weights throughout the year for various classes of sheep relative to performance.

5. Ewe culling policies

The following classes of stock are removed from the main breeding flock annually. They are either slaughtered or placed in a second flock on the property to another breed of ram. The anticipated proportion of animals falling into each of these classes is also shown.

(a) Ewes mated after 34-40 days — 5%
(b) No-lamb ewes — 4%
(c) Assisted ewes and their female progeny — 6%
(d) Lost-lamb ewes — 9%
(e) Ewes which fail to rear lambs to weaning — 2%

Assuming a 5% death rate from mating to lambing, on the average, approximately 30% of ewes are either unproductive or demonstrate undesirable performance characteristics.

6. Female selection policies

The ewes in the main breeding flock each year consist of 2 groups.

(a) Those females which successfully reared lambs to weaning the previous year less those culled for age, wool and other defects.
(b) Maiden ewes entering the flock for the first time. These are the progeny of ewes and rams selected on performance. The parameters used, in combination, in maiden ewe selection are:
   (i) Hogget oestrus—detected by vasectomized rams.
   (ii) Evenness of wool and freedom from faults. Consideration is now being given to selection based on hogget fleece weight—animals being classed into various weight grades at shearing.
   (iii) Yearling and or premating body weight as follows:
      < 43 kg — culled
      44-46 kg — may qualify if more required
      > 46 kg — qualify for entry into flock

These levels are raised as the programme progresses.

7. Disease control

Disease control programmes including those for parasite control, are based on:

(a) Definition of the disease problems on the property and district patterns.
(b) Economics.

8.   Results

As there are many properties in the scheme and as each property situation is different it is difficult to provide specific data. However, the following general effects have been noted through application of this basic programme.

(a)   Significant reduction in ewe and lamb mortality.

(b)   Improved thrift, particularly of young sheep.

(c)   Increased conception rates to restricted mating programmes.

(d)   The proportion of diseased and genitally unsound rams reduced from a mean of 28% at commencement to a nil incidence.

(e)   No significant effects on the incidence of no-lamb ewes.

(f)   Reduction in the proportions of lost lambs and assisted ewes.

(g)   Increased lamb marking percentages, particularly in maiden ewes by improved total lamb drop and lower incidence of non-productive females.

(h)   Improved ewe hogget weights allowing greater flexibility in selection for entry into main flock.

(i)   Improved wool weights.

(j)   Higher average lamb weights at slaughter and a greater proportion of lambs drafted earlier in the season.

These effects should be evident by the third year of participation. Should a flock fail to respond within this period, other breeding alternatives are examined. The biggest problem during this first 3 year period is maintaining ewe numbers.

The success of the programme depends largely on the owner's enthusiasm to improve the productive output of his flock, the responsiveness of the flock, results and the availability of sires to service the programme.

(b)   Ram breeding programmes

As a consultant whose efforts are primarily directed towards improvement of per head production, one is totally committed to the concept of group breeding. The availability of rams to match the commercial programmes outlined above is cause for concern. The organization, structure and some relative results obtained so far from a breeding company formed by registered Romney breeders, serves as an example of what can be done to overcome this shortage.

1.   Organization

The N.Z. Stud Co. was formed in 1971 and is committed to demonstrate that the productive merit of the Romney breed can be improved. By the use of shareholders' ram breeding flocks as daughter, satellite studs it is becoming a potent force within the framework of the Romney breeding industry.

The company has 24 shareholders and overall, they have approximately 15,000 fully recorded ewes at their disposal, producing 7000 rams annually. All shareholding is equal and all shareholders have equal rights. The activities of the company are controlled by a directorate of 4 and 2 technical advisors—myself and A.E. Henderson, Professor of Wool Science, Lincoln College.

2. Selection policies

At establishment in 1971, shareholders were requested to submit all their ewes aged 3, 4 and 5 years old with the following reproductive combinations, based on lambs weaned. All ewes must have reared a *minimum* of 3 lambs in 2 years.

Age (yr)	Acceptable combinations		
1.5	1	2	2
2.5	2	1	2
3.5	2	2	2
4.5	2	2	2
5.5	2	2	2

Various triplet and quadruplet combinations were also acceptable. The following steps were taken in the selection process.

(a) The records of all nominated ewes were examined, with particular emphasis given not only to the performance of the ewe herself but also that of her dam, grand dams, daughters and half-sisters. Only ewes which demonstrated the following characteristics themselves and their female relations were accepted.

  (i) High live lamb production
  (ii) Freedom from assistance at lambing
  (iii) Above average weaning weights of progeny
  (iv) Above average wool weights of progeny
  (v) High proportion of saleable ram progeny and low rejection rate of ewe progeny

(b) All ewes which qualified were inspected for wool faults by Professor Henderson. Those with unsuitable fleeces were rejected.

(c) These sheep were then inspected by a panel of 5 for structural soundness and breed faults.

From the 5,500 ewes of the 3 age groups above, 189 (3.4%) qualified for entry into the company's central flock for a trial period of one year. Those which failed to rear twins or better to acceptable levels in the trial year, were returned to their original owners. All ram and ewe progeny from unacceptable females were slaughtered.

This method of selection still exists within the company and applies to both shareholders' ewes which enter the flock each year, and to company-bred females. Table 3 shows the numbers of ewes mated annually in the flock and their source.

TABLE 3

NUMBER EWES MATED ANNUALLY AND SOURCE

Source	1971	1972	1973	1974	1975	1976
Company owned	—	113	111	73	68	47
Shareholders	189	95	56	62	21	94
Co. bred 2-tooth ewes	—	—	50	72	59	111
Co. bred 4-tooth ewes	—	—	—	38	50	46
Co. bred 6-tooth ewes	—	—	—	—	22	35
Co. bred 4yr. ewes	—	—	—	—	—	9
Total mated	189	208	217	245	220	342

It is apparent from these data, that in order to maintain present performance levels high culling rates are necessary.

3. Results

Space precludes the presentation of the full results from this flock. Several important aspects are highlighted.

(a) The length of the mating period is 34 days. All ewes which fail to conceive within this period are culled. From 1971-76 the percentage of ewes pregnant within this period ranged from 96-99.

(b) Tables 4 and 5 summarize some of the reproductive data. Table 4 shows the mean reproductive rates over the period 1971-75. All parameters have remained relatively static during this 5 year period. For comparison, mean data for Romney stud flocks in the southern North Island over the same period are entered at the bottom of the table.

TABLE 4

MEAN REPRODUCTIVE RATES

Year	% $\dfrac{\text{Total lambs}}{\text{Ewes lambing}}$	% $\dfrac{\text{Lambs alive}}{\text{Ewes to ram}}$	% $\dfrac{\text{Lambs alive}}{\text{Total lambs}}$
1971	180.6	164.4	92.0
1972	178.9	158.3	92.4
1973	178.0	158.3	92.7
1974	181.0	162.0	92.6
1975	180.0	160.0	93.0
Means for Romney Studs	144.0	110.0	87.0

Table 5 presents the same data by ewe age groups. Asterisks beside each year denotes company bred ewes only. The balance are ewes from shareholders' flocks only.

TABLE 5

REPRODUCTIVE RATES — AGE GROUP DIFFERENCES

(a) 2-tooth ewes

Year	% $\frac{\text{Total lambs}}{\text{Ewes lambing}}$	% $\frac{\text{Lambs alive}}{\text{Ewes to ram}}$	% $\frac{\text{Lambs alive}}{\text{Total lambs}}$
1973*	161.2	142.0	89.9
1974*	164.1	141.2	91.4
1975*	166.1	143.5	89.2
(b) 4-tooth ewes			
1974*	180.0	162.2	95.2
1975*	171.4	160.0	95.2
(c) 6-tooth ewes			
1971	172.0	160.0	93.0
1972	164.4	143.2	91.9
1973	164.2	144.6	93.1
1974	177.2	160.3	92.1
1975*	186.5	173.7	95.6
(d) 4-year ewes			
1971	185.7	170.3	93.2
1972	185.4	169.0	93.4
1973	200.0	174.4	94.4
1974	200.0	178.9	89.5
1975	177.4	164.5	92.7
(e) 5-year ewes			
1971	189.2	168.4	91.4
1972	191.9	178.4	93.0
1973	176.9	155.6	91.3
1974	204.5	181.8	88.9
1975	206.3	170.6	87.9
(f) 6-year ewes			
1971	181.8	163.6	90.0
1972	204.2	184.0	93.9
1973	193.1	182.8	94.6
1974	209.5	185.7	95.1
1975	166.6	166.6	100.0

Year	% $\dfrac{\text{Total lambs}}{\text{Ewes lambing}}$	% $\dfrac{\text{Lambs alive}}{\text{Ewes to ram}}$	% $\dfrac{\text{Lambs alive}}{\text{Total lambs}}$
	(g) 7-year ewes		
1972	171.4	142.9	83.3
1973	200.0	175.0	93.3
1974	207.1	200.0	96.6
1975	220.0	220.0	100.0
	(h) >7-year ewes		
1973	200.0	200.0	100.0
1974	183.3	161.5	95.5
1975	200.0	200.0	100.0

Of particular interest, is the performance of the company bred 2, 4 and 6 tooth ewes. The results demonstrate that high levels of reproductive performance are being obtained, particularly when compared to those from other Romney studs. At this stage in its development, the company's flock has the highest reproductive performance of any Romney stud flock in New Zealand.

(c)   Large and repeatable between-sire differences in reproductive rates are being obtained.

    (i) From the ewes mated to various rams in any one year. An example of the ranges obtained at the 1975 lambing are shown in Table 6. These data suggest a sire effect on ewe fertility. These appear to relate to embryonic survival and to full term survival rates of lambs.

TABLE 6

RANGES IN REPRODUCTIVE RATES — BETWEEN SIRE EFFECTS

% $\dfrac{\text{Total lambs}}{\text{Ewes lambing}}$	% $\dfrac{\text{Lambs alive}}{\text{Ewes to ram}}$	% $\dfrac{\text{Lambs alive}}{\text{Total lambs}}$
171.4-204.3	154.7-182.6	85.2-96.4

    (ii) Between daughters of various sires. Table 7 shows the ranges obtained from the 2 tooth daughters of various sires at the 1975 lambing.

TABLE 7

RANGES IN REPRODUCTIVE RATES — BETWEEN SIRE PROGENY GROUPS

$\%\dfrac{\text{Total lambs}}{\text{Ewes lambing}}$	$\%\dfrac{\text{Lambs alive}}{\text{Ewes to ram}}$	$\%\dfrac{\text{Lambs alive}}{\text{Total lambs}}$
155.5-173.9	117.6-162.5	83.3-97.5

(d)   Significant levels of improvement are being obtained in weaning weights and in post-weaning growth rate gain. Table 8 demonstrates the degree of improvement obtained in weaning weights from 1971-75.

TABLE 8

MEAN WEANING WEIGHTS (kg) — 1971-75

Year	Ram lambs				Ewe lambs			
	S	T	Tr	Q	S	T	Tr	Q
1971	35.0	28.6	29.4	27.3	30.2	29.5	26.5	24.1
1975	36.4	34.1	30.9	31.4	31.8	32.7	28.6	29.5

(e)   Average fleece weights have remained constant but count variation within fleeces has been markedly reduced.

Progeny testing facilities have been established to ensure continuity in sire replacement and both artificial insemination and egg transfer have been used to further disseminate the influence of proven rams and ewes. The stud's operation is geared to breed stud rams of high productive merit for shareholders and other stud breeders. These in turn service the commercial flock programmes described above.

REFERENCES

Anon (1975) "Feeding the ewe". *British Meat and Livestock Comm. Sheep Improvement Service—Tech. Report No. 2.*
Coop, I.E. (1972). *Proc. N.Z.V.A. Sheep Soc. 2nd Seminar* 8.
Quinlivan, T.D. and Martin, C.A. (1971a). *N.Z. Journal of Agricultural Research* 14 : 417.
Quinlivan, T.D. and Martin, C.A. (1971b). *Ibid* 14 : 858.
Quinlivan, T.D. and Martin, C.A. (1971c). *Ibid* 14 : 880.
Quinlivan, T.D. and Martin, C.A. (1971d). *Australian Journal of Agricultural Research* 22 : 497.
Quinlivan, T.D., Martin, C.A., Taylor, W.B. and Cairney, I.M. (1966). *Journal Reproduction and Fertility* II : 379.

IV

REPRODUCTION IN THE EWE

36

LIMITATIONS TO FEMALE REPRODUCTIVE EFFICIENCY

C.H. van NIEKERK
University of Stellenbosch, Republic of South Africa

In order to maximize the production potential of a flock it is essential to increase the reproductive rate towards an ideal of two lambing seasons per year with two lambs per season per ewe. However, there are two major groups of factors which limit this ideal reproductive efficiency namely those which influence the interlambing period and those which influence the fecundity of the ewe. The latter group has received wide attention and is so well documented that this aspect will not be discussed here. This chapter therefore deals with the interlambing period and attention is drawn particularly to factors which influence firstly the post-partum involution of the uterus, secondly the post-partum ovarian activity and thirdly the endocrinology of the post-partum period.

FACTORS WHICH INFLUENCE THE INTERLAMBING PERIOD

Post-partum involution of the uterus

The macroscopic changes which occur during the involution of the uterus after parturition have been described by several authors (Foote, Call & Hulet 1967; Foote & Call 1969; Hunter 1968; Nel & Pretorius 1968; Van Wyk, Van Niekerk & Belonje 1972a, 1972b and Botha 1976). However the only comprehensive histological studies are those of Uren (1935), Van Wyk *et al.* (1972a) and Botha (1976).

The data on the macroscopic changes in the involuting uterus of non-lactating Merino ewes from one to 28 days after parturition have been summarised by Van Wyk *et al.* (1972b). It was shown that uterine mass and volume were closely correlated. Furthermore there was a rapid reduction in uterine mass during the first 12 days to reach a constant mass about 24-28 days post-partum. However Botha (1976) found that both season and suckling affected this time interval. So, for instance, during the non-breeding season (August) uterine mass and volume were constant after 30 days in non-lactating ewes and only after 36 days in lactating ewes. On the other hand during the breeding season (March) there was little difference between non-lactating and lactating ewes but uterine mass and volume in both groups were already constant by 26 days post-partum.

The uterine horns reached their maximum reduction in length by the 20th day post-partum while the largest caruncles had reached their resting size by 16 days (Uren 1935); Van Wyk *et al.* 1972b). Again it was shown by Botha (1976) that this process took less time in March than in August.

If all these macroscopic parameters are taken into consideration it appears that lactation has a limited effect on uterine involution, but that this process is completed far more quickly during March (26 days) than during August (30-34 days).

The comprehensive study by Van Wyk *et al.* (1972a) on the microscopic changes in the placentome before and during parturition, and the uterine involution during 28 days after parturition will be used to explain the basic micromorphological changes in the post-partum caruncle. The histological study of the placentome before and during parturition revealed that shortly before parturition the connective tissue of the proximal areas of the maternal villi and adjacent caruncular tissue becomes hyalinized. The result is a narrowing of the arteries and compression of the veins with vascular congestion in the distal vessels of the maternal villi and the formation of blood baths just under the allanto-chorion. After parturition the foetal villi are withdrawn and together with the foetal membranes are expelled from the uterus. During this process the blood baths rupture and free red cells are found on the surface of the caruncle and endometrium. However the blood vessels in the distal part of the caruncle are still intact and distended with blood. Twenty four hours after parturition the endometrium is still covered with dark autolysed blood cells but by 48 hours this free blood has mainly been evacuated through the cervix leaving a "cleaner" endometrium.

The process of involution of the caruncle occurs in three distinct stages viz.:

Degenerative changes in the most distal portion of the maternal villi commence soon after parturition. This area liquifies completely by the 4th day into a dark brown to black semisolid substance which lies on and between the caruncles and oozes through the cervix.

The central portion of the villi with the distended blood vessels degenerates more slowly to form a black tarry substance. By the 12th day this area has autolysed completely.

The remaining amorphous mass of villi and the band of hyalinized tissue does not liquefy but comes off as a plaque. This occurs between the 16th and 20th day leaving a naked connective tissue surface. The plaques are expelled through the cervix leaving the uterus free from all necrotic material.

Botha (1976) found that both season and lactation affected the time taken for the above process. By 12 days post-partum autolysis had proceeded further in both lactating and non-lactating ewes in March than in August. Furthermore, within each season the process always occurred faster in non-lactating than in lactating ewes. In addition, the uterus was cleared of necrotic tissue and excess fluid by 14 days in the non-lactating ewe during March and only by 18 days in the non-lactating and 22 days in the lactating ewe during August.

As soon as the caruncular plaque loosens from the underlying connective tissue, the uterine epithelium starts growing from the edges over the exposed tissue to completely cover the surface by the 28th day (Van Wyk *et al.* 1972a).

Botha (1976) found that both season and lactation also influenced this caruncular epithelialization. In August in the non-lactating ewe, epithelialization was complete by the 22nd day while the epithelium in the centre of the caruncle had reached its normal height by the 32nd day. In the lactating ewes these two stages were reached by the 26th and 34th day. In March, however, the stages were reached sooner and were found to be the 18th and 26th days for the non-lactating and the 22nd and 30th day for the lactating ewe.

It appears, then, that if all the parameters are taken into account, the uterus of the non-lactating ewe involutes slightly faster than that of the lactating ewe and that involution proceeds faster within the breeding season. Nevertheless it is reasonable to accept that the uterus will have recovered completely in the lactating and non-lactating ewe by about 30 days in March and 34 days in August.

Artificial hastening of uterine involution

Research has been conducted by a few workers to attempt to hasten uterine involution so as to increase the frequency of lambing. Foote, Call & Hulet (1967) used progesterone with or without oestrogen and/or PMSG without significantly effecting involution. Foote & Call (1969) found that on days 10, 17 and 24 the quantity of residual debris was greater in progesterone treated and lactating ewes than in non-lactating ewes, while oestradiol treatment reduced the amount of debris. Van Wyk, Van Niekerk, Belonje & Jöchle (1972) treated ewes which lambed either in Spring or in Autumn with either oestradiol or flumethasone intrauterine 24 hours after parturition. These treatments also had no effect on uterine involution.

It is clear therefore that further study is required to determine whether exogenous substances are capable of promoting uterine involution.

Post-partum ovarian activity and factors which influence it

It is maintained that in both ewes (Mauléon & Dauzier 1965) and cows (Tassel 1967) cyclical ovarian activity commences before parturition and that factors such as prepartum nutrition have an important effect on ovarian activity after parturition (Meaker 1970).

A sound knowledge of the normal morphological and endocrinological changes in the hypophysis and ovary after parturition is essential if one is to study the factors which influence post-partum ovarian activity. Unfortunately present knowledge on this aspect leaves much to be desired as in most studies the number of animals which have been used is too small and the interval between observations too large (Quinlan & Maré 1931; Cole & Miller 1935; Nel & Pretorius 1968; Van Wyk 1971 and Botha 1976).

The visual observation of changes in the ovary by laparotomy has limitations since laparotomies cannot be conducted at short intervals over a period of weeks in the same ewe, and the extent of the influence of the operation on ovarian activity is unknown (Miller & Wiggins 1962, 1964; Mauléon & Dauzier 1965; Vosloo, Hunter & Carstens 1969; Hunter, Belonje & Van Niekerk 1970; and Meaker 1970).

It would appear from present knowledge that the ovaries are relatively inactive at parturition and contain a few small follicles and a degenerated corpus luteum. After about five days post-partum ovarian activity increases markedly and by the 10th day large follicles are present with silent ovulation generally occurring between the 12th and 25th days. Thereafter ovarian activity follows the normal cyclic pattern with ovulation about every 17 days. The second post-partum ovulation can therefore be expected between the 30th and 40th days. The first silent ovulation results in the formation of a corpus luteum which some workers maintain is not as large nor does it secrete as much progesterone as a normal corpus luteum. Botha (1976) has however shown that these corpora lutea do not differ from the normal. This finding was based on morphological and histological examination and confirmed by determining the progesterone content of both the corpus luteum as well as the peripheral plasma.

It appears from the literature that within the breeding season oestrus is associated with the second ovulation in about 75% of ewes within 34 days of parturition. Outside of the breeding season up to five silent ovulations have been observed (Hunter & Lishman 1967b). Nel (1965) studied non-lactating Karakul ewes during the breeding season and found that sexual recovery after parturition followed roughly four phases:

First two weeks—sexual activity negligible.

Second two weeks—oestrus occurs at a comparatively high rate but conception is negligible.

Four to eight weeks—conception takes place with about 50% efficiency.

Most breeds of sheep have a restricted breeding season of a shorter or longer duration. It is generally accepted that if a ewe lambs outside of the breeding season her ovaries will remain inactive until the next breeding season. Botha (1976) has shown however that the ovaries of lactating and non-lactating ewes which lambed in August did not remain inactive during the first 40 days post-partum but showed cyclic changes in volume, mass and follicular size. From the 24th day post-partum the ovaries of non-lactating ewes were more active than those of the lactating ewes and in fact a few ovulations were even observed in this group between the 30th and 40th days. Similarly Van Wyk (1971) found a gradual increase in ovarian follicular fluid (3.03 ± 0.05%) in the first 28 days in Merino ewes which had lambed in October.

However there are large breed differences in the length of the anoestrous season. So, for instance, Van Niekerk & Mulder (1965) have shown that, all other factors being equal, the duration of the post-partum anoestrous period in sheep lambing in August is greatly influenced by the breed involved. The length of the period from parturition to first oestrus for the following breeds of sheep was found to be: Dorper 81 (38-200), Dohne Merino 94.4 (36-202) and Merino 68.7 (43-114) days. These results agree with those for various breeds and cross-breeds by Joubert (1962) and Basson *et al.* (1969) in South Africa and Barker & Wiggins (1964b) in America. The fact that all the Merino ewes used by Van Niekerk & Mulder (1965) returned to their normal cyclic pattern of oestrous behaviour within 43 to 114 days after parturition during the season when cyclic oestrual activity is normally at its lowest clearly demonstrates the amazing reproductive potential of this breed if all the interfering factors are kept to a minimum.

From the literature it appears that a number of factors can affect post-partum ovarian activity. The most important of these are: lactation; the level of nutrition, the stimulus of the ram and post-partum fertility.

In a review, Hunter (1968) summarized the work of most authors on the above factors. Since this review a number of publications have appeared on these factors and their effect on the post-partum period (Le Roux & Nel 1968; Nel & Pretorius 1968; Vosloo *et al.* 1969; Mallampati, Pope & Casida 1969; Meaker 1970; Fletcher 1971 & 1973; George 1973; Gould & Whiteman 1973; Shevah *et al.* 1974 & 1975, Le Roux, Van der Westhuizen & Marais 1975; Timariu, Urseseu & Peteu 1975; Sefidbakht & Movassagh 1975; Botha 1976).

Lactation

Some of the above-mentioned authors maintain that lactation suppresses the clinical manifestation of oestrus but not ovulation. Most, however, find that within the breeding season lactation has no effect on either post-partum oestrus or ovulation. The reason for these differences can possibly be found in differences in total milk production in different breeds or individuals. It is well known that high producing milk cows have a far longer post-partum anoestrus than low producers (Van Niekerk 1975), and the same may be true for the ewe.

On the other hand, when ewes lamb outside of the breeding season it would appear that lactation has a depressing effect on both ovarian activity and the length of the post-partum anoestrous period (Botha 1976). According to Pretorius (1966) lactation even influenced this period in spring (October) in Merino ewes. Non-lactating ewes took 41.4 (30-64) and lactating ewes 86.2 (38-153) days after parturition before first oestrus. The fact that 90% of Merino ewes returned to cyclic oestrous activity before the lambs were weaned at approximately 90 days post-partum during the season of low sexual activity suggests that the effect of both lactation and season can, to a large extent, be overcome by the use of a sufficiently high plane of nutrition.

Level of nutrition

It is a well known fact that feeding plays an important role in the reproductive efficiency of animals. However the effect of nutrition on the post-partum interval to ovulation and oestrus has received scant attention, although it appears that both pre- and post-partum nutrition may effect this interval (Smith 1964; Van Niekerk & Mulder 1965). Vosloo *et al.* (1969) and Meaker (1970) confirm this supposition. In the study of Meaker (1970), nutrition only affected the interval to first post-partum ovulation significantly when a high plane was fed both before and after parturition. Good feeding during pregnancy only or only during lactation was not sufficient to hasten post-partum ovulation significantly. To some extent this agrees with the work of Hunter & Lishman (1967b). They found no significant effect on the post-partum interval to ovulation or oestrus by supplementing the ration with 0.45 kg of maize per ewe daily for 25 days after lambing. However, there was a significant linear regression (P<0.01) between the post-partum interval to first ovulation and the ewe's mass measured 3 days post-partum. The heavier ewes ovulated after a mean interval of about 20 days while the lighter ewes ovulated almost 20 days later. Le Roux & Nel (1968) found that, although supplementary feeding to non-lactating Karakul ewes did not shorten the post-partum interval to first oestrus, it resulted in a shorter interval between parturition and subsequent conception. These examples, then, support the view that factors such as nutrition which affect the condition (mass) of the ewe have an important influence on post-partum ovarian activity. Variation in the plane of nutrition is probably the greatest single source of variation in the duration of post-partum anoestrus within a given breed and season.

The stimulus of the ram

It has been claimed that the association of the sexes shortly before the start of the normal breeding season synchronizes the first oestrus in a high proportion of ewes. Initially a silent ovulation occurs within a few days after the introduction of the rams, while the first oestrus of the season usually occurs about one cycle length later. For this reason the role of the ram as a stimulant for ovulation during the post-partum period also requires consideration.

Hunter (1968) has summarized the findings on this aspect. Evidently the presence of the rams during the post-partum period serves as a positive stimulus only if they are not introduced earlier than 3 weeks after lambing. However Hunter *et al.* (1970) could only demonstrate a ram effect on the post-partum period in ewes which lambed outside of the breeding season. This was because ewes in the presence of rams merely came into oestrus earlier at the beginning of the following breeding season than ewes without rams. This effect is the same as the well documented ram effect on dry ewes early in the breeding season.

Post-partum fertility

Normal oestrus

If all the factors mentioned previously are favourable it can be accepted that oestrus accompanied by ovulation can occur in a high percentage of ewes between 30 to 45 days after parturition, particularly during the normal breeding season. However, it appears that conception is very low, particularly during the first oestrous period (Timariu *et al.* 1975). Munro (1959) found that amongst a large number of ewes, whose lambs had been removed at birth and subsequently ran with a ram for eight weeks, only one conceived although several came into oestrus and accepted service.

The data on non-lactating Karakul ewes reported by Nel (1965) show that conception within the first 4 weeks after parturition is very low (7.7%). The conception rate increased to 30.8% between 30 to 34 days but was only normal after 60 days. On the other hand Schäfer (1963) reported that in non-lactating fat-tailed breeds, 1.5% of the ewes had already conceived before day 20 and the majority of the ewes conceived between 32 and 48 days. Results from a twice-yearly lambing study (Whiteman *et al.* 1972) indicated that the interval from lambing to conception in spring and autumn were 66 and 44 days respectively. Also, 71% of the ewes which lambed in autumn conceived while only 23% conceived when they were rebred in spring. Thus a large percentage of these Dorset and Dorset-Rambouillet ewes did not conceive soon enough after lambing in spring to maintain a successful twice-yearly lambing programme. Land & McClelland (1971) in a similar experiment with a breed, Finn-Dorset, which has a long enough breeding season to mate the same ewes twice in the same breeding season found that 32% of the ewes conceived twice a year. All the ewes carried an average of 3 pregnancies and gave birth to an average of 5.8 lambs per ewe when mated four times in two years.

Early weaning trials by Gould & Whiteman (1973) indicated that the interval from spring lambing to conception was shorter (58 days) when lambs were weaned at 30 days compared with weaning at 70 days of age (62 days). The 58-day interval was still however not short enough to ensure a successful twice-yearly lambing programme.

From the above it is obvious that if two lambing seasons per year are required one will have to resort to hormone therapy. This would help to decrease the time from lambing to first oestrus and increase fertility particularly in spring.

Induced oestrus

Synchronization of the oestrous period of sheep with progestagens and PMSG is widely practised. From the latest reports it seems that the fertility of dry ewes inseminated at the synchronized oestrus is normal both within or outside of the breeding season.

This is unfortunately not the case in the early post-partum ewe. For instance, Van Wyk, Van Niekerk & Belonje (1973) found that when synchronization commenced 20 days post-partum in non-lactating Karakul ewes, there was a very good oestrus response both within (100%) and outside (90%) of the breeding season. However the fertility was only 33% compared with 73% in dry ("barren") ewes inseminated at the same time. There was little variation in the conception rate in both post-partum and "barren" ewes between the two seasons. Shevah *et al.* (1974, 1975) reported more or less the same results in ewes synchronized 21-24 days after parturition.

Thimonier *et al.* (1968) and Cognié, Cornu & Mauléon (1974) reported that irrespective of season, oestrus could be induced in up to 98% of ewes by treating them with intravaginal progestagen sponges and PMSG. Although the fertility was low it increased with time from 10 to 35% as the period from parturition to the beginning of treatment increased from 17 to 25 days. They also found a breed difference as 55% of the Prealpes conceived compared with only 32% of the Ile-de-France ewes.

The unpublished data of Boshoff (1976) indicate a normal fertility (69.7% conception) in 300 non-lactating Karakul ewes sponged about 31 (28-35) days after parturition. This experiment was conducted outside of the breeding season and each ewe received 300 IU PMSG subcutaneously on the day of sponge withdrawal.

Boshoff & Faure (1976) investigated the effect of lactation on the ovulation rate in dry and lactating ewes sponged on the 28th, 38th or 48th day post-partum in autumn and spring. Of the ewes that came into oestrus, they found no difference in the ovulation rate (1.4 ovulations per ewe) between the groups or between seasons. They also found that within the breeding season 90 to 100% of all the ewes treated on the 28th, 38th or 48th days came into oestrus. Outside the breeding season there was also little difference in oestrous response between the lactating and non-lactating ewes, but only 60% of the ewes treated on day 28 came into oestrus compared with about 90% of those treated on days 38 and 48. Results of the conception rates in this experiment are not available but if Boshoff's previous conception figures are considered it appears possible to induce oestrus with a normal conception rate by treating lactating and non-lactating ewes with progestagen impregnated sponges and 300 IU PMSG commencing treatment within 30 to 40 days after parturition both within and outside the breeding season.

The endocrinology of the post-partum period

It is imperative that the genital system must return to normal after parturition before normal fertility can be expected. However it is just as important that the endocrine system responsible for oestrus, ovulation and implantation should return to normal as well.

The published data show that the high plasma progesterone levels found in late pregnancy begin to decline a few days before parturition. This is followed by a sharp rise in the unconjugated oestrogen levels on the day of parturition. The concentrations of both these hormones then decrease sharply during parturition. Progesterone reaches low levels within a few hours while oestrogens reach their lowest levels up to 48 hours after parturition. The LH level remains relatively constant from a few days before to about 30 hours after parturition (Flylling 1970; Stabenfeldt, Drost & Fanti 1972; Chamley, *et al.* 1973; Robertson & Smeaton 1973; Emady, *et al.* 1974). Little work has been done on the LH and progesterone values in the plasma of the post-partum ewe (Foster & Crighton 1973; Botha 1976; Shevah *et al.* 1974, 1975). These data show that throughout the sampling period the lactating ewe has levels of LH similar to the basal values found during the oestrous cycle. The non-lactating ewes had similar levels except that there was a peak on the 17th day post-partum. However none of the ewes showed oestrus during the sampling period (Foster & Crighton 1973). No further data could be found on the plasma concentrations of gonadotrophins or the ovarian hormones during the post-partum period to conception. This leaves a large gap in our understanding of this particular period and means that treatments by means of exogenous hormones to shorten the post-partum period is being done empirically. Further research in this field is therefore imperative before the ideal of twice-yearly lambing can possibly be attained.

CONCLUSIONS AND RECOMMENDATIONS

If all the factors which hinder post-partum uterine involution in the ewe are minimized it can be accepted that nidation can occur within 30 days after parturition. However in a large percentage of ewes, particularly outside the breeding season, ovarian activity has not returned to normal by this early stage. Thus if one wishes to follow a twice-yearly breeding system it will be necessary to select for animals with both a long breeding season and a short post-partum anoestrous period. The large variation in the length of post-partum anoestrus within a given breed suggests that rapid progress may be made here provided that the heritability of this trait is sufficiently high. At present however it appears that if ewes are to be rebred within 30 to 40 days after parturition oestrus and ovulation will have to be induced by exogenous hormones such as progestagens and PMSG. Since progestagens apparently retard uterine involution it is inadvisable to commence treatment before 30 days post-partum. Even so good results can be expected, particularly in a breed with a long breeding season, as this induction will always commence during June and December which will fall within the breeding season.

If the use of exogenous hormones is unacceptable, such as in the extensive breeding sytems, then productivity can be increased by adopting an eight-monthly breeding programme. For instance, Basson *et al.* (1969) have shown that the life-time production of a ewe can be increased by 40% in an eight-monthly compared with a 12-monthly breeding programme.

REFERENCES

Barker, H.B., and Wiggins, E.L. (1964). *Journal of Animal Science* 23 : 973.
Basson, W.D., Van Niekerk, B.D.H., Mulder, A.M., and Cloete, J.G. (1969). *Proceedings of the South African Society of Animal Production* 8 : 149.
Boshoff, D.A., and Faure, A.S. (1976). *South African Journal of Animal Science* 6: 187.
Botha, H.K. (1976). Ph.D. Thesis. University of Stellenbosch.
Chamley, W.A., Brown, J.M., Cerini, M.E., Cumming, I.A., Goding, J.R., Obst, J.M., Williams, A., and Winfield, C. (1973). *Journal of Reproduction and Fertility* 32 : 334.
Cognie, Y., Cornu, C., and Mauléon, P. (1974). Proceeding of the International Symposium on Physio-Pathology of Reproduction and Artificial Insemination in Small Ruminants, Thessaloniki.
Cole, H.H., and Miller, R.F. (1935) *American Journal of Anatomy* 57 : 39.
Cunningham, N.F., Symons, A.M., and Saba, N. (1975). *Journal of Reproduction and Fertility* 45 : 177.
Emady, M., Hardley, J.C., Noakes, D.E., and Arthur, G.H. (1974). *Veterinary Record* 91 : 168.
Fletcher, I.C. (1971). *Animal Behaviour* 19 : 108.
Fletcher, I.C. (1973). *Journal of Reproduction and Fertility* 33 : 293.
Flylling, P. (1970). *Acta Endocrinologica* 65 : 273.
Foote, W.C., Call, J.W., and Hulet, C.V. (1967). *Journal of Animal Science* 26 : 943.
Foote, W.C., and Call, J.W. (1969). *Journal of Animal Science* 29 : 190.
Foster, J.P., and Crighton, D.B. (1973). *Journal of Reproduction and Fertility* 35 : 599.
George, J.M. (1973). *Australian Veterinary Journal* 49 : 242.
Gould, M.B., and Whiteman, J.V. (1973). *Journal of Animal Science* 36 : 1041.
Hunter, G.L., and Lishman, A.W. (1967b). *Journal of Reproduction and Fertility* 14 : 473.
Hunter, G.L. (1968). *Animal Breeding Abstracts* 36 : 347.
Hunter, G.L., Belonje, P.C., and Van Niekerk, C.H. (1970). *Proceedings of the South African Society of Animal Production* 9 : 179.
Joubert, D.M. (1962). *Journal of Reproduction and Fertility* 3 : 41.
Land, R.B., and McClelland, T.H. (1971). *Animal Production* 13 : 637.
Le Roux, P.J., and Nel, J.W. (1968). *Proceedings of the South African Society of Animal Production* 7 :141.
Le Roux, P.J., Van der Westhuizen, V., and Marais, C.B. (1975). *South African Journal of Animal Science* 5 : 101.
Mallampati, S., Pope, A.L., and Casida, L.E. (1971). *Journal of Animal Science* 32 : 673.
Mauléon, P., and Dauzier, L. (1965). *Annales de Biologie animale, Biochimie, Biophysique* 5 : 131.
Meaker, H. J. (1970). M.Sc. Thesis, University of Stellenbosch.
Miller, W.W., and Wiggins, E.L. (1964). *Journal of Animal Science* 23 : 981.
Munro, J. (1959). *Animal Production* 1 : 190.
Nel, J.H. (1965). *Proceedings of the South African Society of Animal Production* 4 : 200.
Nel, J.A., and Pretorius, P.S. (1968). *Proceedings of the South African Society of Animal Production* 7 : 137.
Pretorius, P.S. (1966). *South African Journal of Agricultural Science* 9 : 823.
Quinlan, J., and Mare, G.S. (1931). *Report of the Veterinary Research Unit of South Africa* 17 : 633.
Robertson, H.A., and Smeaton. T.G. (1973). *Journal of Reproduction and Fertility* 35 : 461.
Schäfer, H. (1963). *Züchtungskunde* 35 : 158.
Sefidbakht, W., and Movassagh, H. (1975). *Animal Breeding Abstracts* 43 : 427.

Shevah, Y., Black, W.J.M., Carr, W.R., and Land, R.B. (1974). *Journal of Reproduction and Fertility* **38** : 369.

Shevah, Y., Black, W.J.M., and Land, R.B. (1975). *Journal of Reproduction and Fertility* **45** : 289.

Smith, I.D. (1964). *Australian Veterinary Journal* **40** : 199.

Stabenfeld, G.H., Drost, M., and Franti, C.E. (1972). *Endocrinology* **90** : 144.

Tassel, R. (1967). *British Veterinary Journal* **123** : 550.

Thimonier, J., Mauléon, P., Cognié, Y., and Ortavant, R. (1968). *Annales de Zootechnie* **17** : 257.

Timariu, S., Urseseu, A., and Petcu, D. (1975). *Animal Breeding Abstracts* **43** : 4059.

Uren, A.W. (1935). Technical Bulletin Michigan (State College) Agricultural Experimental Station **144** : 64.

Van Niekerk, C.H. (1975). *South African Journal of Animal Science* **5** : 155.

Van Niekerk, B.D.H., and Mulder, A.M. (1965). *Proceedings of the South African Society of Animal Production* **4** : 205.

Van Wyk, L.C. (1971). M.Sc. Thesis, University of Stellenbosch.

Van Wyk, L.C. Van Niekerk, C.H., Belonje, P.C. (1972a). *Journal of the South African Veterinary Association* **43** : 19.

Van Wyk, L.C., Van Niekerk, C.H., Belonje, P.C. (1972b). *Journal of the South African Veterinary Association* **43** : 29.

Van Wyk, L.C., Van Niekerk, C.H., Belonje, P.C., and Jöhle, W. (1972). *Agroanimalia* **4** : 73.

Vosloo, L.P., Hunter, G.L., and Carstens, J.W. (1969). *Proceedings of the South African Society of Animal Production* **8** : 145.

Whiteman, J.V., Zollinger, W.A., Thrift, F.A., and Gould, M.B. (1972). *Journal of Animal Science* **35** : 836.

37

EMBRYO MORTALITY

T.N. EDEY

University of New England, Armidale, N.S.W., Australia

SUMMARY

Estimates of basal embryo mortality in sheep (i.e. that occurring in the absence of recognized stress) are now available for a wide range of environments and in a number of genotypes. About 20-30% of fertilized eggs are usually lost, though occasionally the loss is much lower. The many ova or embryos which die by day 12 cause no disturbance of the normal cycle length, but those that survive beyond this time prevent regression of the corpus luteum. Rapid elongation of the membranes commences on day 12, and subsequent death results in delayed oestrus because a secreting corpus luteum is maintained until resorption of the membranes is substantially complete. The majority of deaths occur before day 18, and most ewes, even if experiencing a long cycle, will be re-mated within the normal joining period. However, fertility is lower at this mating, due probably to impaired sperm transport, and some ewes remain barren because of delayed embryonic death. A significant part of the basal embryonic loss is inevitable and even desirable in disposing of unfit genetic material arising mainly from mutations and damaged zygotes. Knowledge is lacking as to whether the variation between estimates of basal loss is mainly genetic or environmental in origin.

Embryonic loss can be increased above the basal level by environmental factors such as high temperature, severe short-term undernutrition, disease and perhaps by a general "stress" syndrome. Considerable progress has been made in identifying the conditions under which these factors may be detrimental. In many studies, embryos arising from twin ovulations have proved more vulnerable than singles, though the reason for this is not understood.

Some progress is being made toward understanding why many apparently normal embryos fail. Gradually, biochemical studies of the uterine and tubal fluids are characterizing the maternal environment provided for the ovum and embryo, while *in vitro* culture studies are elucidating the conditions which they can tolerate. There have been some studies relating hormonal changes to variations in the reproductive tract, and a few which indicate the hormonal changes which can occur in response to environmental stresses. However, as these fields are complex, progress will inevitably be slow.

INTRODUCTION

Embryo mortality is normally taken to mean the deaths of fertilized ova and embryos up to the end of implantation—about day 40 in the sheep. It is the main source of loss during pregnancy, deaths during the foetal period usually being few. Because most of the deaths occur sufficiently early in pregnancy to allow at least one more service before the rams are removed, embryo mortality does not usually cause a dramatic fall in lambing percentages; rather, it delays lambing, increases its time distribution, reduces twinning rates and leaves a few barren ewes. Thus its effects tend to be insidious rather than dramatic, and it tends to be accepted as an inevitable part of the reproductive process. This chapter will consider the incidence and nature of embryonic death, and the prospects for reducing its effects in breeding flocks.

BASAL EMBRYO MORTALITY

When embryo mortality is assessed in a group of ewes under good husbandry conditions, it is assumed that the losses after fertilization are not associated with any particular environmental factor, but are due to some innate failure of the cleaving ovum or of the embryo to develop, or alternatively, of the maternal utero-tubal system to support their development. This can be termed the basal loss of embryos. Under certain circumstances, it is possible to detect increased mortality due to detrimental environmental or genetic factors, and this is termed induced embryonic loss.

The many estimates of basal embryo mortality, and the procedures used to make them were summarized in an earlier review (Edey 1969). Under a wide range of conditions, and with many breeds, a high proportion of the estimates of loss of fertilized ova fall in the range 20-30%. However, occasional reliable estimates fall well below this level and encourage the hope that certain genotypes or management systems could yield reduced losses.

It has been argued persuasively by Bishop (1964) that much basal embryo mortality is unavoidable and should be regarded as a normal way of eliminating unfit genotypes in each generation. His thesis is that the fault often lies in the male or female gametes or arises by mutation during development, rather than being attributable to the maternal environment. In this context it is known that a considerable number of human abortuses have abnormal karyotypes (Review by Biggers 1969). The relative importance of such inevitable deaths in sheep is unknown and the early stage at which most deaths occur would make investigations difficult.

EMBRYONIC DEVELOPMENT—THE NORMAL PROCESS

Taking the day of onset of oestrus as day zero, ovulation and fertilization usually occur on day one and the cleaving ovum passes into the uterus on day four. Migration of embryos between the horns of the uterus apparently occurs after day 10 but not later than day 14 (Abenes and Woody 1971). Rowson and Moor (1966) reported that the early cleavage stages occur within the zona pellucida but that by day seven a majority of morulae have shed their zonae, and by day nine there is considerable swelling and a prominent embryonic disc is evident. Day 11 blastocysts are readily visible to the naked eye and from about day 12 rapid elongation of the chorionic sac occurs to a mean of about four cm on day 13 and ten cm on day 14. By day 20 the fluid-filled chorion, allantois and amnion distend the uterus and a one cm embryo with a visible heartbeat is present.

The presence of the developing embryo about day 12 prevents regression of the corpus luteum (Moor and Rowson 1966) and permits establishment of the corpus luteum of pregnancy. However, if an embryo dies before day 13 the corpus luteum regresses and the ewe comes into oestrus again about day 17 (Edey 1967) thus giving no evidence that it had conceived. Embryos which die after day 13 extend the secreting life of the corpus luteum (Dooley, Wodzicka-Tomaszewska and Edey 1974) resulting in prolonged oestrous cycles.

THE CAUSES OF EMBRYO DEATH

The last decade has seen a quickening of progress in defining the factors associated with, or causing, increased levels of embryo mortality. Most work has been concentrated on the effects of undernutrition and high temperature but some studies of the effects of disease, genotype, adrenal hyperfunction, ovulation rate and age have also been undertaken.

(1) Nutrition

The relationship between nutrition and embryo survival was reviewed in detail recently (Edey 1976). It should be noted that nutritional treatments have not only a short-term effect which may be reflected in blood nutrient levels, but also long-term effects represented by change in live weight. The inter-relationships between nutrition, live weight and ovulation rate on the one hand, and ovulation rate and embryo mortality on the other, complicates the interpretation of many experiments in which differential treatments, resulting in changes in ovulation rate, are imposed before mating. These will be referred to later.

Regarding high-gain rations, a series of studies reviewed by Casida (1964) suggested inconclusively that high planes of nutrition before and after mating could increase embryo mortality. In a recent study, Cumming *et al.* (1975) concluded that Merino but not Border Leicester x Merino ewes showed higher embryo mortality when fed 200% compared to 100% of a maintenance ration from day two to 16 post-mating.

Experiments involving undernutrition can be placed in three main categories involving: (a) long periods of chronic underfeeding during pregnancy, (b) short periods of severe underfeeding early in pregnancy,

(c) combinations of low and high nutrition in the few weeks before and after mating. In category (a), work by Coop and Clark (1969) and Bennett, Nadin and Axelsen (1970) showed no induced embryo mortality, while Bennett, Axelsen and Chapman (1964) reported a significant drop in the lambing percentage of two-year-olds but not of mature ewes. In category (b), Edey (1966) found that severe undernutrition for seven days during the first three weeks of pregnancy caused the loss of 16% of fertilized ova, but similar treatments in subsequent years gave negative or equivocal results (Edey 1970b). However, Cumming (1972b) Blockey, Cumming and Baxter (1974) and MacKenzie and Edey (1975a) have confirmed that severe short-term undernutrition can cause a measurable level of embryo mortality. It appears from the work of Cumming (1972b) that the week of application is not critical but that increasing the duration of underfeeding from one to three weeks is detrimental to survival. In category (c) experiments, involving high or low nutrition during the few weeks before and after mating, Cumming (1972a) and MacKenzie and Edey (1975b) found that pre-mating undernutrition had little effect on embryo survival but that in ewes which lost weight after mating, embryo mortality was increased.

By using more prolonged or more severe pre-mating treatments, Gunn, Doney and Russel (1972) and Gunn and Doney (1975) produced mating groups which differed widely in body condition. They concluded that regardless of the post-mating treatment, low body condition at mating was detrimental to embryo survival, thus confirming the tentative findings of Edey (1970a, 1970b) and Guerra, Thwaites and Edey (1971). Coop (1962) had earlier noted the relatively high rate of barrenness in ewes of low live weight.

As is desirable in such experiments, the severity of many of the treatments goes well beyond what would be expected under farm conditions. Consequently, it is safe to conclude that in mature ewes, prolonged but moderate feed restriction from about day 20 to day 100 of pregnancy is unlikely to reduce lambing percentages. Two-tooth ewes should be treated with more caution in the light of the results of Bennett, Axelsen and Chapman (1964). Mated ewe lambs have not been studied in this context, but in any case, severe feed restrictions for such animals would be undesirable for general growth reasons. Short-term undernutrition due to handling or transport is rarely likely to equal the severity of the experimental treatments recorded, so while in practice embryo losses from this cause will normally be small, their possible occurrence should not be ignored. The poor performance of ewes in low body condition is of special relevance for overstocked flocks and in drought conditions; such ewes can be expected to have low ovulation rates, above average embryo mortality and a high incidence of barrenness.

In a number of studies on nutrition and embryo mortality, it has been noted that losses are disproportionately high in twin ovulators. In some cases (e.g. Edey 1966; Cumming 1972a, 1972b) twin ovulators lose both embryos while single ovulators escape loss, to the point where more twin than single ovulators are barren.

The discussion, so far, has been concentrated on the effects of the energy component of nutrition. In fact, there is no evidence for an effect of protein level, and amongst the minerals only selenium is known to be involved in embryo mortality. Hartley (1963) reported that in some areas of New Zealand a reduced fertility condition characterized by a high incidence of barren ewes was eliminated by a single dose of selenium. Improved lambing rates, presumably due to decreased embryo mortality were obtained following administration of vitamin E and selenium in selenium-deficient areas of Scotland (Mudd and Mackie 1973).

Amongst specific plants, kale can cause embryonic deaths (Williams, Hill and Alderman 1965) as can the weed *Veratrum californicum* (Van Kampen *et al.* 1969). The severe fall in fertility sometimes associated with the grazing of oestrogenic clovers is not attributable to any measurable degree of embryo mortality.

(2) High Temperature

Continuous high temperatures were shown by Dutt, Ellington and Carlton (1959) and Alliston and Ulberg (1961) to cause the deaths of a high proportion of embryos in early pregnancy, the greatest effect being in the first week after mating. Subsequent work on the application of these hot-room findings to field conditions has shown that allowing a diurnal variation in the hot-room conditions to simulate natural day and night temperatures, reduced embryo mortality from 83% (continuous heating) to 35% (Thwaites 1969). Reducing the duration of heating from the seven or more days used by most previous experiments to three days (the likely duration of a heat-wave) still resulted in considerable mortality, especially where the treatment commenced the day after mating (Thwaites 1971). Thwaites concluded that when diurnal variation was introduced, heat-induced embryo mortality was likely to be of minor importance as a source of reproductive wastage in the sub-tropical sheep breeding areas of Australia. Smith, Bell and Chaneet (1966) reached a similar conclusion, and a field study by Entwistle (1972) in north-central Queensland excluded heat-induced embryo mortality as an important source of loss. On the other hand, Lindsay *et al.* (1975), in a study of 29,500 ewes in south-west Western Australia found some evidence that heat stress at the time of mating may have been effecting embryo survival.

(3) Stress

In the literature relating to embryo mortality, this term has been used somewhat loosely as in heat stress, nutritional stress, climatic stress etc., sometimes without much regard to the mode of action of the stressing treatment. Generally speaking, however, workers have shown an awareness that their treatments are confounded with the effects of factors such as intensive handling, crowding and minor surgery. It has been common practice to turn to the concept of the general adaptation syndrome of Selye (1956), and to look for evidence that stressful treatments are having an effect on the adrenal cortex (e.g. Thwaites 1970; Griffiths, Gunn and Doney 1970). The discovery that injections of adrenocorticotrophic hormone increased embryonic loss in rats (Velardo 1959) engendered the idea that in the sheep, stressors may not merely be causing increased activity of the adrenal cortex, but that high corticosteroid levels could, in fact, be the cause of embryonic deaths. Results of trials of this hypothesis have been equivocal or negative (e.g. Howarth and Hawk 1968; Thwaites 1970; Tilton *et al.* 1972).

Fraser, Ritchie and Fraser (1975) pointed out that stressors evoke many different reactions, not necessarily including an increase of gluco-corticoid output. They offered the following definition of stresses in a veterinary context: "an animal is said to be in a state of stress if it is required to make abnormal or extreme adjustments to its physiology or behaviour in order to cope with adverse aspects of its environment and management". This definition appears to fit the way in which most workers on embryo mortality have used the term, but clearly they must continue to investigate the many pathways by which stressors act and not look to the adrenal as necessarily having any commanding role. Consequently it will be important to evolve experimental techniques which allow the imposition of a chosen stress while minimizing the side effects of handling, surgery and other departures from routine.

(4)   Other Factors

Whether or not ovulation rate influences embryo mortality is an important question in view of the potential which exists to increase ovulation rates through manipulation of live weight and through genetic change. The subject has not yet been adequately examined in the sheep, but there is little convincing evidence that basal embryo mortality is other than random i.e. that losses are higher (or lower) in eggs shed as singles or as twins. Data from superovulated (Robinson 1957) or litter-bearing sheep (Scanlan 1972) showing higher losses with larger litters should be applied with caution to single and twin bearers where uterine capacity is not limiting.

In some circumstances such as certain undernutrition experiments (e.g. Edey 1966; Cumming 1972a, 1972b), induced mortality tends to fall most heavily on the twin ovulators, sometimes as an all-or-none phenomenon, but at other times taking mainly one of two embryos.

The tendency for one embryo to migrate to the opposite uterine horn following a double ovulation on one ovary was seen as a possible source of loss by Casida, Woody and Pope (1966). Their data were re-analysed by Sittman (1972) who concluded that the hypothesis could not be sustained. A detailed study by Doney, Gunn and Smith (1973) led to the conclusion that embryos which migrate to the horn having no corpus luteum have decreased survival rates, but this is more than offset because the non-migrating embryos' chances of survival are enhanced by the consequent removal of uterine crowding effects.

Little is known about genotypic effects on embryo mortality. Estimates of basal loss do not differ widely between breeds (Edey 1969) but the few estimates available could scarcely be expected reliably to expose differences which might exist either between or within breeds. Therefore it is possible that there is a useful degree of genetic variation for this character. In response to nutritional treatments, breed differences in embryo mortality have been noted by Foote *et al.* (1959), Bellows *et al.* (1963) and Cumming *et al.* (1975). There is also a suggestion (Doney and Smith 1968) that inbreeding increases the loss of embryos.

A relatively high level of embryo mortality early in the breeding season was reported by Hulet *et al.* (1956). A similar low fertility syndrome was noted in spring-mated Merino ewes by Oldham, Knight and Lindsay (1976). This was associated with failure of a proportion of ewes to show a quiet ovulation before onset of the breeding season, but whether failure to lamb was due to fertilization failure or embryo mortality is uncertain.

The condition of gametes at the time of fertilization is known to affect embryo mortality rates. With natural mating, ageing of gametes can occur through poor synchronization of mating and ovulation, though this should not be a common occurrence. However, the use of frozen semen (Salamon and Lightfoot 1967) and of semen from heat-stressed rams (Rathore 1968) has been shown to permit fertilization which was followed by a high level of embryo mortality.

Severe cold is sometimes proposed as a cause of embryo mortality (e.g. Theriez, Molenat and Aguer 1972). In the only experiment which tests this possibility, Griffiths, Gunn and Doney (1970) found that housed ewes had higher ovulation rates and a small superiority in embryo survival over unsheltered ewes which were wetted periodically.

There is little evidence that disease plays any significant role in embryo mortality in the sheep. The organisms which can cause reproductive problems in this species almost invariably have their effects in middle or late pregnancy.

THE DIAGNOSIS OF EMBRYONIC DEATH

In the field it is quite difficult to identify embryonic death, since a flock infertility problem may be compounded of failure of oestrus, low ovulation rate, fertilization failure, embryo mortality and poor neo-natal survival. Failure of oestrus can be detected readily by the use of raddled rams, and neo-natal performance assessed by counting dead lambs. Failure of fertilization and embryo mortality are much more difficult to assess and to separate since the assessment of fertilization rate requires either the slaughter of a sample of the ewes about day two or the collection of eggs following laparotomy. These services are rarely available to the farmer, and in their absence he must satisfy himself that ram fertility and mating management were at an acceptable level so that a fertilization rate of say, 85 or 90% can be assumed. Returns to service will then indicate the percentage of failures above this level, which represents the loss from embryonic death. The lengths of the oestrous cycles add evidence as to the time at which death occurred. However, this method takes no account of the effect of twin ovulations. Single ovulators and twin ovulators which lose both embryos will return to service after embryonic deaths but twin ovulators losing only one of the two eggs shed will continue their pregnancy and the loss will not be detected. A fertilization check on day two allows an estimate of ovulation rate to be made; an alternative is to examine the ovaries via an endoscope at some stage of the cycle.

Embryos which die are resorbed in a process which has been studied by Edey (1967), Thwaites (1972) and Sawyer and Knight (1975). The times at which ewes return to service after embryonic deaths at different stages are shown in Table 1.

TABLE 1

ESTIMATES OF THE LENGTHS OF CYCLES
FOLLOWING EMBRYO DEATHS ON DAYS 9 TO 30

	Day on which embryo died									Sources of data
	9	11	13*	15	17†	19	20	25	30	
Day of	17	18	20	25		36				Edey (1967)
return to			20	25	35	40				Thwaites (1972)
service							35	40	45	Sawyer and Knight (1975)

*Cycles significantly prolonged. †Standard errors become larger after this time.

Following embryonic death, a rapid breakdown of membranes and embryos occurs, accompanied by a marked leucocyte invasion of the uterine lumen (Thwaites 1972) and a considerable discharge of debris through the vagina. Vaginal swabs show varying amounts of debris, a large bacterial population and many leucocytes (Sawyer and Knight 1975). A fresh ovulation often occurs before resorption is complete and it is clear from the work of Edey (1972), Sawyer and Knight (1975) and Edey *et al.* (1975) as well as from field evidence (Edey 1970b; MacKenzie and Edey 1975a) that fertility at the oestrus following resorption is seriously depressed. Impaired sperm transport at this time has been demonstrated (Edey *et al.* 1975).

The time at which basal embryo mortality occurs still requires further study. Quinlivan *et al.* (1966) found that much of it was detectable on day

18. The presence of debris at this time would be consistent with deaths occurring between days 14 and 18, and ewes suffering such deaths would have a prolonged cycle and be subject to the reduced fertility just discussed. These are probably the main contributors to the pool of barren ewes. Another large component of embryo mortality occurs too early to affect the subsequent cycle length, and there is a need for further studies of the period between day 2 and 13 to discover when these deaths are occurring.

In the diagnosis of embryonic death, it has been traditional to regard cycles up to 21 days in length as normal. However, in unmated ewes, oestrous cycles exceeding 19 days are unusual (MacKenzie and Edey 1975a), and more accurate indications of the occurrence of embryonic death will be obtained if in mated ewes, cycles of 20 days and longer are taken to indicate deaths occurring after day 12.

PHYSIOLOGICAL MECHANISMS

Establishing the physiological reasons for embryonic death is an immensely complex task. It requires, firstly, an understanding of the normal biochemistry of the uterus and its interactions with the ovum-embryo, and then an assessment of the events leading to failure of the embryo to survive. In recent years there has been a rapid acceleration of research on the endometrium, the uterine and tubal secretions, blastocyst-endometrial relationships and the requirements of the embryo. While a high proportion of the work has been with laboratory animals (e.g. McLaren 1973; Wales 1973; Beier 1974) there is some parallel work in the sheep. The rapid development of hormone assay techniques is allowing the definition of normal hormone patterns in early pregnancy and the hormonal response to some stress factors. For example progesterone insufficiency has not only been excluded as a cause of embryonic death but there is evidence that high levels of plasma progesterone which can occur with low planes of nutrition, may be related to embryonic death (Cumming *et al.* 1971).

While the resources now being focused on uterine-embryo relationships in laboratory animals and humans can be expected to have flow-on benefits for knowledge of the sheep, it would be unrealistic to expect that rapid progress will be made in unravelling the physiological causes of basal and induced embryo mortality.

REFERENCES

Abenes, F.B. and Woody, C.O. (1971) *Journal of Animal Science* **33** : 314.
Alliston, C.W. and Ulberg, L.C. (1961) *Journal of Animal Science* **20** : 608.
Beier, H.M. (1974) *Journal of Reproduction and Fertility* **37** : 221.
Bennett, D., Axelsen, A. and Chapman, H.W. (1964) *Proceedings of the Australian Society of Animal Production* **5** : 70.
Bennett, D., Nadin, J.B. and Axelsen, A. (1970) *Proceedings of the Australian Society of Animal Production* **8** : 362.
Biggers, J.D. (1969) *Journal of Reproduction and Fertility,* Supplement 8 : 27.

Bishop, M.H.W. (1964) *Journal of Reproduction and Fertility* **7** : 383.
Blockey, M.A. de B., Cumming, I.A. and Baxter, R.W. (1974) *Proceedings of the Australian Society of Animal Production* **10** : 265.
Casida, L.E. (1964) Proceedings of the 6th International Congress on Nutrition, Edinburgh, 1963 : 366.
Casida, L.E., Woody, C.O. and Pope, A.L. (1966) *Journal of Animal Science* **25** : 1169.
Coop, I.E. (1962) *New Zealand Journal of Agricultural Research* **5** : 249.
Coop, I.E. and Clark, V.R. (1969) *Journal of Agricultural Science, Cambridge* **73** : 387.
Cumming, I.A. (1972a) *Proceedings of the Australian Society of Animal Protection* **9** : 192.
Cumming, I.A. (1972b) *Proceedings of the Australian Society of Animal Protection* **9** : 199.
Cumming, I.A., Blockey, M.A. de B., Winfield, C.G., Parr, R.A. and Williams, A.H. (1975) *Journal of Agricultural Science, Cambridge* **84** : 559.
Cumming, I.A., Mole, B.J., Obst, J., Blockey, M.A. de B., Winfield, C.G. and Goding, J.R. (1971) *Journal of Reproduction and Fertility* **24** : 146.
Doney, J.M., Gunn, R.G. and Smith, W.F. (1973) *Journal of Reproduction and Fertility* **34** : 363.
Doney, J.M. and Smith, W.F. (1968) *Journal of Reproduction and Fertility* **15** : 277.
Dooley, M., Wodzicka-Tomaszewska, M. and Edey, T.N. (1974) *Journal of Reproduction and Fertility* **36** : 462.

Dutt, R.H., Ellington, E.F. and Carlton, W.W. (1959) *Journal of Animal Science* **18** : 1308.
Edey, T.N. (1966) *Journal of Agricultural Science, Cambridge* **67** : 287.
Edey, T.N. (1967) *Journal of Reproduction and Fertility* **13** : 437.
Edey, T.N. (1969) *Animal Breeding Abstracts* **37** : 173.
Edey, T.N. (1970a) *Journal of Agricultural Science, Cambridge* **74** : 181.
Edey, T.N. (1970b) *Journal of Agricultural Science, Cambridge* **74** : 199.
Edey, T.N. (1972) *Journal of Reproduction and Fertility* **28** : 147.
Edey, T.N. (1976) *Proceedings of the New Zealand Society of Animal Production* **36:** 231.
Edey, T.N., Thwaites, C.J., Pigott, F.A. and O'Shea, T. (1975) *Journal of Reproduction and Fertility* **43** : 485.

Entwistle, K.W. (1972) *Australian Veterinary Journal* **48** : 395.
Foote, W.C., Pope, A.L., Chapman, A.B. and Casida, L.E. (1959) *Journal of Animal Science* **18** : 453.
Fraser, D., Ritchie, J.S.D. and Fraser, A.F. (1975) *British Veterinary Journal* **131** : 653.
Griffiths, J.G., Gunn, R.G. and Doney, J.M. (1970) *Journal of Agricultural Science, Cambridge* **75** : 485.
Guerra, J.C., Thwaites, C.J. and Edey, T.N. (1971) *Journal of Agricultural Science, Cambridge* **76** : 177.
Gunn, R.G., Doney, J.M. and Russel, A.J.F. (1972) *Journal of Agricultural Science, Cambridge* **79** : 19.
Hartley, W.J. (1963) *Proceedings of the New Zealand Society of Animal Production* **23** : 20.
Howarth, B. and Hawk, H.W. (1966) *Journal of Animal Science* **25** : 924.
Hulet, C.V., Voigtlander, H.P., Pope, A.L. and Casida, L.E. (1956) *Journal of Animal Science* **15** : 607.
Lindsay, D.R., Knight, T.W., Smith, J.F. and Oldham, C.M. (1975) *Australian Journal of Agricultural Research* **26** : 189.
MacKenzie, A.J. and Edey, T.N. (1975a) *Journal of Agricultural Science, Cambridge* **84** : 113.
MacKenzie, A.J. and Edey, T.N. (1975b) *Journal of Agricultural Science, Cambridge* **84:** 119.
McLaren, A. (1973) *Journal of Reproduction and Fertility,* Supplement 18 : 159.
Moor, R.M. and Rowson, L.E.A. (1966) *Journal of Endocrinology* **34** : 233.
Mudd, A.J. and Mackie, I.L. (1973) *Veterinary Record* **93** : 197.
Oldham, C.M., Knight, T.W. and Lindsay, D.R. (1976) *Proceedings of the Australian Society of Animal Production* **11** : 129.

Quinlivan, T.D., Martin, C.A., Taylor, W.B. and Cairney, I.M. (1966) *Journal of Reproduction and Fertility* **11** : 379.

Rathore, A.K. (1968) *Proceedings of the Australian Society of Animal Production* **7** : 270.

Robinson, T.J. (1957) In "Progress in the Physiology of Farm Animals" Vol. 3. (Ed. by J. Hammond, Butterworths : London).

Rowson, L.E.A. and Moor, R.M. (1966) *Journal of Anatomy* **100** : 777.

Salamon, S. and Lightfoot, R.J. (1967) *Nature, U.K.* **216** : 194.

Sawyer, G.J. and Knight, T.W. (1975) *Australian Journal of Experimental Agriculture and Animal Husbandry* **15** : 189.

Scanlan, P.F. (1972) Proceedings 7th International Congress on Animal Reproduction and Artificial Insemination, Munich, Summaries p.225.

Selye, H. (1956) "The Stress of Life" (Longmans Green & Co. : London).

Sittmann, K. (1972) *Canadian Journal of Animal Science* **52** : 195.

Smith, I.D., Bell, G.H. and De Chaneet, G. (1966) *Australian Veterinary Journal* **42** : 468.

Theriez, M. Molenat, G. and Auger, D. (1972) Proceedings 7th International Congress on Animal Reproduction and Artificial Insemination, Munich, Summaries p.365.

Thwaites, C.J. (1969) *Journal of Reproduction and Fertility* **19** : 255.

Thwaites, C.J. (1970) *Journal of Reproduction and Fertility* **21** : 95.

Thwaites, C.J. (1971) *Australian Journal of Experimental Agriculture and Animal Husbandry* **11** : 265.

Thwaites, C.J. (1972) *Australian Journal of Biological Sciences* **25** : 597.

Tilton, J.E., Hoffman, R.H., Berg, I.E., Light, M.R. and Buchanan, M.L. (1972) *Journal of Animal Science* **34** : 605.

Van Kampen, K.R., Binns, W., James, L.F. and Balls, L.D. (1969) *American Journal of Veterinary Research* **30** : 517.

Velardo, J.T. (1957) *American Journal of Physiology* **191** : 319.

Wales, R.G. (1973) *Journal of Reproduction and Fertility,* Supplement 18 : 117.

Williams, H.L.L., Hill, R. and Alderman, G. (1965) *British Veterinary Journal* **121** : 2.

38

EVIDENCE AND CONFIRMATION OF LATE EMBRYO LOSS IN A FLOCK OF MERINO EWES

R.N. TYRRELL*, A.R. GLEESON†, B.D. FERGUSON‡,
W.J. O'HALLORAN‡, and R.J. KILGOUR #

SUMMARY

The results of fertility investigations between 1972 and 1974 in a commercial Merino flock are presented. Ewes which failed to lamb represented a major component of reproductive wastage (31% of maidens and 12% of adults, P< 0.01). The factors responsible for these failures were assessed by dividing all ewes into five categories based on whether or not they mated and whether or not they returned to service. Overall, 36% of maidens and 25% of adults returned to service. In each year, the majority of ewes mated once only and had non-return intervals of at least 28 days at the end of joining. Six per cent of non-returning adults failed to lamb (compared with 21% of returning ewes, P< 0.01), but 28% of non-returning maidens failed (36% of returning ewes, n.s.). The high incidence of barrenness amongst non-returning maiden ewes indicated a problem of late embryo loss. The alternative possibility that the ewes had failed to conceive and then entered into anoestrus was dismissed in 1974. Plasma progesterone measurements revealed that none of the non-returning maiden ewes had been in anoestrus at the end of joining. The usefulness of barrenness amongst non-returning ewes as an indicator of late embryo loss is discussed.

INTRODUCTION

Most field studies show that barrenness (i.e. failure to lamb) amongst ewes which have mated accounts for an important part of reproductive wastage in Merino flocks (Lino and Braden, 1968; Chopping and Lindsay, 1970; Knight et al., 1975; Kennedy et al., 1976). The variable nature of this and other fertility problems from property to property has prompted the widely held view that individual property investigations are necessary if satisfactory remedies are to be developed. Yet, the solution of "barren ewe" problems on individual properties often presents a formidable task due to the necessity to obtain precise information about fertilization failure and embryo mortality and, almost inevitably, requires specialized personnel and expensive techniques (Marshall, Beetson and Lightfoot, 1976; Wroth and Lightfoot, 1976; Restall et al., 1976).

* Department of Agriculture, c/o CSIRO, Division of Animal Production, P.O. Box 239, Blacktown, N.S.W. 2148.
† Department of Agriculture, State Office Block, Sydney, N.S.W. 2000.
‡ Department of Agriculture, Wagga Wagga, N.S.W. 2650.
# Department of Agriculture, Tamworth, N.S.W. 2340.

327

There is a pressing need to develop simple techniques for partitioning the causes of prenatal mortality, so that large numbers of flocks can be effectively investigated at reasonable cost. The efficient use of routinely collected mating and lambing information (see Connors, 1971) can partly fulfil this need. The combined effects of early embryo mortality and fertilization failure are reflected in the proportion of ewes which return to service within 19 days of mating (MacKenzie and Edey, 1975) and barrenness amongst ewes which do not return to service is, presumably, indicative of late embryo loss or foetal loss (Morley, 1954).

This chapter presents detailed results of an "on farm" fertility investigation where late embryo loss was diagnosed as an important problem. Barrenness amongst non-returning ewes was used as the initial criterion. In view of the possibility that these ewes had mated, failed to conceive and then entered into anoestrus (Edey, 1969) an early pregnancy diagnosis procedure was developed using plasma progesterone measurements, which also enabled the identification of anoestrous ewes (Turrell, Gleeson and Peter; in preparation). This procedure was then included in the flock measurements in an effort to confirm the original diagnosis.

MATERIALS AND METHODS

(i)   Sheep and management

The flock consisted of Merino ewes of the Inglewood strain and was located near Bombala on the southern tablelands of NSW. The study commenced in 1972 and initially aimed to make a broad examination of the flock's reproductive performance. An experimental flock of 400 ewes was formed for this purpose with approximately equal numbers coming from each of the four age groups between $1\frac{1}{2}$ and $4\frac{1}{2}$ years. The ewes were weighed and joined to eight Merino rams for six weeks commencing early March. The rams were fitted with mating crayons (Radford, Watson and Wood, 1960) and the ewes were individually identified so that fortnightly checks could be made to determine when each ewe was mated. Pregnancy was diagnosed two weeks prior to the start of lambing by subjecting ewes not obviously pregnant on the basis of udder development (Dun, 1963) to further examination with an ultrasonic pregnancy detector. Shortly after lambing had ceased, lactating and non-lactating ewes were identified and these results compared with the pre-lambing diagnosis. Ewes which had been diagnosed pregnant and subsequently found to be non-lactating were considered to have lost their lamb(s).

This procedure was repeated in 1973. In 1974 the maiden ewes were joined separately to the adults, joining was for eight weeks and raddle marks were observed weekly.

(ii) Seasonal conditions

Seasonal conditions varied markedly over the three years. Dry conditions prevailed in 1972 so that handfeeding was necessary and the ewes were in poor condition at the start of joining in 1973. Conditions improved after October 1973 and were not harsh in 1974.

(iii) Experiment to confirm the diagnosis of late embryo loss

At the end of joining in 1974, 68 maiden ewes and 81 adult ewes were diagnosed as pregnant, cycling or anoestrous using plasma progesterone measurements (Tyrrell, Gleeson and Peter, in preparation). This involved taking a blood sample on the last day of joining followed by further samples on the fifth and eleventh days thereafter. Only ewes which mated during the first four weeks of joining without returning to service during the subsequent four weeks were sampled.

(iv) Organizaton and analysis of data

The percentages of barren ewes, ewes losing lambs, ewes rearing twins and lambs marked were calculated each year for the whole flock. Barren ewes and ewes which lambed were separated into maiden and adult age groups in each year and divided according to five mating categories:

1. Ewes which returned to service at least once.
2. Ewes which mated once only and having a minimum non-return period of 28 days, by the end of joining.
3. Ewes which mated once only and having a minimum non-return period of 14 days by the end of joining.
4. Ewes mated once only without opportunity to return to service by the end of joining.
5. Not mated.

Differences in the incidence of barren ewes in the various categories were examined using Chi-square.

RESULTS

Flock fertility each year is summarized in Table 1. A prominent area of reproductive wastage was the consistently high level of barren ewes (16-19% of ewes present at lambing). Between 11 and 24% of ewes which lambed lost all of their lambs and between 13 and 24% of ewes rearing lambs reared twins. The combined effect of these factors was to produce between 64 and 92% of lambs marked to ewes joined.

Overall, significantly more maiden ewes than adults failed to lamb (31% vs 12%, $P < 0.01$), and the association between these failures and the five mating categories can be seen in Table 2. Ewes which failed to mate with the ram (Category 5) were responsible for only 8% of the barren maiden ewes

TABLE 1

OVERALL FLOCK FERTILITY

	1972	1973	1974
Number of ewes joined*	398	372	465
Number of ewes present at lambing*	382	357	417
Barren ewes[1]	18.1	16.2	18.7
Ewes losing all lambs[2]	19.5	10.6	24.3
Ewes rearing twins[3]	17.3	24.0	13.4
Lambs marked[4]	71.6	91.9	63.9

[1] Percentage of ewes present at lambing.
[2] Percentage of ewes which lambed.
[3] Percentage of ewes rearing lambs.
[4] Percentage of ewes joined.
* Differences due to deaths and/or missing at lambing muster.

TABLE 2

NUMBERS OF EWES (AND PERCENTAGE WHICH FAILED TO LAMB) WITHIN EACH MATING CATEGORY

Age	Mating Category*	1972 No (%)	1973 No (%)	1974 No (%)	Total No (%)
Maiden	1	22 (59)	33 (30)	71 (31)	126 (36)
	2	45 (16)	27 (26)	90 (34)	162 (28)
	3	15 (13)	17 (12)	9 (11)	41 (12)
	4	5 (40)	3 (100)	2 (50)	10 (60)
	5	9 (44)	3 (33)	4 (100)	17 (56)
	Total	96 (29)	83 (28)	176 (33)	355 (31)
Adult	1	51 (31)	76 (17)	72 (17)	199 (21)
	2	170 (7)	110 (7)	163 (3)	443 (6)
	3	54 (15)	72 (6)	2 (0)	128 (9)
	4	2 (50)	7 (43)	1 (100)	10 (40)
	5	9 (44)	9 (78)	3 (67)	21 (62)
	Total	286 (14)	274 (13)	241 (8)	801 (12)

* Categories are described fully in Materials and Methods.

and 14% of barren adults and could not be regarded as constituting a major component of the reproductive wastage.

Twenty-four per cent, 40% and 41% of maiden ewes returned to service in 1972, 1973 and 1974 respectively, compared with 19%, 31% and 35% for the adults. Between 30 and 59% of returning maiden ewes failed to lamb. These figures can be compared with the level of barrenness amongst ewes which mated once only and had a minimum non-return period of 28 days (Category 2). Although most ewes in this category would be expected to lamb, this was not the case with the maidens where the incidence of barrenness varied between 16 and 34%. Overall, barrenness amongst these non-returning maiden ewes did not differ significantly from that amongst the returning maiden ewes. The situation was quite different amongst the adult ewes where between 17 and 31% of returning ewes failed to lamb, compared with only 3 to 7% in Category 2 (P< 0.01).

(i)   Experiment to test diagnosis of late embryo loss

Of 68 non-returning maiden ewes sampled, 19 failed to lamb and none of these had been diagnosed as anoestrous at the end of joining (Table 3). Twelve were diagnosed as pregnant at the end of joining while the remaining seven were diagnosed as cycling.

TABLE 3

EXPERIMENT TO CONFIRM LATE EMBRYO LOSS

Age	No. sampled	No. barren	Division of barren ewes according to the classification at the end of joining		
			Pregnant	Non-pregnant	
				Anoestrous	Cycling
Maidens	68	19	12	0	7
Adults	81	2	1	0	1

DISCUSSION

Lamb production on this property has been severely restricted by large proportions of barren ewes and the data analyses outlined have been helpful in partitioning the factors responsible for this wastage. Returns to service amongst both maiden and adult ewes during 1973 and 1974 were higher than normally encountered by the NSW Fertility Service (Connors, 1971). Although returns to service are indicative of early embryo loss and/or fertilization failure it may not follow that these factors alone were also responsible for the high proportion of returning ewes which failed to lamb. In many cases, failure of pregnancy following the return to service may be due to factors completely independent of the initial failure. Quinlivan *et al.* (1966a,b) found that prenatal mortality in Romney ewes mating for their first time was not substantially different to ewes mating as returns to service. On the other hand, reports by Edey (1972) and Sawyer and Knight (1975) show that subsequent fertility is greatly impaired at the oestrus following late embryo death.

The incidence of barrenness amongst non-returning ewes has been shown to be a simple and reliable indicator of late embryo loss. This approach should continue to prove useful for field investigations so long as the possibility of anoestrum amongst non-returning ewes can be discounted. None of the ewes sampled for progesterone measurements at the end of joining in 1974 were in anoestrus. Seven of the 58 maiden ewes sampled came into oestrus during the 12-day sampling period and it is likely that these represent the "first wave" of ewes returning to service after suffering late embryo loss. Edey (1967) and Sawyer and Knight (1975) have shown that ewes which suffer embryo loss at about 20 days post coitus do not return to oestrus until between 9 and 19 days later. Theoretically the ewes in our sample would have had non-return intervals of between 28 and 55 days at the end of joining and should have been starting to return to service by this stage if they had suffered embryo loss about 20 days after mating. Although the progesterone results indicated that 12 of the 19 maiden ewes were pregnant at the end of joining, these ewes had probably lost their embryos some time earlier. Dooley and Wodzicka-Tomaszewska (1972) reported that plasma progesterone levels of ewes in which embryo loss was induced 15 days after mating, were indistinguishable from pregnant ewes for at least 11 days thereafter.

It seems reasonable to conclude that most, if not all, of the non-returning barren ewes sampled had at least reached about 20 days of pregnancy before suffering embryo loss. Few losses would be expected after day 30 (Quinlivan *et al.*, 1966a). It is likely that the problem of embryo loss would apply equally to all ewes in the flock which have conceived, whether it be to a first mating or a return to service.

At this stage it is hard to assess the physiological relevance of the distinction between "early" and "late" embryo loss. However, in practical terms, there is an important difference between the two classes, as ewes which suffer late embryo loss are unlikely to be afforded the opportunity to return to service.

The importance of embryo wastage has been assessed in other studies by subtracting the effects of fertilization failure (estimated in sub-samples at joining) from the pregnancy rate, obtained either during pregnancy (Quinlivan *et al.*, 1966a,b; Restall *et al.*, 1976) or at lambing (Marshall, Beetson and Lightfoot, 1976; Wroth and Lightfoot, 1976) and each approach has its own particular advantages. Apart from its relative simplicity, the approach covered in this paper has the advantage that individual ewes which have suffered embryo loss can be identified and then examined further in order to gain some insight into the cause of the problem. This course of action is now being taken on the present property.

ACKNOWLEDGEMENTS

The help and co-operation of the property owners, Messrs S. and J. Hood is gratefully acknowledged. The project was supported by grants from the Australian Wool Corporation and the Australian Meat Research Committee.

REFERENCES

Chopping, M.H. and Lindsay, D.R. (1970) *Proceedings of the Australian Society of Animal Production* **8** : 312.

Connors, R.W. (1971) *Wool Tec¹ nology and Sheep Breeding* **28** : 55.

Dooley, M. and Wodzicka-Tomaszewska, M. (1972) *Journal of Reproduction and Fertility* **28** : 150.

Dun, R.B. (1963) *Australian Journal of Experimental Agriculture and Animal Husbandry* **3** : 228.

Edey, T.N. (1967) *Journal of Reproduction and Fertility* **13** : 437.

Edey, T.N. (1969) *Animal Breeding Abstracts* **37**(2) : 173.

Edey, T.N. (1972) *Journal of Reproduction and Fertility* **28** : 147.

Kennedy, J.P., Auldist, I.H., Popovic, P.G. and Reynolds, J.A. (1976) *Proceedings of the Australian Society of Animal Production* **11** : 149.

Knight, T.W., Oldham, C.M., Smith, J.F. and Lindsay, D.R. (1975) *Australian Journal of Experimental Agriculture and Animal Husbandry* **15** : 183.

Lino, B.F. and Braden, A.W.H. (1968) *Australian Journal of Experimental Agriculture and Animal Husbandry* **8** : 505.

McKenzie, A.J. and Edey, T.N. (1975) *Journal of Agricultural Science, Cambridge* **84** : 113.

Marshall, T., Beetson, B.R. and Lightfoot, R.J. (1976) *Proceedings of the Australian Society of Animal Production* **11** : 225.

Morley, F.H.W. (1954) *Australian Veterinary Journal* **30** : 125.

Quinlivan, T.D., Martin, C.A., Taylor, W.B. and Cairney, I.M. (1966a) *Journal of Reproduction and Fertility* **11** : 379.

Quinlivan, T.D., Martin, C.A., Taylor, W.B. and Cairney, I.M. (1966b) *Journal of Reproduction and Fertility* **11** : 391.

Radford, H.M., Watson, R.H. and Wood, G.F. (1960) *Australian Veterinary Journal* **36** : 57.

Restall, B.J., Wilkins, J., Kilgour, R.J., Tyrrell, R.N. and Hearnshaw, H. (1976) *Australian Journal of Experimental Agriculture and Animal Husbandry* **80**: 344.

Sawyer, G.J. and Knight, T.W. (1975) *Australian Journal of Experimental Agriculture and Animal Husbandry* **15** : 189.

Wroth, R.H. and Lightfoot, R.J. (1976) *Proceedings of the Australian Society of Animal Production* **11** : 224.

39

RETURNS TO SERVICE, EMBRYONIC MORTALITY AND LAMBING PERFORMANCE OF EWES WITH ONE AND TWO OVULATIONS

R.W. KELLY and A.J. ALLISON
Invermay Agricultural Research Centre, New Zealand

SUMMARY

Data drawn from several experiments in which the ovulation rate in the first 17 days of mating was recorded either by laparoscopy or slaughter of the ewes between days 22-35 of pregnancy, were used to examine the effect of number of ovulations (1 *v.* 2) and position of ovulation (left *v.* right ovary) on returns to service, embryonic mortality and lambing performance of sheep. Neither number or position of ovulation had any significant effect on the percentage of ewes returning to service within 51 days of the commencement of mating. The number of ewes that failed to lamb was not significantly affected by position of ovulation, but there was a significant difference between ewes with one and two ovulations (10.1% *v.* 5.0%, S.E.D. = 1.7%, P < 0.01).

The percentage of ewes with two ovulations that lost one embryo during pregnancy was 32.3% (S.E. = 1.7%), and was not significantly different between ewes with two ovulations on one ovary or one on each ovary. It would appear that transuterine migration does not significantly affect embryo survival.

INTRODUCTION

In New Zealand the lambing percentage (number of lambs born per 100 ewes mated) is mainly limited by the number of multiple births, since barrenness figures are generally less than 6% (Quinlivan and Martin, 1971 a,b,c; Allison, 1975a,b). An important factor in determining the number of multiple births is partial failure of multiple pregnancies. Losses due to the death of one of two embryos during pregnancy can be considerable, although the estimates of loss vary widely. Allison (1975c) found that 28% of 5-6 year old Romney ewes with two corpora lutea had only one viable embryo present at slaughter 22-24 days after mating. In pregnant $2\frac{1}{2}$ year old Romney ewes slaughtered at 18, 30 and 140 days after mating Quinlivan, Martin, Taylor and Cairney (1966) reported that approximately 45% of ewes with two corpora lutea showed evidence of loss of one embryo or foetus.

For a high level of reproductive efficiency a low level of embryonic loss post-fertilization is desirable. The site of ovulation may be one factor which can affect the losses in single and multiple ovulating ewes. In ewes shedding two eggs, Casida, Woody and Pope (1966), Gunn, Doney and Russel (1972), and Doney, Gunn and Smith (1973) have suggested that the mortality is less when there is one ovulation on each ovary than the two ovulations on one ovary, although Edey (1970) failed to confirm this observation. Results are given for the reproductive performance of sheep recorded in several experiments in which the number and position of ovulation (left or right ovary) were noted, the ewes either being slaughtered and number of viable embryos counted, or individually lambed.

MATERIALS AND METHODS

All mating and ovulation data are from ewes marked in the first 17 days of mating. Number and position of ovulation were recorded either by laparoscopy (Kelly and Allison, 1976) or at slaughter of the ewes between days 22-35 of pregnancy. All ewes were $1\frac{1}{2}$-$6\frac{1}{2}$ years old at mating, and were predominantly Romney, although results from experiments involving Coopworth (Romney × Border Leicester), Perendale (Romney × Cheviot) and Merino have been included.

Uteri recovered from slaughtered ewes were dissected the same day, and the contents examined. Viability of embryos was determined with reference to crown-rump length, the development of the chorion and allantois and the presence of blood in the circulatory system. Ewes not slaughtered were individually identified with neck-tags at lambing and flocks were inspected twice daily to record number of lambs born per ewe.

For the ewes with two ovulations, two assumptions were made in analysing the results. Firstly, that fertilization rate was all or none. This may lead to a slight over-estimate of embryonic mortality due to some animals with two ovulations having one egg fertilized and one not fertilized. In our experience this is an extremely rare occurrence. Secondly, that the difference between number of ovulations and number of lambs born was due to embryonic mortality before day 30 of pregnancy, since it is well established that there is little embryonic or foetal loss after day 30 (Robinson, 1951; Quinlivan *et al.,* 1966; Dolling and Nicholson, 1967).

RESULTS

The returns to service in ewes from the lambing flocks, classified according to the number and position of ovulations, are summarized in Table 1. The percentage of ewes returning to service was not significantly affected by number or position of ovulations. The percentage of barren ewes, that is those that did not return to service but which failed to lamb (Table 2), was not significantly affected by the position of ovulation. There was, however, a significant difference in barrenness between ewes with one and two ovulations (10.1% v. 5.0% respectively, t=3.07, P<0.01).

TABLE 1

THE PERCENTAGE OF EWES FROM THE LAMBING FLOCKS WHICH
RETURNED TO SERVICE RELATED TO THE NUMBER AND
POSITION OF OVULATIONS* (NUMBER OF OBSERVATIONS IN
BRACKETS)

Number of ovulations	Position of ovulations	% ewes returning to service ± S.E.
1	Left ovary	7.3±1.8 (15/205)
	Right ovary	7.1±1.5 (21/297)
2	One left/one right	10.1±1.8 (28/278)
	Two left	7.0±2.1 (10/143)
	Two right	7.7±2.1 (13/169)

* Returns to service recorded days 18-51 after the start of mating.

TABLE 2

THE PERCENTAGE OF EWES NOT RETURNING TO SERVICE BUT
WHICH FAILED TO LAMB RELATED TO NUMBER AND POSITION
OF OVULATIONS

Number of ovulations	Position of ovulation	% ewes failing to lamb ± S.E.
1	Left ovary	8.9±2.1 (17/190)
	Right ovary	10.9±1.9 (30/276)
	TOTAL	10.1±1.4 (47/466)
2	One left/one right	5.2±1.4 (13/250)
	Two left	7.5±2.3 (10/133)
	Two right	2.6±1.3 (4/156)
	TOTAL	5.0±0.9 (27/539)

The slaughter and lambing results for ewes with two ovulations that conceived are presented in Table 3. The percentage of ewes which had lost one embryo (slaughter data) and the percentage of ewes which had only one lamb was not significantly different for each of the three positions of ovulations, and so the results were combined. For ewes with two ovulations on one ovary there was a tendency for partial mortality to be less when the ovulations were on the right ovary than the left, but the difference was not significant (27.7% v. 36.4%, $\chi^2_1 = 3.11$, $0.10 > P > 0.05$). Thirty-three per cent of the ewes with one ovulation on each ovary lost one embryo. Overall, the partial failure of twin pregnancies was 32.3% (S.E. = 1.7%).

TABLE 3

EWES WITH TWO OVULATIONS:

THE NUMBER OF ANIMALS WITH ONE OR TWO VIABLE EMBRYOS
AT SLAUGHTER, OR WHICH LAMBED ONE OR TWO LAMBS,
RELATED TO THE POSITION OF THE TWO OVULATIONS
(% VALUES AND S.E.S IN BRACKETS)

Source of data	Number of viable embryos or lambs born per ewe	Positions of ovulations		
		1 left/1 right	2 left	2 right
Slaughter	1	41 (30.1±3.9)	20 (37.7±6.7)	25 (29.1±4.9)
	2	95	33	61
	Total	136	53	86
Lambing	1	83 (35.0±3.1)	44 (35.8±4.3)	41 (27.0±3.6)
	2	154	79	111
	Total	237	123	152
Total	1	124 (33.2±2.4)	64 (36.3±3.6)	66 (27.7±2.9)
	2	249	112	172
	Total	373	176	238

TABLE 4

EWES WITH THREE OVULATIONS:

COMBINED DATA* FOR THE NUMBER OF VIABLE EMBRYOS AT
SLAUGHTER AND NUMBER OF LAMBS BORN PER EWE RELATED
TO THE POSITION OF OVULATION (% VALUES AND S.E.S IN
BRACKETS)

Number of viable embryos or lambs born per ewe	Position of ovulation	
	three on one ovary	one on one ovary and two on the other
1	9 (36.0±9.6)	16 (34.0±6.9)
2	6 (24.0±8.5)	16 (34.0±6.9)
3	10 (40.0±9.8)	15 (32.0±6.8)
TOTAL	25	47

* Slaughter and lambing data were not significantly different.

DISCUSSION

Returns to service were not significantly affected by either number or position of ovulation. Over the last six years observation in many experimental flocks has indicated that almost all ewes which return to service do so at a normal cyclic interval, supporting the contention that fertilization is virtually an all or none process.

Barrenness, although not affected by position of ovulation, was approximately twice as great in ewes with a single ovulation as ewes with two ovulations (10.1% v 5.0%). If the barrenness was due to embryonic death before days 20-30 of pregnancy, most of these ewes should have returned to service in the 51-day mating period, since the return to service interval for embryonic death between days 20-30 is about 15 days (Knight and Sawyer, 1975). It seems likely, therefore, that barrenness was due to embryonic or foetal death after at least day 20 of pregnancy. The factor(s) which causes this loss is unknown.

The partial failure of twin pregnancies in 32% of the ewes with two ovulations is probably not related to the levels of barrenness, since the majority of partial losses occur before or about day 18 of pregnancy (Edey, 1969). Position of ovulation had no significant effect on the amount of partial mortality, indicating that transuterine migration which occurs in the majority of cases when two ovulations are on one ovary (Scanlon, 1972) has little effect on the survival rate of the embryos. The evidence from a limited number of observations of ewes with three ovulations confirms this observation, since there was no difference in embryonic mortality between ewes with three ovulations on one ovary and one ovulation on one ovary and two on the other ovary (Table 4). This result is similar to that obtained by Edey (1970), and contrasts with the work of Casida *et al.* (1966), Gunn *et al.* (1972) and Doney *et al.* (1973) who found differences in embryonic mortality related to position of ovulation.

The magnitude of the loss is substantially less than the 43% and 47% recorded by Quinlivan *et al.* (1966), the 44% recorded by Edey (1970) and the 51% recorded by Casida *et al.* (1966). However, other workers have recorded similar partial losses to ours e.g. 26%—Henning (1939), 26%—Averill (1955), 28%—Allison (1975c). There are many possible explanations for these differences (reviewed by Edey, 1969). Nevertheless, it is obvious that partial failure of twin pregnancies may represent a substantial loss in the reproductive performance of sheep. This is the case in many areas in New Zealand, where at least 50% of mature ewes have multiple ovulations at mating (Averill, 1965; Kelly, unpublished data). It is difficult to believe that much of this loss is unavoidable and needed to eliminate unfit genotypes, as suggested by Bishop (1964). Rather, as Bradford (1972) has concluded, there may well be scope for selection for embryo survival to increase lambing performance and reduce its variability, and so improve the reproductive efficiency of sheep.

REFERENCES

Allison, A.J. (1975a) *New Zealand Journal of Agricultural Research* **18** : 1.
Allison, A.J. (1975b) *New Zealand Journal of Experimental Agriculture* **3** : 161.
Allison, A.J. (1975c) *New Zealand Journal of Agricultural Research* **18** : 101.
Averill, R.L.W. (1955) *Studies in Fertility* **7** : 139.

Averill, R.L.W. (1965) *World Review of Animal Production* **3** : 51.

Bishop, M.W.H. (1964) *Journal of Reproduction and Fertility* **7** : 383.

Bradford, G.E. (1972) *Journal of Reproduction and Fertility, Supplement* **15** : 23.

Casida, L.E., Woody, C.O. and Pope, A.L. (1966) *Journal of Animal Science* **25** : 1169.

Dolling, C.H.S. and Nicholson, A.D. (1967) *Australian Journal of Agricultural Research* **18** : 767.

Doney, J.M., Gunn, R.G. and Smith, W.F. (1973) *Journal of Reproduction and Fertility* **34** : 363.

Edey, T.N. (1969) *Animal Breeding Abstracts* **37** : 173.

Edey, T.N. (1970) *Journal of Agricultural Science, Cambridge* **74** : 199.

Gunn, R.G., Doney, J.M. and Russel, A.J.F. (1972) *Journal of Agricultural Science, Cambridge* **79** : 19.

Henning, W.L. (1939) *Journal of Agricultural Research* **58** : 565.

Kelly, R.W. and Allison, A.J. (1976) *Proceedings of the New Zealand Society of Animal Production* **36**: 240.

Quinlivan, T.D. and Martin, C.A. (1971a,b,c) *New Zealand Journal of Agricultural Research* **14** : (a)417, (b)858, (c)880.

Quinlivan, T.D., Martin, C.A., Taylor, W.B. and Cairney, I.M. (1966) *Journal of Reproduction and Fertility* **11** : 379.

Robinson, T.J. (1951) *Journal of Agricultural Science, Cambridge* **41** : 6.

Sawyer, G.J. and Knight, T.W. (1975) *Australian Journal of Experimental Agricultural and Animal Husbandry* **15** : 189.

Scanlon, P.F. (1972) *Journal of Animal Science* **34** : 791.

40

RAM AND EWE EFFECTS ON FERTILITY IN SHEEP BREEDS AND STRAINS DIFFERING IN REPRODUCTIVE POTENTIAL

S.K. STEPHENSON and G.W. ROTHWELL*

University of New England, Armidale, N.S.W., Australia

SUMMARY

The CSIRO sheep breeding programme, which has produced a high fecundity strain of Merinos, appears to have changed both male and female factors affecting the lambing percentage. In the present study, the high fecundity genotype in Merino rams gave a highly significant improvement in the proportion of ewes that conceived and produced lambs, but had no effect on twinning frequency. Conversely, the same genotype in first cross ewes resulted in an increased twinning frequency without any increase in the proportion of ewes lambing. Breed crosses involving the Border Leicester, Poll Dorset and Merino showed no effect of the Border Leicester genotype on male fertility, but in first cross ewes it was associated with a significant increase in twinning frequency.

The pattern shown by the high fecundity Merino genotype could be explained by an increased production of FSH and LH producing an increase in testicular growth and development and libido in rams and an increase in multiple ovulations in ewes. The lack of any ram effect with the Border Leicester, a breed with the characteristic of high twinning frequency, may be a result of its poor thrift and lack of vigour, particularly in a semi-arid environment with poor nutrition.

INTRODUCTION

The percentage of lambs weaned to ewes mated and the growth of these lambs are two of the most important attributes for prime lamb production. Consequently, lambing performance was a key characteristic in the present experiment designed to investigate the merits of breeds and crosses for lamb production in northern New South Wales.

New South Wales lamb production relies mainly on a three tier crossbreeding system that utilizes Merinos, usually culled for age, which are mated to a Border Leicester ram to produce a first cross ewe with a reputation for increasing the number of lambs weaned and their growth rate to weaning. The prime lambs are produced by mating these Border Leicester/Merino ewes to Dorset rams. The main deficiencies in this system are a lack of sufficient Border Leicester × Merino matings to provide the

---

* Present address: Scientific Poultry Breeders Pty. Ltd., Kellyville, N.S.W. 2153.

required numbers of first cross ewes and the poorer quality cross-bred wool from the prime lamb breeding flock.

There are two main alternatives that might help to improve this system. Firstly, it may be possible to develop a new breed that combines the wool quality and fleece weight of the Merino with the ability to grow cross-bred lambs of a similar standard to those produced by the Border Leicester/Merino ewes. Secondly, if benefits from heterosis (Pattie and Smith, 1964; McGuirk, 1967) suggest the desirability of retaining a three tier cross-breeding system, it may be possible to provide an alternative to the Border Leicester that gives the advantages of fecundity and growth without the associated detrimental effects of wool quality.

The present work was undertaken to provide information about these productive traits in different breed and strain crosses and the feasibility of developing a type of sheep that would increase the returns from a prime lamb breeding system. The experiment was designed so that data were obtained from cross-bred matings utilizing the Merino, Border Leicester and Poll Dorset breeds. These were compared with Merino strain matings utilizing Medium Peppin ewes crossed with rams of the high fecundity and high body weight genotypes developed by CSIRO.

MATERIALS AND METHODS

A flock of 1,400, 3 year old, Medium Peppin Merino ewes was made available in 1968 for this work at the University of New England's Research Station, Warialda, NSW. The rams were obtained from the following sources.

(i) High Fecundity Merino

Four Booroola and eight 'T' group rams were obtained from the Division of Animal Genetics, CSIRO. These high fecundity genotypes were developed by selection based on an index incorporating both litter size and fertility (Turner, 1968, 1969). The Booroola flock was founded in 1959 from a high fecundity, non Peppin Merino commercial stock while the 'T' flock was founded with twin born rams and ewes from the Peppin Merino flock of CSIRO at Gilruth Plains. The response to this selection has been to increase twinning frequency in ewes, leading to a greatly increased proportion of lambs marked and weaned to ewes mated. There has, however, been no apparent effect on fertility, or the ability of the ewes to conceive.

(ii) High Body Weight Merino

Seven rams were obtained also from the Division of Animal Genetics, CSIRO. This genotype was developed by selecting for body weight at 15-16 months of age (Turner, Brooker and Dolling, 1970).

(iii) Poll Dorset

The seven Poll Dorset rams used in this work were from the Marylebone Stud, at Cudal, NSW. This flock was bred and maintained under normal commercial stud management.

(iv) Border Leicester

The 10 Border Leicester rams were from the Uandani Stud, at Young, NSW. This was also a normal commercial stud.

The Merino rams were selected from within their specific flocks, being either the first reserves for the CSIRO matings or older rams that had been used in previous years. The Border Leicesters and Poll Dorsets were good quality flock rams from a multiple birth. No account was taken of relationship in the selection.

With these sheep, crosses were made of (a) different Merino strains, and (b) different breeds according to the pattern shown in Table 1.

TABLE 1

DESIGN OF THE MATINGS

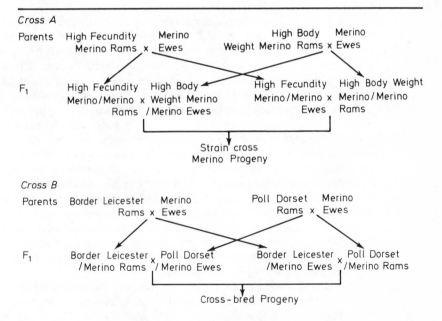

The parental matings were made with a mass mating in May 1968 using 18 to 20 months old rams and rotating the paddocks in sequence. The subsequent lambing was in October. In this year, seven high body weight

Merino and Poll Dorset rams and eight high fecundity Merino and Border Leicester rams were used for their respective matings. In 1969 and 1970 this mating was repeated using the same ewes, but with individual matings using six rams within each group. Each ram was mated with an approximately equal number of ewes, but as far as possible different rams were used in different years. The 1970 mating and lambing were a month earlier than in previous years. F_1 Merinos and cross-breds at 18 and 30 months of age were mated in 1971 and in this instance six rams were mated individually within each of the four mating groups. After mating all ewes were run as a single flock.

The 1968 lambing was carried out with minimum labour while 1970 was a drought year. As a result, these years provided only information for the number of ewe lambs and twins born to individual ewes in the different mating groups. Complete lambing records were available for 1969 and 1971.

RESULTS

As shown in Table 2, the parental matings resulted in a highly significant effect of the ram's genotype on the number of lambs born in relation to ewes mated.

The individual discrepancies between the observed and expected numbers of lambs show that the major contributions to the heterogeneity were made by the superiority of rams of the high fecundity Merino strain and the inferiority of rams of the high body weight Merino strain. The Border Leicester and Poll Dorset rams were intermediate in their effects on lambing percentages. Except in 1969 these variations were fairly consistent although the overall performance was poor owing to a series of dry seasons and a lack of adequate good quality pasture. The 1969 mating was affected by a particularly dry autumn and this may have been responsible for the extremely poor performance of Border Leicester rams in this year. There were no differences in the proportion of twin births, which were uniformly low in all instances, so the differences observed were a result of variations in the proportion of ewes producing lambs. The conclusion is that the ram's genotype affected the proportion of ewes that conceived at any particular mating.

The 1971 mating of first generation cross individuals gave results that were consistent with this pattern and, in addition it showed an effect of the ewe's genotype on the proportion of twin births. The lambing results for the F_1 genotypes are shown in Tables 3 and 4.

Overall, there was significant heterogeneity for the number of lambs born in relation to ewes mated with the cross-bred matings producing more lambs than the Merino matings. The best mating was that of Border Leicester/Merino ewes crossed with Poll Dorset/Merino rams. The two components of this lambing result are the proportion of ewes lambing and the number of lambs produced by those ewes that lambed. The overall χ^2

TABLE 2

THE NUMBER OF EWES MATED, THE NUMBER OF LAMBS BORN AND THE NUMBER OF TWIN BIRTHS IN THE FOUR PARENTAL MATING GROUPS

		Mating groups			Expected number † of lambs if all rams of equal fertility	χ^2 values (3 d.f.)	Twin births
Year	Rams*	Merino ewes mated	Ewe lambs born	Total lambs born			
1968	HWM	350	92		119.75		1
	HFM	350	147		119.72		3
	PD	350	122		119.75	12.70	0
	BL	350	118		119.75	P<0.01	6
1969	HWM	240	66	133	140.5		6
	HFM	240	76	162	140.5	16.78	5
	PD	240	82	163	140.5	P<0.01	7
	BL	240	57	104	140.5		4
1970	HWM	290	58		69.94		Not available drought year
	HFM	316	101		76.21	11.22	
	PD	327	71		78.86	P<0.05	
	BL	311	70		75.00	≅.01	
TOTAL	HWM	880	216		258.6		
	HFM	906	324		266.2		
	PD	917	275		269.4		
	BL	901	245		264.8		

* Abbreviations: HWM=High Body Weight Merino; HFM=High Fecundity Merino; BL=Border Leicester; PD=Poll Dorset.

† Expected numbers of ewe lambs for 1968 and 1970. Expected number of total lambs for 1969. These are expected numbers used to calculate X^2 values.

values for both these components considered individually do not show significant heterogeneity. The results however show significant differences in the proportion of twin births.

These results are summarized in the form of percentages in Table 4.

With the Merino matings the cross of high fecundity Merino/Merino ewes with high body weight Merino/Merino rams resulted in a low proportion of ewes producing lambs, but a relatively high proportion of twin births. Conversely, the reciprocal cross, i.e. the mating of high body weight Merino/Merino ewes with high fecundity Merino/Merino rams, gave a higher proportion of ewes lambing, but no twin births. The overall percentage of lambs born for ewes mated was the same in both crosses.

The cross-bred matings showed a similar proportion of ewes producing lambs in both crosses but the twinning rate was higher when Border

TABLE 3

THE LAMBING RESULTS FROM THE MATING OF FIRST CROSS INDIVIDUALS (1971 ONLY)

| Genotype | | Ewes mated | Ewes lambed | | Number of lambs ($\sigma\sigma+\varphi\varphi$) | | | Birth type for ewes lambed | | χ^2 for heterogeneity within each mating (1 d.f.) |
Ewe	Ram		Observed	Expected for ewes mated	Observed	Expected Ewes mated	Expected Ewes lambed	Singles	Twins	
BL/Merino	PD/Merino	106	75	66.8	102	79.9	89.5	48	27	5.29 $P<0.05$
PD/Merino	BL/Merino	141	94	88.9	113	106.0	112.2	75	19	
HFM/Merino	HWM/Merino	128	67	80.7	79	96.2	79.9	55	12	10.27 $P<0.05$ (corrected for continuity)
HWM/Merino	HFM/Merino	101	64	63.7	64	76.0	76.4	64	0	
		476	300		358					
			$\chi^2=3.62$ 3 d.f. ($p\cong0.3$)		$\chi^2=11.7$ ($P<0.01$)	$\chi^2=3.75$ ($P\cong0.3$)				

TABLE 4

LAMBING PERCENTAGE RESULTS FROM THE MATING OF FIRST
CROSS INDIVIDUALS (1971 ONLY)

Genotype		% of ewes lambing	% lambs for ewes mated	% lambs for ewes lambing	Birth type for ewes lambed	
Ewe	Ram				% Single	% Twins
BL/Merino	PD/Merino	71	96	136	64	36
PD/Merino	BL/Merino	67	80	120	80	20
HFM/Merino	HWM/Merino	52	62	118	82	18
HWM/Merino	HFM/Merino	63	63	100	100	0
	TOTAL AV. %	63.2	75.2	119	81.5	18.5

Leicester/Merino ewes were crossed with Poll Dorset/Merino rams. This
gave the Border Leicester/Merino ewe a distinct advantage in the total number of lambs produced for ewes mated.

DISCUSSION

Since the Merino high fecundity genotypes that were used in this work
were bred by selecting ewes for twinning ability and fertility (Turner, 1968;
1969) the ram effects are a correlated response to this selection.

The Merino mating showed that the ram's genotype had a direct effect on
the proportion of ewes that produced a lamb, but it had no effect on the
proportion of multiple births. The improvement in the proportion of ewes
lambing obtained with the high fecundity rams could have been a result of
lower embryonic mortality, although there is little evidence from other
sources to support this interpretation. Bindon, Ch'ang and Turner (1971)
showed no differences in embryonic survival between 'T' group ewes and
those selected against twinning ability and fertility ('O' group). In a similar
comparison involving transfer of fertilized eggs between 'T' and 'O' group
ewes, Trounson and Moore (1972) found a lower embryonic survival with
'T' group ewes, but a higher survival of 'T' genotype embryos.

The alternative explanation is that the high fecundity Merino genotype
has increased the ram's fertility. Field observations indicate that these rams
were more active and mated a higher proportion of their ewes in the first cycle than those belonging to other genotypes in the experiment. This suggestion also agrees with the results obtained by Land (1973) at Edinburgh with
experiments involving the Finnish Landrace. Land's hypothesis to explain
such results was that the primary effect of the high fecundity genotype in the
Finnish Landrace was to increase the production rates of follicle stimulating
hormone (FSH) and luteinizing hormone (LH). These hormones then increased both testicular growth and development and libido in rams and

multiple ovulations in ewes. Bindon (personal communication) has confirmed that Booroola and 'T' group rams do have a faster rate of testis growth than random bred Merinos.

The effects of the high fecundity ('T' group) genotype on twinning frequency and multiple ovulations in ewes have been well documented in the literature (Packham and Triffitt, 1966; Bindon, Ch'ang and Turner, 1971; Trounson and Moore, 1972). In the present experiment ewes bred from high fecundity rams also gave a higher twinning frequency than those bred from high body weight rams.

The cross-bred F_1 matings showed that rams of both genotypes gave a similar proportion of ewes lambing. The Border Leicester/Merino ewes, however, gave a considerably higher frequency of twin births than the Poll Dorset/Merino ewes. This is consistent with the fact that the Border Leicester has one of the highest fecundity levels for any British sheep breed (Thomson and Aitken, 1959). Under Australian conditions, the pure-bred Border Leicester appears to suffer from poor thrift and adaptation to the environment (Trounson and Roberts, 1970; Fogarty, 1971), but the twinning ability is well expressed in the first cross Border Leicester/Merino ewe, leading to the suggestion that there is a large "component of heterosis" for this characteristic (McGuirk, 1967).

Land's hypothesis suggests that there should, in addition, be an increase in fertility with matings involving the Border Leicester rams. The fact that this did not occur may also be explained by the poor thrift and lack of vigour of this genotype. Such a suggestion is supported by the very poor performance of the Border Leicester rams in the parental matings of 1969, a year with extremely dry conditions at mating.

The results of this work add support to previous conclusions concerning the detrimental effects of the poor thrift and adaptation of the Border Leicester to some parts of the Australian environment. An improvement in the fertility of Border Leicester rams or their equivalent in the crossing stratification would help to increase the supply of first cross ewes and make their breeding more attractive to sheep producers. At the next level in the stratification, the Border Leicester/Merino ewes had good fertility, together with the highest proportion of twin births. Since these features are extremely important for prime lamb production the results suggest that in many environments a stratified cross-breeding system is probably superior to our Merino strain cross.

ACKNOWLEDGEMENTS

We are indebted to the Division of Animal Genetics, CSIRO and particularly to Dr H.N. Turner for making available the rams from their experimental Merino flocks. We are also very grateful to Mr P.L.H. Cowper for field and technical assistance. The work was made possible by the financial assistance of the Australian Meat Research Committee.

REFERENCES

Bindon, B.M., Ch'ang, T.S. and Turner, Helen N. (1971) *Australian Journal of Agricultural Research* **22** : 809.

Fogarty, N.M. (1971) *Wool Technology and Sheep Breeding* **18**(2) : 20.

Land, R.B. (1973) Annual Report. Animal Breeding Research Organisation. pp.7-12. Agricultural Research Council. Edinburgh EH93JQ.

McGuirk, B.J. (1967) *Wool Technology and Sheep Breeding* **14**(2) : 73.

Packham, A. and Triffitt, L.K. (1966) *Australian Journal of Agricultural Research* **17** : 515.

Pattie, W.A. and Smith, M.D. (1964) *Australian Journal of Experimental Agriculture and Animal Husbandry* **4** : 80.

Thomson, W. and Aitken, F.C. (1959) Technical Communication No. 20. Commonwealth Bureau of Animal Nutrition; Edinburgh.

Trounson, A.O. and Moore, N.W. (1972) *Australian Journal of Agricultural Research* **23** : 851.

Trounson, A.O. and Roberts, E.M. (1970) *Proceedings of the Australian Society of Animal Production* **8** : 326.

Turner, Helen N. (1968) *Proceedings of the Symposium on Physiology and Reproduction.* United States Sheep Development Program Stillwater, Oklahoma.

Turner, Helen N. (1969) *Animal Breeding Abstracts* **37** : 545.

Turner, Helen N., Brooker, N.G. and Dolling, C.H.S. (1970) *Australian Journal of Agricultural Research* **21** : 955.

41

OVULATION AND LAMBING RESULTS WITH BOOROOLA AND BOOROOLA-CROSS MERINO EWES IN WESTERN AUSTRALIA

D.E. ROBERTSON

Muresk Agricultural College, Western Australian Institute of Technology

SUMMARY

Rams from the Merino Booroola strain, which has been selected for high fertility and fecundity, were mated to unselected Merino ewes in commercial flocks to produce Booroola-cross ewes for comparison with ordinary Merino ewes born in the same flocks, and with Booroola ewes born on another property.

The Booroola and Booroola-cross ewes had similar numbers of lambs born per ewe mated (1.19 and 1.36 as 2-year-olds, and 1.58 and 1.54 as 3-year-olds), significantly more than the control ewes (0.74 as 2-year-olds and 0.92 as 3-year-olds). Most of the increase arose from the higher incidence of multiple births in the daughters of the Booroola rams.

Ovulation rates, assessed by laparoscopic examination, were 1.95 and 1.81 in maiden Booroola and Booroola-cross ewes respectively and 1.06 in contemporary control ewes.

These results suggest that the use of Booroola rams could be the simplest and most effective means yet available of rapidly achieving high lambing percentages in Australian Merino flocks.

INTRODUCTION

Reproductive efficiency of Australian Merino ewes is low. In Western Australia the average flock produces less than 70 lambs per 100 ewes mated. This limits both the income from the sale of stock and the scope for genetic improvement by selection.

Many attempts have been made by breeders and scientists to find ways of increasing Australian Merino lambing percentages. The most successful work to date has been that by the Seears brothers, breeders near Cooma in New South Wales, who, during the 1940s and 1950s developed by selection for increased litter size, a strain of high fecundity Merinos. This strain, now called the Booroola, was taken over and further selected for maximum litter size at birth by the CSIRO Division of Animal Genetics. Its performance and response to selection has been described by Turner (1968) and Turner and Young (1969). McGuirk (1976) quoted figures from CSIRO workers which showed their Booroolas had an average of 1.94 lambs born per ewe mated.

Ewes and rams sold by CSIRO in 1969 were used to establish a Booroola flock at Gidgegannup in Western Australia. This flock is also selected for maximum litter size. Lambing results with these sheep in Western Australia over the past seven years, have been similar to those recorded by CSIRO in New South Wales.

There is a very large gap between the lambing performances of the Booroolas and unselected Merinos. It is important to know how much of the Booroola's superiority would appear in crosses with other Merinos. If the crosses performed well, breeders could readily use the Booroolas to achieve higher lambing percentages.

This experiment was designed to determine whether ewe progeny from the mating of Booroola rams with unselected Merino ewes would have more lambs than ordinary Merino ewes.

MATERIALS AND METHODS

In 1972 four Booroola rams from the Gidgegannup flock were sent to a co-operating farmer who mated them with a randomly chosen flock of 200 of his own Merino ewes. Seventy-three ewe lambs from this mating, with a randomly drawn control group of 40 ewe lambs of the same age but sired by the farmer's own rams, were transferred at weaning to Gidgegannup where they have since run with Booroola ewes of the same age.

A similar procedure was followed in 1973 and 1974 with different rams being used each year and going to different farms. Thus twelve Booroola rams and three farm flocks have been sampled.

The first flock sampled was of the Peppin strain; the second, mixed Peppin and South Australian strains; and the third, the Collinsville strain.

No control ewes could be obtained in 1973 so only the 1972 and 1974 drafts of Booroola crosses were accompanied by appropriate control groups.

After weaning, each annual intake of ewes has run continuously with Booroola ewes of the same age and all ewes have been treated alike. At mating, Booroolas, crosses and controls have been given equal exposure to the same Booroola rams. The ewes were first mated at 18 or 19 months of age with the exception of the 1973-born ewes which were mated when eight months old.

At lambing, the ewes were held in small two to four hectare paddocks and were closely observed at least twice daily so that the lambs borne by each ewe could be identified and recorded. Lambing results presented here include only lambs born in the first five weeks of lambing. Ewes lambing after that time have been treated as "not lambing".

In 1976, the laparoscopic technique described by Oldham *et al.* (1975) was used to count the corpora lutea on the ovaries of the 1974-born 18-month-old ewes, then being mated for the first time.

RESULTS

Table 1 shows the distributions of litter sizes and the mean numbers of lambs born to the 1972-born ewes. The Booroolas and their crosses had similar numbers in both years, having about 70% more lambs than the unselected control ewes. This difference in favour of the Booroolas and Booroola crosses arose principally because nearly half of these ewes had multiple births while only one pair of twins was born to the control ewes in two years. The difference also stems in part from the greater incidence of dry ewes in the control group.

TABLE 1

LAMBS BORN TO 1972-BORN EWES WHEN TWO AND THREE YEARS OLD

Genotypes of ewes	Numbers of ewes	Percentages of ewes with the following numbers of lambs					Mean numbers of lambs born per ewe mated ± standard errors
		0	1	2	3	4	
1974 lambing when ewes 2 years old							
Booroola	27	18.5	44.4	37.0	0	0	1.19±.14
Booroola-cross	73	9.6	45.2	45.2	0	0	1.36±.08
Unselected	38	26.3	73.7	0	0	0	0.74±.07
1975 lambing when ewes 3 years old							
Booroola	24	4.2	45.8	41.7	4.2	4.2	1.58±.17
Booroola-cross	72	4.2	44.4	44.4	4.2	0	1.54±.08
Unselected	39	10.2	87.2	2.6	0	0	0.92±.06

In Table 2 are the results following the mating of the 1973-born ewes when they were mated as 8 month old lambs. The two groups produced very similar results with half of the ewes lambing and about 40% of these having multiple births.

No controls were available for this age group.

TABLE 2

LAMBS BORN TO 1973-BORN EWE LAMBS LAMBING WHEN 13 MONTHS OLD

Genotypes of ewes	Numbers of ewes	Percentages of ewes with the following numbers of lambs					Mean numbers of lambs born per ewe mated ± standard errors
		0	1	2	3	4	
Booroola	37	48.6	32.4	16.2	0	2.7	0.76±.15
Booroola-cross	82	47.6	31.7	20.7	0	0	0.73±.09

Ovulation rates estimated from counts of the corpora lutea on the ovaries of the 1974-born ewes when they were mated for the first time at 18 months of age, are shown in Table 3.

Again, the Booroola and Booroola-cross ewes could not be separated but they showed a significant advantage over the control ewes. They had this advantage because a high proportion had multiple ovulations while very few of the control ewes shed more than one egg.

TABLE 3

OVULATION RATES ASSESSED BY LAPAROSCOPIC EXAMINATION IN 1974-BORN MAIDEN EWES WHEN 18 MONTHS OLD

Genotypes of ewes	Numbers of ewes	Percentages of ewes with the following numbers of corpora lutea					Mean ovulation rates per ewe mated ± standard errors
		0	1	2	3	4	
Booroola	21	0	14.3	76.2	9.5	0	1.95±.11
Booroola-cross	68	5.9	23.5	54.4	16.2	0	1.81±.09
Unselected	31	3.2	87.1	9.7	0	0	1.06±.06

DISCUSSION

The Booroola-cross ewes have substantially outstripped the control ewes in ovulation rate and in lambing results. They appear to have equalled the performance of their Booroola contemporaries. It is tempting to suggest that heterosis is evident here. However, it must be remembered that the Booroola ewes were apart from the others until weaning and furthermore were borne and reared by Booroola mothers, many as twins, while the crosses and controls were reared by ordinary Merino ewes, mostly as singles. Hence the apparent superiority of the Booroola-cross over the expected mid-parent mean could arise from hybrid vigour or from differences in early-life environments. To resolve whether heterosis was involved it would be necessary to rear the three genotypes in a common environment and to remove the maternal effects.

The question of heterosis is of scientific interest but is not of immediate practical importance. What is significant to sheep breeders is the big difference between the Booroola crosses and ordinary Merino ewes. Comparisons between these two genotypes in this experiment are valid. They demonstrate that Merino lambing percentages could be lifted dramatically within a single generation by the use of Booroola sires.

The gains reported here considerably exceed the responses to lupin feeding reported by Lightfoot and Marshall (1974) and by Knight, Oldham and Lindsay (1975). They are also more attractive than the gains expected from selection for fecundity in an ordinary Merino flock without access to more prolific sheep. Assuming that the rate of improvement reported by Turner (1968) could be sustained for a long period, it might take 30 years of laborious and expensive selection effort to lift another strain to equal the prolificacy already available in Booroola crosses.

The results presented here strongly suggest that crossing with the Booroola strain offers the easiest and most effective means yet available of rapidly achieving high lambing percentages in Australian Merino flocks.

ACKNOWLEDGEMENT

I am grateful to Messrs. C.M. Oldham and P. Gheradi of the University of Western Australia for the ovulation rate assessments recorded in Table 3.

REFERENCES

Lightfoot, R.J. and Marshall, T. (1974) *Journal of Agriculture Western Australia*, **15**: 29.

McGuirk, B.J. (1976) *Proceedings of the Australian Society of Animal Production* **11**: 93.

Oldham, C.M., Knight, T.W. and Lindsay, D.R. (1976) *Australian Journal of Experimental Agriculture and Animal Husbandry* **16**: 24.

Knight, T.W., Oldham, C.M. and Lindsay, D.R. (1975) *Australian Journal of Agricultural Research* **26**: 567.

Turner, Helen Newton (1968) In "Proceedings of US Sheep Development Program Symposium on Physiology of Reproduction in Sheep" ed. G. Scott, Stillwater, Oklahoma.

Turner, Helen Newton and Young, S.S.Y. (1969) Quantitative Genetics in Sheep Breeding. Macmillan, Australia.

42

LUPIN STUBBLES AND EWE FERTILITY

T. MARSHALL, R.J. LIGHTFOOT, K.P. CROKER and
J.G. ALLEN

Department of Agriculture, South Perth, Western Australia

SUMMARY

An experiment was conducted to investigate the effects of rate of stocking and selenium supplementation on the fertility of ewes grazing sweet lupin stubble (*Lupinus angustifolius* L. cv. Unicrop).

Compared to controls (run on dry subterranean clover based pastures) ewes stocked on lupin stubble (for 35 days commencing 14 days prior to joining) at either 15, 30, 45 and 60/ha produced more lambs. Fertility was depressed among ewes stocked at 75/ha (per cent ewes lambing = 76 versus 83, 86, 80, 72 and 66; per cent ewes twinning = 9 versus 17, 26, 18, 19 and 4 for controls and stubble treatments respectively). Liver damage as a result of the disease "Lupinosis" increased with increasing stocking rate. Deaths due to Lupinosis occurred among ewes stocked at 75/ha but not at the lower stocking rates. Selenium supplementation (25 mg, oral) did not affect any of the parameters measured. The practical implications of the findings are discussed.

INTRODUCTION

Considerable areas of sweet lupins are now grown in south Western Australia for grain production, the areas having increased markedly in recent years (27 000, 44 000, 63 000, 129 000 ha for the seasons 1971/72 to 1974/75 respectively, Anon., 1975). A complementary sheep grazing enterprise is conducted on many properties in the areas involved and it is common practice to graze the lupin stubbles after harvest to fatten sheep for sale.

Marshall and Lightfoot (1974) have shown that ewes which grazed sweet lupin stubble from 35 days before the rams were joined, and throughout the joining period, had higher ovulation rates than similar control ewes grazed on either dry subterranean clover based pastures or oaten stubble. Subsequent data from that work (presented by Marshall, 1974) also showed that more ewes lambed and had twins in the lupin stubble group. The ewes in that study grazed the stubbles for a relatively long period (77 days) and were run at a low stocking rate (12/ha). It has subsequently been demonstrated that lupin stubbles can carry high stocking rates during the summer months (Croker, Allen, Lester and Guthrie, 1975). However, when sheep graze lupin stubbles there is an ever present risk of Lupinosis (Gardiner, 1967) frequently complicated with white muscle disease in young stock. It is likely that this risk could be lessened if the grazing period was kept relatively short.

The work reported in this chapter was undertaken to determine if fertility responses could be obtained when ewes grazed lupin stubbles at high stocking rates for a short period before and during joining. As a secondary aim the effects of selenium administration among ewes grazing these stubbles were investigated.

MATERIALS AND METHODS

The work was conducted at the Wongan Hills Research Station situated approximately 150 km north-east of Perth. The station experiences a Mediterranean climate with approximately 75% of mean annual rainfall (380 mm) falling in the period May to September. In December 1974, immediately following harvest, a uniform area of a commercial sweet lupin crop (*Lupinus angustifolius* L. cv. Unicrop, yield one tonne/ha of harvested grain) was selected and subdivided into five plots. The areas of individual plots were adjusted so that groups of between 74 and 79 ewes would graze at either 15, 30, 45, 60 or 75/ha. An additional area of 16 ha of dry subterranean clover based pasture was slected at the same time in an adjacent paddock to be grazed by 80 "control" ewes.

Experimental groups were randomly selected on December 30, 1974 from a flock of mature, medium wool Merinos and the ewes then "teased" with 1% vasectomized rams for 14 days prior to the experimental grazing. Each plot was sampled for residual grain (10 quadrats, 100 cm × 10 cm/plot) immediately before grazing commenced. The ewes were introduced to their respective plots on January 23, 1975 and one half, selected at random from within each treatment, were given a single oral dose of 25 mg of selenium as sodium selenite. Entire rams fitted with marking harnesses and crayons were introduced on February 6, 1975 (3/plot) for a 42-day joining. Each seven days liveweights were recorded, service records obtained and rams rotated between ewe groups. The crayon colours were changed each 14 days.

Deaths due to Lupinosis occurred among ewes grazed at 75/ha on days 28, 30, 31 and 33 of grazing. No deaths occurred on any other treatment. All sheep were removed from their treatment plots on day 35 of grazing (February 27, 1975) and were bulked and run as one flock on normal dry pasture. Immediately following removal from the treatment plots, liver biopsy samples were obtained from 20 ewes (10 "selenium+", 10 "selenium−") from within each treatment. The samples underwent histopathological examination and the degree of damage scored as described by Croker *et al.* (1975).

The ewes ran as one flock until lambing when they were again divided into treatment groups to permit the collection of individual ewe lambing records.

RESULTS

Overall, a mean of 485 ± 56 kg/ha of residual lupin grain was found, there being considerable variability in estimates both between plots and between samples within plots. Approximately half the grain on the ground was still in the pods suggesting that pod drop as well as pod shattering contributed to the post harvest residual grain. Based on the above samplings, estimated lupin grain availabilities for the 35-day grazing period were 32.3, 16.2, 10.8, 8.1 and 6.5 kg/head for the 15, 30, 45, 60 and 75 ewes/ha treatments respectively.

Ewes stocked at 15/ha continued to gain weight throughout the stubble grazing period. Those stocked at higher rates gained less weight initially, and in all cases began to lose weight before the end of the grazing period. There were significant ($P < 0.05$) changes in liveweight between control ewes and ewes in the 15, 30, 45 and 60/ha treatments during the stubble grazing period (liveweight changes were -2.7, $+2.5$, $+1.5$, $+1.1$, -0.3, -1.9 kg for control and 15, 30, 45, 60, 75/ha treatments respectively). Selenium supplementation did not significantly affect liveweight changes within the groups (overall liveweight changes were $+0.2$ and -0.1 kg for selenium+ and selenium− treatments respectively).

There was a highly significant ($P < 0.01$) and positive relationship between stocking rate and mean liver damage score (Table 1) but no effect of selenium on this parameter (mean scores were 4.7 and 4.9 for selenium+ and selenium− treatments respectively).

The proportions of ewes lambing (Table 1) tended to be higher among those grazing the lupin stubbles at 15, 30 or 45/ha (mean 83%) than among control ewes (76%). A higher proportion ($P < 0.05$) of ewes grazed at 15, 30, 45 and 60/ha had twins (mean 20%) than did control ewes (12%).

TABLE 1

MEAN LIVER DAMAGE SCORE AND PER CENT EWES LAMBING
AND TWINNING

Treatment	Liver damage score	% ewes lambing [†]	% ewes twinning [†]
Control	0.6	76 (79) [‡]	9
15 ewes/ha	3.1	83 (70)	17
30 ewes/ha	2.9	86 (69)	26
45 ewes/ha	4.2	80 (74)	18
60 ewes/ha	5.3	72 (72)	19
75 ewes/ha	7.2	66 (73)	4

† Per cent of ewes present at lambing.
‡ Number of ewes.

There was no effect of selenium supplementation on the proportion of ewes lambing or twinning (per cent ewes lambing were 77, 76 and per cent ewes twinning were 13, 17 for selenium+ and selenium− respectively).

DISCUSSION

The data confirm the observations of Croker *et al.* (1975) that lupin stubbles can be a valuable summer feed for sheep and that the severity of liver damage characteristic of Lupinosis increases with increasing stocking rate. It is interesting to note that even at low stocking rates there was considerable liver damage relative to controls, though neither liveweight change nor fertility were obviously affected. This suggests that, for practical purposes, liver damage score *per sé* may be too sensitive a measure of Lupinosis. Selenium supplementation had no effect on liveweight change, fertility or liver score.

The present study has also confirmed the earlier findings of Marshall and Lightfoot (1974) that the fertility of ewes can be increased by grazing lupin stubbles immediately before and during joining. In addition it has shown that fertility responses can be obtained at higher stocking rates (15-45 ewes/ha), with a shorter pre-joining duration of grazing (14 days), than employed in the original study (12 ewes/ha, 35 days pre-joining grazing). In the latter instance, this observation is in keeping with parallel studies on feeding lupin grain (Marshall and Lightfoot, 1974; Lightfoot, Marshall and Croker, 1976) where responses have been obtained with 7 to 14 days pre-joining supplementation at 250 g/head/day.

In this experiment approximately one-third of the total grain produced by the crop remained on the ground following harvesting. While the amount of residual grain in lupin stubbles will be affected by many factors (e.g. cultivar, seasonal conditions and time of harvesting) and may vary considerably, the levels measured in this experiment appear to fall within the expected range (H.M. Fisher, personal communication).

Highest fertility (112% lambs born) was obtained in this experiment among ewes stocked at 30/ha. Lupin grain availability at this stocking rate was estimated to be 16.2 kg/head for the 35-day grazing period. As a rough approximation, 16 kg of Unicrop lupin grain/ha is equivalent to 10 grains/m^2. This suggests a rule of thumb method for estimating stocking rates on Unicrop lupin stubbles which would result in maximum fertility responses. Each 10 grains/m^2 should represent 35 days grazing for one ewe for each ha of stubble. Further work is required to verify these estimates.

An additional issue to consider is the risk of Lupinosis which increases with increasing grazing intensity. Thus, in stubbles with high concentrations of residual grain the final choice of stocking rate may need to be a compromise between grain availability and Lupinosis risk.

ACKNOWLEDGEMENTS

Acknowledgement is made to Messrs. V.E. Gartner (Manager), A.E. Ralph, P.J. Coppin and C.R. Mouritzen of the Wongan Hills Research Station for day-to-day running of the experiment and to Messrs. G.R. McMullen, T.J. Johnson and R.B. Guthrie for skilled technical assistance. The work was supported by a grant from the Wool Research Trust Fund.

REFERENCES

Anonymous (1975) *Commonwealth Bureau of Census and Statistics Western Australian Office. Agricultural Census: Principal Statistics, Season 1974-75.*

Croker, K.P., Allen, J.G., Lester, C.R. and Guthrie, R.B. (1975) *Journal of Agriculture Western Australia* **16** : 84.

Gardiner, M.R. (1967) *Advances in Veterinary Science* **11** : 85.

Lightfoot, R.J., Marshall, T. and Croker, K.P. (1976) *Proceedings of the Australian Society of Animal Production* **11** : 5P.

Marshall, T. (1974) *Proceedings of a Seminar: Sheep Fertility. Recent Research and its Application in Western Australia.* (University of Western Australia: Crawley).

Marshall, T. and Lightfoot, R.J. (1974) *Journal of Agriculture Western Australia* **15** : 31.

43

THE FERTILITY OF MERINO EWES GRAZED ON STANDING, UNHARVESTED SWEET LUPINS PRIOR TO AND DURING JOINING

K.P. CROKER, R.J. LIGHTFOOT and T. MARSHALL

Western Australian Department of Agriculture, Perth, Australia

SUMMARY

The fertility of Merino ewe flocks grazed on 'failed' sweet lupin (*Lupinus angustifolius* L. cv. Unicrop) crops for either 14 or 0 days prior to and during 32 days of joining was compared with that of a control flock grazed on dry annual pasture for the same period.

The occurrence of first service was the same in all groups, but the numbers of ewes that returned to service differed significantly ($P<0.01$) being reduced in the ewes that grazed lupins. When compared with the controls there was a significant increase in both the proportion of ewes lambing ($P<0.05$) and ewes with twins ($P<0.001$) for the treatments grazed on lupins. In addition, lambing in these ewes was more concentrated.

The results indicate that a period of seven days grazing of lupins before the start of joining is required to obtain a substantial increase in the lambing performance above that of similar ewes grazed on dry annual pasture.

INTRODUCTION

The high proportion of ewes which mate, but fail to lamb is primarily responsible for the low lamb marking percentages obtained in Western Australia (Knight, Oldham, Smith and Lindsay 1975). However, ovulation rates in Merino ewes, which make up approximately 90% of the ewe flock, are also very low in Western Australia (Knight *et al.* 1975; Lindsay, Knight, Smith and Oldham 1975; Marshall, Beetson and Lightfoot 1976; Wroth and Lightfoot 1976) and this has been shown to be related to the dry ewe problem (Lindsay *et al.* 1975).

In a number of experiments in south Western Australia it has been clearly demonstrated that ovulation rates and lambing percentages can be increased when ewes are supplemented with sweet lupin grain immediately before and during the first oestrous cycle of the joining period (Lightfoot and Marshall 1974, 1975; Marshall and Lightfoot 1974; Knight, Oldham and Lindsay 1975; Lightfoot, Marshall and Croker 1976). Arnold and Charlick (1976) have also shown that the fertility of ewes grazed on a normal standing sweet lupin crop prior to and during joining is higher than in equivalent ewes joined on sub clover based pastures.

The experiment described here was conducted to investigate the fertility of ewes grazed on standing sweet lupin crops for various periods prior to and during joining.

MATERIALS AND METHOD

The experiment was carried out at Badgingarra Agricultural Research Station, which is situated 180 km north of Perth. The climate is Mediterranean with some 75% of the annual average rain (460 mm) falling between May and September.

Two adjacent 5 ha plots sown to sweet lupins (*Lupinus angustifolius* L. cv. Unicrop) but considered unsuitable for harvesting (low yield) were available for the experiment. For control purposes an adjacent area of pasture (10 ha) based on serradella (*Ornithopis compressus* L.) was selected. The total dry matter present was estimated.

In early December 1974, 300 Merino ewes were randomly allocated to three equal sized groups. Each group consisted of forty-two $2\frac{1}{2}$, forty-two $3\frac{1}{2}$ and $4\frac{1}{2}$, and sixteen $6\frac{1}{2}$ and $7\frac{1}{2}$-year-old ewes which were individually tagged for identification, given an injection of anthelmintic (Nilverm ICI) and received a Co bullet (ICI). They were then introduced to a dry annual pasture.

In mid January 1975 half of each group of ewes was drenched with 10 mg of Selenium (as sodium selenite) (Terra) and this was repeated six weeks later. After the application of the first drench, one group (L_{14}) was placed on a lupin plot (20/ha) while a second group (C) was placed onto the pasture plot (10/ha). Fourteen days later the third group of ewes (L_0) was introduced to the other lupin plot (20/ha) and five harnessed Merino rams were added to each flock.

The rams were joined with the ewes for 42 days. Service records were obtained and the rams rotated between ewe flocks at seven day intervals. Crayon colours were changed at 14 day intervals.

Grazing of the experimental plots was terminated 32 days after the start of joining because the L_{14} lupin plot had insufficient feed for the remainder of the joining period. Subsequently, the ewes were run as one flock on sub clover based pastures until approximately 14 days prior to the expected start of lambing. At this time the flock was divided into treatment groups, the ewes side numbered and observed daily for individual lambing records.

The results were subjected to chi square analyses.

RESULTS

There was an average of 2508 kg DM/ha on the lupin plots of whch 450 kg was grain. The pasture plot yielded 3506 kg DM/ha.

There were no differences between treatments in the incidence of first services and the proportions of ewes served at least once throughout joining were 98, 99 and 97% for C, L_0 and L_{14}, respectively. However, the numbers of ewes that returned to first service did differ significantly (P< 0.01), the proportions being 31, 25 and 11% for C, L_0 and L_{14}, respectively.

The lambing results are recorded in Table 1. When compared with the controls, there was a significant increase in the proportion of ewes lambing for the treatments grazed on lupins (P< 0.05). The proportion of ewes that had twin births also was higher in these ewes (P< 0.001). The highest levels of both parameters were in the L_{14} ewes. The differences between groups were accentuated after the first 21 days of lambing because by this stage nearly all the ewes (89%) in the L_{14} group had lambed.

TABLE 1

THE PROPORTION OF EWES THAT LAMBED, AND WITH MULTIPLE BIRTHS, AFTER THREE WEEKS AND AT THE COMPLETION OF LAMBING

Treatment	Lambing results				
	After 3 weeks		Total (6 weeks)		
	Ewes lambing†	Ewes twinning‡	Ewes lambing†	Ewes twinning‡	Lambs marked†
	%	%	%	%	%
C	61	7	80	8	76
L_0	82	19	89	18	84
L_{14}	92	37	95	36	109
x^2	<0.001	<0.001	<0.05	<0.001	

† of ewes joined ‡ of ewes lambing.

During each of the first three weeks of lambing 10, 4 and 8% respectively of the control (C) ewes that lambed had twin pregnancies. Similarly, results of the L_0 and L_{14} treatments were 9, 30, 17 and 39, 42 and 25%, respectively.

In the L_0 and L_{14} flocks there were no differences between age groups in the proportions of ewes that lambed. However, there was a significant increase in the proportion of lambing ewes which had twin births with increase in age of the ewes (20, 28 and 46% for $2\frac{1}{2}$, $3\frac{1}{2}+4\frac{1}{2}$ and $6\frac{1}{2}+7\frac{1}{2}$ years, respectively, P< 0.05). When all ewes were considered the relationship between ewe age and fecundity was accentuated (P< 0.001) because only the oldest ewes had twins in the control group.

Irrespective of feed, or duration of grazing, treatment of ewes with Se did not affect fertility (91 versus 93% ewes lambing for Se treated and non-treated ewes, respectively).

DISCUSSION

The results of this experiment confirm that grazing unharvested lupins prior to and during joining can increase ewe fertility (Arnold and Charlick,

1976). The improved lambing performance was associated with both increases in the proportion of ewes lambing and in the incidence of twins, similar to the responses obtained when ewes were supplemented with lupin grain only (Lightfoot and Marshall, 1975).

The rapid lamb drop in the lupin fed groups was unexpected. This, in association with the decreased rate of return to service, suggests that embryo survival was enhanced by the intake of lupin material over most of the joining period. Comparable evidence of a rapid lambing among lupin fed ewes does not appear to have been reported previously, although in other experiments (Lightfoot and Marshall 1975; Lightfoot, Marshall and Croker 1976) lupin feeding usually ceased after the first 17 days of joining.

The different rates of twinning obtained for the treatment groups over the first three weeks of lambing also illustrated the need for intake of lupins prior to joining for maximum response. In addition, these figures confirmed that the feeding of lupins for as short a period as seven days could markedly improve ewe fertility (Lightfoot, Marshall and Croker 1976; Lindsay 1976).

ACKNOWLEDGEMENTS

The authors wish to acknowledge the co-operation of Mr. R. Randall, Manager, and the staff, at Badgingarra Research Station during the experiment, in particular Mr. H.A. Abrahams for the detailed lambing observations. In addition, Mr. R.B. Guthrie is thanked for his skilled technical assistance. The project received financial assistance from the Wool Research Trust Fund.

REFERENCES

Arnold, G.W. and Charlick, A.J. (1976) *Proceedings of the Australian Society of Animal Production* **11** : 233.

Knight, T.W., Oldham, C.M. and Lindsay, D.R. (1975) *Australian Journal of Agricultural Research* **26** : 567.

Knight, T.W., Oldham, C.M., Smith, J.F. and Lindsay, D.R. (1975) *Australian Journal of Experimental Agriculture and Animal Husbandry* **15** : 183.

Lightfoot, R.J. and Marshall, T. (1974) *Journal of Agriculture Western Australia* **15** : 29.

Lightfoot, R.J. and Marshall, T. (1975) *Proceedings of the 7th Annual Conference, Australian Society for Reproductive Biology,* Abstract 36. Abstract in *Journal of Reproduction and Fertility* (1976) **46** : 518.

Lightfoot, R.J., Marshall, T. and Croker, K.P. (1976) *Proceedings of the Australian Society of Animal Production* **11** : 5P.

Lindsay, D.R. (1976) *Proceedings of the Australian Society of Animal Production* **11** : 217.

Lindsay, D.R., Knight, T.W., Smith, J.F. and Oldham, C.M. (1975) *Australian Journal of Agricultural Research* **26** : 189.

Marshall, T. and Lightfoot, R.J. (1974) *Journal of Agriculture Western Australia* **15** : 31.

Marshall, T., Beetson, B.R. and Lightfoot, R.J. (1976) *Proceedings of the Australian Society of Animal Production* **11** : 229.

Wroth, R.W. and Lightfoot, R.J. (1976) *Proceedings of the Australian Society of Animal Production* **11** : 225.

44

AGE OF EWE AND RESPONSE TO LUPINS: EFFECT OF LUPIN SUPPLEMENTATION ON OVULATION RATE

T. MARSHALL, K.P. CROKER and R.J. LIGHTFOOT
Department of Agriculture, Western Australia

SUMMARY

An experiment of factorial design was conducted to investigate the effects of age of ewe ($1\frac{1}{2}$, $2\frac{1}{2}$, $3\frac{1}{2}$, $4\frac{1}{2}$ years), rate of lupin grain supplementation (0, 250, 500 g/head/day) and duration of supplementation prior to pre-joining (7 versus 14 days with, in each case, supplements continued for the first 17 days of joining) on incidence of service and ovulation rate.

Significantly less ($P< 0.05$) $1\frac{1}{2}$ and $2\frac{1}{2}$ year old ewes than $3\frac{1}{2}$ and $4\frac{1}{2}$ year old ewes were served both during the first 17 days of joining and throughout the joining period. Lupin supplementation did not effect service.

There was a significant ($P< 0.05$) interaction between ewe age and rate of lupin grain supplementation on ovulation rate. Maiden ($1\frac{1}{2}$ year old) ewes failed to respond to increasing levels of lupin grain supplementation, $2\frac{1}{2}$ year old ewes responded only when 500 g/head/day was fed from 14 days before joining, while $3\frac{1}{2}$ and $4\frac{1}{2}$ year olds showed dramatic responses irrespective of rate or duration of supplementation.

INTRODUCTION

It is well established that "maiden" (nulliparous $1\frac{1}{2}$ year old) ewes are less fertile than older ewes. Knight, Oldham, Smith and Lindsay (1975), working with Merino flocks in south Western Australia found that a reason for this was the low ovulation rate of the young ewes. Ovulation rate in "old" ewes can be increased by feeding sweet lupin grain at joining (Lightfoot and Marshall 1974; Marshall and Lightfoot 1974; Knight, Oldham and Lindsay 1975). Data from two recent experiments conducted by our group (Lightfoot and Marshall 1975; Lightfoot, Marshall and Croker 1976) showed that $2\frac{1}{2}$ year old ewes had significantly lower ovulation rates than older ewes even when supplemented with lupins. In one of these experiments (Lightfoot and Marshall 1975) there was a significant interaction between age of ewe and rate of lupin grain supplementation. Neither experiment, however, examined $1\frac{1}{2}$ year old ewes. The information reported here relates to an experiment conducted to investigate further the relationship between age and ovulation rate response among ewes supplemented with lupin grain.

MATERIALS AND METHODS

The investigation was conducted during January and February, 1976 at the Merredin Agricultural Research Station situated approximately 260 km east of Perth. The region experiences a Mediterranean climate with approximately 75% of mean annual rainfall (330 mm) falling between May and September. The experiment was of factorial design ($4 \times 3 \times 2$, n=25, N=600) and examined the effects of age of ewe at joining ($1\frac{1}{2}$, $2\frac{1}{2}$, $3\frac{1}{2}$, $4\frac{1}{2}$ years—medium wool Merino), rate of sweet lupin grain supplementation (0, 250, 500 g/head/day—*Lupinus albus* L. cv. Ultra, 34% crude protein) and pre-joining duration of supplementation (7, 14 days), on incidence of service and ovulation rate.

The ewes were randomly allocated, within age groups, to treatments on January 21, 1976 following 14 days exposure to vasectomized rams (1 ram/100 ewes). The treatment groups were then introduced to three equal subdivisions (23.5 ha each) of a wheat stubble paddock according to their rate and duration of feeding schedule. The lupin supplements were fed from troughs on the ground thrice weekly (Monday, Wednesday and Friday) commencing either January 21 (14 day treatments) or January 28 (7 day treatments). Six entire rams fitted with marking harnesses and crayons were introduced to each group on February 4. Throughout the feeding period each ewe group was randomly rotated between the stubble subdivisions on a weekly basis. The ram groups remained in the subdivisions in which they had been originally placed during this period. At the end of the supplementation period (February 21) all ewes and rams were run together. Crayon colours were changed on days 14 and 28 of joining and the rams were removed at the completion of a 42 day joining period.

Live weights of all ewes were taken 14 days before joining commenced, at the start of joining, at the end of the supplementation and at the end of the joining. Service records were obtained on days 7, 14, 17, 28, 35 and 42 of joining. Half of the ewes served within each treatment during days 1 to 17 of joining were selected at random and examined by laparoscopy to determine ovulation rates.

RESULTS

(a) Liveweight changes

All groups gained weight during the lupin supplementation period (21.1.1976 to 21.2.1976). There was a significant ($P < 0.05$) linear effect of age on liveweight increase (liveweight increase = 3.3, 4.5, 6.1 and 6.4 kg for $1\frac{1}{2}$, $2\frac{1}{2}$, $3\frac{1}{2}$ and $4\frac{1}{2}$ year old ewes, respectively). There was no significant effect of rate of lupin supplementation on weight gain (4.6, 5.1 and 5.2 kg for 0, 250 and 500 g/head/day treatments, respectively) but ewes supplemented for 14 days gained more liveweight (P< 0.05) than those supplemented for 7 days (5.7 versus 4.7 kg respectively).

(b)   Incidence of service

Significantly less (P< 0.05) 1½ and 2½ compared with 3½ and 4½ year old ewes were served, both during the first 17 days of joining (83, 81, 89 and 88% for 1½, 2½, 3½ and 4½ year old ewes, respectively), and throughout the whole of the joining period although in the latter instance the effect was marginal (93, 90, 96 and 95% respectively). This factor was not affected by either rate or pre-joining duration of lupin supplementation.

There were no treatment effects on the proportion of ewes returning to first service.

(c)   Ovulation rate

Despite highly significant (P< 0.001) main effects of both age and rate of lupin grain supplementation on ovulation rate these two factors interacted (Table 1, P< 0.05). Maiden ewes did not respond at all to lupin supplementation. The 2½ year old ewes only responded when 500 g/head/day was fed from 14 days before joining. In contrast, the 3½ and 4½ year old ewes responded to all lupin supplementation treatments with no significant differences between either rate (250 versus 500 g/head/day) or duration of supplementation.

TABLE 1

RELATIONSHIP BETWEEN AGE OF EWE AND RATE AND
DURATION OF LUPIN GRAIN SUPPLEMENTATION
ON OVULATION RATE

Rate of Lupin (g/head/day)	Age of Ewe (years)				Duration (days)		Overall
	1½	2½	3½	4½	7	14	
0	1.00	1.00	1.08	1.04	1.04	1.02	1.03
250	1.00	1.05	1.24	1.32	1.11	1.21	1.16
500	1.00	1.24	1.14	1.30	1.12	1.23	1.17
Overall	1.00 (63)*	1.09 (65)	1.15 (72)	1.27 (71)	1.09 (135)	1.15 (136)	1.12 (271)

*No. of ewes examined.

DISCUSSION

The finding that fewer 1½ and 2½ year old ewes were served, both in the first 17 days of joining and throughout the entire joining period, differs from the related study with "maiden" ewes by Knight *et al.* (1975) but agrees with studies on $2\frac{1}{2}$ year olds by Marshall, Beetson and Lightfoot (1976) and Wroth and Lightfoot (1976). Reasons for the variability between studies are not clear but are likely to be related to environmental factors affecting growth and development. The results obtained in the present study indicate that a failure for $1\frac{1}{2}$ and $2\frac{1}{2}$ year old ewes to be served could be a significant area of wastage, even when ewes are "teased" prior to the introduction of entire rams. In addition, the results suggest that lupin grain supplementation will not be effective in overcoming this deficiency.

All groups made substantial liveweight gains during the period of lupin supplementation. This reflected, however, changes in the quantity of base pasture on offer as the ewes were moved from a sparsely covered pasture to the freshly harvested wheat stubble for the period of lupin supplementation and joining. This factor could also be responsible for the differences in liveweight gains found between the 7 and 14 day pre-joining supplementation groups. Liveweight increases were similar however, for both the lupin supplemented and the control groups which suggests that liveweight change *per se* was not responsible for the increases in ovulation rate observed. Knight, Oldham and Lindsay (1975) and Lightfoot, Marshall and Croker (1976) also reported increases in ovulation rate independent of changes in liveweight among ewes fed lupin grain.

The data presented here confirms previous reports that maiden Merinos (Knight *et al.* 1975) and $2\frac{1}{2}$ year old ewes (Wroth and Lightfoot 1976) may have lower ovulation rates than older ewes. Of more importance, however, lupin supplementation increased ovulation rates among older ewes, but the $1\frac{1}{2}$ year old ewes did not respond, and $2\frac{1}{2}$ year old ewes only responded when high rates were fed for relatively long periods. This confirms previous indications with $2\frac{1}{2}$ year old ewes (Lightfoot and Marshall 1975; Lightfoot, Marshall and Croker 1976) and adds new evidence that the comparatively unresponsive condition is even more pronounced among $1\frac{1}{2}$ year olds. The practical implication of this work is, that compared with older ewes, it will be relatively uneconomic to supplement $1\frac{1}{2}$ to $2\frac{1}{2}$ year old animals with lupin grain at joining to increase ovulation rates and fertility.

More work is required to understand the reasons for the failure to stimulate higher ovulation rates among young ewes. It is important to determine if they are totally unresponsive to lupin grain supplementation, or whether they have a higher threshold requirement.

ACKNOWLEDGEMENTS

The authors gratefully acknowledge the help and assistance given by Messrs. W.M. Booth (Manager), K.J.I. Burchall and S.M. Crook of the Merredin Agricultural Research Station and the skilled technical assistance of Messrs. G.R. McMullen and T.J. Johnson. The work was supported by a grant from the Wool Research Trust Fund.

REFERENCES

Knight, T.W., Oldham, C.M. and Lindsay, D.R. (1975) *Australian Journal of Agricultural Research* **26** : 567.

Knight, T.W., Oldham, C.M., Smith, J.F. and Lindsay, D.R. (1975) *Australian Journal of Experimental Agriculture and Animal Husbandry* **15** : 183.

Lightfoot, R.J. and Marshall, T. (1974) *Journal of Agriculture Western Australia* **15** : 29.

Lightfoot, R.J. and Marshall, T. (1975) *Proceedings of the 7th Annual Conference, Australian Society for Reproductive Biology Abstract 36*. Abstract in *Journal of Reproduction and Fertility* (1976) **46** : 518.

Lightfoot, R.J., Marshall, T. and Croker, K.P. (1976) *Proceedings of the Australian Society of Animal Production* **11** : 5P.

Marshall, T. and Lightfoot, R.J. (1974) *Journal of Agriculture Western Australia* **15** : 31.

Marshall, T., Beetson, B.R. and Lightfoot, R.J. (1976) *Proceedings of the Australian Society of Animal Production* **11** : 229.

Wroth, R.H. and Lightfoot, R.J. (1976) *Proceedings of the Australian Society of Animal Production* **11** : 225.

45

CLOVER DISEASE IN WESTERN AUSTRALIA

N.R. ADAMS
Division of Animal Health, CSIRO, Nedlands, Western Australia

SUMMARY

Clover disease infertility occurs in ewes after prolonged grazing on oestrogenic subterranean clover, and results from a permanent dysfunction of the cervix. It is different from the temporary infertility observed in ewes and cows grazing oestrogenic pasture around the time of mating. The pathogenesis of clover infertility is not fully understood. Pathological, physiological and statistical investigations suggest that clover infertility is widespread in the higher rainfall areas of south Western Australia.

Current work supports the contention that infertility in the higher rainfall areas is associated with an increase in returns to service by the ewe, and a higher proportion of ewes failing to lamb. The oestrogenicity of pasture was found to be related linearly to the proportion of oestrogenic subterranean clover in the pasture. However, spot estimates of pasture oestrogenicity were not related either to the average annual rainfall or the reproductive performance of ewes on the property. There is a need for methods to study the oestrogenicity of pastures over a prolonged period.

INTRODUCTION

The climate in the southern part of Western Australia is characterized by wet winters and hot, dry summers. Pasture production is therefore almost entirely dependent upon annuals, with subterranean clover (*Trifolium subterraneum* L.) as the legume. The exploitation of this plant, beginning around 1930, has had a dramatic effect on pasture productivity in the State.

It is unfortunate that Dwalganup, the first cultivar of subterranean clover to be widely sown, contained high levels of oestrogenic isoflavones. The newly planted clover-dominant pastures offered grazing animals massive levels of phytooestrogens, with sheep ingesting up to 20 g of isoflavone phytooestrogens daily. The resulting syndrome of dystocia, uterine prolapse and infertility in ewes, and urinary obstruction in wethers, was called clover disease. This severe syndrome is now uncommon, but a residual problem is still widespread in Western Australia.

THE CURRENT EXTENT OF CLOVER INFERTILITY

Both the higher proportion of grass in pastures and the planting of low-oestrogen cultivars of subterranean clover, have led to a marked reduction in the severity of clover disease on commercial pastures. This has made diagnosis of the residual levels of clover infertility more difficult, but all available evidence indicates that subclinical clover infertility occurs extensively in ewes in the higher rainfall areas of south Western Australia.

Both macroscopic and microscopic cysts in the uterus and cervix are widespread among ewes from the higher rainfall area (Adams 1975). The prevalence of cysts in the reproductive tract has been associated with infertility from clover disease (Davies and Nairn, 1964). Thus it is likely that the common occurrence of cysts reflects widespread clover infertility in this area.

Lamb marking rates are generally better in the lower rainfall wheatbelt areas of Western Australia than in the higher rainfall areas. Prior to the widespread plantings of subterranean clover in the higher rainfall areas during the 1940s, the opposite was true (Lightfoot, 1972). Wroth and Lightfoot (1976) and Marshall, Beetson and Lightfoot (1976) have shown that the main difference between ewes in these two areas is the depressed efficiency of ovum fertilization now found in ewes in the higher rainfall areas, associated with the presence of oestrogenic clover. There was no significant difference between the two areas in ovulation rate.

Data obtained from a wider but less detailed survey (Pearson, 1969) also indicate that ewes in the higher rainfall areas have similar twinning rates, but a higher proportion of ewes fail to bear a lamb, when compared with ewes in the drier non-clover area. It has been estimated that clover infertility causes a loss of 500,000 lambs annually in Western Australia alone (Lightfoot, 1972).

CHARACTER OF CLOVER INFERTILITY

Progressive and prolonged infertility was recognized as part of the clover disease syndrome by Bennetts, Underwood and Shier (1946). The sheep grazed green, oestrogenic subterranean clover during the winter, and were infertile when mated on dry, non-oestrogenic pasture during the summer. The infertility observed in these ewes is different from that seen in ewes which graze green, oestrogenic pasture around the time of mating.

(a)   Infertility due to concurrent oestrogens

Infertility due to the intake of phytooestrogens around the time of mating may be distinguished from clover disease, or post-oestrogen infertility, by the stage of the reproductive process at which loss occurs. The infertility observed with concurrent oestrogen exposures is expressed primarily as a reduction in ovarian function. The ovulation rate is reduced, so that fewer ewes come into oestrus, and also fewer ewes bear twin lambs (Lightfoot and Wroth, 1974). The efficiency of fertilization is also somewhat decreased. This may result from decreased sperm transport, or altered ovum transport. The effects on embryonic mortality have not been fully defined. A similar form of infertility, characterized mainly by altered ovarian function has been observed in cows grazing oestrogenic pasture (Thain, 1966).

(b) Post-oestrogen infertility

The infertility associated with clover disease occurs after prolonged oestrogen ingestion and is accompanied by normal or even slightly increased ovarian function and ovulation rates. This post-oestrogen infertility is manifested primarily as a reduced efficiency of ovum fertilization. The ovum transport rate does not appear to be altered, but embryonic death may be slightly more common in affected ewes. This form of infertility has not been described in the cow.

Lightfoot, Croker and Neil (1967) found that fertilization failure, the main abnormality in ewes with clover infertility, was due primarily to a failure of sperm to establish themselves in the cervix. Fewer sperm were available to migrate up to the oviduct, and fertilization efficiency was reduced. Smith (1971) showed that the change in cervical function was due to the production of abnormal "watery" mucus by infertile ewes. This mucus is less efficient at guiding sperm migration *in vitro* (Adams 1976), and has a reduced spinbarkheit. The spinbarkheit is the length to which a thread of mucus can be drawn out between two glass slides. It is related to those factors which direct sperm migration, and is relatively constant over the period of oestrus. Studies on commercial flocks have shown that spinbarkheit may be used as a diagnostic tool in ewes with low levels of clover infertility (Adams, unpublished). It is this form of infertility which is important in Western Australia.

RELATIONSHIP OF CLOVER INFERTILITY AND OESTROGENIC PASTURE

Clover infertility has been studied experimentally by exposing ewes to relatively pure stands of oestrogenic clover, or by varying the composition of a pasture at a single site. In order to obtain an idea of factors controlling pasture oestrogenicity and clover infertility under commercial conditions, flocks on 11 properties in 9 shires in south Western Australia have been studied. These properties all had a high level of management, but varied in clover infertility status.

(a) Experimental methods

Ewes from 4 properties which had oestrogenic pasture had previously been observed to have depressed fertilization rates, or a high level of endometrial cysts. Fertilization rates on properties A, B and C (Table 1) were 72%, 81%, and 84% respectively (Wroth and Lightfoot, 1976). Macroscopic cysts were found in the uterus or cervix of 52 of 114 ewes (57.9%) from property D at the abattoirs (Adams, unpublished). These properties were classified as clover affected (Properties A-D, Table 1). Three other properties had pastures substantially free of subterranean clover, and ewes from these had fertilization rates between 95% and 98%. These properties were classified as free of clover infertility (Properties I-K, Table 1). The clover infertility status of flocks on the other 4 properties was unknown. Flocks of between 280 and 660 ewes aged between 4 and 6 years old were studied on each property under normal management conditions during the 1974-75 breeding season. Return-to-service rates, and the proportion of mated ewes failing to lamb in each flock were determined by the methods described by Lindsay *et al.* (1975).

In winter a paddock selected by the owner as typical of that grazed by the ewes throughout their lifetime was examined. The proportion of subterranean clover in the total plant material available was estimated by 3 independent observers traversing the paddock, and the proportion of the more highly oestrogenic cultivars (Dwalganup, Yarloop and Geraldton) was also estimated. Nine wether hoggets which had previously run on non-oestrogenic pasture were placed on each pasture at this time and the change in their teat length measured over a 14 day period.

(b)   Results and discussion

TABLE 1

PASTURE AND SHEEP REPRODUCTION CHARACTERISTICS FROM 11 PROPERTIES DURING 1974-75 IN SOUTH WESTERN AUSTRALIA

Property	Shire	Average annual rainfall (mm)	Assessed clover infertility status	% return to service of ewes	%of mated ewes failing to lamb	% oestrogenic sub-clover in pasture	mean increase of wether teat length on pasture (cm)
A	West Arthur	635	+	51.2	23.0	30	.53
B	West Arthur	508	+	37.2	30.9	15	.16
C	West Arthur	584	+	40.5	22.1	15	.28
D	Moora	508	+	53.5	31.3	37	.29
E	Boyup Brook	610	?	44.1	15.6	15	.23
F	Wagin	508	?	33.5	26.8	24	.30
G	Cunderdin	381	?	37.7	7.1	17	.48
H	Dowerin	330	?	28.4	19.7	48	.58
I	Kulin	381	—	30.9	18.1	7	.37
J	Bruce Rock	330	—	29.5	14.6	4	.28
K	Merredin	305	—	40.5	16.0	0	.01

It can be seen from Table 1 that both return-to-service rates and the proportion of mated ewes failing to bear a lamb, were higher in flocks on clover affected properties than on properties with little oestrogenic clover. Unpublished studies with cervical mucus spinbarkheit on all the properties also indicated that both return-to-service rates and the proportion of ewes failing to lamb could be related to cervical mucus differences, and thus to variations in clover infertility in the flocks.

Return-to-service rates and the proportion of ewes failing to lamb were lower on properties with lower average annual rainfall. Results from the properties of unknown clover history also appeared to be related to their geographical location, so that there was a correlation over all the properties between the average annual rainfall on a property and the return-to-service rate recorded (r=0.654, P<0.05). This conclusion agrees with the finding of Pearson (1969) that fewer ewes fail to lamb in the drier wheatbelt areas, and the studies of Wroth and Lightfoot (1976) and Marshall, Beetson and Lightfoot (1976) which concluded that return-to-service rates of ewes were higher in the higher rainfall clover areas.

Most previous studies on the oestrogenicity of pastures have been carried out on clover-dominant plots. Since sheep graze pasture species selectively, bioassay methods used experimentally may give a different result when applied to pastures found on commercial properties, since these are not normally clover dominant. Although results obtained in the present work were variable, the proportion of oestrogenic subterranean clover in the pasture was linearly related with the increase in teat length of wethers running on the pasture (r=0.666, P<0.05). Thus selective grazing of either grass or clover by the wethers did not seem to be a serious problem. The relatively high teat length increases recorded on properties I and J may be due to either variability in the teat length response, or to selective grazing of the clover by the wethers.

A surprising finding was the variability of pasture oestrogenicity, when compared with geographical location. Neither the proportion of oestrogenic clover in the pasture nor the teat length increase of wethers on the pasture was correlated with the average annual rainfall, return-to-service rates, or the proportion of ewes failing to lamb. The wether teat length response was also unrelated to previously determined clover infertility status.

This finding accentuates the obvious difficulties in extrapolating results obtained on one paddock at one point in time to the lifetime intake of phytooestrogen by ewes on a property. The relative proportions of grass and clover in a pasture change markedly both within a season, and between seasons. Furthermore, the average annual growing season is 1.5 times longer on the clover affected properties (Properties A-D) than on properties I-K, so that ewes would be exposed to green oestrogenic clover for a longer period. It is obvious that there is an overall relationship between cumulative intake of phytooestrogen by a ewe, and the resulting infertility. However, detailed, long term studies are needed to relate the rate and duration of phytooestrogen intake with the degree of clover infertility resulting.

The hazards to reproductive performance from mating ruminants on oestrogenic pasture have been well recognized. Suitable methods for determining the oestrogenicity of pastures have been developed, and it appears that these may also be useful on the type of pasture found on commercial properties. However, as shown in the present study, a single bioassay of pasture may be misleading in assessing the likelihood of post-oestrogen clover infertility in sheep which have grazed the pasture. A simple method for determining cumulative oestrogenicity of a pasture would be of great assistance in studying post-oestrogen clover infertility.

ACKNOWLEDGEMENTS

The help of Dr. W.J. Collins and Dr. R.C. Rossiter with estimates of pasture composition is gratefully acknowledged.

REFERENCES

Adams, N.R. (1975) *Australian Veterinary Journal* **51** : 351.

Adams, N.R. (1976) *Research in Veterinary Science* **22**: 216.

Bennetts, H.W., Underwood, E.J. and Shier, F.L. (1946) *Australian Veterinary Journal* **22** : 2.

Davies, H.L. and Nairn, M. (1964) *Proceedings of the Australian Society of Animal Production* **5** : 62.

Lightfoot, R.J. (1972) *Journal of Agriculture Western Australia* **13** : 102.

Lightfoot, R.J., Croker, K.P. and Neil, H.G. (1967) *Australian Journal of Agricultural Research* **18** : 755.

Lightfoot, R.J. and Wroth, R.H. (1974) *Proceedings of the Australian Society of Animal Production* **10** : 130.

Lindsay, D.R., Knight, T.W., Smith, J.F. and Oldham, C.M. (1975) *Australian Journal of Agricultural Research* **26** : 189.

Marshall, T., Beetson, B.R. and Lightfoot, R.J. (1976) *Proceedings of the Australian Society of Animal Production* **11** : 229.

Pearson, J.R. (1969) Lambs (Wesfarmers Technical Services: Perth).

Smith, J.F. (1971) *Australian Journal of Agricultural Research* **22** : 513.

Thain, R.I. (1966) *Australian Veterinary Journal* **42** : 199.

Wroth, R.H. and Lightfoot, R.J. (1976) *Proceedings of the Australian Society of Animal Production* **11** : 225.

46

GENETIC ASPECTS OF PUBERTY IN MERINO EWES

M.L. TIERNEY

University of New South Wales, Sydney, Australia

SUMMARY

Ewes representing a number of Merino strains were run together in a single environment in south-west New South Wales. Observations were carried out on the ewes from weaning until their first joining at approximately 20 months of age, to determine the age and live weight at which they first exhibited oestrus. A number of body, yearling fleece, and reproductive traits were also measured to investigate any correlations between these and the first exhibition of oestrus.

Age at first exhibition of oestrus, and whether or not ewes had exhibited oestrus by certain ages appeared to be determined principally by the live weight of the ewe. Differences between strains with respect to exhibition of first oestrus could usually be explained by differences in live weights of the various strains. However, some strains, particularly those selected for high fertility, appeared to have a potential for exhibiting oestrus earlier than would have been expected by their live weights.

It was not possible to determine accurate estimates of the genetic parameters associated with age and live weight at first oestrus. However, the heritability of live weight at first oestrus (0.61 ± 0.44) and the genetic correlation between live weight at first oestrus and live weight at 15 months of age (0.86 ± 0.30) appeared to be the two most significant parameters estimated.

A further study, on a higher plane of nutrition, revealed a significant strain effect and a significant interaction between strain and level of nutrition, for the percentages of ewes exhibiting oestrus by seven and 15 months of age, respectively. These effects could be explained largely in terms of the live weight of the ewes and the periods of the year during which they first exhibited oestrus.

INTRODUCTION

The age at which a ewe can first be mated is of considerable practical importance both from the viewpoint of increasing the lifetime performance of the ewe and from the probable relationship between earliness of sexual activity in the ewe lamb and a generally higher level of reproductive efficiency in the adult ewe (Dyrmundsson, 1973).

Hafez (1952) and Dyrmundsson (1973) have both presented extensive reviews concerning the incidence of first oestrus in ewes. From this latter review the mean age at first oestrus is seen to range from 163 days for Finnish Landrace x Polled Dorset ewes in Scotland to 900 days for Hungarian Merino ewes in Hungary. Similar wide variation is seen to exist in the mean live weight at first oestrus, which is reported to range from 28.8 kg for Corriedale ewes in Uruguay to 54.7 kg for Rambouillet ewes in the USA.

A number of authors have reported considerable variability in the incidence of first oestrus within a particular breed (e.g. Hafez, 1953; Ch'ang and Raeside, 1957; Basson, van Niekerk and Mulder, 1970; Wiggins, Miller and Barker, 1970). The purpose of this study was to examine genetic differences with respect to puberty in a number of strains of Merino ewes, with puberty being defined as the age at which a ewe first exhibits behavioural oestrus.

MATERIALS AND METHODS

(a)　Sheep involved

The sheep in the study were kept at the University of New South Wales Hay Field Station in south-west New South Wales. The foundation flocks established at the Hay Field Station and the environment at that location have been described by Jackson and Roberts (1970).

Five groups which were the progeny of these original foundation flocks were studied. These are referred to as SA1 and SA2 representing two studs of the South Australian Merino strain, SAP representing one stud of the South Australian x Peppin Merino strain, and MP1 and MP2 representing two studs of the Medium Peppin Merino strain. In addition, four other groups were included: C, representing a commercial South Australian strain flock, F1 and F2, two flocks of the Medium Peppin strain which had been selected for increased fertility, and F3, a flock of SA1 ewes mated to Medium Peppin rams selected for increased fertility.

The ewes available for study in this trial were the ewes born and surviving in 1966, 1967 and 1968 in each of the above groups. These totalled 298 in 1966, 148 in 1967 and 222 in 1968.

(b)　Observations and measurements

The ewes were kept together with vasectomized rams fitted with raddle harnesses (Radford, Watson and Wood, 1960) from weaning until they were mated with entire rams, at approximately 20 months of age. This enabled the age at first exhibition of oestrus to be determined.

In addition, live weights were recorded monthly, and a number of fleece and body traits were recorded at shearing at approximately 15 months of age. These characters included face cover score, folds score, greasy fleece weight, fleece washing yield, clean fleece weight, fleece quality number, fleece character, fleece handle, fleece colour, staple length, crimps per inch, fibre diameter, fleece density and fleece value.

Further, using the technique of pelvic endoscopy (Roberts, 1968) the ovulation rate of the ewes was determined at two oestrous cycles just prior to mating, and the joining and lambing performances of the ewes at this mating were also recorded.

All these traits were recorded to allow any relationships between any of these traits and exhibition of first oestrus to be determined.

The method of correlation between half-sibs was used to determine the genetic parameters associated with first oestrus.

(c) Supplementary trial

As an extension to this original trial, a further trial was initiated in December 1969 to examine more closely the effect of level of nutrition on the exhibition of first oestrus, and the possible interactions between level of nutrition and strain of ewe, using three of the strains of ewes that had been involved in the initial study.

Fifty spring-born ewes of each of the SA2, SAP, and MP2 types were purchased from properties with environments as similar as possible to that at Hay, and transferred to the Hay Field Station.

The ewes of each type were randomized on a live weight basis into high and normal nutrition groups. The high group was kept on irrigated natural pasture and had access to grain oats fed *ad lib*. The normal group was kept on the same type of pasture, but the stocking rate was adjusted to maintain the live weights of the ewes in this group at approximately 20% below those in the high group.

The ewes were observed for exhibition of oestrus from January 1970 (five months of age) until December 1970 (15 months of age) and live weights were recorded every two weeks.

RESULTS

(a) Live weights of ewes and percentages exhibiting oestrus

In the initial study, no ewes in any group in any year had exhibited oestrus by the end of their first autumn season, at about 10 months of age. The mean live weights of the ewes at this age were 20.3 kg for the 1966 ewes, 21.2 kg for the 1967 ewes and 28.6 kg for the 1968 ewes. The mean live weights of the heaviest group in each of these years (the SA2 group) were 21.9 kg, 23.7 kg and 29.7 kg respectively.

By 15 months of age, the percentage of ewes exhibiting oestrus in each group, and the mean live weights of the groups are presented in Table 1. Only one of the 1966 ewes had exhibited oestrus by this age, while there were significant percentages in other years.

By 20 months of age greater percentages of all ewes had exhibited oestrus and there were only small differences between groups, with the exception that the high fertility groups performed better than the other groups. Of the

TABLE 1

PERCENTAGE OF EWES EXHIBITING OESTRUS BY 15 MONTHS OF
AGE, AND THE MEAN LIVE WEIGHTS OF THE EWES AT THAT
AGE

Group	Born 1966		Born 1967		Born 1968	
	Percentage exhibiting oestrus	Live weight (kg)	Percentage exhibiting oestrus	Live weight (kg)	Percentage exhibiting oestrus	Live weight (kg)
SA1	0.0	37.5	26.6	33.8	34.8	33.8
SA2	0.0	41.4	33.4	39.1	11.1	37.6
SAP	0.0	40.9	—	—	14.0	35.8
MP1	0.0	30.2	0.0	29.1	29.6	29.6
MP2	0.0	30.7	11.9	33.6	14.6	33.3
C	2.6	28.9	28.0	33.5	21.0	36.2
F1	—	—	43.0	35.0	40.0	35.5
F2	—	—	7.0	29.6	—	—
F3	0.0	30.5	17.0	33.4	35.0	34.1

ewes born in 1966, 85% of those in the high fertility groups had exhibited oestrus by 20 months of age compared to a mean 46.8% for all the other groups. The corresponding figures for the 1967 ewes were 85.3% and 64.4% respectively, and for the 1968 ewes were 93.3% and 79.9% respectively.

(b) Age and live weight at first oestrus

Mean ages and live weights at first oestrus for the various groups were only calculated for those ewes which had exhibited oestrus by the end of their first mating period, at 20 months of age. These are listed in Table 2.

There were no significant differences with respect to age at first oestrus for the ewes born in 1966 or 1967. For those born in 1968, the SA1, F1 and F3 groups were significantly ($P<0.05$) younger at first oestrus than the other groups.

With weight at first oestrus, the MP2, C and F3 groups were lighter than the others for ewes born in 1966. For ewes born in 1967 the SA2 and F1 groups were heavier than the rest, while for those born in 1968 the SA2 and SAP groups were heavier than the others and the MP1 group was lighter. These differences were significant at the 5% level.

(c) Relationship of first oestrus with other characters

Differences between ewes that had and had not exhibited oestrus by 20 months of age were examined for all the body and fleece traits that were measured. Consistent differences were only found with respect to live

TABLE 2

MEAN AGE AND LIVE WEIGHT AT FIRST OESTRUS FOR EWES
THAT HAD EXHIBITED OESTRUS BY 20 MONTHS OF AGE

Group	Born 1966		Born 1967		Born 1968	
	Mean age at first oestrus (days)	Mean live weight at first oestrus (kg)	Mean age at first oestrus (days)	Mean live weight at first oestrus (kg)	Mean age at first oestrus (days)	Mean live weight at first oestrus (kg)
SA1	560	32.2	511	40.0	494	36.7
SA2	560	35.4	521	45.3	542	42.4
SAP	560	36.0	—	—	531	41.2
MP1	580	32.7	561	36.8	529	33.1
MP2	545	31.2	548	38.0	529	37.8
C	N.A.	30.5	509	39.7	518	39.3
F1	—	—	511	42.0	502	37.9
F2	—	—	N.A.	36.9	—	—
F3	542	31.2	546	39.2	506	38.3

weight at 15 months of age, and associated characters such as greasy and clean fleece weights.

The only consistently significant phenotypic correlation found between either live weight or age at first oestrus and any of the other characters was between live weight at 15 months and live weight at first oestrus. The correlation coefficients for this character ranged from 0.67 to 0.99 for the various groups.

Of the reproductive characters measured, the only significant result was a higher overall level of fertility and for the F1, F2 and F3 groups than for the other groups. This was reflected in higher percentages of ewes ovulating in each year in the fertility groups (88% *v.* 65% for ewes born in 1966; 74% *v.* 56% for those born in 1967; 81% *v.* 74% for those born in 1968). The differences were significant for the 1966 and 1967 ewes ($P < 0.05$) but not for those born in 1968.

In addition, the ovulation rates per ewe ovulating were significantly higher for the fertility groups than for the other groups in each of the three years ($P < 0.05$). The rates were 1.13 *v.* 1.00 for the 1966-born ewes, 1.20 *v.* 1.08 for the 1967-born ewes and 1.24 *v.* 1.14 for the 1968-born ewes.

(d) Estimation of genetic parameters

Due to the low numbers of ewes available for the study, it was necessary to pool all the data over all groups and years to obtain any estimates of genetic parameters with reasonable standard errors. The estimates of heritability obtained from these pooled data were 0.11 ± 0.43 for age at first oestrus, 0.61 ± 0.44 for live weight at first oestrus and 0.06 ± 0.24 for the exhibition of oestrus by 20 months of age.

Of the genetic correlations estimated, the only ones for which the standard errors were low in relation to the value estimated were for live weight at first oestrus with live weight at 15 months of age (0.86 ± 0.30), body folds score (0.85 ± 0.25), and greasy fleece weight (0.70 ± 0.46); for age at first oestrus with total folds score (0.86 ± 0.30); and for exhibition of oestrus by 20 months of age with: live weight at 15 months (0.64 ± 0.50), fleece quality number (-0.75 ± 0.30), fleece character score (0.86 ± 0.26) and staple length (0.68 ± 0.50).

(e) Supplementary trial

In the supplementary trial, to examine the effect of nutrition and possible interactions between level of nutrition and the strains involved, the results obtained were as presented in Table 3.

Analyses of variance of the percentages of ewes exhibiting oestrus revealed a significant strain effect at seven months of age and a significant strain x level of nutrition interaction at 15 months of age. With live weight, both strain and level of nutrition effects were significant at both ages, and there were no interactions. All significant effects were at the 5% level.

TABLE 3

PERCENTAGES OF EWES UNDER TWO LEVELS OF NUTRITION
EXHIBITING OESTRUS BY SEVEN AND FIFTEEN MONTHS OF AGE,
AND MEAN LIVE WEIGHTS AT THESE AGES

Group		Percentage of ewes exhibiting oestrus by 7 months	Mean live weight at 7 months (kg)	Percentage of ewes exhibiting oestrus by 15 months	Mean live weight at 15 months (kg)
Strain	Level of nutrition				
SA2	High	66.7	40.2	95.8	43.6
SA2	Normal	56.5	33.9	56.5	38.6
SAP	High	4.0	32.7	20.0	41.2
SAP	Normal	4.0	26.3	4.0	34.5
MP2	High	0.0	30.6	0.0	39.3
MP2	Normal	4.0	24.4	4.0	33.0

DISCUSSION

The lack of exhibition of oestrus by any ewes at 10 months of age can be explained in terms of the live weights of the ewes at that age. The review by

Dyrmundsson (1973) reports no incidences of mean live weights at first oestrus as low as those recorded here at 10 months of age.

At 15 and 18 months of age, the percentages of ewes exhibiting oestrus were fairly closely related to the live weights of the ewes, with the exception that in most cases the high fertility groups had more ewes exhibiting oestrus than would have been expected from their live weights. This relationship between early puberty and overall fertility has also been suggested by other authors including Spencer *et al.* (1942) and Wiggins (1955).

It is of interest that although the ewes born in 1966 were of a similar weight at 15 months of age to those born in 1967 and 1968, only one ewe born in 1966 had exhibited oestrus by that age. This could possibly be explained by the growth patterns of the ewes, since the ewes born in 1966 had made very little gain in live weight from weaning to 13 months of age, whereas those born in the other two years gained steadily throughout that period.

The mean ages at first oestrus reported here are considerably above the minimum values that would have been expected from other reports. In contrast, the mean live weights at first oestrus are within the ranges usually reported. This reflects the fact that due to the drought conditions which prevailed throughout most of this trial, it had been live weight and not age, that had been the restraining influence limiting first exhibition of oestrus.

The estimates obtained for the genetic parameters associated with first oestrus in this study can be taken only as guides to the values which might have been obtained from data involving a greater number of animals.

The considerably lower estimate of heritability obtained for age at first oestrus compared with that for live weight at first oestrus would be expected from the greater variability between groups for live weight at first oestrus than for age. A higher estimate for the heritability of age at first oestrus may have been obtained if the ewes had been sufficiently well grown to have exhibited oestrus in their first breeding season.

The significant genetic correlations reported may be explained in terms of the genetic correlation between folds score and fertility (Turner and Young, 1969), and the high phenotypic correlation between live weight at first oestrus and live weight at 15 months of age.

The supplementary trial confirmed that it was possible for ewes of the SA2 strain to exhibit oestrus by seven months of age provided they were sufficiently well grown.

A strain by level of nutrition interaction was demonstrated. By 15 months of age there was a difference in the percentage of ewes exhibiting oestrus between the high and normal nutrition groups for both the SA2 and SAP strains, but not for the MP2 strain in which, in fact, virtually no ewes had exhibited oestrus.

Watson and Gamble (1961) have reported that with medium wool Merino sheep in Australia, first oestrus occurs, with few exceptions, between November and May. In the supplementary trial the ewes were seven months of age in early May and 15 months in early December. Since both the SA2 and SAP high nutrition groups had ewes coming into oestrus between May and December it would appear that under good growth conditions this seasonality of first oestrus does not apply to the same extent in these strains as in the MP2 strain. In fact, in both the SA2 and SAP groups ewes were still exhibiting first oestrus in June, while in the SAP group first oestrus activity began again in October.

ACKNOWLEDGEMENTS

I wish to thank Associate Professor E.M. Roberts and Dr. J.W. James for their invaluable assistance in all respects of this project. I wish also to thank the staff of the University of New South Wales Hay Field Station for their care of the experimental animals.

The project was supported throughout by an Australian Wool Board Post-graduate Scholarship.

REFERENCES

Basson, W.D., van Niekerk, B.D.H. and Mulder, A.M. (1970) *Proceedings of the South African Society of Animal Production* **9** : 171.

Ch'ang, T.S. and Raeside, J.I. (1957) *Proceedings of the New Zealand Society of Animal Production* **17** : 80.

Dyrmundsson, O.R. (1973) *Animal Breeding Abstracts* **41** : 273.

Hafez, E.S.E. (1952) *Journal of Agricultural Science, Cambridge* **42** : 189.

Hafez, E.S.E. (1953) *Empire Journal of Experimental Agriculture* **21** : 217.

Jackson, N. and Roberts, E.M. (1970) *Australian Journal of Agricultural Research* **21** : 815.

Radford, H.M., Watson, R.H. and Wood, G.F. (1960) *Australian Veterinary Journal* **36** : 57.

Roberts, E.M. (1968) *Proceedings of the Australian Society of Animal Production* **7** : 192.

Spencer, D.A., Schott, R.G., Phillips, R.W. and Aune, B. (1942) *Journal of Animal Science* **1** : 27.

Turner, Helen N. and Young, S.S.Y. (1969) "Quantitative Genetics in Sheep Breeding" (Macmillan of Australia: South Melbourne).

Watson, R.H. and Gamble, L.C. (1961) *Australian Journal of Agricultural Research* **12** : 124.

Wiggins, E.L. (1955) *Journal of Animal Science* **14** : 1260 (Abstract).

Wiggins, E.L., Miller, W.W. and Barker, H.B. (1970) *Journal of Animal Science* **30** : 974.

47

ASSESSMENT OF NEW AND TRADITIONAL TECHNIQUES OF SELECTION FOR REPRODUCTION RATE

B.M. BINDON and L.R. PIPER

CSIRO, Division of Animal Production, Armidale, N.S.W., Australia

SUMMARY

Sheep reproduction rate is a function of fertility (the number of ewes conceiving), fecundity (the number of offspring per pregnancy) and lamb survival (broadly the number of offspring surviving the birth process). This chapter assesses the problems of direct selection for these components and reviews indirect approaches under investigation.

Estimates of the parameters required for predicting the likely success of direct selection for the above components come either from analyses of randomly-bred flocks or from long term experiments in which selection was based on a combination of fertility and fecundity. Selection has never been attempted for sheep fertility, fecundity or survival alone. Jointly, these two sources of data allow the following conclusions:

> Fertility is a character of low repeatability (0.08-0.09) and heritability (0.03-0.10) and the rate of improvement from direct selection is likely to be very slow.

> Fecundity has higher repeatability and heritability than fertility. Direct selection for the combined components of fertility and fecundity has resulted in increases of 0.011 to 0.023 lambs per ewe per year in two experiments.

> The heritability of viability in the Merino appears low (0.02) and there are no estimates of within flock genetic variation in mothering ability.

> A number of other factors contribute to the difficulty of increasing reproduction rate by direct selection: The characters are sex-limited and may be measured only in the ewe. Seasonal and other "environmental" effects (age and birth-type of ewe) affect performance and further increase the difficulty of identifying superior replacements. Fertility and fecundity require the male for their expression and this undoubtedly contributes to their low repeatability. Survival is a complex character because both viability of the foetus, the ewe's maternal behaviour and milk production contribute.

The physiology of reproduction is being studied in the hope that there may be measurements which are highly correlated with the components of reproduction rate but which do not suffer from the same limitations. The following characteristics and their relation to the components of reproduction rate are assessed in this chapter.

> Ovulation rate. Several measures may be made in one year, it is not confounded by direct male effects and it has higher repeatability than either fertility or fecundity.
>
> Number of oestrous cycles. This character is easily measured before the first joining. It has been shown to be genetically correlated with reproduction rate of the Romney and has a high heritability in Merinos.
>
> Plasma LH and FSH. These hormones lie at the basis of the components of reproduction rate. Their potential advantage rests in the fact that their measurement may be made in both ram and ewe lambs as early as the first month of life.
>
> Testis growth. This is a simple measurement which may directly assist the identification of superior males in the first year of life. Early results of selection for testis growth are promising.

The above approaches could be readily applied in group breeding schemes but may be beyond the scope of individual farmer's flocks.

INTRODUCTION

Low reproduction rate is a major factor limiting productivity of sheep enterprises in Australia and elsewhere. It assumes special significance after the national flock has been depleted by natural disasters and rural recessions. It is also an important factor limiting the success of breeding schemes of all types (national, group, individual farmer) since low reproduction rate limits genetic progress in all characters under selection.

Consistently reliable non-genetic or "environmental" methods for increasing reproduction rate are not available. Interest in the genetic approach to improvement therefore continues. Major reviews of the work in this area have been prepared by Turner (1969), Bradford (1972) and Land (1974). The purpose of this chapter is to re-assess the problems involved in traditional methods of selection for reproduction rate and to evaluate new approaches resulting from recent research in sheep reproductive biology. The results presented are related to within-breed improvement, although it is acknowledged that cross-breeding is responsible for substantial increases in fertility.

TRADITIONAL METHODS OF SELECTION FOR REPRODUCTION RATE

(a)   Components of reproduction rate

Reproduction rate is a complex character. Even when presented in a

TABLE 1

COMPONENTS OF REPRODUCTION RATE

Component and definition	Major contributory factors		
	Ewe	Ram	Lamb
Fertility (number of ewes conceiving)	Number of oestrous cycles	Sexual drive, sperm fertility	—
Fecundity (number of lambs per pregnancy)	Ovulation rate; uterine capacity	Fertilization rate	—
Survival (number of lambs surviving to 7 days age)	Parturition; post natal behaviour; milk production	—	Birth weight; viability

simplified way such as in Table 1 it is clear that variation in gross reproductive performance (e.g. lambs weaned per ewe joined per year) may be the result of variation arising from the ewe, the ram, the lamb or all three.

(b) Direct selection for fertility and fecundity

Data from randomly-bred flocks and two long term selection experiments (Wallace, 1958; Turner, 1962) have been reviewed by Turner (1969) with a view to assessing the likely rate of progress from direct selection for various components of reproduction rate. More recently, McGuirk (1976) reported progress in a third selected flock and discussed some consequences of the observed reproduction rate increase in this and other flocks.

In Table 2 an attempt has been made to summarize the range of estimates available for repeatability and heritability of fertility, fecundity and survival. This has been done by drawing from the reviews of Turner (1969) and McGuirk (1976), and adding data for repeatability of failure to lamb in Merinos (B.J. McGuirk, unpublished) and heritability of lamb viability in

TABLE 2

ESTIMATES OF GENETIC PARAMETERS FOR COMPONENTS OF REPRODUCTION RATE

Component	Repeatability (range)	Heritability (range)	Likely annual genetic gain*
Fertility	0.08-0.09	0.03-0.10	Small
Fecundity	0.04-0.28	0.04-0.26	0.011-0.023
Survival			
—lamb viability		0.02	Nil
—mothering ability	0.0 -0.10	No estimate	No estimate

* Lambs per ewe per year

Merinos (L.R. Piper, unpublished). Table 2 also contains an estimate of the likely rate of genetic progress if selection were applied to each character.

It should be stated that it appears that selection has never been applied to fertility, fecundity, mothering ability or survival alone. (The long term experiments of Wallace (1958), Turner (1962) and McGuirk (1976) yielded a positive response in terms of fecundity but selection was actually applied to a combination of fertility and fecundity.) There are, therefore, few estimates of realized heritability.

The data of Table 2 suggest that with the exception of fecundity, direct selection for the individual components of reproduction rate is unlikely to yield gains that are useful in individual or even group breeding schemes. The rate of response in the 'B' flock of the CSIRO experiment (see Turner, 1969) appears higher than the upper limits given in Table 2 (Turner, unpublished) but this flock had exceptional fecundity when selection commenced. It is not known whether the high initial fecundity is responsible for the apparently higher rate of selection response.

However, greater progress is likely to be achieved if selection is based on an index combining information on all or several of the components and, though it is perhaps obvious, the trait "lambs born per ewe joined" is unlikely to be the optimum index.

Genetic parameter considerations aside, there are a number of other factors which contribute to the difficulty of increasing reproduction rate by direct selection.

(i)   The characters (other than lamb viability) under selection are sex-limited and may be measured only in the ewe.

(ii) All components are influenced by age of the ewe (see Turner, 1969 for review). This may mean it is necessary to ignore a ewe's first lambing record and to assess her fertility and fecundity over several later lambings.

(iii) Fertility and fecundity are influenced by a ewe's birth-type. In the Merino, for example, Piper and McGuirk (1976) have shown that in some flocks, ewes born as twins have lower reproductive performance than singles until age 4 years. The "environmental" penalty of being born a twin thus mitigates against that animal leaving offspring for future generations.

(iv) Both fertility and fecundity require a contribution from the ram for their expression. A ewe may thus be dry or have a litter size lower than her potential simply because of infertility of the ram to which she was joined.

(v)   Although it is perhaps self evident, low reproduction rate itself limits the amount of selection pressure that can be applied: in flocks with the greatest need for improvement selection differential for reproduction rate is likely to be lowest.

NEW APPROACHES TO SELECTION FOR REPRODUCTION RATE

The difficulties of improving reproduction rate by direct selection have stimulated the study of other indirect characteristics that might obviate some of the problems outlined above. The studies incorporate some recent advances in reproductive biology and the philosophy of the approach has been described several times by Land (1974, 1976). Before reviewing these investigations it should be stressed, yet again, that indirect selection will only improve rate of genetic gain if the heritability of the indirect character is higher than the direct one and if the genetic correlation between them is high. Some additional benefit may arise where indirect selection can be practised earlier in life or at less expense than direct selection.

(a) Ovulation rate

It is now possible to obtain repeated estimates of ovulation rate in sheep without impairing reproductive performance using the technique of endoscopy. Until recently only one or perhaps two estimates were possible by mid-ventral laparotomy. The reasons for choosing ovulation rate as a criterion for selection have been outlined by Bradford (1972) and Hanrahan (1974). Based on analysis of ovulation rates and egg transfer experiments between breeds and in the selection experiment of Turner (1962) within the Merino it was predicted that an increase in litter size would follow an increase in ovulation rate.

Theoretical studies by Hanrahan (1974), led to the prediction that at least a two-fold increase in rate of genetic gain per generation would be achieved by selection on ovulation rate rather than litter size. These conclusions rest partly on the results of egg transfer experiments which indicate that uterine capacity is not limiting and partly on the assumption that the genetic correlation between early and subsequent ovulation rate is close to unity. The conclusions are nonetheless interesting and warrant further attention.

(i) Repeatability of ovulation rate

Repeatability of ovulation rate has been assessed in the low, medium and high fecundity flocks of the selection experiment of Turner (1962) as well as in a random bred control flock (these are respectively the 'O', 'T', 'B' and 'C' flocks). Estimates have been made of the repeatability of ovulation rate between successive cycles ('C', 'O', 'T' and 'B' flocks) in the one year and between years ('T' flock only). The latter estimate is derived from a new selection experiment using 'T' ewes in which selection is being applied directly for ovulation rate just prior to joining. All ewes are examined each year.

TABLE 3

REPEATABILITY OF OVULATION RATE IN MERINOS

Flock	Basis of estimate	Repeatability	d.f.
'C', 'O', 'T', 'B'	Between successive cycles in one year	0.69±0.04	171
'T'	Between years	0.24±0.08	124

The results are shown in Table 3. Repeatability of ovulation rate both within and between years is high and by comparison with the results in Table 2, much higher than for the components of reproduction.

These results confirm the high repeatability of ovulation rate in ewes of the 'B' flock which have a range (1 to 6) of ovulation rates larger than in most Merinos (Bindon, 1975).

Further estimates of repeatability are available from the work of Hanrahan (1976) in sheep breeds of different fecundity in Ireland. The results in Table 4 demonstrate the quite high repeatability of ovulation rate in Finnsheep (mean ovulation rate = 3.8) and the High Fertility flock (mean ovulation rate = 2.6). Shown also in Table 4 are the corresponding estimates of repeatability of litter size in these flocks. These data lend further support to the conclusion that ovulation rate is a character of higher repeatability than litter size.

(ii)   Heritability of ovulation rate

There are no comprehensive estimates for heritability of this character. Preliminary analyses by Hanrahan (1976) yield a value of 0.35 for Finn-sheep based on 47 d.f. for sires and 2.4 daughters per sire. A similar estimate for the Galway breed yielded a negative value for heritability. Estimates of this parameter should be derived from random bred Merino flocks in Australia.

(iii)   Uterine capacity and the relation between ovulation rate and litter size

It is important to know if increases in ovulation rate are likely to be accompanied by increases in litter size. The subject has been reviewed by Bradford (1972) and Land (1974) whose conclusions suggest that in terms of their effect on the number of lambs born, an increase in ovulation rate is four times as important as an increase in uterine capacity. More convincing evidence comes from egg transfer experiments which demonstrate that ewes with low and high ovulation rates have similar uterine capacity. An example comes from recent work by Hanrahan and Quirke (1976), summarized in Table 5. It is clear from this study that when two or three eggs are transferred to a low fecundity ewe (Galway) their survival rate is only slightly lower than that after transfer to a high fecundity (Finn) ewe.

TABLE 4

REPEATABILITY OF OVULATION RATE AND LITTER SIZE IN IRISH
SHEEP BREEDS
(FROM HANRAHAN 1976)

Breed	Ovulation rate		Litter size[†]	
	r	d.f.	r	d.f.
Finnsheep	0.66**	109	0.07	188
Galway	0.15*	135	0.20**	707
Fingalway	0.15*	79	0.06	99
High Fertility	0.78**	32	0.13	79

† Pooled estimates based on ewe age combinations 2 and 3, 2 and 4, 3 and 4.
* ($\underline{P}$< 0.05)   **($\underline{P}$< 0.01)

TABLE 5

LITTER SIZE FOLLOWING EGG TRANSFER TO SHEEP
GENETICALLY DIFFERENT IN FECUNDITY
(FROM HANRAHAN AND QUIRKE 1976)

Recipient breed	Ewes with 3 eggs		Ewes with 2 eggs	
	No. of ewes	Litter size	No. of ewes	Litter size
Finnsheep	48	2.02	15	1.75
Galway	49	1.89	12	1.50
Fingalway	34	2.00	7	1.80

Analysis of the relation between ovulation rate and litter size leads to the general conclusion that loss of potential embryos increases with ovulation rate. In the Irish study, for example (Hanrahan, 1976) it has been shown that because of a curvilinear relationship between ovulation rate and litter size in Finnsheep, further increases in ovulation rate may actually lead to a decrease in litter size. For other low fecundity breeds, however, there is a linear relation between the two characters. If the same conclusions apply to the Merino it seems safe to say that increases in litter size would follow selection for ovulation rate.

As an indirect character for selection ovulation rate has the advantages:

(i) It is easily measured and more than one estimate could be derived in a particular breeding season. The technique could be usefully applied in group breeding schemes.

(ii) Ovulation rate has high repeatability and initial estimates of heritability are also high at least in high fecundity sheep.

(iii) The character does not require the male for its expression. All ewes should have a record each year.

(iv)   Indications are that at the level of fecundity of the average Merino increased litter size should accompany increases in ovulation rate.

(b)   Number of oestrous cycles

The number of times a ewe shows oestrus is, like fertility or fecundity, a reflection of ovarian activity. There is much evidence from between breed comparisons to show that frequency of oestrus is correlated with high fecundity (see Land, 1974 for review). Other evidence comes from the study of frequency of oestrus and ovulation in the Merino flocks of the Turner (1962) selection experiments. In Figure 1, for example, the frequency of ovulation throughout the year in the low ('O' flock) and high ('B' flock) fecundity flocks is presented. Additional data indicate that the 'B' flock ewes show oestrus about 40% more often than 'O' flock ewes. High fecundity is correlated with the frequency of oestrus and ovulation.

Recent analyses (Bindon and Piper, unpublished) indicate that 'B' flock ewes have a higher total number of cycles to first joining than 'O' flock ewes and also a lower age at first oestrus. In these same data the heritability of total oestrous records to first joining was 0.05 ± 0.01 (120 d.f. for sires with an average of 8.06 daughters per sire). Frequency of oestrus to first joining has

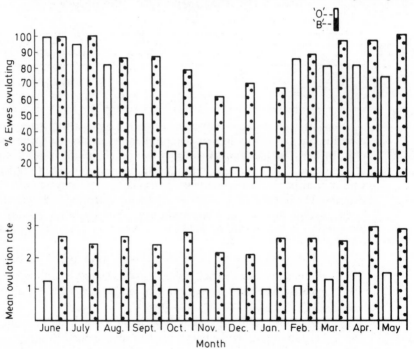

Figure 1—Frequency of ovulation and ovulation rate in ewes genetically low ('O' flock) and high ('B' flock) in fecundity.

been found to be genetically correlated ($r_g = 0.5 - 0.9$) with reproduction rate in the New Zealand Romney (Ch'ang and Rae, 1972) and this character might therefore be considered as a useful indirect selection criterion. There are no reported results from selection experiments to further evaluate this approach.

(c) Plasma luteinizing hormone (LH) and follicle stimulating hormone (FSH)

There is little doubt that gonadotrophic hormones are the major determinants of the number of eggs shed. The spectacular (if rather variable) increases in ovulation rate that follow injection of sheep with exogenous gonadotrophins attest to this fact. With the advent of methods of measuring LH and FSH in plasma it was logical to attempt to relate genetic (and phenotypic) differences, particularly in fecundity, to the level of endogenous LH and FSH. In adult ewes to date there are no convincing data relating quantitative differences in LH or FSH to ovulation rate, fertility or litter size. In any event it is so much easier in the adult to measure the real thing—i.e. ovulation rate itself. A more important advantage of FSH and LH measurements is that they may be made early in life and the same hormones are measurable in ram and ewe lambs.

The major investigations in this field have been reviewed by Land (1974). A serious deficiency in the work is that it is based largely on between-breed comparisons or on flocks in the selection experiment of Turner (1962), rather than on randomly bred flocks that might yield some estimates of genetic parameters necessary to evaluate the approach in genetic terms. Be this as it may there are a number of useful conclusions that may be drawn at this stage.

- (i) Breeds (Finnsheep, Romanov) or flocks (CSIRO 'B' and 'T' Merino) of high fecundity have higher levels of plasma LH than breeds of low fecundity compared in the same environment. This difference appears as early as the first month of life.

- (ii) A single measurement of LH is an unreliable indicator of LH "status" since this hormone fluctuates widely from hour to hour.

- (iii) At least in the 'O', 'T' and 'B' Merino flocks neither a single LH value or repeated LH measurements around day 30 of life in ewe lambs has been shown to be positively correlated with litter size at first and second lambing. The LH differences appear to be unrelated to subsequent reproductive performance.

- (iv) In one study (B.M. Bindon and T.S. Ch'ang, unpublished) in which LH levels were measured on young lambs four times at 30-minute intervals an estimate of heritability of less than 0.1 was recorded. Even if the character were related to reproductive rate it would be difficult to alter by selection.

(v) It is possible to obviate the technical difficulties associated with measuring LH in young animals. Both injections of oestrogen (Land, 1974; Bindon, Ch'ang and Evans, 1974) and synthetic gonadotrophic releasing hormone (Gn RH) have been used to over-ride endogenous LH fluctuations to improve the discrimination between LH status of different genetic groups. This approach is far too laborious for practical consideration.

(vi) Measurement of FSH is relatively more complicated. Early results (e.g. Findlay and Bindon, 1976) however, reveal that ewe lambs from the 'B' and 'T' flocks have higher FSH than low fecundity flocks and that FSH levels have a much higher between measurement repeatability than LH. These points are illustrated in Table 6 and Figure 2.

TABLE 6

MEAN FSH LEVELS (DAY 30) OF LAMBS FROM MERINO FLOCKS GENETICALLY DIFFERENT IN FECUNDITY
(FROM FINDLAY AND BINDON 1976)

Flock	Incidence of multiple births %	Mean plasma FSH (ng/ml)	
		Male lambs	Female lambs
'C'	0-10	17.8± 2.8	42.0±16.6
'O'	0-10	17.0 1.3	26.8 2.2
'T'	50-60	17.7 1.4	111.0 73.5
'B'	70-80	22.2 4.8	159.1 77.6

The main disappointment with FSH is that the differences appear only in ewe lambs, whereas the greatest need is for an indirect selection character that could be applied to the young ram lamb. Further studies in which FSH measurements have been made fortnightly for four months (B.M. Bindon and J.K. Findlay, unpublished) failed to demonstrate ages at which FSH might be used to discriminate between males.

There remains, of course, the possibility that FSH differences may bear no relation to subsequent fertility or fecundity. There are insufficient data to draw a conclusion on this point.

(d)   Testis growth

The search for a character in the ram that might be correlated with the components of reproduction in the female led Land (1973) to study testis growth in ram lambs from breeds and crosses different in fecundity. This seems to be a most sensible approach since the measurement of testis growth would appear to obviate almost all of the difficulties associated with the characters discussed so far. That is:

(i) It may be measured directly in the male.

(ii) It may be measured early in life.

(iii) Its measurement is simple and its heritability and genetic correlation with ovulation rate are high in the mouse. Initial heritability estimates for sheep also appear high.

(iv) It is known to be directly dependent on FSH and LH and in this sense is the counterpart of ovulation rate in the ewe.

The relation between testis growth and fecundity has been examined in the Merino by measuring monthly testis diameter in rams from the CSIRO 'O', 'T' and 'B' flocks during the ages 4 to 18 months. Data averaged from two lamb drops have been summarized in Figure 3 which shows the increase in testis diameter relative to body weight. While the slopes of the regression lines do not differ between flocks the difference in elevation are significant (P< 0.05). It should be pointed out that testis measurements commenced at 4 months of age which is likely to coincide with the latter part of the rapid testis growth phase studied by Land (1973). One must stress, however that only limited conclusions can be drawn from these and similar data since sequential measurements on the same animals are likely to be highly

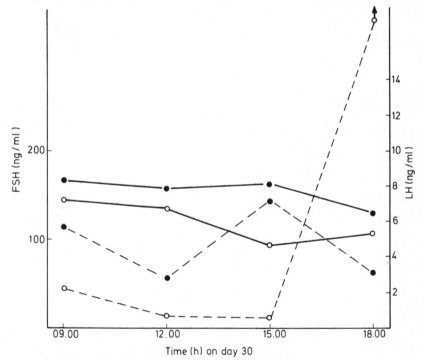

Figure 2—Sequential measurements of plasma LH (o---o) (•---•) and FSH (o—o; •—•) in lambs from the CSIRO 'B' flock.

correlated and thus over-estimate the precision of the regression coefficients and the differences between them.

Direct selection for ram testis weight is now in progress in the UK (Land,
1976). Selection is based on a pooled estimate of testis diameter from
measurements at 6, 8, 10, 12 and 14 weeks of age. Preliminary results have
shown encouraging progress in testis size and in components of reproduction in ewes. In the line selected for high testis weight, for example, ewes
showed a significantly earlier onset of the breeding season. Effects on ovulation rate have been partly confounded by differences in body weights of the
divergent lines. A summary of the ovulation rate differences between the
high and low lines in the early stages of this experiment is reproduced in
Figure 4 by kind permission of R.B. Land. The data show a 15% change in
ovulation rate in four years. Early indications are that this difference is
reflected also in a corresponding difference in litter size.

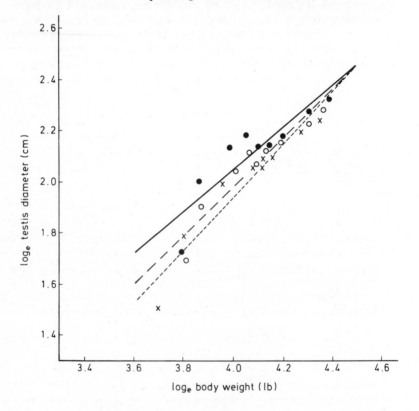

Figure 3—Relation between testis diameter and body weight in rams of the C.S.I.R.O. 'O'
(×---×), 'T' (●—●) and 'B' (○--○) flocks. Measurements were made monthly between 4 and 12 months of age.

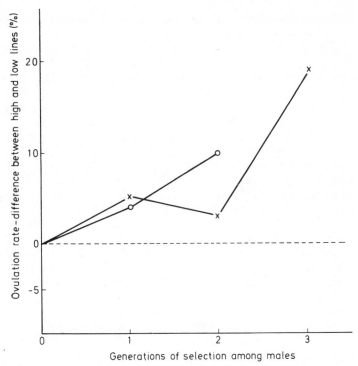

Figure 4—Percentage change in ovulation rate of ewes aged 7 months (× — ×) and 19 months (○ — ○) following 3 and 2 generations of divergent selection for testis growth. (From Land 1976.)

DISCUSSION

This chapter outlined the limitations to traditional methods of selection for components of reproduction rate. Based on present estimates of genetic progress only selection for fecundity could be considered as worthwhile for individual farmers or group breeding programmes.

Although not dealt with in this chapter the increase in reproduction rate from cross-breeding should be mentioned. Relative to progress possible by direct selection, quite spectacular improvement in fertility and fecundity may be achieved. Within the Merino breed similar benefits arise from "cross-breeding" with the CSIRO high fecundity 'B' Merino (e.g. Piper *et al.*, 1977). Apart from the initial benefits a single cross with the 'B' Merino would lead to increased variation in litter size. Such increases in variation may enhance attempts to further increase reproduction rate by direct selection.

There are few estimates of genetic parameters for lamb survival. Preliminary evidence suggests that not much can be done to improve lamb viability in the Merino. The scope for genetic improvement of the maternal side of lamb mortality is unknown.

Research in Australia and elsewhere has highlighted a number of indirect approaches to selection for reproduction rate. Most of the approaches are interesting in terms of reproductive biology but not all are useful in genetic terms. One problem is that the data are generally drawn from experimental populations that cannot yield genetic parameters required for their evaluation.

At this stage measurement of gonadotrophic hormones looks unlikely to yield useful methods of selecting between animals of potentially low and high reproductive performance. There is some merit in further investigation of frequency of oestrus as a useful alternate measure of ovarian activity. A selection experiment could profitably be started along these lines.

Available evidence reviewed suggests that in the adult ewe, ovulation rate itself is likely to be a useful indirect selection approach. This could be especially relevant to the Australian Merino where average ovulation rate is at a level where maximum benefit should result. There should be no great difficulty in incorporating such an approach into group breeding schemes, if selection for reproduction rate is regarded as important. Frequency of oestrus in young ewes also appears to be a worthwhile approach to indirect selection for reproductive performance.

Selection for testis growth is under practical evaluation in the UK and early results are promising. Similar evaluation in Australian sheep populations should perhaps be undertaken.

ACKNOWLEDGEMENTS

The authors acknowledge the co-operation of Dr J.P. Hanrahan, Agricultural Institute, Belcare Co. Galway and Dr R.B. Land, ABRO Edinburgh in allowing presentation of their results in this paper. The assistance of computing staff at CSIRO North Ryde is also gratefully acknowledged. The work was supported by the Australian Wool Corporation and the Austrralian Meat Research Committee.

REFERENCES

Bindon, B.M. (1975) *Journal of Reproduction and Fertility* **44** : 325.

Bindon, B.M., Ch'ang, T.S. and Evans, R.E. (1974) *Journal of Reproduction and Fertility* **36** : 477.

Bradford, G.E. (1972) *Journal of Reproduction and Fertility* Supplement **15** : 23.

Ch'ang, T.S. and Rae, A.L. (1972) *Australian Journal of Agricultural Research* **23** : 149.

Findlay, J.K. and Bindon, B.M. (1976) *Journal of Reproduction and Fertility* **46** : 515.

Hanrahan, J.P. (1974) *Proceedings of the First World Congress of Genetics Applied to Livestock Production* **III** : 1033.

Hanrahan, J.P. (1976) Personal communication.

Hanrahan, J.P. and Quirke, J.F. (1976) *Animal Production* **22**: 162 (abstr.).

Land, R.B. (1973) *Nature (London)* **241** : 208.

Land, R.B. (1974) *Animal Breeding Abstracts* **42** : 155.

Land, R.B. (1976) Personal communication.

McGuirk, B.J. (1976) *Proceedings of the Australian Society of Animal Production* **11**: 93.

Piper, L.R., Allison, A.J., Bindon, B.M., Gheradi, P., Kelly, R.W., Killeen, I.D., Lindsay, D.R., Oldham, C., Robertson, D. and Stevenson, J.R. (1977) *Animal Breeding Absracts* **45,** No. 2818.

Piper, L.R. and McGuirk, B.J. (1976) *Proceedings of the Australian Society of Animal Production* **11** : 105.

Turner, Helen Newton (1962) *Wool Technology and Sheep Breeding* **9** : 19.

Turner, Helen Newton (1969) *Animal Breeding Abstracts* **37** : 545.

Wallace, L.R. (1958) *New Zealand Journal of Agricultural Research* **97** : 545.

48

SYNCHRONIZATION OF OVULATION

I.A. CUMMING
Department of Agriculture, Werribee, Victoria, Australia

SUMMARY

The accepted role of synchronization programmes in large scale sheep farming is to provide reliably and economically a manageable number of ewes for artificial insemination on any one day and to reduce the number of days on which artificial insemination must be carried out. Additional benefits may accrue from the partial control of lambing dates which follow such a programme. Synchronization programmes can induce oestrus and/or ovulation four to six weeks prior to the normal start of the breeding season. Superovulation techniques have been used in synchronization programmes to increase the proportion of ewes lambing twins. A challenging opportunity exists for graziers to integrate management techniques with synchronization programmes so as to exploit maximal ovulation rates and embryo survival. However, no fail safe system for synchronizing ewes which is suitable for large scale sheep farming has yet been thoroughly tested.

Techniques to synchronize oestrus and/or ovulation in ewes can be placed in 3 major categories: firstly there are techniques to initiate ovulatory activity during the anoestrous season, secondly techniques to induce luteolysis and finally techniques to prevent ovulation until removal of treatment.

Recent advances in endocrinology have given additional hope that synchronization programmes may incorporate hormone treatments capable of accurately timing ovulation.

Management techniques may in the future be used to greater effect if synchronization is practicable. Increases in ovulation rate can be achieved by feeding supplements, by manipulating ewe live weights, or by injecting ewes with ovulatory stimulants. With the continual improvement in our knowledge of reproductive physiology one may predict considerable increases in our ability to achieve higher conception rates in synchronization programmes. As conception rates improve economic and management considerations will finally determine the acceptance or otherwise of synchronisation systems for the artificial insemination programmes on properties running large flocks.

INTRODUCTION

Increasingly farmers, geneticists and economists see advantages in using artificial insemination within breeding programmes. For artificial insemination to be accepted in large scale sheep farming an economical method must be found to supply a reliable and manageable number of ewes for artificial insemination on any one day and to reduce the number of days for each property on which inseminations will be carried out. Thus there is a need for an acceptable synchronization programme. While undertaking such a programme increasing numbers of farmers are seeking methods to increase both the fertility and fecundity of their ewe flocks.

Additional benefits, such as partial control of lambing dates may flow from synchronization. Conceptions may occur four to six weeks prior to the normal start of the breeding season with, as a result, an earlier lamb crop.

It is the aim of this chapter to develop an understanding of the physiological mechanisms which influence the effectiveness of synchronization programmes. Techniques will be discussed which may be integrated with the synchronization programmes to increase the proportion of ewes lambing twins or decrease the proportion of ewes not lambing. To achieve these aims a brief resume of current knowledge of hormonal control of the oestrous cycle and ovulation is first given.

THE OESTROUS CYCLE AND OVULATION

The first indications that a primordial follicle has commenced development are changes which occur in the oocyte. These changes then initiate all divisions of the granulosa cells which surround the oocyte. This initial follicular growth is unlikely to be induced by pituitary hormones. However, the future controlled development of these follicles appear to be under at least partial control of the two pituitary hormones, Luteinizing Hormone (LH) and Follicle Stimulating Hormone (FSH). These hormones are known as the Gonadotrophins. The development of the follicles involves a multiplication in the number of granulosa cells and with this a gradual development of a vascular layer of thecal cells (Figure 1). As the follicle increases in size it vacuolates and the antrum fills with follicular fluid. The follicle is now known as a Graafian follicle. The Graafian follicle is the major site for the production of oestrogens. Twenty four hours prior to ovulation LH is released from the pituitary in a surge and this induces the final preparation of the follicle for ovulation. The oocyte resumes its latent meiosis. At ovulation the ovum floats free of the follicular antrum and is swept into the tubes by ciliated cells on a tissue which surrounds each ovary. Once the ovum is shed the ruptured follicle reduces in size and the granulosa cells transform into lutein cells and this tissue becomes infiltrated with capillaries. This newly formed body, named the corpus luteum (CL), is responsible for the production of progesterone.

Progesterone is secreted in large amounts by the CL. It is a steroid hormone which maintains pregnancy and controls the length of oestrous cycles.

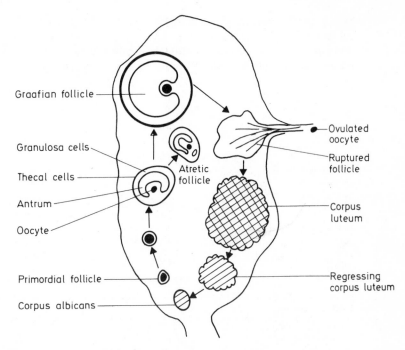

Graafian follicle

Granulosa cells

Thecal cells

Antrum

Oocyte

Primordial follicle

Corpus albicans

Atretic follicle

Ovulated oocyte

Ruptured follicle

Corpus luteum

Regressing corpus luteum

Figure 1 — Diagrammatic representation of an ovary showing follicular development, ovulation and formation and regression of the corpus luteum.

Whilst progresterone levels are elevated, oestrus and ovulation do not occur. Progesterone also affects the development of the mammary gland. Exogenous progesterone can be administered to a ewe by intra-muscular injection or by slow release from either subcutaneous implants or intravaginal pessaries.

During the normal oestrous cycle the CL secretes progesterone until Day 15 or 16. A day later the ewe shows oestrous behaviour and will ovulate about the end of oestrus (Figure 2). The ewe will not come into oestrus if progesterone/progestagen levels are elevated (either because of endogenous production of progesterone from the CL or because of exogenous progesterone/progestagen treatment). However if the CL regresses earlier the ewe will return to oestrus earlier. Progestagens are compounds with actions like those of progesterone. Depending on the compound they can be administered orally in stock feeds, in the water, intravaginally in pessaries, by subcutaneous implants or by injection.

The endogenous control of the CL lifespan is thought to be exerted by the uterus. Uterine tissue under the influence of progesterone and oestrogens will produce the hormone Prostaglandin $F_{2\alpha}$ ($PGF_{2\alpha}$) which when released from the uterus in sufficient quantity and over a required period of time will cause the complete regression of the CL. This process of regression is termed luteolysis (Figure 2). Following luteolysis, ovulation and oestrus can occur. If the animal becomes pregnant luteolysis does not take place and the CL persists until the end of pregnancy.

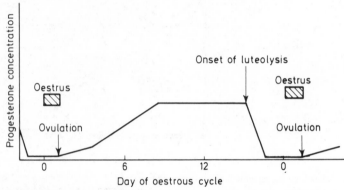

Figure 2 — Diagrammatic representation of progesterone concentration in peripheral blood throughout the oestrous cycle.

Prostaglandin $F_{2\alpha}$ is very rapidly cleared from the ewe's blood by the lungs. If exogenous $PGF_{2\alpha}$ is administered to induce luteolysis then it must be given either as a very large injection into the muscle so that sufficient will pass through the lungs and thus reach the ovaries or in smaller amounts placed directly into the uterus. Neither system has potential application in the sheep industry for the first method would be too costly and the second would be impossible without recourse to surgery. An alternative method has been devised whereby the chemical structure of the $PGF_{2\alpha}$ molecule has been slightly changed and the new chemical produced has retained the luteolytic properties of $PGF_{2\alpha}$ but is not subject to the extremely rapid clearance by the lungs. Such an analogue is Cloprostenol (formerly ICI 80996) which is an effective luteolysin in sheep and cattle. However, ewes should be in the mid- to late-luteal phase of the oestrous cycle if the treatment is to reliably induce luteolysis.

As progesterone levels fall as a result of luteolysis the Graafian follicle/s secretes increasing amounts of oestrogens (Figure 3). This increasing oestrogen level, in the absence now of progesterone, is thought to evoke a release of a LH-releasing hormone from the hypothalamus (Figure 4) and this immediately results in a surge release of LH (with concomitant release of FSH) from the pituitary. In addition to evoking this hormone release the oestrogen brings the animal into oestrus. Ovulation occurs about 24 hours after the onset of the preovulatory LH surge.

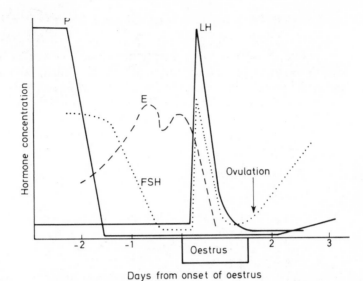

Figure 3 — Diagrammatic representation of hormonal changes in peripheral blood during the period from onset of luteolysis to ovulation. P = progesterone; E = oestrogen; FSH = follicle stimulating hormone; LH = luteinizing hormone.

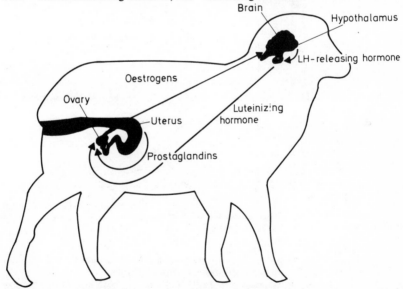

Figure 4 — Prostaglandins released from the uterus stops progesterone production from the ovary. Growth of follicles follows and oestrogens released from the ovaries travel to the brain where they cause LH-releasing hormone to travel from the hypothalamus to the pituitary. The releasing hormone induces luteinizing hormone (LH) to be released from the pituitary. LH acts on the ovary to cause follicles to ovulate.

CONTROLLING FOLLICULAR DEVELOPMENT

Follicular development may be a continuous process in which a number of primordial follicles develop daily or almost hourly. Obviously only a small proportion of these develop into Graafian follicles and ovulate. For the majority, development is interrupted and atresia sets in, but some follicles escape atresia and develop to the point at which they ovulate.

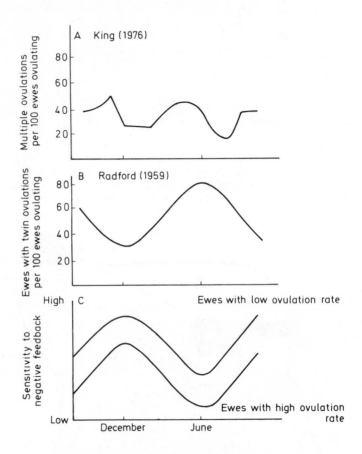

Figure 5 —A Pooled monthly percentage of multiple ovulations from 3 years in largely Polwarth and Corriedale ewes from Tasmania.
B Relationship ($P= 0.571-0.246 \sin \theta -0.015 \cos \theta$) between observed incidence of twin ovulations in ovulating Merino ewes held at constant body weight in Victoria.
C Hypothesised changes in sensitivity to negative feedback of the hypothalamic-pituitary axis to oestrogens which regulates the peripheral concentration of follicle stimulating hormone.

The development of a follicle to ovulation may take about two weeks. However, it is known that the number of follicles which eventually ovulate can be determined as late as Day 14 of the oestrous cycle. After Day 14 no increase in number of follicles which could ovulate appears possible. As early as Day 11 exogenous gonadotrophin treatment can increase follicular numbers with resultant increases in ovulation rate (Cumming and McDonald, 1967).

Cumming and Findlay (1976) have shown that when one ovary is removed the peripheral level of FSH increases for a short period. This increased level in FSH may be the signal for increased follicular development. Certainly the time period in which the FSH levels were elevated was short and the levels only increased by 25-30%. If indeed this change is responsible for the observed ovarian compensation (Land, 1973) then the degree of sophistication in the control mechanism is apparent. Such a mechanism would require a very close feedback relationship between the ovaries and the hypothalamic-pituitary axis. This feedback relationship may be further affected by environmental factors such as season of the year. If the sensitivity of the hypothalamic-pituitary axis to oestrogens varies throughout the year with high sensitivity about the summer solstice and low sensitivity six months later (Figure 5C) such a control system could explain the low proportion of ovulating ewes which have twin ovulations at the time of the summer solstice.

The ovary secretes varying levels of oestrogens throughout the oestrous cycle depending upon the number and stage of development of the follicles. As negative feedback controls gonadotrophin levels and assuming the sensitivity of the hypothalamic-pituitary axis varies as described above, these oestrogens would lower FSH levels to a greater degree about the summer solstice. With lower FSH levels there may then be lower ovulation rates.

However appealing such an hypothesis appears, this control mechanism must be associated with a second system controlling the process of ovulation rather than the development of Graafian follicles. This second system is responsible for positive feedback. Specifically for ovulation to take place there must be positive feedback, i.e. the release of LH and FSH in a surge evoked by elevated oestrogen levels (in the absence of progesterone). The absence of positive feedback could be responsible for anoestrus.

Ovulation rates in ewes which commence their breeding season before the summer solstice frequently show a pattern of falling ovulation rates until about the solstice and then an increase as the breeding season progresses. This pattern has been described by King (1976) for Polwarth and Corriedale ewes studied in northern Tasmania (Figure 5A), by Radford (1959) for Merino ewes studied in southern Victoria (Figure 5B) and by Cumming, Rizzoli, Baxter and White (1976) for 3 of 4 Border Leicester × Merino flocks studies in Victoria (Figure 6). King and Radford both recognized the relationship in the ewes which were ovulating as opposed to the pattern seen in ovulation rate changes for the total flock. Perhaps the absence of positive feedback suppresses the clear expression of the negative feedback system during the late anoestrous/early oestrous portion of the year. However, ewes not showing ovulatory activity until after the summer solstice have lowest ovulation rates at the start of the breeding season.

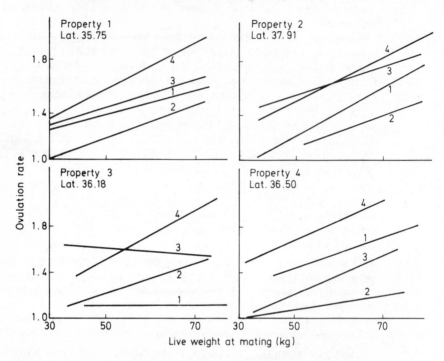

Figure 6 — Association of live weight at mating and ovulation rate in Border Leicester × Merino ewes studied in four commercial flocks in Victoria each on four separate occasions:

Mating Period	Property			
	1	2	3	4
1	15.11 — 6.12.73	29.11 — 20.12.73	6.11 — 27.11.73	8.11 — 29.11.73
2	3. 1 — 24. 1.74	17.1 — 7. 2.74	27.12 — 15. 1.74	27.12 — 17. 1.73
3	14. 2 — 7. 3.74	28. 2 — 21. 3.74	5. 2 — 26. 2.74	7. 2 — 28. 2.73
4	28. 3 — 18. 4.74	11. 4 — 2. 5.74	19. 3 — 9. 4.74	21. 3 — 11. 4.73

Exogenous gonadotrophins, of which Pregnant Mare Serum Gonadotrophin (PMSG) is a widely used example, can be given to the ewe about Day 12 of the oestrous cycle and can stimulate follicular growth and subsequently the number of ovulations. However, response to such treatment varies considerably and this variation considerably limits the application of this method of increasing fecundity. Nevertheless, PMSG is used to increase ovulation rates in progestagen treated ewes in a number of commercial situations. For example Robinson (1974) describes the use of this technique as used in France. Additional benefits from using the exogenous gonadotrophin include the possible stimulation of follicular growth during the anoestrous season of the year when, otherwise, the ovaries would be quiescent with minimal follicular growth and ovulation might well not occur following progestagen treatment. However, ewes with higher ovulation rates do not always have higher FSH levels (Findlay and Cumming, 1976a). Either ewes that ovulate at higher rates, have ovaries inherently capable of higher ovulation rates or a pituitary-hypothalamic axis which selectively releases the gonadotrophins (FSH and LH) in such a pattern and/or at such a level as to produce the higher ovulation rates.

A novel approach to the question of controlling follicular development is that developed by Findlay and Cumming (1976b). An analogue of a FSH/LH-releasing hormone (Hoe 766, Hoechst) was used to manipulate the animal's own pituitary function. This particular LH–RH analogue evokes a release of FSH and LH during the luteal phase of the oestrous cycle. When given at about Day 12, considerable increases in mean ovulation rates were achieved without any of the 'super' responses often seen with use of PMSG. In fact no ewe had more than 2 ovulations. This approach has been applied in a large scale field programme in Western Australia in conjunction with a synchronization programme. The grazier merely injected all his ewes once on either Day 12 or Day 13 with the releasing hormone analogue and ovulation rates were increased by up to 23% depending on the day of injection and the synchronization treatment used (Findlay, Cumming and Fairnie 1977).

An exciting advance has recently taken place in understanding the relationship between nutrition and ovulation rate. Feeding high protein supplements such as lupin grain (Knight, Oldham and Lindsay 1975) and soya bean meal (Davis and Cumming 1977) result in significant increases in ovulation rates. These responses may be greater in certain environments and seasons than in others (Lightfoot and Marshall 1974; Rizzoli, Baxter, Reeve and Cumming 1976). In Western Australia responses in ovulation rate can be large and can occur within 6 days of commencement of feeding the lupin supplement (Lindsay 1976). In such an environment it is likely that lupin grain could be more economically used if fed only for the week or say 10 days prior to ovulation, and such a feeding programme could be integrated with a synchronization programme. However, in other environments, possibly where protein levels in the pasture at a similar mating time are at higher levels than found in Western Australia, such a programme would not result in comparable responses. It is of interest to note that ewes supplemented with lupin grain have higher FSH concentrations in plasma than in unsupplemented ewes five days before oestrus when compared to levels 1-2 days before oestrus (Brien, Baxter, Findlay and Cumming 1976).

TIMING OF OVULATION

Ovulation occurs 24 hours after the onset of the preovulatory surge of LH (Cumming, Buckmaster, Blockey, Goding, Winfield and Baxter 1973) and the coincident surge of FSH (Salamonsen, Jonas, Burger, Buckmaster, Chamley, Cumming, Findlay and Goding 1973). This interval appears to be constant both in maiden or mature ewes, whether progestagen treated or naturally cyclic. However, recent evidence from cattle studies suggests that 'stress' may extend this interval (Cumming, Baxter, White and Findlay 1976).

The preovulatory surge of LH often does not start immediately at the onset of oestrus as is frequently believed (Goding, Catt, Brown, Kaltenbach, Cumming and Mole 1969). In the extreme, situations exist where these two events are known to be completely dissociated. For example, the first ovulation of the breeding season is not accompanied by overt oestrus. There is an accumulation of evidence that ovulation and oestrus are only related coincidentally. Additionally it is known that in young ewes, lactating ewes or ewes undergoing nutritional restriction oestrus expression can disappear while ovulation continues.

In the progestagen treated ewe the pre-oestrus elevation in oestrogen levels and the LH surge can commence earlier with respect to the onset of oestrus than in the untreated ewe (Cumming, Blockey, Brown, Catt, Goding and Kaltenbach 1970). While ovulation and oestrus are not usually related it is desirable that the two events coincide in order to achieve maximum fertility.

In 1967 Robinson wrote that the aim of a synchronization programme in the breeding season is that ovulation must be 'blocked' in a simple manner and then effectively released at will. "If natural service is to be used, oestrus must accompany this release." Robinson recognized that this may not be essential if artificial insemination were to be used provided that the time of ovulation could be controlled with sufficient accuracy.

Evidence is accumulating that fertility is likely to be highest in ewes which show oestrus (Fairnie, Cumming and Martin 1976a; 1976b).

Reproductive physiologists have shown in detail that as LH levels increase in the blood, pituitary stores of LH rapidly fall. Further, they have shown that surges of LH of similar magnitude result from:

1. *Injections of LH-releasing hormone (Cumming, Buckmaster, Cerini, Cerini, Chamley, Findlay and Goding 1972) or of its analogues (Findlay and Cumming 1976b)*

This releasing hormone is the chemical messenger from the hypothalamus which controls the release of LH from the pituitary; such injections bring about an immediate release of LH.

2. *Injections of oestrogen*

These injections bring about a release of LH some 10-18 hours later if progesterone levels are low or negligible. The oestrogens may evoke a release of LH-releasing hormone and this releasing hormone in turn results in the release of LH.

The preovulatory surge of LH can be mimicked by injections of Human Chorionic Gonadotrophin (HCG). After injection with HCG nearly all ewes ovulate between 22 and 26 hours later.

As it is desirable to inseminate about the time of ovulation to achieve maximum fertilization rates and subsequent viability of the conceptus, it may be of value to time ovulation and then inseminate about this time. Unfortunately, there is no strict time relationship from onset of oestrus to ovulation, and, further it is not practical to accurately time onset of oestrus on a large farm scale beyond daily checks for mating marks made during the previous twenty four hours. Accurate timing of oestrous onset requires continuous observation of mating behaviour patterns and this is only feasible within a research programme. Thus a practical method of timing ovulation will likely not use behavioural indexes but will involve exogenous hormone treatment. This treatment may be either the mimicking of the preovulatory surge of LH by using LH or HCG or the evoking of an LH surge by injecting an oestrogen or releasing hormone.

Such a theoretical approach may have many advantages. But any attempt to time ovulation must be timed to coincide with the development of follicles in the ovary and preparation of the reproductive tract to receive the ova and spermatozoa.

All synchronization methods result in a variable interval from treatment to onset of oestrus or to ovulation. With current knowledge there appears to be on the one hand little or nothing which can be done to 'slow down' the first ewes which show oestrus following treatment. On the other hand, it may be possible to shorten the time interval of the ewes showing longer intervals from treatment to onset of oestrus or to ovulation. Such arguments thus suggest a reduction in the spread in the timing of ovulation may be possible and with this reduction will come an increased knowledge of when ovulation is likely to occur for a proportion of the flock.

Evidence from studies with dairy heifers suggests that use of an LH-releasing hormone can improve conception rates. Cows injected with LH-releasing hormone 52 hours after $PGF_{2\alpha}$ analogue and inseminated 26 hours later had equivalent conception rates to heifers not receiving any LH-releasing hormone after $PGF_{2\alpha}$ analogue treatment and inseminated both 72 and 96 hours later. This double inseminated group and the LH-releasing hormone treated heifers had higher conception rates than heifers inseminated once 72 hours after $PGF_{2\alpha}$ analogue treatment (Lawson and Cumming 1976).

Preliminary studies in the ewe (Fairnie and Cumming 1976) have no identified such an advantage in using LH-releasing hormone. Nevertheless, knowing the complexities it is hardly surprising that the first testing of such an hypothesis was equivocal.

SYNCHRONIZATION TECHNIQUES

Techniques to synchronize ewes can be placed into 3 major categories; firstly by initiating ovulatory activity, secondly by effecting luteolysis and thirdly by treating ewes to prevent ovulation until the treatment is removed.

1.   Initiating ovulatory activity

Immediately prior to the start of the breeding season ovulation can be induced by a number of methods. However, this first ovulation will not be associated with overt oestrus except if the ewe has been exposed to progesterone/progestagen treatment shortly before the ovulation. Thus a ewe will normally first express overt oestrus immediately before the second ovulation of the breeding season or at the first ovulation only if the ewe has undergone a period of exogenous progesterone or progestogen treatment (Figure 7A).

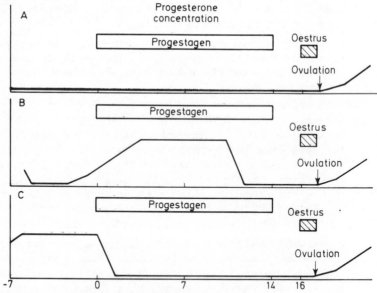

Figure 7 — Diagrammatic representation of onset of oestrus, ovulation and progesterone concentration in peripheral blood with progestagen treatment commencing
A — shortly before the start of the breeding season
B — during the early luteal phase of the oestrous cycle, and
C — during the late luteal phase of the oestrus cycle.

(a) *The ram effect.* One method often of considerable value in achieving partial oestrous synchronization is that of introducing rams to the ewe flock shortly before the ewes are expected to commence their breeding season. This joining can stimulate a considerable proportion of the flock to ovulate within a few days. However, this ovulation is at a "silent oestrus". The ewes then express overt oestrus at the following ovulation (Figure 8). If the ewes have already commenced ovulatory activity or are in deep anoestrus the ram effect will not operate.

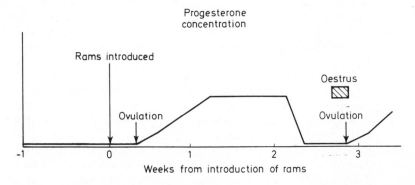

Figure 8 — Diagrammatic representation of the "ram effect" showing ovulation, oestrus and progesterone concentration in peripheral blood following the introduction of rams to the ewe before the start of the breeding season.

A recent communication from Western Australia (Oldham 1976) suggests that although a large proportion of a Merino flock can be induced to ovulate in October and show overt oestrus, as the breeding season progresses an increasing number of ewes fail to demonstrate oestrus following each progressive oestrous cycle until that portion of the year is reached when it is generally considered to coincide with the normal breeding season.

The ram effect has distinct advantages in that it is potentially cheap and requires a minimum of labour. The partial synchrony may be valuable if follow-up luteolysin treatments are to be used. About two weeks after rams are introduced a large proportion of the flock could be predicted to be in the luteal phase of the oestrous cycle. During this luteal phase the corpora lutea can be destroyed by luteolysins such as the $PGF_{2\alpha}$ analogue Cloprostenol (formerly ICI 80996). Another advantage may be the advancement achieved in the expected lambing dates.

If an artificial insemination programme is timed to coincide with the first peak in oestrous activity induced by the ram effect a large proportion of the flock may be inseminated within a week, perhaps sufficient animals to restrict the inseminations to just this one week if the grazier wishes to inseminate only a certain number of the ewes.

(b)   *Progesterone/progestagen treatment.* Progesterone/progestagen treatment can be used to induce oestrus at the first ovulation of the breeding season. The withdrawal of the treatment in the few weeks before ewes are entering the breeding season enables follicular growth to proceed and oestrus and ovulation to take place (Figure 7A).

Such treatments will ensure a higher level of synchronization than is possible using the ram effect (Gordon 1975). In addition this treatment can be used in conjunction with exogenous hormone treatments to increase ovulation rates. A method currently receiving acceptance by a number of European producers is that of injecting ewes with PMSG (Robinson 1974). PMSG tends to increase ovulation rates when given about the time of stopping progestagen treatment. Its use also increases the number of ewes showing oestrus immediately after progestagen treatment and tends to increase conception rates (Robinson and Smith 1967).

(c)   *Photoperiod manipulation.* An increasing interest is being shown in the possibility of inducing oestrous activity by manipulating the hours of darkness a ewe is subjected to each day. However, it is known that latitude, breed, previous management and probably other factors not yet recognized influence a ewe's response to changes in photoperiod. Much research will have to be carried out to assess the potential benefits of such an approach. Results from a limited number of experiments suggest that additional benefits may be had in addition to inducing oestrous activity. These include a degree of synchronization and an increase in ovulation rates (Dunstan, 1976).

2.   *Ovulation control using progesterone/progestagens*

A great deal of research, both basic and applied, has been done studying inhibition and control of ovulation in the ewe (e.g. Robinson 1967). Daily injections of progesterone for 12-14 days inhibit oestrus and ovulation with oestrus generally occurring 2-3 days after the last injection. Obviously such a system has a high labour demand and, from this consideration alone, would be unacceptable for use in commercial sheep farming. Progestagens have been manufactured for oral administration. The need for precise treatments to individuals within large flocks precluded these systems from commercial use.

Vaginal pessaries and subcutaneous implants provide a progesterone/progestagen cover over a 14 day insertion period (Figure 7). Following withdrawal the oestrogens produced from the developing follicles stimulate oestrus and ovulation. The interval from ceasing treatment to the onset of oestrus varies with season, age and breed of ewe and the type and amount of progesterone/progestagen used.

Much developmental work was carried out in Australia and France to develop the Cronolone impregnated intravaginal sponge (e.g. Robinson 1967). However, it is not widely used in Australia today. It is interesting to speculate as to why this may have occurred for the technique is used commercially in Ireland (Gordon 1975) and France (Robinson 1974). It has been suggested by Bindon (1974) that the technique would be a practical reality in the Australian sheep industry today if the appropriate research had received continuing support during the period when economic factors in the industry made the method unattractive. Whatever may be the reason/s for the general non-acceptance of the technique in Australia, the recent developments in commercial artificial insemination programmes suggest that the 'sponge' method should be reassessed. Perhaps it should be used in conjunction with hormonal treatments to accurately time ovulation. Thus, with appropriately timed inseminations, fertility may be increased with a high degree of reliability.

A particularly valuable use for progestagens is immediately before and about the beginning of the breeding season. At this time treatment will usually initiate oestrous activity. Techniques involving the use of luteolysis (see below) cannot be used reliably at this time for there is no guarantee that a CL will be present.

Whether or not 'sponges' or subcutaneous implants are used there is a high labour input at insertion and withdrawal. Both methods have their peculiar problems. Implants, particularly, can become dislodged and lost if incorrectly inserted and a small percentage may become surrounded by localized infections. Sponges can be unpleasant at withdrawal and may be difficult for farmers to dislodge if the tapes break from the sponge.

It is generally believed that the major problem with such treatments is low fertilization. Thus, particularly in artificial insemination programmes, it is imperative that semen quality is at a maximum and that the quantity of semen used is sufficient. There is accumulating evidence with the use of progestagens/progesterone of imbalance in hormone levels or of mistiming of endocrinological events. The evidence includes lowered incidence of overt oestrus and a relatively early release of LH compared to the onset of oestrus. Detailed research using recently developed techniques should now be used to identify methods to overcome these limitations.

3. Luteolysins

Prostaglandins were detected in human semen more than 30 years ago when it was found that semen caused uterine muscle to contract but little additional knowledge of these compounds was gained for the next twenty years. In the late 1950s the chemical structures of most prostaglandins were defined. In addition to their potent effects on smooth muscle they affect the reproductive, digestive, nervous and blood systems and they have further roles in the lungs and kidneys. The potential for using prostaglandins to synchronize oestrus in farm animals arises from their ability to control the length of the oestrous cycle and thereby control the time of ovulation. Premature destruction of the CL during the oestrous cycle results in the onset of oestrus and ovulation about three days later (Figure 9).

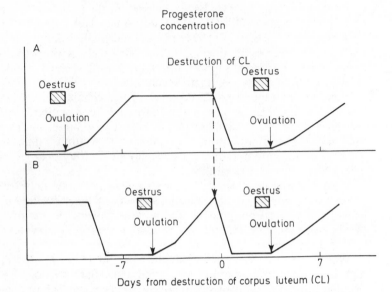

Progesterone concentration

Figure 9 — Diagrammatic representation of onset of oestrus, ovulation and progesterone concentration in peripheral blood with destruction of the corpus luteum occurring
A — towards the end of the luteal phase of the oestrous cycle, and
B — towards the beginning of the mid-luteal phase of the oestrous cycle.

$PGF_{2\alpha}$ was identified as the hormone produced by the uterus which causes the demise of the CL in sheep, goats and cattle. This provided an effective means of controlling the oestrous cycles of these species. Infusions of $PGF_{2\alpha}$ into the ovarian artery, the uterine vein or the uterus itself during mid-cycle caused luteal regression, followed by oestrus and ovulation. Intramuscular injections of $PGF_{2\alpha}$ at dosages similar to those which were effective locally, did not cause luteal regression. Potent enzymes in the lungs destroyed the $PGF_{2\alpha}$ before it could reach the ovary and CL. Very high dose rates of $PGF_{2\alpha}$, however, induced luteal regression when given by intramuscular injection.

In 1973 Inskeep wrote, "Surely no research worker who has suffered disappointing conception rates in animals in which ovulation was controlled by progestin treatment can fail to be excited by the possibility that prostaglandins can be used to control ovulation in farm animals. . . . While the potential uses of prostaglandins are exciting, there is no substitute for hard data to determine their real value in livestock management".

The cost of effective intramuscular injections of $PGF_{2\alpha}$ in large numbers of sheep was prohibitive. In addition luteal regression could not be induced reliably in the sheep by $PGF_{2\alpha}$ treatment before the 5th day of the oestrous cycle.

To enable prostaglandins to be used intramuscularly a number of drug companies have developed analogues of $PGF_{2\alpha}$ which are not destroyed as rapidly by the lungs. These could therefore be given at low doses and yet reach their site of action in the ovary.

During 1975-76 $PGF_{2\alpha}$ analogues have been commercially released for use with cattle and horses. One such analogue, Cloprostenol (formerly ICI 80996), which is effective in cattle, is currently being field tested in a number of sheep flocks during synchronization programmes (Fairnie, Cumming and Martin 1976a; 1976b).

Fairnie, Cumming and Martin (1976a) used a double treatment regime of Cloprostenol (125 μg i.m.) by first injecting ewes 10 days after oestrus with a second injection given 14 or 15 days later. All ewes were inseminated 64-70 hours after the second injection whether they were detected in oestrus or not following this injection. This synchronization programme was followed by 'acceptable' fertility with 41% of ewes lambing to the one insemination. Similar fertility has been seen in paddock matings in a similar environment over a similar time of year (Knight, Oldham, Lindsay and Smith 1975). The proportion of ewes lambing was similar in a control (i.e. no Cloprostenol treatment) group (52%) and in those ewes which were detected in oestrus from the double treatment group (53%). These results are encouraging.

Commercial programmes involving artificial insemination of ewes synchronised with luteolysins requires decisions on the following:

1. Should all ewes be inseminated whether or not they are detected in oestrus following treatment;

2. Should a single or double injection regime be used;

3. Should insemination be at a strict time in relation to treatment and/or to oestrous expression;

4. Should additional hormonal treatments (e.g. LH-releasing hormone) be used to reduce the time interval over which ovulation occurs;

5. When should vasectomized rams be introduced in respect to proposed treatment dates;

6. If double treatment regimes are to be used, how widely spaced should they be;

7. Over how many days should an artificial insemination programme be conducted bearing in mind the considerable day to day variations which can occur in fertility;

8. Should a single or double insemination be used particularly if frozen semen is to be used.

It is not the intent of this chapter to give recipes which take into account all those points listed above—and there are probably many others not listed. Unfortunately there is at present far too little of the hard data available which would be required in the framing of recommendations.

THE FUTURE

Assuming an increasing demand for synchronization programmes and artificial insemination, research will continue into understanding the delicately balanced and intricately interwoven physiological events which influence fecundity. Much of this understanding has been made possible only by the development of sophisticated techniques. Techniques to study concentrations in peripheral blood of most hormones believed to be of importance in reproduction have been developed during the past ten years. Recently techniques have been developed to study hormone receptors at the site of action of these hormones.

No doubt within a few years scientists will understand more completely such questions as why a ewe has either single or twin ovulations, how follicular development can be programmed so that ovulation takes place at a set interval from luteolysis or progestagen treatment and why within artificial insemination programmes is oestrous expression correlated with improved fertility.

There have been a number of recent developments in understanding the reproductive processes of the sheep. Much of the potential commercial application of this knowledge to synchronize ovulation and maximize fecundity is still in the early stages of development. Unfortunately many systems which have looked valuable to the research scientist have not been as successful as hoped when tested on a large scale or in commercial flocks.

In large flocks improvement in reproductive performance at artificial insemination will continue to come from better farm and animal management and improved synchronization programmes. There is tremendous scope for such improvement.

REFERENCES

Bindon, B.M. (1974) *Proceedings of the Australian Society of Animal Production* 10 : 232.
Brien, F.D., Baxter, R.W., Findlay, J.K. and Cumming, I.A. (1976) *Proceedings of the Australian Society of Animal Production* 11 : 237.
Cumming, I.A., Baxter, R.W., White, M.B. and Findlay, J.K. (1976) *Proceedings of the VIIIth International Congress on Animal Reproduction and Artificial Insemination* Vol. III, 453.
Cumming, I.A., Blockey, M.A.deB., Brown, J.M., Catt, K.J., Goding, J.R. and Kaltenbach, C.C. (1970) *Proceedings of the Australian Society of Animal Production* 8 : 383.
Cumming, I.A., Buckmaster, J.M., Blockey, M.A.deB., Goding, J.R., Winfield, C.G. and Baxter, R.W. (1973) *Biology of Reproduction* 9 : 24.
Cumming, I.A., Buckmaster, J.M., Cerini, J.C., Cerini, M.E.D., Chamley, W.A. Findlay, J.K. and Goding, J.R. (1972) *Neuroendocrinology* 10 : 338.
Cumming, I.A. and Findlay, J.K. (1976) *Journal of Reproduction and Fertility* 46 : 516.
Cumming, I.A. and McDonald, M.F. (1967) *New Zealand Journal of Agricultural Research* 10 : 226.
Cumming, I.A., Rizzoli, D.J., Baxter, R.W. and White, D.H. (1976) Personal communication.
Davis, I.F. and Cumming, I.A. (1977) *Animal Breeding Abstracts* 45, 3276.
Dunstan, E.J. (1976) Personal communication.
Fairnie, I.J. and Cumming, I.A. (1976) Personal communication.
Fairnie, I.J., Cumming, I.A. and Martin, E.R. (1976a) *Proceedings of the Australian Society of Animal Production* 11 : 133.

Fairnie, I.J., Cumming, I.A. and Martin, E.R. (1976b) *Proceedings of the 53rd Annual Conference of the Australian Veterinary Association* 186.

Findlay, J.K. and Cumming, I.A. (1976a) *Biology of Reproduction* **15**: 115.

Findlay, J.K. and Cumming, I.A. (1976b) *Biology of Reproduction* **15**: 335.

Findlay, J.K., Cumming, I.A. and Fairnie, I.J. (1977) *Animal Breeding Abstracts* **45**: 2811.

Goding, J.R., Catt, K.J., Brown, J.M., Kaltenbach, C.C., Cumming, I.A. and Mole, B.J. (1969) *Endocrinology* **85** : 133.

Gordon, I. (1975) *Proceedings of the IIIrd World Conference on Animal Production* 473.

King, C.F. (1976) *Proceedings of the Australian Society of Animal Production* **11** : 121.

Knight, T.W., Oldham, C.M. and Lindsay, D.R. (1975) *Australian Journal of Agricultural Research* **26** : 567.

Knight, T.W., Oldham, C.M., Lindsay, D.R. and Smith, J.F. (1975) *Australian Journal of Agricultural Research* **26** : 577.

Land, R.B. (1973) *Journal of Reproduction and Fertility* **33** : 99.

Lawson, R.A.S. and Cumming, I.A. (1976) *Proceedings of the VIIIth International Congress on Animal Reproduction and Artificial Insemination* Vol. III, 484.

Lightfoot, R.J. and Marshall, T. (1974) *Journal of Agriculture Western Australia* **15** : 29.

Lindsay, D.R. (1976) *Proceedings of the Australian Society of Animal Production* **11** : 217.

Oldham, C.M. (1976) Personal communication.

Radford, H.M. (1959) *Australian Journal of Agricultural Research* **10** : 377.

Robinson, T.J. (1974) *Proceedings of the Australian Society of Animal Production* **10** : 250.

Robinson, T.J. (1967) In "The control of the ovarian cycle in the sheep", Edited by T.J. Robinson 237.

Robinson, T.J. and Smith, J.F. (1967) In "The control of the ovarian cycle in the sheep", Edited by T.J. Robinson (Sydney University Press) 144.

Rizzoli, D.J., Baxter, R.W., Reeve, J.L. and Cumming, I.A. (1976) *Journal of Reproduction and Fertility.* **46** : 518.

Salamonsen, L.A., Jonas, H.A., Burger, H.G., Buckmaster, J.M., Chamley, W.A., Cumming, I.A., Findlay, J.K. and Goding, J.R. (1973) *Endocrinology* **93** : 610.

49

CONTROLLED BREEDING OF SHEEP AND GOATS

T.J. ROBINSON
University of Sydney, N.S.W., Australia

INTRODUCTION

In 1975, more than 400,000 breeding ewes and a smaller number of goat-does were brought into a system of controlled breeding in Europe, with France as the centre, an increase of some 50% over 1974 and of some 400% over 1972. Countries into which this system is spreading include Spain, Italy, Holland, Germany, Greece, Israel, Sardinia, and a number of Eastern European countries. A separate development has been in Eire where more than 20,000 ewes per annum are now used for early lamb production.

This is the first major application of a system of controlled animal breeding and it hinges upon the development, some 12 years ago in Sydney, of a simple method of controlling the time of mating and ovulation. This chapter gives the background to this development. Others will deal with more recent attempts to improve upon it.

The key to controlled breeding is a simple device used for the long-term administration of progesterone or one of its more active analogues (Robinson, 1965). It is a plastic sponge, with a drawstring attached, impregnated with a suitable progestagen, which is inserted into the vagina and left in place for 12-16 days. The hormone is absorbed through the vaginal wall and enters the circulation. In the cyclic animal, it takes over the role of the normal corpus luteum after it regresses, so preventing ovulation and at the same time conditioning the central nervous system (CNS) to respond to the events associated with the subsequent controlled ovulation. In the anoestrous animal, it conditions the reproductive tract and the CNS to respond in a normal physiological manner to stimulation by a suitable gonadotrophin such as PMSG given on cessation of treatment. In both cases the sponge is withdrawn, the progestagen inhibition of pituitary action is removed, and ovulation occurs some 72 hours later.

DEVELOPMENT OF THE METHOD FOR CONTROL OF OVULATION

Background

There were four major preliminary steps. The first was the demonstration by O'Mary, Pope and Casida in 1950 that daily injection of progesterone inhibited ovulation in the ewe and that ovulation occurred 2-4 days after cessation of treatment.

The second important step was the independent demonstration two years later by different groups in France, the USA, and in Melbourne that the use of progesterone before PMSG induced oestrus with ovulation in the anoestrous ewe (Dauzier and Wintenberger, 1952; Dutt, 1952, 1953; Robinson, 1952, 1954a). It had long been known from the work of Cole and Miller (1933, 1934) that ovulation, but not accompanying oestrus, could be induced in the anoestrous ewe with PMSG. If a second injection followed some 16 days later, when the resultant corpus luteum was waning, oestrus frequently accompanied the second induced ovulation. It was also known that the first ovulation at puberty and the breeding season usually was unaccompanied by oestrus. Clearly the progesterone from the preceding corpus luteum played some role in the manifestation of oestrus.

The third step was the elucidation of this phenomenon. Extensive studies in Australia during the mid 1950s showed that normal oestrus in sheep was dependent upon a progesterone-oestrogen interaction (Robinson, 1954b,c, 1955a,b, Robinson, Moore and Binet, 1956; Moore and Robinson, 1957a,b). Parallel studies in ovariectomized and in entire anoestrous ewes led to the following conclusions:

(1)　A period of progesterone influence prior to oestrogen is essential for a normal oestrous response—behavioural and vaginal.

(2)　The relationship between progesterone and oestrogen is precise, with response greatly affected by duration of progesterone pre-treatment, dose of progesterone, dose of oestrogen, and time of injection of oestrogen relative to cessation of progesterone.

(3)　The behavioural response is particularly precise.

(4)　The amount of oestrogen produced by the ovary is very small— equivalent to the injection of some 20 μg of oestradiol (OD).

The fourth step was the use of the progesterone-oestrogen interaction as a bioassay for progestagens (Shelton, 1965). The importance of this assay stemmed from earlier observations that long acting forms of progestagen, while effective in blocking oestrus, were relatively ineffective in producing a synchronized oestrus (Robinson, 1960, 1962). A number of such compounds became available in the 1950s as a spin-off in the development of the human oral contraceptive. During this period American efforts towards synchronization of oestrus concentrated upon incorporation of such compounds in the feed. Other workers, in particular Lamond (1964), concentrated on studies of the interaction between season and management on the response to injected progesterone, with or without PMSG. Others went in a different direction. It was reasoned that a short-acting progestagen, either progesterone itself or a more active analogue, was needed which could be incorporated into an "artificial corpus luteum" allowing it to be inserted and removed at will. Insertion and removal had to be simple and the materials inexpensive. There were two logical methods of insertion, subcutaneous or intravaginal, of which the latter was particularly appealing on the ground of simplicity. A short acting, highly potent progestagen which could be absorbed through the vaginal wall was needed. The progesterone-oestrogen interaction in spayed ewes was used as a test in the search for such a progestagen.

The Search for Suitable Progestagens

In 1959 G.D. Searle and Co. in Chicago agreed to supply a range of experimental progestagens. A screening programme was then commenced by

TABLE 1

OESTRUS RESPONSE TO ODB OF SPAYED EWES PRETREATED WITH PROGESTERONE OR WITH SC-9880

(from Shelton, Robinson and Holst 1967)

Notes: (1) *Progestagen priming period — 12 days*
(2) *Interval progestagen-oestrogen — 2 days*
(3) *Dose/day — progesterone, 10 mg; SC-9880, 0.4 mg*

Main effect	Total ewes	Positive responses	
		Behavioural oestrus	Vaginal oestrus
A. **Standardization to Progesterone**			
Test Number			
1	72	27	17
2	72	27	37
$\underline{P}$		N.S.	< 0.001
Dose of ODB (μg)			
10.0	48	2	10
15.6	48	18	16
24.3	48	34	28
$\underline{P}$ linear		< 0.001	< 0.01
$\underline{P}$ quadratic		N.S.	N.S.
B. **Comparison of SC-9880 with progesterone**			
Type of Progestagen			
SC-9880	72	36	Not recorded
Progesterone	72	35	
$\underline{P}$		N.S.	
Dose of oestradiol benzoate (μg)			
10.0	48	5	
15.6	48	27	
24.3	48	39	
$\underline{P}$ linear		< 0.001	
$\underline{P}$ quadratic		N.S.	

Shelton (1964, 1965). The plan was to screen in spayed ewes and then to test in entire animals. Spayed ewes were used for initial screening because a test could be conducted every 16 or 17 days all year round, using the same animals, whereas work with entire ewes was restricted to the breeding season and each animal could be used but once. The object was to find a highly potent progestagen which, in its reaction pattern, exhibited the same relationship with oestrogen as did progesterone itself. The behavioural response was chosen because of its precision and repeatability (Table 1A). The details of the sheep bioassay are readily available (Shelton, Robinson and Holst, 1967; Robinson, 1967) and need no repetition here. Suffice it is to say that tests were conducted initially on 11 and later on several more progestagens over a period of some three years in a flock of 160 spayed ewes. Each bioassay gave information about the potency of the progestagen under test and on the duration of its activity.

Finally one promising steroid, SC-9880, which is a halogenated short-acting progestagen, now known as FGA or Cronolone was selected, which showed an activity profile indistinguishable from that of progesterone (Table 1B) except that it was some 25 times more active.

Of the then well known progestagens, Provera (MAP) and Chlormadinone (CAP), the former showed a reaction pattern similar to SC-9880 but it was slightly longer acting and half as potent and CAP, while highly potent, was considerably longer acting.

Promising new progestagens were also evaluated in entire ewes for their capacity to block ovulation and to release it after cessation of treatment. These tests confirmed the predictive value of the spayed ewe test.

Intravaginal Application of Progestagens

During this testing period, absorption of three progestagens through the vagina of spayed and entire ewes was studied (Shelton and Robinson, 1967; Shelton and Moore, 1967). Physiological quantities of progesterone and of some other progestagens, but not of SC-9880, were absorbed if administered intravaginally in propylene glycol daily, and fertility at the controlled oestrus was high. Gauze and cotton wool tampons soaked in a propylene glycol solution were relatively ineffective for longer term administration.

By the end of 1963 the following stage had been reached:

(1)   A short acting progestagen, SC-9880 was available, which, when given intramuscularly, was highly effective for conditioning for oestrus and subsequent release of ovulation.

(2)   SC-9880 was inactive when given intravaginally, daily in propylene glycol, but some progestagens were inactive when administered in this manner.

Development of the Intravaginal Sponge Technique

In an effort to find a more suitable means of administering the progestagens by the vaginal route attention was focused on the possibility of using cylindrical pessaries made of polyurethane sponge material. Pessaries of different sizes (diameter 2.5 cm or 3.2 cm) and textures were prepared and threaded with a linen cord. Solutions in ethanol of progesterone, SC-9880, and a mixture of both, were made up so that 5 ml contained the required dose of steroid—either 30 or 50 mg SC-9880 or 500 or 800 mg progesterone. This volume was then pipetted onto each sponge which was suspended by its string and the alcohol allowed to evaporate off.

The two initial experiments conducted in early 1964 (Robinson, 1964, 1965) were highly successful (Table 2)—too successful in fact because faults which became apparent later were masked. The sponges were inserted either subcutaneously or intravaginally for the synchronization of oestrus in ewes in the normal breeding season. Intravaginal retention was better for the larger (3.2 cm diameter) sponges than for the smaller (2.5 cm) and for the firmer of the two types of sponge. Intravaginal insertion was at least equally as effective as subcutaneous in suppressing oestrus and in initiating a fertile oestrus on cessation of treatment. SC-9880 appeared to be

TABLE 2

SUMMARY OF RESULTS OF FIRST TESTS CONDUCTED USING PROGESTAGEN IMPREGNATED POLYURETHANE SPONGES FOR THE CONTROL OF OESTRUS IN EWES

(from Robinson 1965)

Progestagens — Progesterone and SC-9880

Site of implantation	Number of ewes			
	(a) Treated	(b) With sponges intact	(c) With synchronized oestrus	(d) Lambed of (c)
Test 1 Natural mating				
Subcutaneous	36	36	26	18
Intravaginal	36	30	29	22
Test 2 Artificial insemination				
Intravaginal	106	97	67*	54
Total	178	163	122	94

* Majority of failures were in ewes treated with progesterone.

generally more effective than progesterone, particularly in the second test in which 71% of all treated ewes lambed to artificial insemination after SC-9880 treatment as compared with only 31% of those treated with progesterone (P< 0.001).

Steroids Tested in Intravaginal Sponges

Five progestagens, previously screened, were then tested in intravaginal sponges for the control of ovulation and oestrus in 930 cyclic Merino ewes (Robinson, Moore, Holst and Smith, 1967). The most useful steroids proved to be SC-9880 (at doses of 10-40 mg), Provera (20-80 mg), and one other, SC-9022 (not< 80 mg). The time of onset of oestrus following the use of SC-9880 was earlier and appeared to be more predictable than that following the use of Provera, while SC-9022 required a higher dose. Consequently all subsequent work, and that of the French workers, has been confined to the use of Cronolone.

TESTING OF THE METHOD

Breeding in Anoestrus

In late 1964, the possibility of using intravaginal sponges impregnated with progestagen, in conjunction with PMSG, was examined in the dry anoestrous cross-bred ewe and, a month or two later, with or without PMSG, in Dorset Horn ewes approaching the breeding season. Despite the highly promising results (Moore and Holst, 1967; Robinson and Smith, 1967a), this is all that was done with the anoestrous animal or with PMSG.

Synchronization of Oestrus for Artificial Insemination in the Breeding Season

Large scale testing commenced with Merino ewes at "Cocketgedong", Urana, New South Wales, in October 1964. Although in the conventionally accepted trough of breeding activity, spring mating was the practice at this station. Three tests were conducted, involving 3,388 ewes. SC-9880 was in short supply and was used only in 536 ewes (30 mg/ewe) while progesterone (500 mg/ewe) was used on the remaining 2,852. Overall results obtained with progesterone were much worse than with SC-9880 due largely to excessive loss of sponges (32.8% in Test 1 for which full data were available) and to a poor oestrous response (Table 3). The lambing rate to a single insemination of ewes in oestrus was comparable to that following SC-9880, namely 45%.

TABLE 3

PERCENTAGE EWES IN OESTRUS, INSEMINATED AND LAMBED
FOLLOWING INTRAVAGINAL TREATMENT WITH SC-9880 OR
PROGESTERONE

(from Robinson and Moore 1967)

	No. of ewes		Percentages of ewes available			
	Total	Avail-able*	% In oestrus	Insem-inated	Lambed	
					1st oestrus	1st & 2nd oestrus
SC-9880						
Inseminated at oestrus	284	270	87.0	87.0	45.6	65.8
Inseminated at fixed time	252	233	—	100.0	45.1	75.1
PROGESTERONE (Test 1 only)						
Inseminated at oestrus	1067	699	61.7	61.7	27.5	41.6
Inseminated at fixed time	357	230	—	100.0	36.5	64.4

* After loss of sponges, deaths, etc.

Such a conception rate was low by normally accepted standards but caused little concern because of the scale of the operation and the consequent necessity to work quickly and to use the available rams to excess. Indeed the possibility of developing a planned programme of artificial insemination at a set time was becoming real.

There followed a massive programme of field testing involving 9,552 Merino ewes on nine properties in three Australian States (Robinson, Salamon, Moore and Smith, 1967). Not yet being aware of the importance of dose of Cronolone and desiring to make maximum use of limited resources, sponges containing 10 mg steroid were used in 5,072 of the treated ewes. The result was a disaster! Oestrus and ovulation were well synchronized but fertility to artificial insemination was appalling, ranging from 4.0 to 41.2%. As the magnitude of the disaster became apparent sponges containing 20 and 40 mg Cronolone were introduced. There was an immediate improvement but the general pattern of results remained poor.

THE REASONS FOR FAILURE

General

In retrospect it is easy to be wise. Conditions could not have been worse: it was a shocking drought year, the programme was conducted in mid-summer when temperatures were exceedingly high, large numbers of sheep were involved so the operations were rushed, ewes were stressed, and rams over-worked. Semen was extended to—and beyond— the limit used in any nor-mal AI programme. As the programme advanced into the autumn the results improved due to a combination of factors—season, size of in-dividual operation, increased dose of SC-9880. However by that time it was realized that there were many details which needed clarification. Fortu-nately many of the faults *were* details—correct dose of progestagen, type of sponge and progestagen release characteristics, avoidance of stress on ewes—and on operators—preparation of rams, correct use of teaser rams, to name the most obvious. Dose of SC-9880 was implicated early, but it was not simply a case of amount in the sponge. Commercially made sponges were becoming available containing 30 mg SC-9880 but they were relatively ineffective, behaving like laboratory made sponges containing 10 mg. Although the nature of the polyurethane sponge affected rate of absorption (Morgan, Lack and Robinson, 1967) it was also found that the method of impregnation was important. SC-9880 in the early commercial Syncromate (Searle) sponges was applied in only 1 ml ethanol injected into the centre of the sponge. As a consequence it was concentrated in a small region in the centre of the sponge and was poorly absorbed. Reconstitution with 5 ml ethanol ensured even distribution of the hormone and consequent effectiveness.

Hence dose of steroid, type of sponge, and method of preparation were all important in determining the dose of progestagen available to the animal, and hence subsequent fertility. But this was not the whole answer. There was—and still is—a specific problem of subfertility following pro-gestagen treatment, to which attention was now turned.

Specific

This was soon shown to be an impairment of fertilization (Moore, Quinlivan, Robinson and Smith, 1967). This was not due to variation in time of ovulation relative to that of withdrawal of sponges and onset of oestrus (Robinson and Smith, 1967b) or to faulty ovum transport and capacity for fertilization. Dose of Cronolone was involved, with the critical dose in the sponge being of the order of 20 mg. The use of diluted semen markedly exacerbated the problem. Then it was shown by Quinlivan and Robinson (1967, 1969) that there was a marked impairment in the pattern of sperm transport in the reproductive tract of the progestagen treated ewe (Figure 1). The problem could be overwhelmed by the use of large numbers of spermatozoa in a dense inseminate.

EARLY CONCLUSIONS

This was more or less where the matter was left. The sheep industry of Australia appeared to be on the verge of collapse and there was no commer-cial interest in artificial insemination or controlled or intensified breeding,

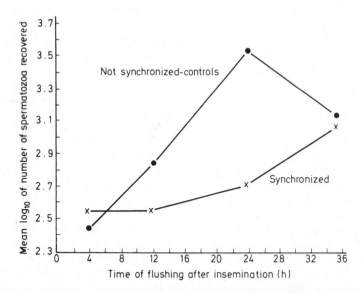

Figure 1—Estimated mean numbers of spermatozoa recovered from the Fallopian tubes 4, 12, 24 and 36 hours after the insemination of control ewes or ewes treated with intravaginal SC-9880 for synchronization of oestrus (modified from Quinlivan and Robinson 1967).

nor was there any interest by grant awarding bodies. The use of PMSG was out of the question on grounds of availability and costs. The commercial application of the Syncromate technique had failed on the grounds of lack of demand, inadequate servicing facilities and technical failure due largely to incorrect preparation of the sponges.

In 1970 Robinson published the results of the last field trials and concluded that under conditions prevailing in Australia the technique of the use of intravaginal sponges impregnated with Cronolone alone has been taken as far as possible. Controlled artificial insemination on a fairly large scale in the normal breeding season on a fixed time basis can be carried out with two inseminations at a half-day interval, and a 60% lambing rate can be expected following insemination with milk diluted semen containing 200×10^6 spermatozoa (Robinson, 1970).

The use of $2 \times 200 \times 10^6$ spermatozoa per ewe was not practicable under most Australian conditions and so, in practice, sheep were normally inseminated at the second oestrus following withdrawal of sponges. Detection of oestrus was necessary but a single insemination only was needed and semen could be more greatly extended. Present use is limited to several thousand sheep a year in Australia.

CURRENT RESEARCH

Since 1968 limited resources have been concentrated towards a solution of the problem of impaired fertilization. Quinlivan's work first implicated sperm transport to the Fallopian tubes. Subsequently Allison and Robinson (1970) confirmed that failure of fertilization is due to the impaired pattern of sperm transport. This is due to endocrine imbalance (Allison and Robinson, 1972; Croker, Robinson and Shelton, 1975a), and the seat of the trouble is the cervix. Allison (1972) was the first to show that passage through the cervix is critically affected by the relative levels of progesterone and oestrogen. This was confirmed by Croker, Robinson and Shelton (1975b) who came to the important conclusion that the pattern of distribution of spermatozoa through the cervix shortly after insemination is highly predictive of the numbers in the Fallopian tubes some 20-24 hours later and available for fertilization. Studies on sperm transport through the cervix have been conducted in an attempt to relate this pattern to the internal endocrine status of the animal. In addition, taking a lead from the French the value of PMSG is currently being reassessed. Finally, the fundamental nature of the post-partum period is being studied using a battery of modern techniques, again with the progestagen sponge as a cornerstone. Suffice it to say that the six-month barrier has already been broken with a number of sheep, that is, some ewes have lambed twice within 12 months and with no apparent ill effects. These are early days, but at least it can be done.

DEVELOPMENTS ABROAD

General

Most of what has been said is history. Some of it is very recent history but none the less its recitation is often worth while. Applied biology is only of use if it *is* applied—and applied it has been. The nature of this application and the reasons for its rather "spotty" application, have already been documented in papers to the Australian and New Zealand Societies of Animal Production (Robinson, 1974a,b) and require no replication here. Suffice it to say that strong economic incentives coupled with academic astuteness and the machinery for implementation of innovations have been instrumental in the application of controlled breeding in France and in Eire.

France

The incentive to produce extra lambs in the EEC is great. A single lamb born out of season is worth some $12 to the farmer more than is a lamb born in season, while twins are worth an extra $20. If two lambings per year can be obtained—and this is common in some small flocks—the gross return per ewe can exceed $125 per annum. This incentive, coupled with the carefully controlled programme of developmental research of INRA, has resulted in a dramatic increase in the numbers of ewes treated as the value of treatment has been recognized and the means for its implementation have been established (Table 4). The technique is so well established that a workshop symposium was held in Montpellier in France in February 1975 in order to evaluate present progress and to formulate plans for its expansion into other European countries (Le Provost and Mauleon, 1975).

TABLE 4

CONTROLLED SHEEP BREEDING IN FRANCE

(from Le Provost and Mauleon 1975)

Year	Number of ewes treated
1966	Several hundred
1967	3,500
1968	17,500
1972	100,000
1974	300,000
1975	>400,000

The French (INRA) system is highly sophisticated and is continually evolving. In its most advanced form it incorporates a whole range of innovations including controlled time of ovulation, artificial insemination at a predetermined time, use of frozen semen, early pregnancy diagnosis, controlled parturition, and the breeding of lambs at 6-7 months of age. At an experimental level it also includes genetic parameters such as cross breeding and the introduction of the highly prolific Finnish Landrace (nee Russian Romanov) associated with batch production of lambs every 7 weeks on a year-round basis.

Eire

In Eire, Professor Gordon has applied the Cronolone impregnated intravaginal sponge/PMSG technique for the advancement of the breeding season on an ever-increasing scale and with consistently improving results on several hundred private farms over the past 9 or 10 years. Extension of the programme has now passed into the hands of the Department of Agriculture and it is an accepted practice. In 1971/72 some 10,000 ewes passed through Gordon's hands with results improving every year (Gordon, 1973).

Scotland

Mention must be made of the programme at the Rowett Research Institute where a "hot house" system of lamb production has been in operation for several years (Ørskov and Robinson, 1972). The system is highly intensive and involves year-round housing in a light controlled environment. Breeding activity is regulated by photoperiod (Yeates, 1949). The light year is reduced from 365 to 205 days so that ewes exhibit a short breeding season every 7 months. Superimposed upon this is synchronization of oestrus using Cronolone impregnated intravaginal sponges in order to use the short breeding season to maximum advantage. Genetic parameters are introduced in that highly fecund Finnish Landrace × Dorset Horn ewes are bred to Suffolk rams to have high prolificacy plus maximum hybrid vigour. The lambing efficiency of the flock is extraordinary (Table 5).

TABLE 5

CONCEPTION RATES AND LITTER SIZES IN EWES WITH A
LAMBING INTERVAL OF 205 DAYS

(from Ørskov and Robinson 1972)

Group	No. of ewes	Previous date of lambing	Date of induced oestrus	% Conception at induced oestrus and 1st return	Mean litter size
1	48	18 March-12 April	21 May	93.7	2.3
2	48	24 June-18 July	29 Aug	97.9	2.1

CONCLUSIONS

One is entitled to ask why controlled sheep breeding, pioneered in Australia, has been neglected here but exploited elsewhere. The reasons are manifold, but three are paramount, namely lack of economic incentive, failure to fully appreciate the benefits and value of the technique, and the absence of a national agency such as INRA to implement such an innovation.

A large measure of the French success is due to the recognition of the several physiological states of animals subjected to treatment. This is summed up by Mauleon (1975) in the following words: "Au cours de ces années et même actuellement, les personnes décues ont été celles qui ont cru trouver la méthode universelle alors que c'est et doit être une *méthode raisonnée*" ("During these years and even now, those deceived have been those who have believed in finding a universal method when it is and must be *reasoned method*"). A factorial situation, representing 48 potentially different physiological states is recognized (Table 6).

Each of the possible combinations is considered separately in relation to factors such as the size of sponge, dose of Cronolone, dose of PMSG, number of spermatozoa used for insemination and whether fresh or frozen semen is to be used. By dint of recognizing the limitations of the technique and compensating for them, programmes satisfactory for all breeds, types and ages of sheep for all seasons and for all regions of France have been developed and put into practice (Le Provost and Mauleon, 1975).

TABLE 6

EXTERNAL AND INTERNAL FACTORS WHICH INTERACT WITH A CONTROLLED BREEDING PROGRAMME

External					
Season	(4)	Breeding season	Early anoestrus	Mid anoestrus	Late anoestrus
Internal					
Age	(3)	Pubertal lamb	Maiden ewe		Mature ewe
Lactational state	(4)	Dry	Postpartum dry	Postpartum suckled	Postpartum milked

The progestagen impregnated sponge, with or without associated PMSG treatment, has been widely used elsewhere on an experimental or pilot commercial basis, but generally speaking, on an *ad hoc* basis. Progestagens other than Cronolone—notable progesterone and Provera—have been incorporated into sponges and have been used in a manner similar to that worked out in detail for Cronolone but there have been no *systematic* studies other than those detailed above and so results continue to be variable and, in some hands, unreliable. Research and development of this new technique must go hand in hand in each environment into which it is introduced. It is not a simple commercial "product" like the oral contraceptive which can be marketed for more or less universal use. Mauleon (1975) summed up the situation perfectly when he pointed out that the potential value of the intravaginal sponge technique is revealed only by *precise* determination of the manner in which it is to be used. The same situation will apply to other techniques as they become available—as indeed they will. There is no panacea in controlled animal breeding; there is no "méthode universelle".

REFERENCES

Allison, A.J. (1972) *Journal of Reproduction and Fertility* **31** : 415.

Allison, A.J. and Robinson, T.J. (1970) *Journal of Reproduction and Fertility* **22** : 515.

Allison, A.J. and Robinson, T.J. (1972) *Journal of Reproduction and Fertility* **31** : 215.

Cole, H.H. and Miller, R.F. (1933) *American Journal of Physiology* **104** : 165.

Cole, H.H. and Miller, R.F. (1934) *Anatomical Record* **58** : Supplement 56.

Croker, K.P., Robinson, T.J. and Shelton, J.N. (1975a) *Journal of Reproduction and Fertility* **43** : 405.

Croker, K.P., Robinson, T.J. and Shelton, J.N. (1975b) *Journal of Reproduction and Fertility* **44** : 11.

Dauzier, L. and Wintenberger, S. (1952) *Annales de Zootechnie (Paris)* **4** : 49.

Dutt, R.H. (1952) *Journal of Animal Science* **11** : 792.

Dutt, R.H. (1953) *Journal of Animal Science* **12** : 515.

Gordon, I. (1973) *Third World Conference on Animal Production.* Pre-Conference Volume **3** : 34.

Lamond, D.R. (1964) *Animal Breeding Abstracts* **32** : 269.

Le Provost, F. and Mauleon, P. (eds.) (1975) In "Maitrise des Cycles Sexuels chez les Ovins", Proceedings of Meeting at Montpellier, France, February 1975 (Searle: Paris).

Mauleon, P. (1975) In "Maitrise des Cycles Sexuels chez les Ovins", p.4, Proceedings of Meeting at Montpellier, France, February 1975 (Searle: Paris).

Moore, N.W. and Holst, P.J. (1967) In "The Control of the Ovarian Cycle in the Sheep", p.133 (Sydney University Press: Sydney).

Moore, N.W., Quinlivan, T.D., Robinson, T.J. and Smith, J.F. (1967) In "The Control of the Ovarian Cycle in the Sheep", p.169 (Sydney University Press: Sydney).

Moore, N.W. and Robinson, T.J. (1957a) *Journal of Endocrinology* **14** : 297.

Moore, N.W. and Robinson, T.J. (1957b) *Journal of Endocrinology* **15** : 360.

Morgan, J., Lack, R.E. and Robinson, T.J. (1967) In "The Control of the Ovarian Cycle in the Sheep", p.195 (Sydney University Press: Sydney).

O'Mary, C.C., Pope, A.L. and Casida, L.E. (1950) *Journal of Animal Science* **9** : 499.

Ørskov, E.R. and Robinson, J.J. (1972) Rowett Research Institute Annual Report **28** : 116.

Quinlivan, T.D. and Robinson, T.J. (1967) In "The Control of the Ovarian Cycle in the Sheep", p.177 (Sydney University Press: Sydney).

Quinlivan, T.D. and Robinson, T.J. (1969) *Journal of Reproduction and Fertility* **19** : 73.

Robinson, T.J. (1952) *Nature (London)* **170** : 373.

Robinson, T.J. (1954a) *Journal of Endocrinology* **10** : 117.

Robinson, T.J. (1954b) *Nature (London)* **173** : 878.

Robinson, T.J. (1954c) *Endocrinology* **55** : 403.

Robinson, T.J. (1955a) *Journal of Endocrinology* **12** : 163.

Robinson, T.J. (1955b) *Journal of Agricultural Science* , Cambridge **46** : 37.

Robinson, T.J. (1960) *Proceedings of the New Zealand Society of Animal Production* **20** : 42.

Robinson, T.J. (1962) *Journal of Endocrinology* **24** : 33.

Robinson, T.J. (1964) *Proceedings of the Australian Society of Animal Production* **5** : 47.

Robinson, T.J. (1965) *Nature (London)* **206** : 39.

Robinson, T.J. (1967) In "The Control of the Ovarian Cycle in the Sheep" (Sydney University Press: Sydney).

Robinson, T.J. (1970) *Australian Journal of Agricultural Research* **21** : 793.

Robinson, T.J. (1974a) *Proceedings of the Australian Society of Animal Production* **10** : 250.

Robinson, T.J. (1974b) *Proceedings of the New Zealand Society of Animal Production* **34** : 37.

Robinson, T.J. and Moore, N.W. (1967) In "The Control of the Ovarian Cycle in the Sheep", p.116 (Sydney University Press: Sydney).

Robinson, T.J., Moore, N.W. and Binet, F.E. (1956) *Journal of Endocrinology* **14** : 1.

Robinson, T.J., Moore, N.W., Holst, P.J. and Smith, J.F. (1967) In "The Control of the Ovarian Cycle in the Sheep", p.76 (Sydney University Press: Sydney).

Robinson, T.J., Salamon, S., Moore, N.W. and Smith, J.F. (1967) In "The Control of the Ovarian Cycle in the Sheep", p.208 (Sydney University Press: Sydney).

Robinson, T.J. and Smith, J.F. (1967a) In "The Control of the Ovarian Cycle in the Sheep", p.144 (Sydney University Press: Sydney).

Robinson, T.J. and Smith, J.F. (1967b) In "The Control of the Ovarian Cycle in the Sheep", p.158 (Sydney University Press: Sydney).

Shelton, J.N. (1964) *Proceedings of the Australian Society of Animal Production* **5** : 43.

Shelton, J.N. (1965) *Nature (London)* **206** . 156.

Shelton, J.N. and Moore, N.W. (1967) In "The Control of the Ovarian Cycle in the Sheep", p.59 (Sydney University Press: Sydney).

Shelton, J.N. and Robinson, T.J. (1967) In "The Control of the Ovarian Cycle in the Sheep", p.48 (Sydney University Press: Sydney).

Shelton, J.N., Robinson, T.J. and Holst, P.J. (1967) In "The Control of the Ovarian Cycle in the Sheep", p.14 (Sydney University Press: Sydney).

Yeates, N.T.M. (1949) *Journal of Agricultural Science* , Cambridge **39** : 1.

50

MANIPULATION OF THE BREEDING CYCLE

P. MAULEON

I.N.R.A.—Station de Physiologie de la Reproduction, Nouzilly, France

SUMMARY

The technique of vaginal pessaries for the control of the oestrous cycle in the ewe is now well developed in France (500,000 ewes in 1976). The precise method of its application has been studied. This concerns standardization of supporting material, quality and assay of PMSG, and conditions of mating or AI. Two alternative methods of administration have been attempted, oral or subcutaneous implants, with similar results. The technique has greater possibilities than that used at present, particularly in terms of lamb production: 3.2 lambs per ewe per year. However, breeds of sheep with a marked sexual season show low fertility at the induced oestrus in spring, particularly if they are suckling two lambs. The suppression of lactation and use of artificial rearing is an efficient practical solution. The major difficulties encountered are fertilization and embryonic mortality. A hypofunctional corpus luteum develops after ovulation with consequently low progesterone concentrations at 10 days post-conception. Many endocrine parameters are depressed due to an interaction between post-partum × lactation × presence of young and season. No solution has yet been found by modifications in treatment. Different hypotheses concerning this physiological problem are presented.

INTRODUCTION

The desire of French sheep breeders to control the oestrous cycle stems from the structure of sheep farming and the system of production of prime lambs in this country.

Techniques using progestagens and PMSG allow oestrus to be induced regardless of time of year and type of animal (lambs, lactating or dry ewes) and permit oestrus to be synchronized and prolificacy to be increased. The techniques for the control of the sexual cycle of the ewe have developed in France for these three reasons since 1966, when several hundreds of sheep were treated, until 1974 when the number increased to 300,000. It is likely to be more than 500,000 in 1976, that is, 7% of the national breeding flock.

This chapter will be concerned with 3 aspects of the subject: (1) a short history of the development of such techniques, (2) a brief resumé of the improvements produced in France during the last 10 years, and above all of the concept for its use in the future, (3) the physiological problems remaining to be resolved and the current state of research efforts to solve them.

HISTORICAL SUMMARY OF THE TECHNIQUE

Three dates have marked the development of an hormonal treatment resulting in planned oestrus, ovulation and pregnancy in the sheep:

● 1948-1952: progesterone, which inhibits ovulation during the entire period of its administration, is obligatory to synchronize and induce oestrus and ovulation (Dutt and Casida, 1948).

● 1964: the method of administration via the vagina facilitates the application of small doses of progestagens during the two weeks required for treatment. Using the induction of oestrous behaviour in the castrated ewe as a model, the progestagen FGA was chosen from among the numerous progestagens produced by the chemical industry as being that with the most interesting physiological characteristics (Robinson, 1964).

● 1966-1968: after a period of infatuation, the technique was rapidly abandoned throughout the world because of the low fertility at the first post-treatment oestrus.

The work of I.N.R.A. was to search for the know-how to use the technique, and to educate breeders and technicians so that its use could be rationalized.

Actually, 3 different motives exist for French sheep breeders as stated previously and methods are chosen and adapted according to the particular breeding systems.

IMPROVEMENTS IN THE TECHNIQUES FOR THE CONTROL OF THE CYCLE

(a)   In the recent past

The variations were simple and concerned the vaginal method of administration of progestagens but they could be applied to all other methods of administration. They contributed to the standardization of the manufacture of sponges (supporting material, size, distribution of progestagen) and quality of PMSG (a single batch per year, assay by different methods). This latter precaution is essential if the dose recommendations are to have a practical value. Different doses are advised depending on breed, age, time of year and the physiological state of the ewe. They are notified and better adapted to the flock as the breeders' advisers acquire knowledge of the result for a given year.

However, irregularities in the fertility rates observed at induced oestrus during the first trials are for a large part dependent on the number of spermatozoa available at the time of fertilization of the oocyte. The endocrine balance created by the progestagens is unfavourable for the transport and survival of spermatozoa (Quinlivan and Robinson, 1969). It is possible to correct this fault by improvements in the conditions of mating: nutritional preparation and behaviour of the rams, adaptation of the number of rams to the number of ewes in oestrus (one ram for 6 ewes) or on the other hand a staggering of oestrous periods by using two periods of sponge insertion separated by several days, and two durations (12 and 16 days). This latter method means that one ram can be used for a group of 48 to 64 ewes, with good fertility (Prudhon, Galinder and Reboul, 1975). A variety of ways of organizing the mating period were devised by breeders once they realized the importance of the male factor in the fertility results at the induced oestrus. The technique of artificial insemination developed by Colas has in fact obviated the idea of poor fertility associated with oestrus controlled by progestagens.

(b) Present time

Other methods of administration of progestagens have been the object of experimentation either because the insertion of pessaries presents difficulties in certain types of females (lambs) or because of sanitary problems at the vaginal level (flocks in the Mediterranean area).

It also appears that the method of administration of progestagens influences the dispersion of spermatozoa at the level of the vagina: it would be less rapid after administration of the progestagen *per os* than if the product had been inserted into the vagina (Hawk and Conley, 1971). The unfavourable influence of the progestagens on the transport of gametes in the genital tract occurs at the cervical level; it could be decreased in cases where the progestagen was administered by another route. Two other methods of administration have been the subject of numerous trials:

(i) Oral administration

Contrary to the results of authors who used oral administration of MAP (Lindsay *et al.*, 1967), Cognié (1972) found that daily administration *per os* of 6 to 8 mg of FGA gave a good synchronization of oestrus at the end of treatment (Table 1).

TABLE 1

OESTROUS SYNCHRONIZATION AFTER VARIOUS ROUTES OF
PROGESTAGEN ADMINISTRATION

| | Interval between the end of treatment (12 days) and onset of oestrus (hours) | | | | | |
| | 24 | 36 | 48 | 60 | 72 | 84 |
	% of treated animals					
8 mg FGA/day (oral route)			29	54	4	4
40 mg FGA (vaginal pessary)	9	40	41	8	2	
3 mg SC 21009 Norgestomet (subcutaneous implant)	51	49				

In all cases, PMSG injection at the last day of treatment.

(From Cognié 1972; Martin *et al.* 1976.)

Results were strictly comparable with those obtained after vaginal administration of the same product with the maximum number of ewes in oestrus within 48 hours. The fertility of lactating ewes after two inseminations was identical if not superior to that obtained with vaginal sponges during comparable trials in respect of season, type of animal and conditions of insemination (64-67% *v.* 53-59%).

(ii) Subcutaneous Norgestomet implants

The fertility after this type of administration is identical, when matings are controlled or carried out during the entire period of oestrus. To obtain a normal fertility after a single artificial insemination at this induced oestrus, it is necessary to advance the time of insemination by 6 hours compared to the FGA treatment (50 *v.* 56 h). This is because of a more rapid appearance of oestrus after the treatment, as noted in Table 1 and a more rapid ovulation after the beginning of oestrus (27 *v.* 32 h) (Martin *et al.*, 1976; Cognié, Mariana and Thimonier, 1970).

Thus it appears that if the principles noted above (necessity of PMSG, large numbers of spermatozoa, reduced time of treatment) are adhered to and the time of insemination is altered more because of the differences in metabolism of the compounds rather than their method of administration, several progestagen treatments have identical value. Practical consideration of local problems, methods of administration and the relative prices of the different progestagens (bearing in mind the relative doses necessary to effect complete inhibition of the hypophysis) are the factors which determine the choice of one particular treatment rather than another.

French sheep breeders have used the method for the control of the cycle to resolve particular problems stemming from their system of flock management. From the point of view of an increase in lamb production, this method offers undisputed possibilities (Table 2).

TABLE 2
IMPROVEMENT IN REPRODUCTIVE PERFORMANCE OF CHARMOISE EWES USING OVULATION CONTROL DURING THE EARLY POST-PARTUM PERIOD

	1968	1969	1970	1971	1972	1973	1974	1975
No. of lambs reared per 100 ewes/year	90	87	162	159	148	179	149	160 320*†
No. of lamb crops per 100 lambs/year	91	84 (before controlled breeding)	159	139	142	142	125	135 200†

* Alive 5 days after birth.
† Results for Romanov × Charmoise F₁ cross-bred. (From Louhault, Ramblière and Cornu 1976.)

A move towards intensive sheep breeding and production methods comparable with those seen for pigs and poultry would appear to be a logical progression in France, especially if sheep are to remain competitive with alternative forms of livestock production. Obvious benefits in terms of organization of work (Thimonier *et al.*, 1975) offered by this technique may well determine its development, especially in the light of ever increasing costs of production. If the technique does fail to form the basis of a twice-a-year lambing system it will be because of the poor fertility of lactating ewes when inseminated earlier than 70 days post-partum. This latter problem is exaggerated in the meat breeds which exhibit a much more marked seasonality of reproduction.

PROBLEMS RESULTING FROM TWICE YEARLY LAMBING

The major difficulties encountered in those breeds of sheep with a marked sexual season results from the combination of two unfavourable physiological conditions. The interference of post-partum anoestrus with seasonal anoestrus is shown by the fact that the duration of post-partum anoestrus is a function of the interval between lambing date and the beginning of seasonal anoestrus (Mauleon and Dauzier, 1965). This is confirmed by the differences in fertility at an induced oestrus between the Merinos d'Arles and Merinos x Romanov breeds with month of treatment during seasonal anoestrus.

The induction of oestrus and ovulation during anoestrus does not itself constitute a problem as long as the dose of PMSG is modified according to breed and season. In contrast, the conception rate at the induced oestrus is variable in spring and is always lower than in autumn under similar situations. Conception rate at the induced oestrus increases the longer the interval from parturition to artificial insemination (Cognié *et al.*, 1974). The difference in fertility between AI performed 43 or 70 days after lambing is greater in some breeds than others. Conception rates are low (27-45%) in lactating Ile-de-France ewes and much higher in Prealpes or Romanov-cross ewes (45-60%) when inseminated 43 days post-partum.

One solution to this problem is therefore to choose judiciously a breed for more frequent lambing which will respond to existing hormonal treatments. Alternatively, two other approaches can be taken with existing breeds to try to overcome the problems encountered at 45 days post-partum:

(a) Elimination of one of the unfavourable factors

This can only be lactation, since season is a factor imposed by the choice of an accelerated lambing rhythm.

The restoration of cyclical activity after parturition is strongly dependent on lactation. Prolactin is released in large quantities during lactation.

TABLE 3

%CONCEPTION RATE AT INDUCED OESTRUS OF AUTUMN AND
SPRING LAMBING EWES. INFLUENCE OF BREED AND
INTERVAL PARTURITION — AI

	Interval parturition — AI (days)	Breed			
		Ile-de-France		Préalpes	
Autumn	43	45	(89)	60	(60)
	70	71	(55)	68	(41)
Spring	43	27	(426)	45	(178)
	70	46	(76)	70	(54)

( ) No. of treated ewes

(From Cognié, Cornu and Mauleon 1974.)

TABLE 4

%CONCEPTION RATE AT INDUCED OESTRUS OF AUTUMN AND
SPRING LAMBING EWES. INFLUENCE OF NUMBER OF LAMBS
SUCKLED AND INTERVAL PARTURITION — AI

	No. of lambs suckled	Interval parturition — AI (days)			
		43		70	
Autumn	1	53	(77)	78	(92)
	2	43	(42)	57	(33)[*]
Spring	1	37	(312)	65	(70)
	2	28	(186)[*]	53	(39)

( ) No. of treated ewes
* $P < 0.01$

(From Cognié, Cornu and Mauleon 1974.)

After denervation of the mammary gland, the prolactin peaks observed during milking and suckling disappear (Kann and Martinet, 1975). The suppression of the prolactin peaks by either surgical suppression of the reflex suckling stimulus or pharmacological inhibition using 2-bromo-α ergocryptin results in an earlier return to cyclical ovarian activity (Kann *et al.*, 1975).

The influence of lactation persists when oestrus is induced by progestagen-PMSG treatment. Regardless of the breed, season and interval from parturition to AI, the conception rate is lower when the ewe suckles two lambs instead of one (Table 4). The complete suppression of lactation achieved by removing lambs at birth improves conception rates by 20% at an induced oestrus, compared with the lactating ewe suckling a single lamb (Cognié *et al.*, 1974).

One further benefit of weaning lambs at birth is that any animal failing to conceive to the treatment insemination will show a return to oestrus. An improvement in fertility can also be achieved by weaning lambs 30 or more days after parturition. A conception rate of 64% was obtained in Aragonese ewes when the lambs were removed at birth compared with 58% when weaning took place at the time of sponge insertion 50 days post-partum. The comparative figure for lactating ewes was 38% (Cognié, personal communication). Ile-de-France ewes treated with progestagen-PMSG in spring which suckle a lamb for 30 days have identical conception rates when inseminated 45 days post-partum to those which have been dry since lambing (88.7% *v.* 87%). Fertile returns-to-oestrus appear in the cycle following the induced cycle in both cases but not at this time of the year if the ewes are lactating at the time of treatment.

The early weaning of lambs modifies the endocrine balance of the post-partum period. This is a valid field solution, since artificial feeding of the lambs is technically possible. An increase of the order of 30 to 50% in the overall fecundity of ewes subjected to an accelerated lambing rhythm is obtained by early weaning.

(b)   Adapt the hormonal treatment to the endocrine status of the lactating ewe lambing in the non-breeding season

(i)   Biological parameters responsible for low fertility

To obtain a high percentage of ewes ovulating after a progestagen-PMSG treatment, it is necessary to use a higher dose of PMSG for lactating animals than for those weaned at lambing or for dry ewes (Thimonier *et al.*, 1968). This results in a greater variation in the number of ovulations in lactating animals and a higher number of ovulations greater than four. The spread of ovulations occurs over a long time (Signoret and Cognié, 1975). These are sufficient reasons for a low fertility in such ewes: the fertilization rate is lower because the eggs are of variable age (Cognié *et al.*, 1975). Difficulties continue during embryonic development. A larger number of eggs at 7 days show a delay in development. It appears that they die prior to Day 12 since they are incapable of assuring the embryonic signal which brings about the maintenance of the corpus luteum, and they return to oestrus at 16 days post AI. In addition, a greater embryonic mortality between 18 and 50 days of gestation occurs in such lactating ewes in the non-breeding season (Thimonier, Cornu and Terqui, 1975) (Table 5).

Data suggest that the quality of the ovulation is inadequate after treatment because there are too many eggs of variable age, often from immature follicles. In fact, 2-day fertilized ova collected from lactating ewes treated with progestagen-PMSG, and transferred to dry ewes, die prematurely and gestation is established less often (50% *v.* 67 or 68%) than if eggs from dry ewes are transferred to other dry ewes or even to lactating ewes.

TABLE 5

ACCURACY OF PREGNANCY DIAGNOSIS IN NURSING EWES

	1st semester winter & spring		2nd semester summer & autumn	
	Préalpes	Ile-de-France	Préalpes	Ile-de-France
% ewes lambing/ presumed pregnant*	80.4 (110)	58.8 (178)	92.0 (112)	74.0 (213)

ACCURACY OF PREGNANCY DIAGNOSIS IN EWES IN DIFFERENT PHYSIOLOGICAL STATES

	Nursing	Dry	Nursing	Dry
% ewes lambing/ presumed pregnant*	74.3 (761)	91.4 (321)	87.4 (662)	91.8 (416)

* High plasma progesterone determination at 18 days after A I

(From Thimonier, Cornu and Terqui 1975.)

The quality of ovulation appears inadequate because the induced corpora lutea do not function normally in that either the synthesis or the release of progesterone is impaired. In fact, for an equal number of corpora lutea, the plasma concentration of progesterone between 10 and 14 days after AI is higher in pregnant dry ewes than in pregnant lactating ewes (Cognié, Barreto and Saumande, 1975). These low levels of progesterone may explain the elevated early embryonic mortality.

The foregoing discussion points to difficulties at the ovarian level being responsible for the problems in lactating ewes. However, this ignores the fact that uterine involution is as important as ovarian function for normal fertility to be achieved. The poor viability of fertilized ova transferred from the uterus of a dry ewe to that of a lactating ewe (Cognié, Barreto and Saumande, 1975) serves as a reminder of the importance of the uterus and its involvement for embryo survival and transport of gametes. In order to correct the latter it is necessary to use both more inseminations and a greater concentration of spermatozoa per inseminate in lactating compared with dry ewes.

(ii) The endocrine responses of lactating ewes in the non-breeding season

The absence of ovulation may result from either insufficient gonadotrophic hormones liberated by the hypophysis or from a poor response by the ovary to the induced hormonal discharge.

Gonadotrophic insufficiency may result from a reduced ability of the pituitary to synthesize or liberate hormones, and seems to be more asso- ciated with a seasonal effect than with post-partum *per se*. Certainly, under the effect of lactation there is a decreased pituitary content of FSH 2 days post-partum (Mallampati, 1970) but the pituitary FSH and LH contents of dry ewes during anoestrus is already very low (Thimonier and Mauleon, 1969). There is also a decrease in the total quantity of LH released by the hypophysis of lactating ewes during seasonal anoestrus after the classic treatment for induction of ovulation (Pelletier and Thimonier, 1973; Cognié and Pelletier, 1976). Both the maximum concentration and duration of release are lower after LRF injection (Pelletier, 1974), but this is not the case in ewes injected with LRF at the beginning of the breeding season, whether they be hyper- or hypoprolactinaemic (Kann *et al.*, 1975).

The insufficiency of gonadotrophin release may be due to the progestagen treatment itself. It appears that LRF is present in sufficient quantities even though hypothalamic LRF activity increases as a function of the interval post-partum. During seasonal anoestrus there is no lack of LRF. A progestagen treatment of 12 days certainly has a dose-dependent depressive effect on the LRF activity of the hypothalamus of the lactating ewe, but no effect in dry ewes (Pelletier and Thimonier, 1972).

The low levels of oestrogens which precede the induced ovulation in lactating animals (Cognié, Barreto and Saumande, 1975) suggest a poor ovarian response although the response to endogenous gonadotrophic hormones can be improved after a number of stimulations with oestrogens (Kann *et al.*, 1975). There is an inhibitory influence of prolactin on ovarian steroid synthesis of other species and this may be so in the lactating ewe.

(iii)   Modifications to treatment resulting from the preceding results

Several applications of the preceding physiological data have been attempted.

• Improvement in the synthesis of gonadotrophic hormones: females were submitted to an artificial light regime (Mauleon and Rougeot, 1962), so that the two inseminations for twice-yearly lambing were carried out in April and November under the photoperiodic conditions of autumn. The ewes were synchronized by progestagen treatment. Fertility remained low in April in lactating ewes, but was high in November (25 and 75% respectively). However, under these photoperiodic conditions, returns to oestrus of animals not pregnant to the induced oestrus occurred in April as they do in November. With no photoperiod control, these returns are never observed in April (Cognié, personal communication).

Several trials have been attempted to decrease the dose of progestagens impregnated into the vaginal sponges for lactating ewes treated during the anoestrous period (Ile-de-France (20 *v.* 40 mg)), but these have not resulted in increased fertility. In animals with complete ovarian inactivity, there is a tendency towards a favourable effect of a short progestagen treatment (6 *v.* 12 days) which merits further investigation.

• Improvement in the release of gonadotrophic hormone: in seasonally anoestrous lactating ewes, progestagen treatment is not followed by spontaneous release of LH. There is no influence of the interval from sponge removal to PMSG injection on the magnitude of the LH peak if the interval is less than 24 h, although there is a reduced LH release if PMSG is given after 24 h (Pelletier and Thimonier, 1975). No studies on the fertility of sheep after such modifications in treatment have been undertaken.

To avoid excessive superovulation which may result from the standard dose of PMSG used after progestagen treatment in these females, an injection of 50 μg of oestradiol benzoate has been given 24 h after sponge removal to compensate for a reduction in the dose of PMSG. The number of ovulations induced is in fact lower, and less variable but the fertility is low. The same is true if treatment consists of a low dose of PMSG (400 or 750 IU) and followed 24 h later with two injections of 200 μg GnRH 4 hours apart. In the latter case, the synchronization of ovulation is perfect but the quality of the eggs liberated is poor. Ovulation takes place in immature follicles. Several combinations of PMSG, oestradiol and GnRH have been studied without an improvement in fertility (Pelletier, Cognié and Thimonier, personal communication).

• Acceleration of uterine involution by oestrogens soon after parturition (Table 6).

It appears that this treatment increased the number of ewes diagnosed pregnant, but much of the benefit was lost due to a high incidence of subsequent embryonic mortality. It is possible that corpus luteum function remains impaired (Cognié, unpublished) although the state of the uterus was in fact improved, thus leading to an increase in the level of fertilization.

It is certain that the interaction of post-partum state × lactation × season creates a particularly unfavourable hormonal environment. Control of the seasonal factor by means of light control should prove the best means of improving the fertility of treated animals. Any hormonal treatment is likely to be too complex to be of any practical value.

TABLE 6

FERTILITY RATE AT INDUCED OESTRUS IN SUCKLING EWES FOLLOWING PROGESTAGEN PMSG TREATMENT COMMENCING 24 DAYS AFTER PARTURITION

		Ewes pregnant (estimated by progesterone: pregnancy diagnosis at day 18) %	Ewes lambing %
Autumn	Control*	47	29
	Treated*	61	44
Spring	Control	34	29
	Treated	48	37

* Treated ewes 50μg ODB Day 20 p.p.   (From Cognié, unpublished data.)
Control ewes no ODB.

ACKNOWLEDGEMENT

The skilful assistance of Dr. Meredith Lemon for the English correction is gratefully acknowledged.

REFERENCES

Cognié, Y. (1972) *VIIth International Congress of Animal Reproduction and Artificial Insemination,* Munich **2** : 925

Cognié, Y., Hernandez-Barreto, M. and Saumande, J. (1975) *Annales de Biologie Animale, Biochimie, Biophysique* **15** : 329.

Cognié, Y., Mariana, J.C. and Thimonier, J. (1970) *Annales de Biologie Animale, Biochimie, Biophysique* **10** : 15.

Cognié, Y., Cornu, C. and Mauleon, P. (1974) *Symposium of Physiopathology of Reproduction and Artificial Insemination,* Thessaloniki, 33.

Cognié, Y. and Pelletier, J. (1976) *Annales de Biologie Animale, Biochimie Biophysique* **16:** 521.

Dutt, R.H. and Casida, L.E. (1948) *Endocrinology* **43** : 208.

Hawk, H.W. and Conley, H.H. (1971) *Journal of Reproduction and Fertility* **27** : 339.

Kann, G. and Martinet, J. (1975) *Nature* **257** : 63.

Kann, G., Carpentier, M.C., Meusnier, C., Schirar, A. and Martinet, J. (1975) *Journées de la Recherche Ovine et Caprine en France, Paris,* 272.

Lauferon, M. (1975) *Journees "Maitrise des Cycles Sexuels"* (Searle), Montpellier, 9.

Lindsay, D.R., Moore, N.W., Robinson, T.J., Salamon, S. and Shelton, J.N. (1967) *In: The Control of the Ovarian Cycle in the Sheep, p.3 ed. Robinson, T.J. (Sydney University Press).*

Mallampati, R.J. (1970) *Dissertation Abstracts* **31** : 6.

Martin, E., Cognié, Y., Espinosa, E. and Juaristi, J.L. (1976) *VIIIth International Congress on Animal Reproduction and Artificial Insemination,* Krakow Abstracts No. 168.

Mauleon, P. and Rougeot, J. (1962) *Annales de Biologie Animale, Biochimie, Biophysique* **2** : 209.

Mauleon, P. and Dauzier, L. (1965) *Annales de Biologie Animale, Biochimie, Biophysique* **5** : 131.

Pelletier, J. (1974) *Compte-rendus de l'Acad*émie *des Sciences de Paris* **279** : 179.

Pelletier, J. and Thimonier, J. (1972) *Journal of Reproduction and Fertility* **31** : 496.

Pelletier, J. and Thimonier, J. (1973) *Journal of Reproduction and Fertility* **33** : 311.

Pelletier, J. and Thimonier, J. (1975) *Annales de Biologie Animale, Biochimie, Biophysique* **15** : 131.

Prudhon, M., Galindez, F. and Reboul, G. (1975) *Journées de la Recherche Ovine et Caprine en France, Paris,* 293.

Quinlivan, T.D. and Robinson, T.J. (1969) *Journal of Reproduction and Fertility* **19** : 73.

Robinson, T.J. (1964) *Proceedings of the Australian Society of Animal Production* **8** : 47.

Signoret, J.P. and Cognié, Y. (1975) *Annales de Biologie Animale, Biochimie, Biophysique* **15** : 205.

Thimonier, J., Mauleon, P., Cognié, Y. and Ortavant, R. (1968) *Annales de Zootechnie* **17** : 257.

Thimonier, J. and Mauleon, P. (1969) *Annales de Biologie Animale, Biochimie, Biophysique* **9** : 233.

Thimonier, J., Cornu, C. and Terqui, M. (1975) *Journées de la Recherche Ovine et Caprine en France, Paris,* 232.

51

USE OF A PROSTAGLANDIN ANALOGUE (ICI 80996) FOR THE SYNCHRONIZATON OF OESTRUS AND LAMBING IN MERINO EWES

R.J. LIGHTFOOT, K.P. CROKER and T. MARSHALL

Western Australian Department of Agriculture, Perth, Australia

SUMMARY

Two experiments were conducted to examine the use of a synthetic prostaglandin analogue (ICI 80996) for the synchronization of oestrus and lambing in Merino ewes. Factors studied were injection procedure (1 versus 2 injections), dose of ICI 80996 (range, 30-150 μg i/m) and ram/ewe ratio (5 versus 10%) during a seven day joining period.

Joining after a single injection proved unsatisfactory as only partial synchronization was achieved and the proportion of ewes lambing to those served was very low (26%). Two injections, 10 days apart, improved synchronization (85-90% ewes served) but among ewes mated at this oestrus lambing rate was depressed by 19 to 25% (P< 0.001) when compared with that resulting if mating was delayed approximately 17 days until the subsequent "natural" oestrus. With the double injection procedure there were no effects of dose of ICI 80996 on either oestrous synchronization or fertility, but ewe lambing rates were improved by approximately 12% for those joined at the higher ram percentage.

INTRODUCTION

A variety of techniques have been used to induce synchronization of behavioural oestrus in the ewe. Those involving drugs or hormones include a multitude of progesterone or progestagen treatments administered either orally, by injection, subcutaneous implant or intra-vaginal pessary (Lammond, 1964; Robinson, 1967; Cumming, 1979), to more recently the use of prostaglandin $F_{2\alpha}$ (Douglas and Ginther, 1973; Hawk, 1973) and related synthetic analogues (Hearnshaw, Restall, Nancarrow and Mattner, 1974; Fairnie, Cumming and Martin, 1976).

A problem common with the hormone procedures used to synchronize oestrus in the sheep has been reduced fertility at the oestrus immediately following treatment. Despite this problem some methods, in particular the vaginal pessary technique developed by Robinson (1967), have experienced comparatively widespread adoption. In many situations mat ng can be delayed until the second oestrus following treatment when fertili> quite normal, though synchronization is less precise.

The present investigation was conducted to examine a recently developed prostaglandin analogue (ICI 80996), and in particular, to test over a range of doses whether fertility was depressed at the synchronized oestrus.

MATERIALS AND METHODS

Two experiments were conducted, Experiment 1 at the Mt. Barker Agricultural Research Station (approximately 320 km south of Perth) commencing December 1974, and Experiment 2 at the Badgingarra Agricultural Research Station (approximately 180 km north of Perth) commencing December 1975. The climate at both locations is of the Mediterranean type with approximately 75% of the mean annual rainfall (700 mm Mt. Barker, 460 mm Badgingarra) falling between the months of May and September. Pastures are based on subterranean clover (cvs. Mt. Barker/Woogenellup) with volunteer annual herbs and grasses comprising some 50 to 75% of available forage.

Experiment 1 was of factorial design $(4 \times 3 \times 2; n = 21, N = 504)$ and examined the effects of dose of ICI 80996 (0, 40, 80, 120 μg i/m), injection schedule (i. single injection on day -1; ii. two injections, days -11 and -1; iii. two injections, days -27 and -17) and ram/ewe ratio at joining (5%, 10%) on oestrous synchronization and lambing.

Experiment 2, also of factorial design $(6 \times 2 \times 2; n = 24, N = 576)$, examined dose of ICI 80996 (0, 30, 60, 90, 120, 150 μg i/m), injection schedule (ii. two injections, days -10 and -1; iii. two injections, days -27 and -17) and ram/ewe ratio at joining (5%, 10%).

Experimental animals were drawn from flocks of Merino ewes (Expt. 1, 3.5 year old Peppin; Expt. 2, mixed age Murray) and were allocated at random to treatments. The prostaglandin analogue (ICI 80996), initially in vials of 2 ml containing 500 μg, was diluted with sterile water until the required dose was contained in 1 ml. Control ewes (0 dose) were injected with water only. All injections were timed relative to the first day of joining = day 1; (no day 0).

Prior to joining, the ewes ran as single flocks with harnessed vasectomized rams. Throughout the 7-day joining to fertile Merino rams (Expt. 1, 4-10 February, 1975; Expt. 2, 8-14 January, 1976) the ewes were run as two flocks according to the ram/ewe ratio treatments and service records were collected daily.

For lambing, the ewes were divided into treatment groups, side numbered and observed daily to collect individual lambing records. Statistical significance of main effects and interactions was examined by analysis of Chi Square.

RESULTS

(a) Experiment 1

(i) Incidence of oestrus

There was a highly significant ($P < 0.001$) effect of dose on the proportion of ewes served (Table 1). Only 44% of control ewes (0 dose) were served compared with 89% of ewes injected with the active compound. In the latter group (40, 80 and 120 μg) there were no significant differences between doses over the range examined.

TABLE 1

EXPERIMENT 1: SERVICE AND LAMBING RESULTS

Treatment	% ewes served/joined	% ewes lambed/served
1. Dose of ICI 80996 (n = 73 – 124*)		
(i) 0 μg i/m	44	55
(ii) 40 μg i/m	85	45
(iii) 80 μg i/m	88	37
(iv) 120 μg i/m	93	50
P	< 0.001	n.s.
2. Injection schedule** (n = 115 – 124*)		
(i) 1 injection: Day – 1[†]	84	26
(ii) 2 injections: Days –11 and –1[†]	90	40
(iii) 2 injections: Days –27 and –17[†]	92	65
P	n.s.	< 0.001
3. Ram/ewe ratio (n = 213 – 217*)		
(i) 5% (1 ram : 20 ewes)	83	39
(ii) 10% (1 ram : 10 ewes)	79	51
P	n.s.	< 0.05

* Number of ewes joined: variation due to accidental exposure to fertile rams prior to experimental joining

** Excludes 0 dose (control ewes)

† No. of days before rams were joined

There was an interaction ($P < 0.01$) between dose and injection schedule due primarily to the inclusion of control ewes for between schedule comparisons. When the zero dose was excluded (Table 1), however, the interaction was not apparent and there were no significant differences between the three injection schedules studied. It was noticed, however, that the result for schedule (i) (84% ewes served) was somewhat higher than the mean of approximately 70% indicated by two earlier observations of response to single injections on days –27 and –11.

Following either a single injection, or the second of a double injection series, most ewes were in oestrus during the ensuing 36 to 72 hours. Within the range studied there appeared to be little effect of dose on this relationship. Pooling results for schedules (i) and (ii), 31% of treated ewes were detected at the first draft approximately 48 hours post injection, increasing cumulatively to 79% by 72 hours, 84% by 96 hours and 84% by 120 hours. Comparable figures for schedule (iii) were 8, 27, 73 and 85% ewes served after 1, 2, 3 and 4 days of joining respectively.

There was no effect of ram/ewe ratio on service incidence.

(ii)   Lambing

Fifty-five per cent of control (0 dose) ewes served during the 7-day joining period subsequently lambed. Fertility among "treated" ewes tended to be lower (mean 44%) with no apparent effect of dose. Considerable variation in fertility occurred due to injection schedule (Table 1). A significantly lower ($P < 0.001$) proportion of ewes lambed after a single injection (Schedule i) compared with the double injection procedures (Schedules ii and iii). In the latter instance, using a double injection technique, lambing was significantly ($P < 0.001$) higher if mating was delayed until the second (natural) oestrus (Schedule iii), rather than joining immediately after the second injection (Schedule ii).

Although there was no effect on service incidence, the lambing rate was significantly higher ($P < 0.05$) for ewes joined at 10%, compared with 5% rams.

(b)   Experiment 2

(i)   Incidence of oestrus

There was a highly significant effect ($P < 0.001$) of dose on the proportion of ewes served to ewes joined during the 7-day mating period. Forty-four per cent of control ewes (0 dose) were served compared with 85, 84, 81, 80 and 85% for ewes injected with either 30, 60, 90, 120 or 150 μg, ICI 80996 respectively.

Among ewes injected with the active compound, there were no significant differences between either doses or injection schedules (Schedule ii, 85%; Schedule iii, 81%), nor were there any significant interactions. Among these ewes the incidence of service was more concentrated for those subject to injection schedule (ii), in which the cumulative distribution for ewes served over successive days (1-7) of joining was 0, 41, 89, 97, 99, 100% compared with 4, 26, 68, 86, 90, 94 and 100% for schedule (iii).

A slightly higher proportion of ewes joined to 10% rams were served when compared with the result for 5% rams (79 versus 74%) but the difference was not significant.

(ii) Lambing

On overall results (Table 2) there were no effects of dose, but fertility was depressed (50 versus 69%, P< 0.001) among ewes injected with ICI 80996 and joined at the first oestrus after treatment (Schedule ii) compared with those joined approximately one cycle later (Schedule iii). Lambing in the latter instance (69%) appeared normal when compared with overall results (71%) for control (0 dose) ewes.

TABLE 2

EXPERIMENT 2: PER CENT EWES LAMBING OF EWES SERVED

Dose of ICI 80996 μg i/m	Injection schedule		Overall
	(ii) Days −10 and −1	(iii) Days −27 and −17	
0	67 (18) †	75 (24)	71
30	48 (40)	73 (40)	60
60	44 (43)	76 (38)	59
90	56 (41)	72 (36)	64
120	51 (37)	51 (39)	51
150	51 (41)	71 (41)	61
Overall	51 (220)	69 (218)	60

† (No. of ewes served)

Of the ewes served a significantly higher proportion subsequently lambed following joining at 10%, compared with 5% rams (66 versus 54% respectively; P< 0.01).

.DISCUSSION

The results presented herein demonstrate that with appropriate technology the synthetic prostaglandin analogue ICI 80996 is an effective agent for the synchronization of oestrus in the breeding ewe. In particular the present work corroborates earlier evidence (Fairnie, Cumming and Martin, 1976) that a series of two, rather than a single injection is necessary for good results. Within the limits of the range examined (30 to 150 μg) dose rate did not appear to be critical, either for oestrous synchronization, or for conception rate to service at that oestrus. With regard to oestrous response, ICI 80996 appears similar in this respect to a related compound, ICI 79939, reported by Hearnshaw *et al.* (1974) to be effective at approximately 30μg.

In common with other synchronization techniques the present work clearly demonstrated that fertility is depressed in ewes mated at the first oestrus after treatment with ICI 80996. The physiological basis of this phenomenon remains to be elucidated but Hawk (1973) has reported a change in the pattern of uterine contractions following treatment of ewes with prostaglandin $F_{2\alpha}$. By delaying joining until the subsequent (natural) oestrus the present study has shown that fertility returns to control levels, but synchronization of oestrus is less precise. Such reduced efficiency in the regulation of oestrus is of consequence with regard to possible use of the compound in artificial breeding programmes where insemination on a time basis, as used by Fairnie, Cumming and Martin (1976), is desirable. Double insemination may overcome the problem, or alternatively, ewes can be joined with entire rams as adopted here. In this case the ram concentration must be high as although oestrous detection was little effected, lambing rates were significantly higher for ewes joined at 10%, compared with 5% rams.

In the most successful treatments used here (Schedule iii, 30 to 150 μg, 10% rams), and pooling results for both experiments, 147 (84%) of 176 treated ewes were served during the 7-day joining. Of these, 109 (74% of ewes served, 62% of ewes joined) subsequently lambed. Despite this degree of success, use of drugs or hormones to synchronize oestrus in Australia's extensive sheep-farming enterprise requires careful economic justification. Advantages in terms of the more widespread use of superior genotypes in artificial breeding programmes, improved control of nutrition and husbandry during mating (e.g. lupin feeding, Lightfoot, Marshall and Croker, 1976) and lambing, and more uniform liveweight in lambs are possible, but often difficult to cost in real cash terms.

ACKNOWLEDGEMENTS

The authors wish to thank the managers of the Research Stations, Mr. D. Rowe (Mt. Barker) and Mr. R. Randall (Badgingarra) for their co-operation throughout the trials, and Messrs. T. Johnson, B. Guthrie, D. Ryall and H. Jellicoe for skilled technical assistance. The prostaglandin analogue was provided by ICI (Australia) Ltd. through courtesy of Mr. E.P. Meyer.

REFERENCES

Cumming, I.A. (1979) In: *Sheep Breeding.* Ed. by Tomes, G.J., Robertson, D.E. and Lightfoot, R.J. p. 403. London: Butterworths.

Douglas, R.H. and Ginther, O.J. (1973) *Journal of Animal Science* **37** : 990.

Fairnie, I.J., Cumming, I.A. and Martin, E.R. (1976) *Proceedings of the Australian Society of Animal Production* **11** : 133.

Hawk, H.W. (1973) *Journal of Animal Science* **37** : 1380.

Hearnshaw, H., Restall, B.J., Nancarrow, C.D. and Mattner, P.E. (1974) *Proceedings of the Australian Society of Animal Production* **10** : 242.

Lammond, D.R. (1964) *Animal Breeding Abstracts* **32** : 269.

Lightfoot, R.J., Marshall, T. and Croker, K.P. (1976) *Proceedings of the Australian Society of Animal Production* **11** : 5P.

Robinson, T.J. (1967) *The Control of the Ovarian Cycle in the Sheep.* Sydney University Press.

52

CULTURE, STORAGE AND TRANSFER OF SHEEP EMBRYOS

N.W. MOORE
University of Sydney, Camden, N.S.W., Australia

The development of effective methods of frozen preservation of embryos of farm animals would have marked application, particularly in the cheap and rapid transport of flocks and herds, and in the retention of genetic material for future use and study. Methods for the storage of mouse embryos in liquid nitrogen ($-196°C$) have been developed (Whittingham, Leibo and Mazur 1972; Wilmut 1972) and the survival and development of frozen/thawed embryos following transfer of recipients has been reported to be comparable with that of un-stored embryos (Whittingham 1974). Within farm animals there are reports of successful frozen storage of sheep (Willadsen *et al.* 1976), goat (Bilton and Moore 1976a) and cattle embryos (Wilmut and Rowson 1973; Bilton and Moore 1976b) but generally less than 10% of stored embryos developed into normal young following transfer to recipient females, or too few embryos were involved to provide meaningful results.

In this laboratory it was considered necessary as part of the studies directed toward frozen storage of embryos, to develop and refine a number of procedures.

Induction of superovulation

In the ewe, superovulation can be induced by a variety of gonadotrophins, particularly PMSG and pituitary extracts, but high ovulation rates are frequently associated with complete failure of fertilization. When gonadotrophins are given following treatment with progesterone marked control over the time of oestrus and ovulation is achieved (Moore 1970a), but the problem of failure of fertilization remains. It would seem that the cervix is involved in the failure of fertilization for it has been demonstrated that high rates of fertilization following superovulation can be achieved by the direct deposition of semen into the uterine horns (Trounson and Moore 1974). Where precise control over the time of ovulation has been achieved by the use of an equine anterior pituitary extract (HAP) and daily injections of progesterone (Moore 1970a) surgical inseminations can be carried out without reference to the detection of oestrus.

Precise control of time of ovulation coupled with surgical insemination has allowed the harvesting, at pre-determined times, of large numbers of embryos of known age.

In vitro culture of sheep embryos

Successful methods of culture of sheep embryos have been developed (Moore 1970b; Moor and Cragle 1971; Moore and Spry 1972) and they have been based on the use of balanced salt solutions, enriched with ovine serum, or albumin of ovine (or bovine) origin. Sheep embryos, in common with those of a number of species hesitate at a particular stage of development (Moore 1973) and require specific conditions in culture to progress through these stages of development. In the ewe, the "block" to continued development in culture occurs around the 8-cell stage of development. Incubation under reduced oxygen tension has been shown to enable embryos to progress through and beyond the 8-cell stage (Tervit, Whittingham and Rowson 1972; Trounson 1974).

Apparently normal development in culture provides a reliable indication of the potential for further development *in vivo* and hence culture allows the rapid assessment of viability of embryos after they have been exposed to a variety of treatments.

Low temperature liquid storage

A number of attempts have been made to store sheep embryos at around 5-10°C and lambs have been born following the transfer of stored embryos to recipient ewes (Averill and Rowson 1959; Kardymowicz *et al.* 1963; Moore and Bilton 1973). Development, as evidenced by continued cleavage, is arrested during storage but the duration of storage is limited to a few days. The duration of storage without appreciable loss in viability, would seem to be influenced by storage temperature. Kardymowicz and Kremer (1971) stored embryos at 10°C and reported the birth of 11 lambs following the transfer of 15 embryos stored for 5 days. Later work (Kardymowicz 1972) suggested that storage at 10°C without marked loss in viability might be extended to 10 days. When temperatures around 5°C are employed viability might be lost more rapidly. Moore and Bilton (1973) found that only one of seven embryos stored for 2 days at 5°C survived following transfer and in further studies (Moore and Bilton 1976) in which the duration of storage at 5°C was extended to 12 days the proportion of embryos showing continued development in culture decreased after 3 days storage.

Most studies have been carried out on 3-day, and older, embryos of 8-cells to early blastocyst. However, the studies of Moore and Bilton (1973) would suggest that 2-day embryos (1, 2 and 4-cells) survive storage at 5°C equally as well as 5-day embryos (morulae of 20 or more cells).

Frozen storage

The success which has been achieved in the frozen storage of mouse embryos is, to a large degree, due to two factors. First, large numbers of embryos of known age can be collected on any given day thus allowing the setting up of critical experiments involving sufficient numbers of embryos to provide meaningful results. Second, a short period of gestation, together with the development of effective methods of *in vitro* culture of mouse embryos allows the rapid assessment of viability of embryos after storage.

Effective methods of culture of sheep embryos are now available, and modifications, particularly surgical insemination, have increased the reliability of harvesting large numbers of embryos at pre-determined times. These techniques can now be profitably exploited in studies on frozen sheep embryos.

Factors most likely to influence the survival of embryos during freezing and thawing are medium, choice and method of use of cryoprotectant, rates of freezing and thawing and stage development of embryos at the time of freezing.

Phosphate buffered saline, e.g., Dulbecco phosphate buffer at pH 7.4-7.6, is the medium most commonly used in the frozen storage of all mammalian embryos so far studied. In this laboratory around 20% serum of the species under test is added to Dulbecco phosphate buffer. A number of compounds with reputed cryoprotectant characteristics are available, but only dimethylsulphoxide (DMSO), glycerol and to a lesser degree polyvinyl pyrrolidone (PVP) have been examined. Willadsen *et al.* (1976) successfully stored sheep embryos in media containing 1.5 M DMSO, others have obtained apparently normal development in culture and lambs from embryos frozen in media containing 1.0-2.0 M DMSO (Bilton and Moore 1976c). However, the procedures used in the addition and removal of DMSO can markedly influence the subsequent survival of embryos (Table 1).

TABLE 1

DEVELOPMENT OF 6 DAY SHEEP EMBRYOS IN CULTURE FOLLOWING EXPOSURE TO 1.5 M DIMETHYLSULPHOXIDE (DMSO) ADDED AND DILUTED AT 37°C OR 5°C

TEMPERATURE (°C)		NUMBER OF EMBRYOS		
Addition of DMSO	Dilution of DMSO	Developed in Culture	Failed to Develop in Culture	Total
37	37	8	7	15
37	5	0	15	15
5	37	12	3	15
5	5	0	15	15
Total:		20	40	60

Cultured for 24-36 hours.

1.5 M DMSO was added at either 37° or 5°C to 6-day embryos held in Dulbecco phosphate buffer containing 25% sheep serum (DB+25% S). The embryos were then held at 5°C for 30 min., or cooled to 5°C after the addition of DMSO and held there for 30 min. DMSO was removed by dilution at 5° or 37°C and the embryos were then cultured in DB+25% S for 60 hours. Irrespective of temperature of addition, dilution of DMSO at 5°C had a disastrous effect on subsequent viability. It may be that dilution at 5°C may not effectively remove DMSO from the embryos with a subsequent deleterious effect upon their viability. In all subsequent studies DMSO has been added and diluted in a stepwise procedure (0.25M DMSO interval between steps) carried out at 30°C.

A wide range of freezing and thawing rates have been used. It would seem that relatively slow rates of freezing, of the order of 0.1 to 0.3°C/min. are required, but thawing rates may not be so critical. When freezing rates of 0.15 to 0.3°C/min. (measured over the range of around 0°C to −60°C) have been used subsequent development both *in vivo* and *in vitro* has been achieved with thawing rates (measured over the same range) as low as 2.2°C/min. (Bilton and Moore 1976c) and as high as 360°C/min. (Willadsen *et al.* 1976).

The success which has been achieved to date in the frozen preservation of sheep embryos is encouraging, but many more critical studies are required before the technique can be evaluated for practical purposes.

REFERENCES

Averill, R.L.W. and Rowson, L.E.A. (1959). *Journal of agricultural Science* **52** : 392.

Bilton, R.J. and Moore, N.W. (1976a). *Australian Journal of biological Sciences* **29**: 125.

Bilton, R.J. and Moore, N.W. (1976b). *Journal of Reproduction and Fertility* **46** : 537.

Bilton, R.J. and Moore, N.W. (1976c). *Proceedings of the Australian Society for Reproductive Biology*, Brisbane, August, 1976.

Kardymowicz, O. (1972). *Proceedings VII International Congress on Animal Reproduction and Artificial Insemination, Munich*, Summaries, 55.

Kardymowicz, M., Kardymowicz, O., Kuhl, W. and Lada, A. (1963). *Acta biologica Cracoviensa, Series Zooiogia* **6** : 31.

Kardymowicz, O. and Kremer, M. (1971). *Acta biologica Cracoviensa, Series Zoologia* **14** : 64.

Moor, R.M. and Cragle, R.G. (1971). *Journal of Reproduction and Fertility* **27** : 401.

Moore, N.W. (1970a). *Journal of Endocrinology* **46** : 121.

Moore, N.W. (1970b). *Australian Journal of biological Sciences* **23** : 721.

Moore, N.W. (1973). *Journal of Reproduction and Fertility, Supplement No. 18* : 111.

Moore, N.W. and Bilton, R.J. (1973). *Australian Journal of biological Sciences* **26** : 1421.

Moore, N.W. and Bilton, R.J. (1976). *Proceedings of the VIIIth International Congress on Animal Reproduction and Artificial Insemination, Crakow* Vol. III: 306.

Moore, N.W., and Spry, G.A. (1972). *Journal of Reproduction and Fertility* **28** : 139.

Tervit, H.R., Whittingham, D.G. and Rowson, L.E.A. (1972). *Journal of Reproduction and Fertility* **30** : 493.

Trounson, A.O. (1974). Ph. D. Thesis, University of Sydney.

Trounson, A.O. and Moore, N.W. (1974). *Australian Journal of biological Sciences* **27** : 310.

Whittingham, D.G. (1974). *Journal of Reproduction and Fertility* **37** : 159.

Whittingham, D.G., Leibo, S.P. and Mazur, P. (1972). *Science, New York,* **178** : 411.
Willadsen, S.M., Polge, C., Rowson, L.E.A. and Moor, R.M. (1976). *Journal of Reproduction and Fertility* **46** : 151.
Wilmut, I. (1972). *Life Sciences* **II** : 1071.
Wilmut, I. and Rowson, L.E.A. (1973). *Veterinary Record* **92** : 686.

53

OVARIAN STEROID SECRETION IN SHEEP DURING ANOESTRUS

R.J. SCARAMUZZI and D.T. BAIRD

M.R.C. Unit of Reproductive Biology, Edinburgh, U.K.

SUMMARY

The secretion rate of oestradiol and the peripheral concentration of luteinizing hormone (LH) were measured during anoestrus in six Merino cross ewes. The basal concentration of LH remained low (< 0.5 ng/ml NIH LH-S-14) with episodic releases of LH occurring with a mean frequency of one per 5 h. The basal secretion rate of oestradiol was about 0.5 ng/min. Each release of LH was followed by a rapid increase in secretion of oestradiol. Maximum secretion was 5.24 ng/min and was reached 25 minutes after LH release. Oestradiol secretion was not maintained and had returned to basal values 2 h after LH release. These results show that follicles which develop during anoestrus have an apparently normal capacity to secrete oestradiol. The data suggest that anoestrus results from an altered sensitivity of the hypothalamic-pituitary axis to a normal oestrogen stimulus from the ovary.

INTRODUCTION

The ovary of the anoestrous ewe is characterized by the absence of corpora lutea and the presence of numerous follicles which often reach 5 mm in diameter (Cole and Miller, 1935; Turnbull, unpublished results). The peripheral concentrations of the main ovarian steroids oestradiol and progesterone are very low during anoestrus (Yuthasastrakosol, Palmer and Howland, 1975). However direct ovarian secretion of oestradiol and androstenedione does take place during anoestrus (Martensz *et al.,* 1976) in variable but sufficient amount to maintain negative feedback. Ovariectomy during anoestrus results in a rapid rise in peripheral LH concentrations (Reeves, O'Donnell and Denorscia, 1972).

The concentrations of LH in peripheral plasma (Scaramuzzi and Martensz, 1975; Yuthasastrakosol, Palmer and Howland, 1975) and the pituitary gland (Robertson and Hutchinson, 1962; Roche *et al.,* 1970) during anoestrus have been measured. The pituitary concentration is similar to that seen during the mid-luteal phase whereas the plasma concentrations of LH differ markedly between anoestrus and the luteal phase (days 4-6) of the oestrous cycle (Scaramuzzi and Martensz, 1975). During anoestrus in Welsh Mountain ewes LH is released in large episodic bursts with a frequency of one per 15 h. During the luteal phase episodic LH discharges have a higher frequency but a lower amplitude and are followed by a rapid increase in oestradiol secretion (Scaramuzzi and Martensz, 1975; Baird, Swanston and Scaramuzzi, 1976). These results indicate that during the oestrous cycle the ovary, like the testis is responsive to short term episodic discharges of LH. The following experiment was conducted to determine if the ovary of the anoestrous ewe remained responsive to infrequent short term changes in peripheral LH concentration.

MATERIALS AND METHODS

(a) Experimental animals

Six Merino cross ewes in which the reproductive tract had been autotransplanted to a more accessible site in the neck (Baird *et al.*, 1976a) were used in the experiments described. These ewes show oestrus at regular intervals and the pattern of steroids secreted during the oestrous cycle is normal (Baird *et al.*, 1976b). The experiments were conducted in June 1975 (mid anoestrus) three months after the last recorded oestrus of the preceding breeding season, and well before the onset of the next breeding season in late September 1975.

(b) Collection of utero-ovarian venous blood

On the day preceding blood collection both jugular veins were cannulated. On the side of the utero-ovarian transplant a 100 cm silastic catheter (i.d. 2.0 mm, o.d. 3.2 mm Dow Corning Corporation, Midland, Michigan, USA) was placed in the jugular vein so that its tip lay near the site of the anastomosis of the utero-ovarian and jugular veins (Baird *et al.*, 1976b). Two intravenous catheters were also placed about 15 cm apart in the opposite jugular (Braunula, 18 G size 2, Armour Pharmaceutical Co. Ltd., Eastbourne). After cannulation the ewes were placed in metabolism cages (Baird *et al.*, 1976b). Twenty-five timed (to provide an estimate of ovarian blood flow) samples of ovarian venous blood (25 ml) were collected every 10 minutes between 1100 and 1500 h or between 1200 and 1600 h. The blood was collected into sterile heparinized universal containers which were centrifuged immediately (1,500 g for 20 minutes). A 10 ml aliquot of plasma was then removed, using a sterile pipette, and stored at −20°C until assay. The remaining plasma and blood cells were returned to the ewe under gravity via the proximal cannula in the contralateral jugular vein. In this way the blood volume of the ewe was never depleted more than 150 ml. Midway between the collection of each ovarian venous sample a 4 ml sample of blood was taken from the distal catheter in the contralateral jugular vein for the assay of LH. The drip which returned blood cells to the ewe was closed while the blood sample was collected. The plasma was removed and stored at −20°C until assay.

(c)   Analytical methods

Oestradiol and LH were measured by radioimmunoassay. Oestradiol was measured in duplicate extracts of ovarian venous plasma using the method and antiserum described by Baird, Swanston and Scaramuzzi (1976). Luteinizing hormone was assayed in triplicate in jugular venous plasma (Martensz *et al.*, 1976), and the results are expressed as ng/ml of NIH-LH-S14. The packed cell volume was measured hourly starting with the first ovarian venous blood sample. The determinations were carried out using a micro haematocrit centrifuge.

(d)   Calculation of secretion rates

The secretion rate of oestradiol was calculated as follows:

$$\text{Secretion Rate (ng/min)} = \frac{\text{plasma concentration (pg/ml)}}{1000} \times$$

$$\text{Ovarian Blood Flow (ml/min)} \times \frac{\text{haematocrit (\%)}}{100}$$

No allowance was made for the volume of plasma in the packed red cells or for the proportion (about 10%) of plasma oestradiol adsorbed onto the surface of the erythrocytes (Baird *et al.*, 1968).

RESULTS

The secretion rate of oestradiol and the concentration of LH varied considerably within and between individual ewes. Because of this variation individual results are presented in Figures 1, 2 and 3. Luteinizing hormone concentrations remained very low (< 0.5 ng/ml) during the period of sampling. However episodic discharges of LH occurred five times in the six ewes over a total of 24 h of sampling, a mean frequency of approximately one discharge per 5 h. The mean amplitude (peak height) of the LH discharges was 6.0 ± 0.3 ng/ml (M±SEM; n=5). Each discharge of LH was followed by a rapid return to basal concentrations usually within an hour (Figure 4). The basal secretion rate of oestradiol was also very low at about 0.5 ng/min. However each episodic discharge of LH was followed by a rapid increase in the secretion of oestradiol. The maximum secretion rate was 5.24 ± 0.84 ng/min (M ± SEM; n=5) and was reached 25 minutes after LH discharge (Figure 4). Oestradiol secretion was not maintained and had returned to basal values by about 2 h after LH discharge. The data centred about the episodic discharges of LH is presented in Figure 4, and illustrates the temporal relationship between LH and oestradiol. The mean overall ovarian blood flow, which also includes blood from the uterus and the skin in the jugular-carotid loop varied significantly between, but not within animals. The mean ovarian blood flow within animals ranged from 19.40 ± 0.46 to 51.76 ± 0.67 ml/min (M±SEM; n=25).

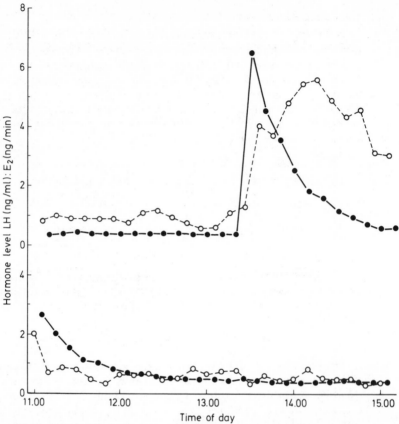

Figure 1.—The secretion rate (ng/min) of oestradiol o-----o and the peripheral concentration of LH (ng/ml) ●——● during mid anoestrus in two ewes bearing utero-ovarian autotransplants. Ovarian venous blood samples were collected every 10 min starting at 1105 h (top) or 1100 h (bottom) and continuing for 4 h. Jugular venous blood samples were taken midway between the ovarian venous blood samples commencing 5 min after the first ovarian venous blood sample.

DISCUSSION

The present data shows quite clearly that during anoestrus LH is released as a series of infrequent episodic discharges, confirming earlier observations (Scaramuzzi and Martensz, 1975; Yuthasastrakosol, Palmer and Howland, 1975). The pattern of LH release seen during anoestrus differs in frequency and amplitude from that seen during the luteal phase of the oestrous cycle (Scaramuzzi and Martensz, 1975; Baird, Swanston and Scaramuzzi, 1976). These data suggest that differences in hypothalamic-pituitary functions in anoestrous and breeding ewes are quantitative rather than qualitative. Early attempts to find differences in hypothalamic-pituitary function at different times of the year were unsuccessful (Beck and Reeves, 1973) however a quantitative change in sensitivity of the ovariec-

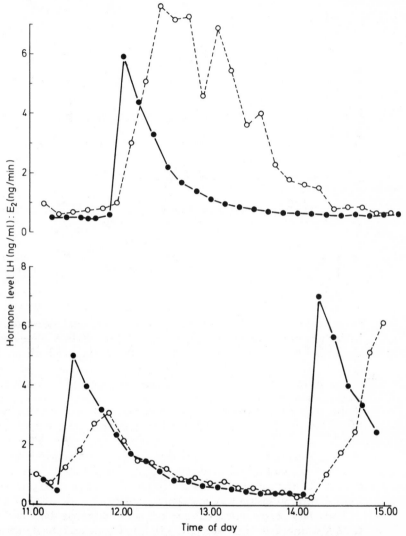

Figure 2. —The secretion rate (ng/min) of oestradiol o-----o and the peripheral concentration of LH (ng/ml) ●——● during mid anoestrus in two ewes bearing utero-ovarian autotransplants. Ovarian venous blood sample were collected every 10 min starting at 1105 h (top) or 1100 h (bottom) and continuing for 4 h. Jugular venous blood samples were taken midway between the ovarian venous blood samples commencing 5 min after the first ovarian venous blood sample.

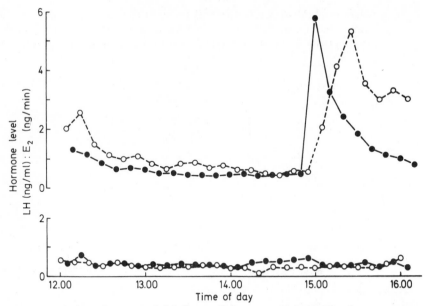

Figure 3.—The secretion rate (ng/min) of oestradiol o-----o and the peripheral concentration of LH (ng/ml) •——• during mid anoestrus in two ewes bearing utero-ovarian autotransplants. Ovarian venous blood samples were collected every 10 min starting at 1205 h (top) or 1200 h (bottom) and continuing for 4 h. Jugular venous blood samples were taken midway between the ovarian venous blood samples commencing 5 min after the first ovarian venous blood sample.

tomized ewe to exogenous oestrogen given at different times of the year has recently been reported (Land, Wheeler and Carr, 1976). These results plus the present data suggest that small quantitative changes in both negative and positive feedback take place during seasonal anoestrus in the ewe. It remains to be determined if these changes are a cause or a consequence of seasonal anoestrus.

The frequency of LH discharges (approximately one per 5 h) seen in the Merino cross ewes used in the present study was greater than previously seen in Welsh Mountain ewes (Scaramuzzi and Martensz, 1975) and probably reflects a breed difference, perhaps related to differences in the "depth" of anoestrus. During the breeding season ewes of both breeds have a frequency of LH discharge of about one per 2 h (Scaramuzzi and Martensz, 1975; Baird, Swanston and Scaramuzzi, 1976).

The ovary of the anoestrous ewe contains numerous small and medium sized follicles (Cole and Miller, 1935). These follicles often reach 5 mm in diameter before becoming atretic (Turnbull, unpublished observations). Virtually all of the oestradiol secreted by the sheep ovary comes from the largest non-atretic follicle (Baird and Scaramuzzi, 1976). Preliminary studies (Martensz *et al.*, 1976) had shown that the ovary of the anoestrous ewe was capable of secreting oestradiol. The present results confirm the

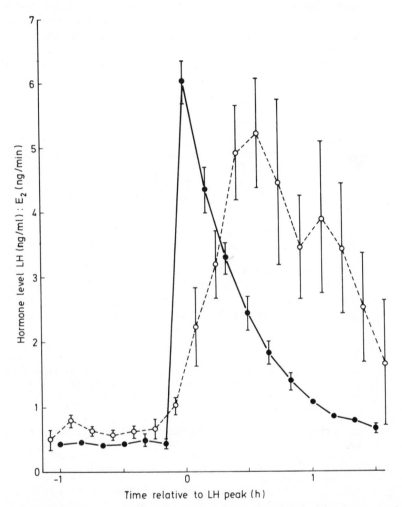

Figure 4.—The secretion rate (ng/min) of oestradiol o-----o and the peripheral concentration of LH (ng/ml) ●——● in six ewes with utero-ovarian autotransplants. The values have been grouped around the peak concentration of LH in jugular venous plasma and each point represents the mean ±SEM of between three and five observations.

preliminary observations and show that follicles present in the ovaries of anoestrous ewes are highly responsive to endogenous LH. The variability of oestradiol secretion during anoestrus (Martensz *et al.*, 1976) is most likely related to the infrequent episodic nature of LH release. Oestradiol secretion from follicles on the ovaries of either anoestrous or non-pregnant breeding ewes appears quantitatively similar (Figure 4; Baird, Swanston and Scaramuzzi, 1976) however biologically important differences may still exist. Nevertheless the present data do show that follicles which develop during anoestrus have a quantitatively normal capacity to secrete oestradiol. The results support the concept of an altered responsiveness of the hypothalamic-pituitary system to oestrogen as a major contributing factor to seasonal anoestrus. The patterns of LH release seen in the present study suggest that during anoestrus the negative feedback loop between LH and oestradiol has *increased* its sensitivity to oestradiol, while the positive feedback loop has *decreased* its sensitivity to oestradiol (Land, Wheeler and Carr, 1976). This suggestion requires additional detailed experimental testing. If, these two key feedback loops which regulate ovulation have shifted their respective sensitivities to oestrogen in opposite directions, the prospect of a simple reliable method for inducing regular oestrous cycles in anoestrous ewes becomes extremely remote.

ACKNOWLEDGEMENTS

We wish to thank Messrs W.G. Davidson, D.W. Davidson and I. Swanston for excellent technical assistance. The purified preparations of sheep LH used in the LH assays were supplied by the Endocrine Study Section of the National Institute of Health, USA.

REFERENCES

Baird, D.T., Goding, J.R., Ichikawa, Y. and McCracken, J.A. (1968) *Journal of En docrinology* **42** : 283.

Baird, D.T., Land, R.B., Scaramuzzi, R.J. and Wheeler, A.G. (1976a) *Proceedings of the Royal Society, London. Section B Biological Sciences* **192** : 463.

Baird, D.T., Land, R.B., Scaramuzzi, R.J. and Wheeler, A.G. (1976b) *Journal of Endocrinology* **69** : 276.

Baird, D.T. and Scaramuzzi, R.J. (1976) *Acta Endocrinologica (Copenhagen)* **83**: 402.

Baird, D.T., Swanston, I. and Scaramuzzi, R.J. (1976) *Endocrinology* **98**: 1490.

Beck, T.W. and Reeves, J.J. (1973) *Journal of Animal Science* **36**: 566.

Cole, H.H. and Miller, R.F. (1935) *American Journal of Anatomy* **57**: 39.

Land, R.B., Wheeler, A.G. and Carr, W.R. (1976) *Annales de Biologie Animale Biochimie Biophysique* **16**: 521.

Martensz, N.D., Baird, D.T., Scaramuzzi, R.J. and Van Look, P.F.A. (1976) *Journal of Endocrinology* **69** : 227.

Reeves, J.J., O'Donnell, D.A. and Denorscia, F. (1972) *Journal of Animal Science* **35** : 73.

Robertson, H.A. and Hutchinson, J.S.M. (1962) *Journal of Endocrinology* **24** : 143.

Roche, J.F., Foster, D.L., Karsch, F.J., Cook, B. and Dziuk, P.J. (1970) *Endocrinology* **88** : 568.

Scaramuzzi, R.J. and Martensz, N.D. (1975) "Immunisation with Hormones in Reproduction Research" p.141 (ed. E. Nieschalag) (North Holland Publishing Company: Amsterdam, The Netherlands).

Yuthasastrakosol, P., Palmer, W.M. and Howland, B.E. (1975) *Journal of Reproduction and Fertility* **43** : 57.

V

MALE REPRODUCTION AND ARTIFICIAL INSEMINATION

54

MATING BEHAVIOUR IN SHEEP

D.R. LINDSAY

Department of Animal Science & Production, University of Western Australia

INTRODUCTION

Sexual behaviour is undoubtedly an essential component of the reproductive process in sheep but it has been studied far less intensively than the more conventional physiological aspects of reproduction. There are two strong reasons for this.

First, a great deal of the research effort in reproductive physiology has been directed towards the elimination of the behavioural component whose variability and capriciousness have been a stumbling block to progress in the control of reproduction. For example, in synchronizing oestrus for artificial insemination the aim has been to inseminate ewes successfully without reference to the onset or nature of behavioural oestrus. Similarly, semen may be collected from rams without considering normal sexual behaviour patterns of rams and in artificial insemination the intricate process by which ewes and rams find each other and mate is of little consequence.

Second, behaviour is often difficult to describe and analyse quantitatively. Sexual behaviour is no exception in this regard. In some cases the characteristic measured is trivial and in other cases the measurements are unrealistic. Thus the development of meaningful measurements of sexual behaviour are of prime concern to workers in this field.

It remains a fact that an overwhelming majority of the world's sheep reproduce without artificial control and the efficiency with which they do so is strongly influenced by inate and learned behavioural patterns.

Measurement of Behaviour in Males

In animal production terms the ability of the ram to disseminate gametes among ewes is the most important behavioural attribute. This is a function of the number of services he performs in unit time and the number of available ewes which he covers.

The number of services in unit time is most usually called libido, although other, less useful, definitions of libido such as "reaction time" (period between introduction to the ewe and first service) or "latent period" (time between successive services), or even subjective assessments of vigour, are occasionally used. Mattner, Braden and George (1971) have shown that there is a correlation between the number of times a ram serves an oestrus ewe in a pen during a series of 20 minute tests and his performance under paddock mating conditions. The "20-minute libido test" may therefore be a useful guide to field performance. Measurements of so-called dexterity such as mounts per service (Pepelko and Clegg, 1965; Mattner and Braden, 1975) seem to be of little practical value. Mounting several times prior to ejaculation seems to be part of normal foreplay in some rams but not in others and the correlation between number of mounts without ejaculation and number of ejaculations in unit time is very poor (Knight, 1967).

The willingness and ability of rams to seek out new ewes in oestrus and the ability to discriminate between oestrus and non-oestrous ewes are the other behavioural characteristics which have important reproductive implications. Pepelko and Clegg (1965) demonstrated clearly that rams are attracted by ewes freshly in oestrus, or ewes which have not been recently served, in preference to those which they have already been serving. It is unlikely then, in the field, that rams will attend one ewe for a long period to the exclusion of others. Rams can detect ewes in oestrus by olfaction (Lindsay, 1965a). When the olfactory bulbs are removed from rams their ability to discriminate between oestrus and non-oestrous ewes is lost entirely but simply by nudging ewes on a trial-and-error basis, and presumably using the stance of the ewe as a criterion, successful identification is achieved. Even rams which can smell distinguish a proportion of ewes in this manner (Lindsay, 1965a). Rams appear to smell the vulva of ewes and often demonstrate "flehmen", a characteristic curling back of the upper lip with the head erect. This is not demonstrated uniquely in response to the smell of oestrus ewes as it can be elicited by touching the nose of a ram with a swab containing urine from a non-oestrous ewe, a wether or another ram.

Measurement of Behaviour in Females

To permit the maximum opportunity for fertilization a ewe must not only accept the male about the time of ovulation but she must be in oestrus long enough to be mated, preferably several times. Under field conditions where rams are working continuously the amount of semen from one ejaculate is probably well below that necessary for optimum chances of fertilization. Knight and Lindsay (1973) showed that the more rams with which a ewe mates the greater the chance she has of conceiving. Therefore behavioural elements which contribute to multiple mating are desirable for successful reproduction. The length of oestrus and the attractiveness of ewes to rams are important competitive attributes.

The length of oestrus is comparatively easy to measure by observation and shows important variability which may be correlated with reproductive performance. Maiden ewes remain in oestrus for a shorter period than older ewes (McKenzie and Terrill, 1937; Lambourne, 1956). Lambourne suggests that this fact may contribute to the lower lambing performance of maiden ewes. Fletcher and Lindsay (1971) have shown that the mean length of oestrus varies with season, the longest oestrus coinciding with the height of the breeding season.

Variations among ewes in attractiveness to rams may also contribute to variations in the amount of multiple mating. Hafez (1951) suggested a number of criteria which are supposed to make ewes more or less attractive to rams but this characteristic is difficult to express quantitatively. One promising technique has been developed by Bell (1974) who ranked individuals within groups of oestrus ewes for attractiveness according to the order in which they were chosen by a ram released among them. This method is repeatable and allows comparisons among ewes of different categories.

Hormonal Control of Reproductive Behaviour

In rams as well as most other mammalian species the development of sexual activity is under the influence of male hormones. Pre-pubertal castration eliminates practically all adult male sexual behaviour. After puberty the role of hormones is less clear. Rams castrated post-pubertally lose sexual activity slowly but variation in libido is not attributable to variation in male hormone levels (Knight, 1973; Mattner and Braden, 1975). Therapeutic doses of luteinizing hormone or testosterone have not produced detectable differences in libido of rams and adult rams with no apparent desire to mate have not been induced to do so by such injections. Therefore in rams it appears likely that the ability to perform many services in unit time is inate and within wide limits is not influenced by hormones.

On the other hand, in females reproductive behaviour is strongly under hormonal control. Robinson (1954) showed that oestrogen induces behavioural oestrus in ovariectomized ewes and that ewes are more sensitive to oestrogen if they have been under the influence of progesterone for several days before oestrogen is administered. Other steroids, such as testosterone and androstenedione, can induce oestrus but at such high doses that they probably have little physiological relevance (Lindsay and Scaramuzzi, 1969). The behavioural response to oestrogen is very precise and is quantitatively related to the dose of oestrogen given (Fletcher and Lindsay, 1971; Scaramuzzi, Lindsay and Shelton, 1971) and is a sensitive bioassay for oestrogens in sheep. Ewes are more responsive to oestrogen in the breeding season than in the non-breeding season and Merino ewes vary less in this respect than sheep of British breeds (Fletcher, Allison and Lindsay, 1971). Acceptance of the male and male seeking behaviour are both controlled by the same hormones but the latter response requires higher levels of oestrogen for its expression (Lindsay and Fletcher, 1972).

It is clear from the work with spayed ewes that hormone regimes for the control of ovulation which interfere with the production and action of oestrogen or which fail to allow a progestational phase of at least 6 days prior to ovulation (Scaramuzzi, 1968), will interfere with the normal expression of oestrus behaviour.

Mating Behaviour in the Flock

Most of the literature on mating behaviour in sheep describes observations and experiments done in pen and animal house situations. Extrapolation to the field is not always valid but on the other hand, field studies are difficult because direct observation of animals is seldom possible. Research workers have been obliged to rely heavily on ram-mating harnesses and crayons (Radford *et al.* 1960) or modifications of them (Knight and Lindsay, 1973) for clues about mating performance. More precise measurement of field behaviour awaits the introduction of remote sensors and recorders of individual mounting and mating in undisturbed flocks.

An example of the divergence between pen and field behaviour is the role of dominance in the mating behaviour of rams. In confined areas a dominant ram can suppress drastically the performance of rams subordinate to him (Lambourne, 1956; Edgar, 1961; Hulet *et al.* 1962a; Lindsay *et al.* 1976). In the field there is little evidence that a dominant ram has any significant adverse effect on the performance of subordinates (Lindsay and Robinson, 1961; Mattner, Braden and Turnbull, 1967). This is particularly the case when many ewes are in oestrus simultaneously. In fact, Lindsay and Ellsmore (1968) showed that individual rams marked significantly more ewes when in competition with two other rams than when working alone.

By tethering rams and thus curtailing their ability to seek out oestrus ewes Inkster (1957) demonstrated that the ewe can play an active part in finding a male with which she can mate. Lindsay (1965b) showed that when a number of ewes are in oestrus simultaneously a number of loosely defined harems develop around individual males and it is to these that the ewes move at oestrus. The most clearly formed harems are around the most dominant rams (Mattner *et al.* 1967). There may often be competition among ewes for the ram (Hulet *et al.* 1962b; Banks, 1964) but this seldom leads to the exclusion of individual females unless their oestrus period is very short. The attractiveness of ewes to rams changes during oestrus (Bell, 1974) and rams are stimulated by new oestrus ewes (Pepelko and Clegg, 1965) so that the dominant competitive advantage of certain ewes is probably overridden by the ram who chooses those most attractive to him. An exception appears to be the case of maiden ewes in competition with old females (Lindsay and Robinson, 1961). The oestrus of maiden ewes is shorter and they are subordinate to the older ewes (Lambourne, 1956).

The ram's contribution to partner seeking behaviour has been shown in tethering experiments not to be indispensible but despite harem formation and migratory activity of ewes the ram probably initiates most contact. Lindsay (1966) showed that once a ewe found a ram she generally remained with him during oestrus even though under some experimental conditions the ram was not permitted to mate with her. On the other hand, Knight and Lindsay (1973) showed that under extensive flock conditions many ewes mated with several rams during joining indicating that rams actively seek new partners when allowed to do so.

How Important is Mating Behaviour to the Production of Lambs?

The overall contribution of variation in mating behaviour patterns and interactions to reproductive performance is generally speculative. However, there are a number of areas currently under study where mating behaviour has a clear influence on reproductive performance.

Mattner, Braden and George (1971) and Walkley and Barber (1976) have shown that in certain cases a high proportion of rams, especially young rams, show no sexual activity at all. They suggest that by using a pen libido test such rams can be eliminated before joining. However, some rams, particularly among those which are seldom handled, do not work in the unfamiliar surroundings of a pen (T.W. Knight and C.M. Oldham, unpublished data) and could be mistakenly culled. A more satisfactory test where rams are not accustomed to handling is to coat the rumps of oestrus ewes with copious quantities of raddle and to turn suspect rams out with them overnight in an open, familiar area. Rams whose briskets are unmarked by raddle after 12-24 hours can be considered to be non-workers.

The overall importance of libido of rams in flocks is equivocal. The average joining proportion in Australia is one ram per 40-50 ewes and at this proportion there is little evidence that reproduction has been influenced by lack of ability to cover the available ewes. In New Zealand Haughey (1959) and Allison (1975) have shown that with proportions as low as one ram to 250 ewes failure to cover the ewes was still not a problem. On the other hand, Fowler (personal communication) suggests that with Merino sheep ratios of less than one ram per 100 ewes will result in insufficient coverage. Mattner *et al.* (1967) don't agree with this view but individual rams vary greatly in libido and the conflicting results may reflect this. When the sperm output of individual rams is increased by high protein supplements (Oldham *et al.*, 1976) allowing a smaller proportion of rams to be joined, the libido of the rams may become a limiting factor. This may be particularly true with young rams which mate significantly less frequently than older rams (Croker and Lindsay, 1970). Resolution of the minimum, safe ratio of rams to ewes which will result in full coverage of the ewes is therefore essential. Under commercial conditions at least 3-4 rams need to be joined per flock to compensate for variability among individuals. To be meaningful experiments into ram/ewe ratios should include ram groups of this size.

The mating behaviour of ewes can be significantly modified by the influence of rams. Schinckel (1954) showed that the sudden introduction of rams before the normal breeding season will induce ewes to commence cycling. In British breeds and Merino cross-bred sheep this advancement of the breeding season seldom exceeds 6 weeks but C.M. Oldham (personal communication) has shown that in the Merino, in which unstimulated flock-breeding activity extends only from late January to May, sudden introduction of rams at any time of the year will stimulate breeding activity in 100% of the flock. Ewes ovulate within a few days of the introduction of the rams but since in most cases this is a silent heat (Oldham *et al.* 1976) mating activity does not begin until the next cycle, or about 20 days after the rams have been joined. Thus in Merino sheep which are bred out-of-season vasectomized teaser rams joined for 2 weeks before the entire rams will lead to a concentrated mating and, subsequently, lambing.

Rams also influence the length of the oestrus period. Ewes remain on heat for longer if rams are absent or present only intermittently than if rams are constantly present (Parsons and Hunter, 1967; Fletcher and Lindsay, 1971). In addition, the presence of rams advances the time of the surge of LH relative to the commencement of oestrus and the time of ovulation (Lindsay *et al.* 1975; Signoret, 1976). Thus in the absence of rams ovulation can be delayed for as long as 12 hours, a mechanism which gives the ewe greater opportunity to find a ram.

In artificial insemination programmes the most common behavioural problem is that of missed cycles. Where oestrus has been artificially synchronized this problem may be due to slight hormonal imbalances, discussed above, or to the fact that all ewes are on heat over a very short period. A very high proportion of vasectomized teaser rams relative to that necessary under conditions of random oestrus would be necessary to cover all available ewes. However, missed cycles are often a problem in artificial insemination programmes even when oestrus is natural and random. In this case the problem has often been ascribed to the stress of mustering and drafting sheep daily. Although stress has been shown under certain conditions to induce ovulation (Braden and Moule, 1964; Lang, 1964), there is no information on its effects on oestrus, so this supposition remains unverified.

REFERENCES

Allison, A.J. (1975) *New Zealand Journal of Agriculture Research* **18** : 1.
Banks, E.M. (1964) *Behaviour* **23** : 249.
Bell, J. (1974) Honours Thesis, University of Western Australia.
Braden, A.W.H. and Moule, G.R. (1964) *Australian Journal of Agricultural Research* **15** : 937.
Croker, K.P. and Lindsay, D.R. (1970) *Journal of Agriculture, Western Australia* **11** : 189.
Edgar, D.G. (1961) Proceedings of the Ruakura Farmers Conference Week 6.
Fletcher, I.C., Allison, A.J. and Lindsay, D.R. (1971) *Journal of Endocrinology* **50** : 539.
Fletcher, I.C. and Lindsay, D.R. (1971) *Journal of Reproduction and Fertility* **25** : 253.
Hafez, E.S.E. (1951) *Nature, London* **167** : 777.
Haughey, K.G. (1959) Sheepfarming Annual, New Zealand, p.17.
Hulet, C.V., Ercanbrack, S.K., Blackwell, R.L., Price, D.A. and Wilson, L.O. (1962a). *Journal of Animal Science* **21** : 865.
Hulet, C.V., Blackwell, R.L., Ercanbrack, S.K., Price, D.A. and Wilson, L.O. (1962b) *Journal of Animal Science* **21** : 870.

Inkster, I.J. (1957) New Zealand Sheep Farming Annual, p.163.

Knight, T.W. (1967) Honours Thesis, University of Western Australia.

Knight, T.W. (1973) *Australian Journal of Agricultural Research* **24** : 573.

Knight, T.W. and Lindsay, D.R. (1973) *Australian Journal of Agricultural Research* **24** : 579.

Lambourne, L.J. (1956) Proceedings of the Ruakura Farmers Conference Week, p.16.

Lang, D.R. (1964) Proceedings of the Australian Society of Animal Production **5** : 53.

Lindsay, D.R. (1965a) *Animal Behaviour* **13** : 75.

Lindsay, D.R. (1965b) *Animal Behaviour* **14** : 73.

Lindsay, D.R. (1966) *Animal Behaviour* **14** : 419.

Lindsay, D.R., Cognie, Y., Pelletier, J. and Signoret, J.P. (1975) *Physiology and Behaviour* **15** : 423.

Lindsay, D.R., Dunsmore, D., Williams, J. and Syme, G. (1976) *Animal Behaviour* **24**: 818

Lindsay, D.R. and Ellsmore, J. (1968) *Australian Journal Experimental Agricultural and Animal Husbandry* **8** : 649.

Lindsay, D.R. and Fletcher, I.C. (1972) *Animal Behaviour* **20** : 452.

Lindsay, D.R., Gherardi, P.B. and Oldham, C.M. (1976) Proceedings of International Congress on Sheep Breeding, Western Australian Institute of Technology, Perth.

Lindsay, D.R. and Robinson, T.J. (1961) *Journal of Agricultural Science* **55** : 141.

Lindsay, D.R. and Scaramuzzi (1969) *Journal of Endocrinology* **45** : 549.

Mattner, P.E. and Braden, A.W.H. (1975) *Australian Journal of Experimental Agriculture and Animal Husbandry* **15** : 330.

Mattner, P.E., Braden, A.W.H. and George, J.M. (1971) *Australian Journal of Experimental Agriculture and Animal Husbandry* **11** : 473.

Mattner, P.E., Braden, A.W.H. and Turnbull, K.E. (1967) *Australian Journal of Experimental Agriculture and Animal Husbandry* **7** : 103.

McKenzie, F.F. and Terrill, C.E. (1937) Research Bulletin, Missouri Agricultural Experiment Station, No. 264.

Oldham, C.M., Knight, T.W. and Lindsay, D.R. (1976) *Proceedings of the Australian Society of Animal Production* **11** : 129.

Oldham, C.M., Adams, N.R., Lindsay, D.R. and MacKintosh, J.B. (1976). Unpublished data.

Parsons, S.D. and Hunter, G.L. (1967) *Journal of Reproduction and Fertility* **14** : 61.

Pepelko, W.E. and Clegg, M.T. (1965) *Animal Behaviour* **13** : 249.

Radford, H.M., Watson, R.H. and Wood, G.F. (1960) *Australian Veterinary Journal* **36** : 57.

Robinson, T.J. (1954) *Endocrinology* **55** : 403.

Scaramuzzi, R.J. (1968) Ph.D. Thesis, University of Sydney.

Scaramuzzi, R.J., Lindsay, D.R. and Shelton, J.N. (1971) *Journal of Endocrinology* **50** : 345.

Schinckel, P.G. (1954) *Australian Journal of Agricultural Research* **5** : 465.

Signoret, J.P. (1976) *Journal of Endocrinology* **64** : 589.

Walkley, J.R.W. and Barber, A.A. (1976) Proceedings of the Australian Society of Animal Production **11** : 141.

55

BEHAVIOURAL PRINCIPLES AS A FOUNDATION FOR SHEEP MANAGEMENT

V.R. SQUIRES
CSIRO, Deniliquin, N.S.W., Australia

SUMMARY

Within the past 10 years there has been a concerted effort to understand the complex subject of behaviour of sheep. Interest has focused upon those aspects of behaviour which have most relevance to sheep raising under commercial conditions, both extensive and intensive.

Behavioural adaptation to a wide variety of environmental and nutritional regimes, which is a common feature among free-ranging sheep, influences reproductive success and lamb survival both of which are matters of considerable commercial concern. In the past the majority of studies on sheep behaviour have been made with few animals under confined conditions and the value of these findings to the behaviour of free-ranging livestock or of intensively managed systems needs to be critically evaluated.

It is shown, from a review of studies on sheep behaviour and ecology, that a knowledge of behaviour is a firm basis for sheep management. Of particular relevance are studies on reproductive behaviour and breeding, grazing behaviour and pasture utilization, defaecating behaviour and parasite control, animal learning, handling and restraint, and indexes of normal behaviour in relation to signs of abnormality associated with disease.

INTRODUCTION

Animal behaviour or ethology is concerned with the observable characteristics of the movements and expressions of animals and the underlying neurophysiologic processes. But our interest is not only with central nervous system mechanisms controlling social behaviour but also with genetically controlled or modified aspects of behaviour and with experiences of the animal early in life which may affect subsequent behaviour. Behaviour is a product—the phenotype if you wish—of a series of environment interactions to which the genotype is exposed during its development.

Ethologists first of all, study a particular animal in its normal activities either alone or interacting with its conspecifics, and all other factors of its environment which may effect the animal. From the study of a single animal one proceeds logically to the study of groups of animals, and from groups to populations and species so that the species-typical behaviour may be described and understood. The total inventory of a species' behaviour repertoire is called an 'ethogram'. We are now only just beginning to piece together an ethogram for the domestic sheep—probably one of the earliest animal species domesticated by man! The importance of having a complete ethogram is so that animal husbandry can work with, rather than against, the species-specific responses.

A sound understanding of the principles of animal behaviour has the most relevance to reproductive behaviour and breeding, grazing behaviour and pasture utilization, defaecating behaviour and parasite control, animal learning, training, handling and restraint, and indexes of normal behaviour in relation to signs of abnormality associated with disease.

REPRODUCTIVE BEHAVIOUR AND BREEDING

Detailed analysis of sexual behaviour in sheep has been until fairly recently confined almost entirely to pen experiments because of the difficulty of obtaining precise information under field conditions. It is difficult to predict from the results of experiments carried out under such confined surroundings what might happen under more extensive conditions, e.g. the dispersal of animals may well influence reproductive efficiency (Nicholls, Fowler and Kearins 1976). Detailed studies of events during the establishment of the ewe-lamb bond (reviewed by Squires 1975a) are critical to the survival of the young and therefore the productivity of the enterprise.

Studies on mating behaviour (Mattner, Braden and Turnbull 1967; Fowler 1975) and on behaviour of ewes during parturition and the post-lambing period (Arnold and Morgan 1975) have added greatly to our knowledge of behaviour of sheep that are reproducing in the field under conditions which are identical to those in which commercial sheep and wool production is practised. For example two forms of ewe behaviour were identified which could directly affect lamb mortality. These were desertion of new-born lambs and pre-lambing maternal interest. Ewe behaviour may also influence lamb behaviour by denying the lamb's access to the udder or by delaying the time taken for the lamb to drink (Arnold and Morgan 1975). A number of other social and environmental factors were found to be important in both mating success and in preventing lamb mortality. A more detailed treatment of these factors can be found in Hafez and Lindsay (1965) and in Squires (1975a).

GRAZING BEHAVIOUR AND PASTURE UTILIZATION

The nutrition of grazing sheep is solely dependent on the animal's ability to gather its own food. Many factors can influence the quantity and quality of plant material eaten by grazing animals (Table 1) and these have been discussed by Arnold (1964). Satisfaction of appetite is determined by interactions of behavioural responses with characteristics of the plant community from which the diet has to be selected, and with environmental factors, particularly weather. Social interactions between flock members can often override other factors in determining choice of grazing area (Squires 1975b).

TABLE 1

SOME FACTORS INFLUENCING GRAZING BEHAVIOUR OF SHEEP

SEASON OF YEAR
BREED, AGE AND DENSITY OF ANIMALS
SOCIAL ORGANIZATION OF THE FLOCK
ENVIRONMENTAL FACTORS (WIND DIRECTION, RAINFALL, DAY LENGTH, TEMPERATURE AND RADIATION)
TYPE AND CONDITION OF VEGETATION
SIZE, SHAPE AND TOPOGRAPHY OF THE PADDOCK
NUMBER AND LOCATION OF WATERING POINTS
NUMBER AND LOCATION OF CAMPING AREAS
AVAILABILITY OF SHADE

DEFAECATING BEHAVIOUR AND PARASITE CONTROL

Strong social cohesion in a flock of sheep can lead to a massive build-up of faecal material at a few selected sites within a pasture (Hilder 1964). Faeces are the principal source of infection or re-infection by internal parasites of grazing animals and the avoidance of contaminated forage may have considerable adaptive significance since it has been shown that parasite larvae migrate only short distances from the faeces.

ANIMAL LEARNING AND RESTRAINT

Learning is an important factor in many facets of sheep behaviour. Animals organize their behaviour through the process of learning and behave according to what they have learned by prior experience. The role of early learning among farm animals is a neglected field. Sheep mortality has been shown to increase when there was no opportunity for lambs to learn diet selection from their dams. Cultural transfer (learned behaviour) from ewes to lambs through successive generations has been shown to be a factor in home range formation among hill country sheep (Hunter 1964). Many learned behaviours are established early in life. 'Critical periods' for learning may be experienced in certain species as the phenomenon of 'imprinting' indicates (Kilgour 1973). The early environment provided by the mother may affect the later behaviour of the offspring. For sheep at pasture the mother-offspring relationship develops freely. But in many intensively managed systems mother and young are often separated shortly after birth. While these practices may be helpful for maximization of individual productivity they may have detrimental effects on the development of normal social relationships.

Handling sheep, both at pasture and in yards, requires knowledge of behaviour. Procedures involved in yarding, transport, and processing sheep at abattoirs could benefit greatly from application of behavioural principles (Kilgour 1971).

ABNORMAL BEHAVIOUR

Without a basic knowledge of normal behaviour, the veterinarian or husbandman is, on many occasions, in a difficult position to make a diagnosis or prognosis. Abnormal behaviour patterns give the veterinarian the basic clinical signs of abnormality (due to disease, trauma or emotional disturbance) upon which he can base a diagnosis. However, without a baseline of what normality is, in terms of overt behavioural responses, his diagnosis will be more difficult to make or at least more laborious and protracted.

The behaviour of animals may be an important factor in the success of intensive husbandry systems. Highly intensive units are characterized by severe restrictions of animal movement, their monotonous environment and a general lack of sensory stimulation of animals. Despite a number of recent reviews on animal adjustment to intensive conditions (see Kilgour 1974) there is urgent need for more information on behaviour under intensive conditions. In general there is a paucity of behavioural information available and indoor husbandry decisions must be made on a trial and error basis.

CONCLUSIONS

The application of ethological principles in animal husbandry can suggest more appropriate ways of keeping domesticated animals and thus, contribute to economic gain. There are many problems in sheep husbandry which could benefit from ethological studies. These break down into four major, economically pertinent categories. First, fertility, reproductive competence and maternal adequacy may be influenced by social and environmental factors such as crowding, stress, and relative social deprivation. Second, productivity in terms of daily weight gain may be influenced by social and environmental factors. Third, the genetic analysis of the inheritance of desirable and undesirable traits and their phenotypic development under different rearing and maintenance regimes should be undertaken. Finally, having searched for and identified those factors that influence productivity the interaction effects of strain or genotype with animal husbandry methods (the "artificial ecology" of farm animals) that may increase or decrease susceptibility to psycho-physiological stress, infectious and metabolic disease should be explored.

REFERENCES

Arnold, G.W. (1964) in: Grazing and Marine and Terrestrial Environments (Ed. D.J. Crisp). Blackwells: Oxford. P.133.
Arnold, G.W. and Morgan, P.D. (1975). *Applied Animal Ethology* **2** : 25.
Fowler, D.G. (1975) *Applied Animal Ethology* **1** : 357.
Hafez, S.E. and Lindsay, D.R. (1965) *Animal Breeding Abstracts* **33** : 1.
Hilder, E.J. (1964) *Proceedings of the Australian Society of Animal Production* **5** : 241.
Hunter, R.F. (1964) in: Grazing in Marine and Terrestrial Environments (Ed. D.J. Crisp). Blackwells: Oxford. p.155.
Kilgour, R. (1971) Thirteenth New Zealand Meat Industry Research Conference Proceedings, p.9.
Kilgour, R. (1973) *New Zealand Journal of Agriculture* **126** : 18.
Kilgour, R. (1974) *Proceedings of the Australian Society of Animal Production* **10** : 286.
Mattner, P.E., Braden, A.W.H. and Turnbull, K.E. (1967) *Australian Journal of Experimental Agriculture and Animal Husbandry* **7** : 103.
Nicholls, P.J., Fowler, D.G. and Kearins, R.D. (1976) *Proceedings of the Australian Society of Animal Production* **11** : 2P.
Squires, V.R. (1975a) *Mammal Review* **5** : 35.
Squires, V.R. (1975b) *Applied Animal Ethology* **1** : 177.

56

SEXUAL ACTIVITY IN MERINO RAMS

I.C. FLETCHER

Agricultural College, Roseworthy, South Australia

SUMMARY

Investigations were carried out with South Australian strong wool Merino rams to assess whether sexual inhibition in young rams is affected by previous contact with ewes, and to determine whether variation in measured sexual activity is related to variation in the number of ewes lambing to single-sire matings.

Sexual activity was recorded for a total of 215 1½ year old rams. Seven rams (3.3%) were unable to achieve ejaculation during three 20-minute periods of contact with oestrous ewes. Only one of 92 rams (1.1%) allowed prior contact with oestrous ewes was sexually inhibited, compared with six of 123 rams (4.9%) which had no previous contact with ewes.

Nineteen rams, with records of sexual activity ranging between 3 and 18 ejaculations in one hour, were mated singly to groups of about 30 ewes for a six-week period. There was no significant relationship between number of ejaculations and number of ewes which lambed. In a further field test, there was no significant relationship between number of ejaculations and number of ewes mated with ewe/ram ratios equivalent to 2.6 and 0.9 rams per 100 ewes.

The implications of these results on the management of young rams and their selection for use are briefly discussed.

INTRODUCTION

Although sexual inhibition and variation in sexual activity have been recorded among a number of different breeds and strains of rams in several countries, at least two aspects of sexual activity have not been resolved.

Mattner, Braden and George (1973) reviewed evidence that the rearing of males in homosexual groups may affect sexual behaviour patterns in sheep and other animals, but any affect of rearing rams in the presence of ewes has not been reported. Among sexually active rams, Wiggins, Terrill and Emik (1953) reported that sexual activity measured by number of ejaculations was correlated with the percentage of ewes which lambed, but Hulet *et. al.* (1962) and Walkley and Barber (1976) found no such relationship.

Investigations of these two aspects of sexual activity in rams are presented in this chapter.

MATERIALS AND METHODS

(a) Experimental Animals

All rams were taken from the "Collrose" Merino stud at Roseworthy Agricultural College, South Australia. The stud is based on the "Bungaree" type of South Australian strong wool Merino, and has been closed to outside introduction since 1953.

(b) Effects of Previous Contact with Oestrous Ewes on Sexual Inhibition in Young Rams

In 1973, 99 rams of mean age 45 weeks were allotted at random, after stratification on the basis of body weight, into three groups of equal size. Rams in one group were taken from the paddock on three separate occasions and allowed contact with oestrous ewes for 24-hour periods at mean age 46, 53 and 59 weeks. At these times rams were individually penned with a single ewe in oestrus but were changed between pens so that they had contact with three or four different ewes during each 24-hour period. The second group of rams were taken from the paddock and put into pens, but without oestrous ewes, at the same times as rams in the first group were allowed contact with ewes. This second treatment was included to account for any effect of penning *per se* on subsequent measurements of sexual activity. The third group of rams remained in the paddock except for routine husbandry operations.

In 1975, 120 rams were allotted at random, after stratification on the basis of body weight, into two groups of equal size. Rams in one group were allowed contact with oestrous ewes, as in 1973, for three 24-hour periods at mean age 46, 53 and 59 weeks. The other groups of rams remained in the paddock without ewe contact.

In both years, the sexual activity of all rams was recorded at a mean age of 76 weeks. Rams were put singly with a group of four or five oestrous ewes in a small pen (about 2 m × 5 m), and records were taken of the numbers of mounts and ejaculations achieved during a 20-minute period of observation. Rams were transferred to a second group of four or five oestrous ewes half-way through the observation, so that they were exposed to a total of nine or ten stimulus animals. Three such 20-minute periods of observation, spaced at weekly intervals, were made for each ram.

Spayed ewes were used as stimulus animals for these observations of sexual activity. Oestrous behaviour was induced by a series of six intramuscular injections of 20 mg progesterone, administered at two-day intervals, followed by a single intramuscular injection of 100 μg oestradiol benzoate administered 48 hours after the last injection of progesterone.

Rams found to be sexually inhibited at 76 weeks of age in November 1973 were tested again in April, August and December 1974.

(c)   Association between Sexual Activity Measured in Pens and Mating Activity in the Field

Nineteen rams, with measured sexual activity ranging between 3 and 18 ejaculations in one hour, were mated singly to groups of about 30 entire ewes for a six-week period during November and December. Six rams were used in 1973, and 13 in 1974. Semen samples taken from the rams before mating were all satisfactory in terms of volume, motility and density. Records were taken of ewes which mated and ewes which subsequently lambed.

In another test, three groups of four rams with low, medium, and high levels of sexual activity (7, 11 or 12, and 16 to 18 ejaculations in one hour respectively), were selected from 18 month old rams tested in 1973. Each group of four rams was put with nine spayed ewes in oestrus in a small (about 2 ha) paddock for a 24-hour period on three different occasions, and with 27 spayed ewes in oestrus for a 24-hour period on one further occasion. On the assumption that one-seventeenth of animals in a flock of unmated entire ewes will be in oestrus on any one day of the breeding season, the use of four rams with 9 and 27 oestrous ewes for 24-hour periods represent ram/ewe ratios of about 2.6 and 0.9 rams per 100 ewes respectively. Rams were fitted with harnesses and marking crayons (with a different crayon colour for each ram in the group), and the number of ewes mated by each ram was recorded from colour marks on the ewes' rumps at the end of each 24-hour period.

RESULTS

(a)   Effects of Previous Contact with Oestrous Ewes on Sexual Inhibition in Young Rams

The percentages of rams which were sexually inhibited (failed to ejaculate) at 76 weeks of age are shown in Table 1. There were no significant differences in sexual activity between rams which had been penned without ewe contact and rams which remained in the paddock in 1973, and results from these two groups have been pooled. For all rams showing sexual activity, the mean value and standard error for the total number of ejaculations in one hour was 11.3 ± 0.3, and the ratio of mounts to ejaculations was 2.4 ± 0.2.

The proportions of sexually inhibited rams did not vary significantly between the two years of observation. There was a lower level of sexual inhibition among rams which had previous contact with ewes in both years, but the difference was not significant. Sexually inhibited rams could not be distinguished from others on any visual appraisal or measurement of body weight or conformation, wool production, horn development, or testis size.

Of the four rams found to be sexually inhibited in November 1973, three were still inhibited in April and August 1974, and two in December 1974.

TABLE 1

PERCENTAGES OF SEXUALLY INHIBITED RAMS AT 76 WEEKS OF AGE

(Total ram numbers are shown in parentheses)

| Previous | Year | | |
Experience	1973	1975	Means
Contact with ewes	0 (33)	1.7 (59)	1.1 (92)
No contact with ewes	6.2 (64)	3.4 (59)	4.9 (123)
Means	4.1 (97)	2.5 (118)	3.3 (215)

Effect of ewe contact: $x^2 = 2.40$, df= 1, NS
Difference between years: $x^2 = 0.42$, df= 1, NS
Interaction: $x^2 = 0.57$, df= 1, NS

(b)   Association between Sexual Activity Measured in Pens and Mating Activity in the Field

All of the 19 rams put with entire ewes for a six-week period mated with all or all but one of their ewes. The association between number of ejaculations in one hour of pen observation and number of ewes which lambed after the six-week mating period is shown in Figure 1. The small positive correlation between number of ejaculations and arcsin root percentage of ewes lambing (r=0.32) was not significant. The correlation between ejaculations and arcsin root percentage of ewes lambing during the first 17 days (an estimate of ewes conceiving at their first oestrus of the mating period) was slightly higher (r=0.39), but still not significant.

The percentages of oestrous spayed ewes mated in the field by rams with varying levels of pen-measured sexual activity are shown in Table 2. The mean percentages of ewes mated by each ram were lower at a ram/ewe ratio equivalent to 0.9 rams per 100 ewes than at 2.6 rams per 100 ewes. However, differences between rams in pen-measured sexual activity were not reflected in the percentage of ewes mated, even when the ram/ewe ratio was equivalent to only 0.9 rams per 100 ewes.

DISCUSSION

A low level of sexual inhibition was a feature of the rams used in this investigation. Only about 5% of rams which had no previous contact with ewes were found to be inhibited, compared with values in the range 10-45% reported by Hulet, Blackwell and Ercanbrack (1964), Mattner, Braden and George (1973), and Walkley and Barber (1976). Also the number of ejaculations in one hour tended to be higher, and the ratio of mounts to

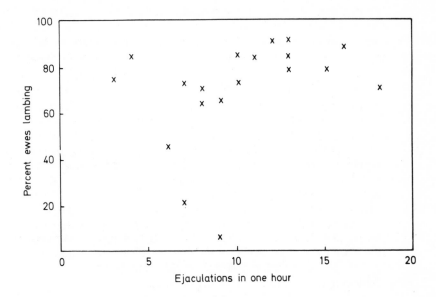

Figure 1—Relationship between ejaculations in one hour and percentage of ewes lambing after a six week mating period.

TABLE 2

PERCENTAGES OF EWES MATED BY RAMS WITH VARYING LEVELS OF PEN-MEASURED SEXUAL ACTIVITY

| Pen-measured sexual activity (no. ejaculations in 1 hour) | No. of rams used | Mean percentages of ewes mated with ram/ewe ratios equivalent to:— | | Means |
		2.6 rams /100 ewes	0.9 rams /100 ewes	
Low (7)	4	88	67	78
Medium (11 to 12)	4	90	60	76
High (16 to 18)	4	87	65	76
Means		88	64	76

ejaculations lower, than values recorded by other authors. Sexual activity may have been greater in this investigation because a high level of stimulus was provided by the use of small pens and ewes with a high intensity of oestrus and by changing ewe groups to reduce stimulus satiation. Further, Mattner, Braden and George (1973) have demonstrated differences between Merino rams of different strains or from different studs, and a low level of sexual inhibition may have been inherent among rams used in this study.

The low level of sexual inhibition among control rams made it difficult to demonstrate any effects of previous contact with ewes. Three separate 24-hour periods with oestrous ewes reduced the level of sexual inhibition from about 5% to only about 1%, but the effect was not statistically significant. Further investigation of the effects on sexual inhibition of rearing young rams with ewes does not seem warranted, except perhaps among flocks with high levels of inhibition. Even if the practice completely eliminated inhibition, the labour and management costs of providing ewe contact for individual young rams, as was done in this investigation, could be prohibitive. The more practicable procedure of providing ewe contact for groups of young rams is less likely to be effective, since sexually inhibited rams may have a naturally low order of dominance (Mattner, Braden and George, 1973), and competition between rams could prevent such animals from achieving ewe contact in a group situation.

The present observation that two or four inhibited rams subsequently developed sexual activity is meaningful only in relation to reports based on larger numbers of animals by Hulet, Blackwell and Ercanbrack (1964) and Mattner, Braden and George (1973), that 83 and 82% respectively of inhibited rams later became active. These results cast doubt on the need to test rams for sexual inhibition before they are used except for single-sire matings, or for multiple-sire matings if ewes are relatively infertile and there is a high proportion of inhibited rams (Mattner, Braden and George, 1973). Culling young rams on the basis of sexual inhibition would obviously be wasteful. Flock rams of no special merit might be culled for sexual inhibition, but more valuable stud animals could be retained with a reasonable expectation that they would subsequently become active.

There was no significant relationship between number of ejaculations measured in a pen test, and ewes lambing after a six-week mating period with a ram/ewe ratio of about 3%. Although lamb production was not measured at lower ram/ewe ratios, the number of ewes mated in the field was still not affected by ram ejaculation number when the ram/ewe ratio was reduced to about 1%. These results, and those of Walkley and Barber (1976), indicate that selecting highly active from less active rams would be of little benefit except, perhaps, if ram/ewe ratios were very low indeed. It is noted that the significant relationship between ejaculation number and per cent ewes lambing reported by Wiggins, Terrill and Emik (1953) was of too small a magnitude ($r = 0.081$) to be useful in predicting the mating performance of young rams from a prior test of sexual activity.

It is concluded that measurement of the sexual activity of young rams, or the development of management techniques to reduce sexual inhibition, cannot be easily justified. For single-sire matings, it would be more practicable to replace rams which show little or no sexual activity during the first one or two weeks of mating, than to test the activity of all rams before they are used. For multiple-sire matings, prior testing of ram activity would be appropriate only when there was evidence or inference that sexual inhibition was a cause of low lamb production. These conclusions would not necessarily be valid if rapid and inexpensive methods of reducing sexual inhibition and of assessing ram activity were available.

ACKNOWLEDGEMENTS

I am grateful to Mr R.E. Brady for technical assistance, and to Dr D.E. Taplin for assistance with experimental work and criticism of the manuscript.

REFERENCES

Hulet, C.V., Blackwell, R.L. and Ercanbrack, S.K. (1964) *Journal of Animal Science* **23** : 1095.

Hulet, C.V., Ercanbrack, S.K., Price, D.A., Blackwell, R.L. and Wilson, L.O. (1962) *Journal of Animal Science* **21** : 857.

Mattner, P.E., Braden, A.W.H. and George, J.M. (1973) *Australian Journal of Experimental Agriculture and Animal Husbandry* **13** : 35.

Walkley, J.R.W. and Barber, A.A. (1976) *Proceedings of the Australian Society of Animal Production* **11** : 141.

Wiggins, E.L., Terrill, C.E. and Emik, L.O. (1953) *Journal of Animal Science* **12** : 684.

57

SEMEN QUALITY AND QUANTITY IN THE RAM

M. COUROT

I.N.R.A.—Station de Physiologie de la Reproduction, Nouzilly, France

SUMMARY

Fertility in the sheep depends on sperm production of the ram as a large number of spermatozoa is a prerequisite for successful breeding. Sperm production varies with testicular development. In the young ram-lamb, the first spermatozoa can be collected around 5-6 months of age. Sperm production increases in quantity and quality with age and ejaculates consistent with normal fertility are produced around 7.5 to 9 months of age according to breed. Adult rams are permanent sperm producers throughout the year but with seasonal variations in sperm quantity and less evident variations in quality. Increasing photoperiod decreases sperm production from the testis in spring but fertility can be maintained provided a sufficient number of spermatozoa of high quality is delivered into the ewes. However, there are some problems with semen deep frozen in spring. Breeding management must take into account the number of ewes joined per ram and the preparation of rams for the breeding period. In good conditions, sheep breeding may be successful throughout the year.

INTRODUCTION

Fertility in the sheep largely depends on sperm production of the ram as a large number of spermatozoa of high quality is a prerequisite for successful natural breeding or artificial insemination. According to the physiological conditions of the ewe, for example natural or induced oestrus, dry or nursing ewe, 250 to 1000×10^6 normal spermatozoa per female are required to have a good chance of fertilization. As the ram normally delivers 3 to 5×10^9 spermatozoa per ejaculate (when not bred too frequently), the number of females which can be served by a ram in natural breeding per day, or inseminated with one ejaculate is fairly low. Moreover, as the ram is a seasonal breeder, he is a low sperm-producer in the non-breeding season. Thus, the problem of sperm quantity and quality in the ram is not a theoretical one, and tends to become critical with the development of intensive sheep farming.

The number and the quality of spermatozoa available for reproduction vary with the efficiency of spermatogenesis (sperm production in the testis), the quality of maturation of sperm in the epididymis and lastly with the frequency of ejaculation, the latter point being largely influenced by sexual activity and breeding management.

The assessment of quantitative sperm production has been thoroughly reviewed by Amann (1970): quantitative measurement of ejaculated semen, measurement of epididymal or gonadal sperm reserves, quantitative histological analysis of the seminiferous epithelium . . . are quite classic tests and need no complementary development. On the other hand, semen quality is much more difficult to assess because of the variability of age and quality of spermatozoa in a semen sample. It can theoretically be defined for ejaculated sperm by its fertilizing ability and the capacity of fertilized eggs to develop as normal embryos. The practical problem is to predict the quality of semen by *in vitro* tests which are indicative of this quality, i.e. significantly correlated with fertility. Several tests have been developed; they are based on (1) the morphological integrity of the sperm cells, head, acrosome, intermediate piece (cytoplasmic droplet) or flagellum either under the light or electron microscope, (2) the viability of cells, motility of raw semen, percentage of motile sperm and motility following dilution of semen, or at thawing after storage of deep frozen semen, maintenance of motility and per cent motile sperm during *in vitro* incubation, percentage of live or dead sperm, and (3) the metabolic activity of spermatozoa, respiration, dehydrogenase activity or fructolysis. The most reliable tests are motility and per cent of motile sperm which are used for selection of the ejaculates to be used as liquid semen or stored as deep frozen semen; in the latter case, survival tests after incubation would help to select the best doses for AI.

In this chapter those factors which affect sperm production in the ram: age, season and general management will be considered.

AGE AND SPERM PRODUCTION

Sperm production is quantitatively dependent on the testicular development. Indeed a positive relationship has been observed between testicular weight, gonadal and extragonadal sperm reserves and sperm production (see review by Amann, 1970).

(a)   Passage of the first spermatozoa

After a period of slow development during 2 to 3 months after birth, the testes of the male lamb develop very rapidly from 3 to 4 months onward (Watson, Sapsford and McCance, 1956; Courot, 1962; Skinner *et al.*, 1968) independently of the season of birth (Courot, de Reviers and Pelletier, 1975). This accelerated gonadal growth is concomitant with inception of spermatogenesis (Courot, 1962). Due to the duration of the cellular processes (Ortavant, 1958) and cellular degenerations which occur in the first spermatogenetic cycles (Courot, 1962), it needs 2 to 3 months to obtain complete spermatogenesis and the release of the first sperm cells. With two more weeks for epididymal transit (Amir and Ortavant, 1968) first spermatozoa can thus be collected around 5 to 6 months of age in Ile-de-France, Prealpes and Romanov ram-lambs (Colas and Zinszner-Pflimlin, 1975; Colas *et al.*, 1975). The age at which first spermatozoa are collected varies with breeds (Watson, Sapsford and McCance, 1956; Louw and Joubert, 1964; Skinner and Rowson, 1968; Colas and Zinszner-Pflimlin, 1975) within breeds, and with body growth. Body growth is more closely correlated to testicular weight than chronological age (Watson, Sapsford and McCance, 1956; Courot, 1962; Skinner *et al.*, 1968; Dyrmundsson and Lees, 1972). Puberty is attained when ram-lambs reach 40-45% of the adult body weight.

(b) Sperm production after puberty

(i) Semen quantity

Testicular growth continues for a long time after puberty but it is seasonally regulated by photoperiodism (Fig. 1; Alberio, 1976) and varies with breed (Land and Carr, 1975; Carr and Land, 1975).

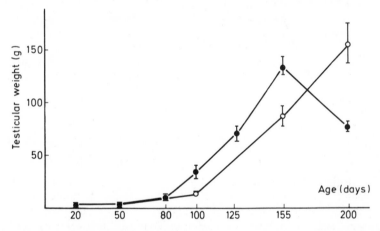

Figure 1 — Testicular development in male lambs born in September (black circles) or February (open circles)

(From Courot, de Reviers and Pelletier 1975)

Thus, sperm production increases after puberty, being also seasonally modulated (Colas *et al.*, 1975; Alberio, 1976). However, quantitative sperm output changes with the sexual behaviour of ram-lambs and the frequency of ejaculations. If one wants to obtain the maximum spermatozoa, the frequency of collection of semen needs to be adapted to the capacity of the animals: 2×1 ejaculate per week from 6 to 10 months, 2×2 from 11 to 17 months and 3×2 from 18 months onward have proved satisfactory with Romanov ram-lambs (Colas *et al.*, 1975). With Ile-de-France and

Prealpes lambs, the same rhythm of 2×1 ejaculate per week between 6 and 11 months showed differences between breeds in total sperm per ejaculate with higher production in the Prealpes than in the Ile-de-France (Fig. 2) especially during the first months of collection. Individual differences within breeds have also been shown which suggest the possibility of selecting animals on their early maturity and perhaps on their sperm productivity in the adult if the latter is related to the performance at a young age as it seems to be in the bull (Hahn, Foote and Seidel, 1969) and in the male goat (Corteel, 1975).

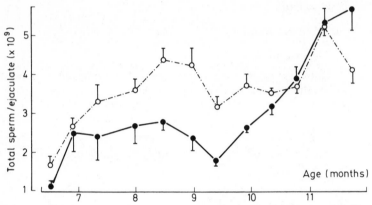

Figure 2 — Sperm production in ram lambs of the Ile-de-France (full line) and Prealpes (dotted line) breeds.
2 x 1 ejaculate/week.
(From Colas and Zinszner-Pflimlin 1975)

Related to sexual experience, one should also mention that yearling lambs well trained for regular collections, at 5 to 6 months old, are better sperm producers (sperm output), when 1.5 years old, than animals of the same age which have never been collected from previously (Colas *et al.*, 1975). In any case, sperm production increases with age; 2.5-year-old rams give significantly more total spermatozoa per ejaculate than 1.5-year-old ram-lambs (Lightfoot, 1968). This increase in sperm production and the increased sexual experience with age would partly explain why older rams (3.5 to 5.5 years old) are more fertile than the 1.5 years old when joined to 50 ewes per ram, the fertility of the latter being higher if joined to 25 instead of 50 ewes (Lightfoot and Smith, 1968).

The increase in sperm production with age could be related with the multiplication of the so-called "reserve" stem spermatogonia (A_0) and their differentiation into "renewing" stem spermatogonia (A_1) which originate sperm formation through spermatogenetic cycles (Hochereau-de Reviers, Loir and Pelletier, 1976).

(ii) Semen quality

The sperm produced at puberty is of poor quality as motility is low, and abnormal and dead sperm are more numerous than in ejaculates of normal

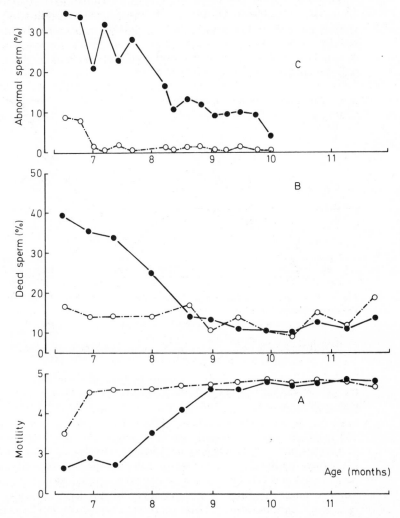

Figure 3 — Sperm quality in ram lambs of the Ile-de-France (full line) and Prealpes (dotted line) breeds.

A — Motility of raw semen

B — % dead sperm (after eosine staining)

C — % abnormal sperm

(From Colas and Zinszner-Pflimlin 1975)

adult rams. This is a general feature in all ram-lambs, but the evolution to a better quality varies with breeds: it is fairly rapid, 1 month, in Prealpes and slower, 3 months, in Ile-de-France lambs (Fig. 3).

This may result from the better efficiency of spermatogenesis with testicular development (Courot, 1962) and probably also from a better epididymal maturation which could be related to the increased androgen secretion which occurs with age (Cotta *et al.*, 1975). Even if spermatozoa can be collected as early as 6 months of age, the quality of ejaculates is only satisfactory at 7.5 months (Prealpes) and 9.5 months (Ile-de-France). Thus, lambs can be placed in progeny-testing programmes at these early ages. The quality of semen is still improving with age as shown by the fertilizing ability of semen from rams of one and two years old (Table 1)

TABLE 1

AGE DEPENDENCE OF FERTILITY IN THE RAM

Age (years)	Ile-de-France	Lacaune
< 1-1.5	45.5 (56)	54.0 (354)
> 2	66.2 (65)	61.5 (311)

2 A.I./ewe 48 and 60h after sponge removal and injection of PMS for oestrous synchronization — 500×10^6 spz./ewe. Results are given in % of ewes lambing; ( ) No. of ewes inseminated.

(Adapted from Colas and Zinszner-Pflimlin 1975.)

SEASON AND SPERM PRODUCTION

In European latitudes, rams are seasonal breeders with maximum sexual activity in summer and autumn (Rouger, 1974). But if animals are regularly trained to mate or to be collected from using an artificial vagina, it is possible to collect sperm each month of the year. Thus, complete spermatogenesis occurs continuously. However, seasonal variations are known in each step of the spermatogenetic processes (Hochereau-de Reviers, Loir and Pelletier, 1976) resulting in a lower and higher quantitative testicular sperm production in spring and autumn respectively (Ortavant, 1958). This is operated by the gradual changes in the photoperiod, the optimum being between 10 and 8 h light in a decreasing photoperiod (Ortavant, 1961). These changes modulate the hypothalamo-hypophyseal activity and the release of gonadotrophic hormones under the regulative influence of androgens secreted by the testis (Pelletier and Ortavant, 1975). Consequently epididymal sperm reserves fluctuate seasonally (Table 2) which will have repercussions only when the frequency of collection is high, 2×2 ejaculates weekly in the adult ram (Colas *et al.*, 1972).

Thus the total number of sperm per ejaculate decreases when daylight increases and *vice versa* (Amir and Volcani, 1965; Smith and Gordon, 1967).

The sensitivity of the ram to the photoperiod occurs early in life. At 24 weeks of age, Ile-de-France lambs stop testicular growth in adverse conditions, i.e. with an increasing photoperiod from 12 weeks of age, however when favourable light conditions occur again, the testes resume their development. The sperm output in 2×2 ejaculates weekly parallels the testicular development (Alberio, 1976).

Semen quality is less affected by seasonal variations than sperm quantity. Motility is indifferent to the light regime (Skinner and van Heerden, 1971; Colas *et al.*, 1972) but sensitive, with breed differences, to elevated temperatures (Lindsay, 1969). Elevated temperatures are also deleterious to spermatozoa: they induce morphological abnormalities and subsequently depress fertilization rate and increase embryonic mortality in the ewes fertilized with such semen (Rathore, 1968). The fertilizing ability of ram liquid semen does not change significantly with seasons (Kalev, Zagorski and Sertev, 1968; Colas, 1975). However, this latter result is an over-estimation because only the best ejaculates, according to the motility of spermatozoa, are used for artificial insemination. Indeed, there is a favourable relationship between semen traits and lambing performance (Goerke, Thrift and Dutt, 1970). In more severe conditions, some seasonal variations in semen quality have been shown: the fertilizing capacity of deep frozen semen is lower for semen frozen in spring than in autumn when both are used on oestrus synchronized ewes either in spring or in autumn with the same number of motile sperm per ewe (Colas and Brice, 1976). Thus, a small seasonal variation also exists in semen quality.

TABLE 2

EFFECT OF SEASON ON THE TESTICULAR AND EPIDIDYMAL SPERM RESERVES IN THE RAM

Number of sperm	Testis ($\times 10^9$)	Cauda epididymis ($\times 10^9$)
Spring	19.6 ± 1.5	37 ± 5.9
Autumn	23.7 ± 1.4	63 ± 5.1

(From Ortavant 1958.)

MANAGEMENT AND SPERM PRODUCTION

Sperm is produced continuously throughout the year, whatever the sexual activity of the ram. When animals do not mate or are not collected from, their spermatozoa are normally voided in the urine (Lino, Braden and Turnbull, 1967). Spermatozoa available for ejaculation come from the ampullae and the distal cauda epididymis; stocks are limited for a given day, thus allowing only a limited number of sperm cells to be released at ejaculation. So the total sperm per ejaculate is decreased by increasing the frequency of collection (Bielanski and Wierzbowski, 1961; Salamon, 1962). This is of importance in AI and also in natural mating because after repeated copulations the number of spermatozoa delivered into the female tract could decrease below the level consistent with normal fertility (Lightfoot, 1968) especially in oestrus induced ewes where the required number of sperm are particularly elevated. Indeed a decrease in the total number of sperm per ejaculate has been observed at the beginning of the breeding period (Mattner and Braden, 1967; Lightfoot, 1968) and a negative correlation has been shown between number of sperm per ejaculate and an increase in the number of females joined each night from 2 to 5 (Schäffer and Matter, 1966) and better fertility was obtained by lowering the ratio of ewes per ram (Lightfoot and Smith, 1968). On the other hand, it needs to be stressed that high ejaculation frequency does not modify sperm quality (Kastyak, 1962; Salamon, 1962) and that if it does result in a lower fertility, it is due only to the delivery of too small a number of spermatozoa into the ewe.

Sperm output can be enlarged by a suitable sexual preparation (false mount, change of teaser ewe . . .) before collection of semen (Petkov, 1969). Pharmacological treatment with oxytocin can also be used for increasing sperm output (Ewy, Bielansky and Zapletal, 1963; Knight, 1974; Voglmayr, 1975). However, one needs to be careful as long term effects of this drug would be harmful to spermatogenesis of the ram (Knight and Lindsay, 1970). A practical way of increasing sperm production is to prepare the rams before the breeding period by a kind of male flushing (Salamon, 1964) having in mind the duration of the spermatogenetic processes and transit being around two months (Ortavant, 1958; Amir and Ortavant, 1968). On the other hand, as spermatogenesis and epididymal maturation are sensitive to elevated temperature (Dutt and Hamm, 1957; Waites and Ortavant, 1968; Rathore, 1968), it is important to maintain rams in a cool environment as far as possible.

CONCLUSION

The ram is a permanent sperm producer whose semen can be used throughout the year provided animals are trained to breed or to be collected from. Careful control of liquid semen quality allows for a good fertility every month of the year. However, some seasonality is observed in sperm quality when using deep frozen semen, with a lower fertility with semen frozen in spring. Ram-lambs are able to ejaculate around 6 months old, but a further 2-2.5 months are necessary for the quality of semen to improve to warrant using them. Thus, young rams can be selected as sperm producers very early and used in progeny testing programmes during their first year thus allowing a fairly rapid genetical improvement. With adult rams utilized as natural breeders, care must be taken of the number of ewes joined per day and/or per week in order to avoid exhaustion. Except for photoperiod control and flushing the male, there is no available practical means of increasing sperm production in the ram.

ACKNOWLEDGEMENT

The skilful assistance of Mr. L. Cahill for the English correction is gratefully acknowledged.

REFERENCES

Alberio, R. (1976) Thesis, University of Paris VI.
Amann, R.P. (1970) "The Testis" (Academic Press: N.Y.) **1** : 433.
Amir, D. and Ortavant, R. (1968) *Annales de Biologie Animale, Biochimie, Biophysique* **8** : 195.
Amir, D. and Volcani, R. (1965) *Journal of Agricultural Science, Cambridge* **64** : 121.
Bielanski, W. and Wierzbowski, S. (1961) *VIth International Congress of Reproduction and Artificial Insemination,* The Hague **2** : 274.
Carr, W.R. and Land, R.B. (1975) *Journal of Reproduction and Fertility* **42** : 325.
Colas, G. (1975) *Annales de Biologie Animale, Biochimie, Biophysique* **15** : 317.
Colas, G. and Brice, G. (1976) *VIIIth International Congress of Reproduction and Artificial Insemination,* Krakow Vol. IV: 977.
Colas, G., Laszczka, A., Brice, G. and Ortavant, R. (1972) *Acta Agraria et Silvestria, series Zootechnica* **12** : 3.
Colas, G., Personnic, D., Courot, M. and Ortavant, R. (1975) *Annales de Zootechnie* **24** : 189.
Colas, G. and Zinszner-Pflimlin, F. (1975) *Journées de la Recherche Ovine et Caprine en France* 235.
Corteel, J.M. (1975) *Journées de la Recherche Ovine et Caprine en France* 4.
Cotta, Y., Terqui, M., Pelletier, J. and Courot, M. (1975) *Compte-rendus de l'Académie des Sciences, Paris* **280** : 1473.
Courot, M. (1962) *Annales de Biologie Animale, Biochimie, Biophysique* **2** : 25.
Courot, M., de Reviers, M.M. and Pelletier, J. (1975) *Annales de Biologie Animale, Biochimie, Biophysique* **15** : 509.
Dutt, R.H. and Hamm, P.T. (1957) *Journal of Animal Science* **16** : 328.
Dyrmundsson, D.R. and Lees, J.L. (1972) *Journal of Agricultural Science, Cambridge* **79** : 83.
Ewy, Z., Bielanski, W. and Zapletal, Z. (1963) *Bulletin de l'Acad*émie Polonaise des Sciences **11** : 145.
Goerke, T.P., Thrift, F.A. and Dutt, R.H. (1970) *Journal of Animal Science* **31** : 445.
Hahn, J., Foote, R.H. and Seidel, G.E. (1969) *Journal of Dairy Science* **52** : 1843.
Hochereau-de Reviers, M.T., Loir, M. and Pelletier, J. (1976) *Journal of Reproduction and Fertility* **46** : 203.
Kalev, G., Zagorski, D. and Sertev, M. (1968) *VIth International Congress of Animal Reproduction and Artificial Insemination,* Munich **1** : 289.
Kastyak, L. (1962) *Zeszyty Nauk W.S.R. (Olsztyn)* **12** : 67.
Knight, T.W. (1974) *Journal of Reproduction and Fertility* **39** : 329.
Knight, T.W. and Lindsay, D.R. (1970) *Journal of Reproduction and Fertility* **21** : 523.
Land, R.B. and Carr, W.R. (1975) *Journal of Reproduction and Fertility* **45** : 495.
Lightfoot, R.J. (1968) *Australian Journal of Agricultural Research* **19** : 1043.
Lightfoot, R.J. and Smith, J.A.C. (1968) *Australian Journal of Agricultural Research* **19** : 1029.
Lindsay, D.R. (1969) *Journal of Reproduction and Fertility* **18** : 1.
Lino, B.F., Braden, A.W.H. and Turnbull, K.E. (1967) *Nature* **213** : 594.
Louw, D.F.J. and Joubert, D.M. (1964) *South African Journal of Agricultural Science* **7** : 509.
Mattner, P.E. and Braden, A.W.H. (1967) *Australian Journal of Experimental Agriculture and Animal Husbandry* **7** : 110.

Ortavant, R. (1958) Thesis, University of Paris. Ed. *Annales de Zootechnie* **8** : 183 and 271 (1959).

Ortavant, R. (1961) *IVth International Congress of Reproduction and Artificial Insemination,* The Hague **2** : 236.

Pelletier, J. and Ortavant, R. (1975) *Acta Endocrinologica* **78** : 435 and 442.

Petkov, E. (1969) Cited by Colas *et al.* (1975).

Rathore, A.K. (1968) *Proceedings of the Australian Society for Animal Production* **7** : 270.

Rouger, Y. (1974) Thesis, University of Rennes.

Salamon, S. (1962) *Australian Journal of Agricultural Research* **13** : 1137.

Salamon, S. (1964) *Australian Journal of Agricultural Research* **15** : 645.

Schäfer, H. and Matter, H.E. (1966) Züchtungskunde **38** : 186.

Skinner, J.D., Booth, W.D., Rowson, L.E.A. and Karg, H. (1968) *Journal of Reproduction and Fertility* **16** : 463.

Skinner, J.D. and Rowson, L.E.A. (1968) *Journal of Reproduction and Fertility* **16** : 479.

Skinner, J.D. and van Heerden, J.A.H. (1971) *South African Journal of Animal Science* **1** : 77.

Smith, P. and Gordon, I. (1967) *Irish Veterinary Journal* **21** : 222.

Voglmayr, J.K. (1975) *Journal of Reproduction and Fertility* **43** : 119.

Waites, G.M.H. and Ortavant, R. (1968) *Annales de Biologie Animale, Biochimie, Biophysique* **8** : 323.

Watson, R.H., Sapsford, C.S. and McCance, I. (1956) *Australian Journal of Agricultural Research* **7** : 574-590.

58

SOME PROBLEMS OF EVALUATING SEMEN QUALITY

R.C. JONES and I.C.A. MARTIN

University of Newcastle and University of Sydney, N.S.W., Australia

SUMMARY

The relationship between the ultrastructure of samples of spermatozoa and their ability to fertilize an ovum is evaluated. Further work is required to describe satisfactorily the structure of a spermatozoon capable of fertilizing an ovum and caution is required in interpreting results of semen processing since some diluents may contain membrane active substances which produce misleading results.

INTRODUCTION

To date no single criterion or group of criteria has proved to be of great value for predicting the fertilizing capacity of semen (Jones, 1971). Consequently, it is desirable to develop new approaches to the assessment of semen quality. This report summarizes a series of studies of the ultrastructure of spermatozoa in which a major objective was to assess whether there is a relationship between ultrastructure and function of spermatozoa in order to provide a basis for developing methods of semen evaluation. The studies have mainly been of processed semen since it is used for artificial insemination more than undiluted semen and because semen processed in various ways produces a wider range of conception rates than is usually obtained with semen from separate males (Linford *et al.*, 1976).

MATERIALS AND METHODS

Three studies of ram semen are described. Experiment 1 is part of an *in vitro* study described by Jones and Martin (1973, Experiment 1) in which semen was diluted 20-fold in glucose-phosphate containing 3 per cent egg yolk and stored at 39°C or 5°C. Experiments 2 and 3 involved the artificial insemination of sheep at Guyra, N.S.W. The general procedures for handling and inseminating semen were described by Jones *et al.* (1969). Where appropriate semen was diluted to a concentration of 900×10^6 spermatozoa/ml in inositol-MES (184 mM inositol, 17 mM fructose, 7 mM sodium chloride, 4 mM potassium chloride and 80 mM MES-Na) containing 5% egg yolk. For storage at 5°C samples were cooled over 1.5 hours. Cervical insemination of 10^8 spermatozoa were carried out and conception rates were determined as non-returns to oestrus within 28-45 days of insemination. Experiment 4 was a laboratory study in which semen was stored in inositol-MES and saline-bicarbonate (115 mM sodium chloride, 5 mM potassium chloride, 1.5 mM calcium chloride, 1 mM magnesium chloride, 5 mM sodium phosphate buffer, 20 mM sodium bicarbonate and 5.6 mM glucose).

Experiment 5 is part of a study carried out at Reading, U.K. (Jones and Stewart, 1976). One ejaculate was collected from each of 6 bulls using an artificial vagina, diluted in a diluent containing 83% v/v skim milk, 10% v/v egg yolk, 7% v/v glycerol, 12.5 mg/ml fructose and antibiotics and frozen in 0.5 ml straws. Each straw contained 20 x 10[6] spermatozoa. Five straws for each ejaculate were used for the electron microscope studies and one was used per insemination after thawing in air following at least 28 days storage at −196°C. The semen was used only for first inseminations and conception rates were determined as non-returns to oestrus within 16 weeks of insemination.

Spermatozoa were prepared for examination with the transmission electron microscope as previously described (Jones, 1973a, 1975). Electron micrographs of areas of the Araldite sections were prepared from each treatment and all sections of spermatozoa through the appropriate plane were scored according to the type and degree of structural change occurring in the head and mid-piece. Usually 100 sections of heads and of mid-pieces were scored per treatment and counts of the frequencies of each type expressed as mean percentages. Heads were scored according to the structure of the plasma membrane and acrosome as corresponding to one of the types shown in Plate 1, Figs. 1 to 9 (Jones and Martin, 1973): the sequence of structural changes shown in Figs. 2 to 5 is referred to as "acrosomal vesiculation" and the sequence shown in Figs. 6 to 9 is referred to as "acrosomal vacuolation" (Jones, 1973b). Mid-pieces were classified according to the structure of the plasma membrane and mitochondria into the types shown in Plate 1, Figs. 10-14 (Jones and Martin, 1973). For summary in this report, several of the classification types have been grouped together in the tables. The statistical significance of differences was determined as in previous studies (Jones *et al.*, 1969; Jones and Martin, 1973).

RESULTS

The results of Experiments 1 to 4 are summarized in Tables 1 and 2. Experiment 1 was carried out to determine whether the observed reduction in fertility caused by processing ram semen (Jones *et al.*, 1969; Lapwood *et al.*, 1972) could be attributed to changes in sperm structure. It was found that storage for 24 hours at 5°C increased the proportion of spermatozoa with swollen acrosomes and concluded that these structural changes accounted, at least in part, for the reduced fertility observed in the insemination studies. In order to confirm this conclusion the basic study was repeated (Experiments 2 and 3) with sub-samples of semen used for artificial insemination. However, it was found that although storage of semen for 1

TABLE 1

THE STRUCTURE OF THE HEADS AND FERTILIZING CAPACITY OF RAM SPERMATOZOA IMMEDIATELY AFTER COLLECTION AND AFTER DILUTION IN GLUCOSE-PHOSPHATE (EXPERIMENT. 1) OR INOSITOL-MES (EXPERIMENTS 2 AND 3) AND STORAGE AT 39°C, 35°C OR 5°C

Results of semen treatments are from n replicates using ejaculates from different rams.

| | Mean percentage of sperm scored as type shown in Plate 1* | | | | | | Conception Rates | |
| | | | | | | | No. ewes insem. | % non-returns |
	1·	2	3-5	6	7	8-9		
Plasmalemma: intact	+	+	+	−	−	−		
broken	−	−	−	+	+	+		
Acrosome: swollen only	−	+	−	−	+	−		
vesiculating	−	−	+	−	−	−		
vascuolating	−	−	−	−	−	+		
Experiment 1 (n=2)								
(a) Undiluted semen**	59.5	0.5	0.0	39.5	0.0	0.5	†	†
Diluted semen:								
(b) 2 hr at 35°C	74.5	0.5	1.5	20.0	0.5	3.0	†	†
(c) 0 hr at 5°C	58.5	17.0	2.0	11.0	7.0	2.5	†	†
(d) 24 hr at 5°C	37.5	24.0	5.0	12.0	18.0	3.5	†	†
P (mean a & b *vs* c & d)		<0.01						
Experiment 2 (n=2)								
(e) Undiluted semen**	73.0	0.0	1.0	18.0	2.0	6.0	14	57
Diluted semen:								
(f) 0 hr at 39°C	76.0	0.0	1.0	13.0	7.0	3.0	15	53
(g) 1 hr at 39°C	82.0	0.5	4.5	10.0	0.0	2.5	14	29
Experiment 3 (n=2)								
(h) Undiluted semen**	76.5	0.0	0.0	20.5	1.0	2.0	23	61
Diluted semen:								
(i) 0 hr at 5°C	87.0	0.5	2.5	6.0	0.0	4.0	22	59
(j) 24 hr at 5°C	87.5	0.0 ·	4.0	1.5	0.0	7.0	20	30
P (mean e, f, h & i *vs* g & j)		>0.05					<0.01	
Experiment 4 (n=3)								
(k) Undiluted semen**	93.3	1.0	1.6	2.0	0.0	2.0	†	†
Diluted semen:								
(l) 1 hr in inositol-MES	87.0	0.7	2.7	3.3	0.0	6.0	†	†
(m) 1 hr in inositol-MES, then 1 hr in saline-bicarbonate	48.7	0.0	35.7	7.0	0.0	8.7	†	†
P (mean k & l *vs* m)		<0.001						

† not determined
* Jones and Martin (1973)
** semen samples fixed as soon as possible after collection.

TABLE 2

THE STRUCTURE OF THE MID-PIECES AND SCORES OF PER CENT
MOTILE SPERMATOZOA IMMEDIATELY UPON COLLECTION OF
RAM SPERMATOZOA AND AFTER DILUTION IN GLUCOSE-
PHOSPHATE (EXPERIMENT 1) OR INOSITOL-MES (EXPERIMENTS 2
AND 3) AND STORAGE AT 39°C, 35°C OR 5°C

Results are means from n replicates using ejaculates from different rams.

Plasma membrane: Mitochondria:	Mean percentage of sperm — scored as type shown in Plate 1*			% motile
	intact		broken	
	Normal or condensed	pale	pale**	
Experiment 1 (n=2)				
(a) Undiluted semen	86.0	9.5	4.5	90
Diluted semen:				
(b) 2 hr at 35°C	85.5	6.0	8.8	†
(c) 0 hr at 5°C	75.0	11.0	14.0	75
(d) 24 hr at 5°C	71.0	15.0	14.0	70
Experiment 2 (n=2)				
(e) Undiluted semen	94.5	0.5	5.0	92
Diluted semen:				
(f) 0 hr at 39°C	92.5	0.0	7.5	85
(g) 1 hr at 39°C	85.0	10.0	5.0	75
Experiment 3 (n=2)				
(h) Undiluted semen	75.0	8.5	16.5	90
Diluted semen:				
(i) 0 hr at 39°C	92.5	3.0	4.5	87
(j) 24 hr at 5°C	89.0	5.0	6.0	75
Experiment 4 (n=3)				
(k) Undiluted semen	95.0	0.3	4.7	90
Diluted semen:				
(l) 1 hr in inositol-MES	94.7	0.0	5.3	70
(m) 1 hr in inositol-MES. then 1 hr in saline-bicarbonate	61.0	0.0	39.0	47
P (mean k & l *vs* m)	<0.001			<0.001

† Not scored.

* Jones and Martin (1973).

** Some spermatozoa (0-6.5%) pooled into this group were scored as having normal or condensed mitochondria (Jones and Martin, 1973, Figs. 13 and 11).

hour at 39°C or 24 hours at 5°C caused a reduction in conception rates, these treatments had no effect on the ultrastructure of spermatozoa. In retrospect it was considered that the absence of an effect of storage on sperm structure may be because the diluent (inositol-MES) was stabilizing the cell membranes and preventing any acrosomal swelling. This proposal was tested in Experiment 4 in which semen diluted to 300 x 10⁶ sperm/ml (9 to 15-fold) in inositol-MES was incubated for 1 hour at 35°C, then concentrated by centrifugation and removal of the supernatant fluid. Half of the samples were fixed (1) and the other half were resuspended in saline-bicarbonate (m) and stored for a further hour before fixation for electron microscopy. Fresh undiluted semen was also prepared (k) for comparison. The results showed that resuspension and incubation of the diluted samples in saline-bicarbonate after storage in inositol-MES caused structural damage to the heads and mid-pieces. Scores of per cent motile spermatozoa were also reduced by treatment m.

A study of deep frozen and thawed bull semen, which was carried out concurrently with the ram semen studies, is reported (Tables 3 and 4) to confirm the relationship between the structure of semen and conception rates following artificial insemination. The average conception rates obtained from bull 6 was lower than for the other bulls and semen from this bull also showed a higher incidence of acrosomal and mid-piece degeneration and a lower score of per cent motile spermatozoa than the average for other bulls.

TABLE 3

EXPERIMENT 5: THE ULTRASTRUCTURE OF THE SPERM HEADS AND THE 16 WEEK NON-RETURN RATES FOLLOWING ARTIFICIAL INSEMINATION OF DEEP-FROZEN AND THAWED SEMEN FROM EACH OF SIX BULLS

		Mean percentage of sperm scored as type shown in Plate 1*						Conception Rate No. cows insem.	% non-returns
		1	2	3-5	6	7	8-9		
Plasmalemma:	intact	+	+	+	−	−	−		
	broken	−	−	−	+	+	+		
Acrosome:	swollen only	−	+	−	−	+	−		
	vesiculating	−	−	+	−	−	−		
	vacuolating	−	−	−	−	−	+		
Bull No. 1		9	31	5	1	12	42	99	60
2		6	19	4	13	6	52	79	63
3		9	15	1	9	12	54	123	59
4		15	14	8	4	4	55	88	64
5		20	30	7	4	8	31	93	63
6		0	0	4	0	2	94	60	15
$\underline{P}$ (mean 1-5 *vs* 6)				<0.05				<0.001	

* Jones and Martin (1973)

TABLE 4

EXPERIMENT 5: THE ULTRASTRUCTURE OF THE HEADS OF BULL
SPERMATOZOA AND SCORES OF PER CENT MOTILE SPERMATOZOA
IN SEMEN SAMPLES AFTER FREEZING AND THAWING

| Bull number | Mean percentage scored as type shown in Plate 1*: | | | % motile |
| | Plasma membrane intact | | Plasma membrane broken | |
	Mitochondria normal**	Mitochondria pale	Mitochondria pale***	
1	44	21	15	60
2	26	52	22	50
3	18	52	30	20
4	16	43	41	40
5	32	42	26	50
6	2	13	85	5
P (mean 1-5 vs 6)	< 0.001			< 0.05

* Jones and Martin (1973)
** A few spermatozoa pooled into this group were scored as possessing condensed mitochondria (Jones and Martin, 1973)
*** A few spermatozoa pooled into this group were scored as possessing normal or condensed mitochondria (Jones and Martin, 1973)

DISCUSSION

The quantitative approach used in these studies to assess the structure of semen is necessary since conception rates vary quantitatively and it is considered that male fertility mainly affects conception rates according to the numbers of spermatozoa in an inseminate and their life span in the female reproductive tract (Jones, 1975). These studies confirm that there is a relationship between structure and function of spermatozoa which may be of use in evaluating semen quality. However, factors of the molecular level must also be considered since ultrastructural changes alone do not explain the reduced fertility due to storage of ram semen at 35°C (Experiment 1) or the reduced fertility due to storage of boar semen at 20°C in glycerol (Jones, 1975). The studies reported in this chapter also indicate that more information is required on the structure of a spermatozoon which is capable of fertilizing an ovum. The problem is demonstrated by Experiment 5 in which the average conception rate for bulls 1 to 5 was 61.6% yet an average of only 11.8% of the spermatozoa were scored as "normal" (Table 3, column 1). Clearly, spermatozoa other than these must be capable of fertilizing an ovum. It is suggested that acrosomal swelling (Table 3, cols. 2 & 7) would not preclude the acrosome reaction and fertilization to occur and that much of the plasma membrane breakage observed (Table 3, cols. 6 & 7) occurs during fixation (Jones, 1973a,c,). When the numbers of spermatozoa in these categories are pooled with those showing no structural change the proportion of spermatozoa potentially capable of fertilization would be 48.2%. It is noteworthy, however, that the susceptibility of spermatozoa to membrane breakage during fixation varies between males within and between species (Jones, 1975) and consequently requires further consideration as a useful criterion for assessing the viability of spermatozoa.

Results presented in this chapter also demonstrate that caution is required in interpreting studies of semen processing techniques as some diluents may act to artificially stabilize cell membranes and so give a misleading impression of the structural integrity of spermatozoa. It is suggested that although the spermatozoa in Experiments 2 and 3 were structurally intact when examined microscopically, Experiment 4 indicates that dilution and storage in the female reproductive tract fluids following artificial insemination would cause cytolysis. Consequently, in future it may be necessary to dilute spermatozoa in a diluent similar in composition to female reproductive tract fluids and to incubate them for a period prior to fixation. The value of such an approach is confirmed by studies which have found that the parameter which will most accurately predict conception rates following insemination of semen is the ability of sperm to survive dilution and incubation at 30 to 37°C (Buckner *et al.*, 1954; Linford *et al.*, 1976).

The studies described in this report were basically to determine the nature of the structural changes which may occur to spermatozoa and whether they can be related to changes in fertility. The ultimate objective (Jones, 1975) is to use the light microscope to evaluate the viability of the head and tail of the same spermatozoon since both parts are necessary for sperm function. Most of the acrosomal changes described in these studies can be recognized with the light microscope (Jones, 1975) which would be used for the evaluation of semen quality once the nature and significance of the acrosomal changes have been determined. Further it has been a consistent finding in these and previous studies (Jones, 1973; Jones and Martin, 1973) that scores of per cent motile spermatozoa are related to counts of per cent undamaged mid-pieces. If the latter can be recognized cytochemically using the light microscope (Jones and Holt, 1974) then it would be possible to assess the structural integrity of the head and tail of the same spermatozoon and count in a sample the proportion of spermatozoa which are structurally intact.

ACKNOWLEDGEMENTS

This work was supported by funds from the Australian Research Grants Committee.

REFERENCES

Buckner, P.J., Willett, E.L. and Bayley, N. (1954). *J. Dairy Sci.* **37** : 1050.

Jones, R.C. (1971). *Nature (Lond.)* **229** : 534.

Jones, R.C. (1973a). *J. Reprod. Fertil.* **33** : 145.

Jones, R.C. (1973b). *J. Reprod. Fertil.* **33** : 113.

Jones, R.C. (1973c). *J. Reprod. Fertil.* **33** : 179.

Jones, R.C. (1975). The Biology of the Male Gamete. Duckett, J.G. and Racey, P.A. (Eds.) p.343.

Jones, R.C. and Holt, W.V. (1974). *J. Reprod. Fertil.* **41** : 159.

Jones, R.C. and Martin, I.C.A. (1973) *J. Reprod. Fertil.* **35** : 311.

Jones, R.C., Martin, I.C.A. and Lapwood, K.R. (1969). *Aust. J. Agric. Res.* **20** : 141.

Jones, R.C. and Stewart, D.L. (1976). Unpublished data.

Lapwood, K.R., Martin, I.C.A. and Entwistle, K.W. (1972). *Aust. J. Agric. Res.* **23**: 457.

Linford, E., Glover, F.A., Bishop, C. and Stewart, D.L. (1976). Unpublished data.

59

FACTORS AFFECTING THE FERTILITY OF DILUTED RAM SEMEN

I.C.A. MARTIN and B.A. RICHARDSON
University of Sydney, N.S.W., Australia

SUMMARY

In 3 field experiments, involving the artificial insemination of 1245, 990 and 1102 ewes respectively, every insemination dose contained a total of 100 million spermatozoa. Ewes were inseminated once and conception rates were estimated from non-returns to oestrus 28 to 45 days after insemination.

Semen was diluted 20- to 30-fold in the first experiment and samples were centrifuged and reconcentrated to give a dose volume of 100 μ1. In the second experiment, semen was diluted 5-fold and dose volumes of 80 to 190 μ1 were used. For the third experiment, dilution rates were about 4- or 12-fold using dose volumes of 110 and 330 μ1 respectively.

Diluents were based on solutions of 184 mM of glucose or inositol and were buffered with 20 mM phosphate or 80 mM MES. Levels of egg yolk of 0.5% or 5% (v/v) were used in diluents in all experiments and, in the third experiment, some diluents contained 0.3% or 3% (w/v) protein from cows' milk.

There were no significant differences in fertility results that could be related to the dose volumes employed. The mixture of inositol and MES was a satisfactory diluent and was stable as a freeze-dried preparation made in advance of the field experiment. Incubation of diluted semen at 39°C for 1 hour before use for AI caused a significant depression of conception rates. A similar drop in fertility occurred when diluted semen was chilled to 5°C. This was followed by a further loss in fertility when chilled samples were stored at 5°C for 24 hrs before AI.

Laboratory studies of the motility and morphology of incubated and chilled specimens of diluted ram semen have not shown any degree of change in the spermatozoa which could be correlated with the loss in fertility, nor have the spermatozoa lost much of their hyaluronidase during these processes.

INTRODUCTION

Lapwood, Martin and Entwistle (1972) showed that, with a single dose of 50 million spermatozoa per ewe inseminated, the fertility of ram semen diluted 10-fold in a buffered glucose solution was improved by centrifugation of the diluted semen to return the concentration of spermatozoa in the

513

dose for artificial insemination (AI) to that of the original semen. This effect did not occur when the semen was diluted in milk. Entwistle and Martin (1972) then used the technique of centrifuging and reconcentrating ram semen diluted 10-fold in buffered glucose to prepare doses of semen containing 50 or 100 million spermatozoa in volumes of 50 or 200 μl. In that experiment, there were no significant differences in fertility between treatments or between diluted and undiluted semen. Martin and Watson (1976) tested the fertility of ram semen diluted 30- to 40- fold and centrifuged to produce insemination doses of 25 or 100 million spermatozoa in volumes of 50 and 250 μl. They found that fertility was significantly reduced by this degree of dilution and by the use of 25 million spermatozoa. The volume of the AI dose had no significant effect on fertility.

Laboratory studies have shown that the addition of fractions extracted from egg yolk to the glucose phosphate diluent improved the survival of ram spermatozoa chilled to 5°C (Watson and Martin, 1975a). More recently, it has been shown that ram spermatozoa survived better during incubation at 39°C or chilled to 5°C after dilution if glucose was replaced by an equimolar amount of inositol in the diluent. In addition, some organic buffers were tested as substitutes for the phosphate buffer usually included in diluents. The response of ram spermatozoa to them closely followed that of bull spermatozoa (Jones and Foote 1972). The compound, 2-(N-morpholino) ethane sulphonic acid (MES), was selected as a better hydrogen ion buffer than phosphate for diluents for ram semen.

The effects on the fertility of ram spermatozoa of diluent composition, incubation at body temperature and cooling to 5°C of diluted semen are reported in this chapter.

MATERIALS AND METHODS

The artificial insemination (AI) experiments were conducted in April-May of 1973, 74 and 75 at Guyra, N.S.W. (30°14'S). The general flock management procedures and techniques of AI were as described by Entwistle and Martin (1972) and Martin and Watson (1976) except that the flocks were only mustered once each day at 0800 hr when oestrous ewes were drafted for AI. All ewes received a single dose of 100 million spermatozoa. Non-returns to oestrus were recorded 28 days after the last day of AI in the experiment (i.e. 28 to 45 day non-returns).

All field experiments followed a basic split-ejaculate design in which a portion of semen from each collection was taken for each treatment including a sample for control inseminations of undiluted semen. Two or three "stages" or "trials" were set up within each experiment to test diluents or processes. Practical difficulties occurred in field AI trials where more than 5 or 6 semen treatments were tested within each semen collection. Hence the sequential approach to testing a number of treatments has been adopted.

In Experiment 1, semen was diluted between 20- and 30- fold and then reconcentrated by centrifugation to give an insemination dose of 100 μ1. The dilution rate for Experiment 2 was 5-fold and the dose volume lay between 80 and 190 μ1. For Experiment 3, semen was diluted to 900 or 300 million spermatozoa/ml, i.e. about 4- and 12- fold dilution, and dose volumes of 110 and 330 μ1 respectively were used for the two concentrations. Semen was diluted 21-fold in Experiment 4.

The composition of the diluents was:—

Glucose-phosphate: 184 mM glucose, 31 mM sodium chloride, 20 mM sodium phosphate buffer and 5 mM potassium chloride

Inositol-phosphate: as for glucose-phosphate but containing 184 mM inositol instead of glucose

Inositol-MES: 184 mM inositol, 7 mM sodium chloride, 4 mM potassium chloride and 80 mM sodium-MES (i.e. 2-(N-morpholino) ethane sulphonic acid, sodium salt titrated to pH 7.0)

Sodium chloride, bicarbonate buffer: 115 mM sodium chloride, 5 mM potassium chloride, 1.5 mM calcium chloride, 1 mM magnesium chloride, 5 mM sodium phosphate buffer and 20 mM sodium bicarbonate.

All diluents contained 17 mM fructose and 500 i.u. each of dihydrostreptomycin sulphate and sodium penicillin-G.

The quantities of egg yolk and non-dialysable fractions of milk added to diluents are shown in Tables 1, 2, 3 and 4. All diluents in Experiment 3 were prepared several weeks in advance of the experiment and freeze-dried. The non-dialysable fraction of milk was prepared by freeze-drying heated skim milk samples after 24 hr dialysis against 15 mM sodium chloride solution.

In Experiment 4, hyaluronidase released into the diluent was measured using the method described by Rogers and Morton (1973) modified to include deproteinization of the samples by perchloric acid treatment and centrifugation. The total hyaluronidase content of spermatozoa was estimated by assaying the level of enzyme in the supernatant obtained from samples of semen which had been twice frozen to $-79°C$ and thawed.

The methods of scoring spermatozoa for motility, percentage motile and changes in the acrosome have been described by Watson and Martin (1975b) and Watson (1975). In Experiment 4, these scores were made after the spermatozoa had been resuspended in sodium chloride, bicarbonate buffer and incubated for 1 hr.

RESULTS

The results of Experiment 1 are summarized in Table 1. In Trial C, magnesium chloride partially replaced the sodium chloride content of the diluent and the bicarbonate buffer similarly replaced part of the phosphate component of the diluent. There were no significant differences in non-return rates between treatments within each of the three trials of different diluents. Over all treatments, fertility was significantly higher in Trial C than in A and B.

TABLE 1

Experiment 1: Non-returns to oestrus of artificially inseminated ewes (Insemination dose of 100 million spermatozoa per ewe; semen diluted 20- to 30-fold and phosphate buffer present in all diluents)

	Treatment	Number of Ewes Inseminated	% Non-return
Diluted Semen			
Composition of diluent			
Egg Yolk (%v/v)			
	Trial A		
0	Glucose	80	58
0.5	Glucose	78	46
0.5	Inositol	82	52
0.5	Sodium chloride	78	50
	Trial B		
0.5	Glucose	84	52
5.0	Glucose	83	48
0.5	Inositol	86	58
5.0	Inositol	82	55
	Trial C		
0.5	Inositol	111	64
0.5	Inositol, 20 mEq magnesium	107	56
0.5	Inositol, 20 mM bicarbonate buffer	108	58
Undiluted Semen (Controls)			
	Trial A	79	57
	B	83	57
	C	104	64

Experiment 2 was divided into two trials (Table 2). In the first, semen was diluted in either inositol-MES or inositol-phosphate and the effects of two levels of egg yolk were tested. No significant differences between treatments were detected. In the second part of the experiment, the buffers were again contrasted and a freeze-dried diluent was tested. This diluent had been freeze-dried 4 to 6 weeks before the experiment and a sample was reconstituted on each day of the AI trial by the addition of distilled water. Neither dilution nor diluents had significant effects on the fertility of the semen. However, chilled storage at 5°C significantly depressed the non-return rate.

TABLE 2

Experiment 2: Non-returns to oestrus of artificially inseminated ewes (Insemination does of 100 million spermatozoa per ewe; semen diluted 5-fold in a diluent based on 185 mM inositol)

Treatment		Number of Ewes Inseminated	% Non-return
Diluted Semen			
Composition of diluent			
Buffer	Egg Yolk ($\%$v/v)		
	Trial A		
MES	0.5	97	42
	5.0	94	50
Phosphate	0.5	96	46
	5.0	100	41
	Trial B		
MES	5.0	87	43
Phosphate	5.0	91	52
Phosphate (freeze-dried diluent)	5.0	88	49
Chilled to 5°C and stored for 24 hr			
MES	5.0	38	24
Phosphate	5.0	39	15
Phosphate (freeze-dried diluent)	5.0	37	30
Undiluted Semen (Controls)			
	Trial A	95	54
	B	128	51

In Experiment 3 (Table 3, Trial A), diluted semen inseminated immediately after dilution was significantly lower in fertility than the undiluted semen. Incubation at 39°C caused a further significant loss in fertility of diluted semen. In Trial B, it was found that a significant loss of fertility had occurred during cooling to 5°C and a further drop in non-return rate was observed when the chilled semen was stored at 5°C for 24 hrs. The factor of dilution rate/dose volume did not have any significant effect on conception rates. Two levels of egg yolk and milk protein were tested in diluents used in the third stage of this experiment (Trial C). The fertility of the diluted incubated samples of semen was not significantly lower than that of the undiluted semen. The response to an increase in content of egg yolk was opposite to that shown to an increase in the level of milk protein and this was significant ($\chi^2 = 5.21$; P < 0.05).

The observations made in Experiment 4 are summarized in Table 4. Hyaluronidase loss was measured at the end of the period in the inositol-MES diluent and motility, percentage motile and acrosome scores were observed after resuspension of the spermatozoa in sodium chloride bicarbonate and incubation for 1 hr. The presence of egg yolk in the diluent significantly reduced loss of hyaluronidase from the spermatozoa and degenerative changes in the structure of the acrosome. The percentage of motile spermatozoa was improved when egg yolk was included in the diluent but the

TABLE 3

Experiment 3: Non-returns to oestrus of artificially inseminated ewes (Insemination dose of 100 million spermatozoa per ewe; semen diluted approximately 4- or 12-fold)

Treatment		Number of Ewes Inseminated	% Non-return
I. *Semen diluted in 5% Egg yolk Inositol-MES*			
Storage conditions before AI	Dose volume (μl)		
	Trial A		
Used immediately after dilution	330	99	51
	110	98	50
Incubated at 39°C for 1 hr	330	98	40
	110	96	42
	Trial B		
Cooled to 5°C in 1.5 hr	330	69	38
	110	69	39
Cooled to 5°C and stored 24 hr	330	60	22
	110	61	25
II. *Semen diluted in Inositol-MES and incubated at 39°C for 1 hr before AI* (Dose volume 110 μl)			
	Trial C		
Type of protein in diluent	Level		
Egg yolk	0.5% v/v	48	69
	5.0% v/v	46	54
Milk	0.3% w/v	47	45
	3.0% w/v	46	63
Undiluted Semen (Controls)			
Trial A		97	64
B		124	67
C		44	59

rate of movement of the spermatozoa (motility score) was not significantly affected by egg yolk. The processes of incubation and storage at 5°C did not have significant effects on the appearance of acrosomes. However, scores of motility and percentage motile fell significantly on storage at 5°C for 24 hr. There was also an increase in loss of hyaluronidase in this period of chilled storage.

TABLE 4

Experiment 4: Characteristics of diluted semen after incubation at 39°C or chilled storage at 5°C

(Results are means from 6 replicates of a 3×4 factorial experimental design)

Treatment	Release of Hyaluronidase (% of Total)	Motility Score (0-4)	% Motile	Acrosome Score (1-4)
A. *Level of egg yolk in diluent (% v/v)*				
0	13.7	3.42	59.0	2.03
0.5	6.7	3.46	61.3	1.96
5.0	6.8	3.43	65.6	1.91
B. *Sample taken from diluted semen*				
Immediately after dilution	7.5	3.67	66.1	1.99
Incubated at 39°C for 1 hr	9.0	3.57	64.7	1.99
Cooled to 5°C in 1.5 hr	8.7	3.36	60.8	1.91
Stored at 5°C for 24 hr	11.1	3.14	56.1	1.99

DISCUSSION

Dilution rates in these experiments have ranged from 5- to 30-fold without causing a major loss in fertility. The related factor of volume of dose for AI also has not had an effect on fertility. Although diluents based on inositol-MES appear to preserve the motility and morphology of ram spermatozoa *in vitro* better than those containing glucose-phosphate or sodium chloride and bicarbonate buffer, fertility results after AI using semen freshly diluted in inositol-MES have not been any better than when the other diluents were used.

The powerful buffering effect of MES was probably valuable in Experiment 3 when samples of diluted semen were incubated for 1 hr before AI. Laboratory experiments comparing diluents containing phosphate or MES buffers have shown the superiority of the latter compound in preserving a high proportion (70% or higher) of motile spermatozoa. The conception rates recorded for diluted, chilled semen were also disappointing as spermatozoa in all samples of chilled semen showed strong progressive motility and, again, at least 70% of the spermatozoa were motile. Therefore, in all stages of the experiments, 70 million, or more, motile spermatozoa would have been present in each dose.

The capacity to fertilize appears to have been deranged before defects in morphology and motility could be detected with the light microscope. However, the changes in morphology of acrosomes and mid-pieces observed in electron micrographs of diluted, incubated or chilled semen samples by Jones and Martin (1973, 1976) have indicated that, though the diluents selected for these AI trials may have been very satisfactory for the preservation of spermatozoa, any benefits may be lost rapidly once the spermatozoa enter the fluids of the female genital tract. The measurement of loss of hyaluronidase from spermatozoa has also shown that the acrosome remained reasonably intact in diluents containing inositol-MES. Research will be continued to determine the rate of loss of the enzyme when the spermatozoa have been removed from diluents of this type.

The composition of the sodium chloride, bicarbonate buffer is similar to the inorganic content of uterine fluid, but semen diluted in this solution did not have better fertility than samples diluted in glucose-phosphate, inositol-phosphate or bicarbonate. Similarly, the incorporation of a relatively high level of magnesium in the diluent had no significant effect even though the level used of this cation was expected to limit changes in acrosomal membranes. The benefit from the use of a milk fraction in part of Experiment 3 may be due to the presence of milk protein or to the calcium which is present in the macromolecules of undenatured casein. Work is in progress to test whether mixtures of egg lipoprotein and milk proteins in the diluent are beneficial to the structure and function of ram spermatozoa.

ACKNOWLEDGEMENTS

We are indebted to Mr D.K. Rainger, "Brockley", Guyra, N.S.W. for generously providing sheep and facilities.

This research was supported by grants from the Australian Wool Corporation and the Australian Research Grants Committee.

REFERENCES

Entwistle, K.W. and Martin, I.C.A. (1972). *Australian Journal of agricultural Research,* **23,** 467.

Jones, R.C. and Foote, R.H. (1972). *Australian Journal of biological Sciences,* **25,** 1047.

Jones, R.C. and Martin, I.C.A. (1973) *Journal of Reproduction and Fertility,* **35,** 311.

Jones, R.C. and Martin, I.C.A. (1976). Proceedings of International Sheep Breeding Congress, Perth, W. Australia, August 23-27.

Lapwood, K.R., Martin, I.C.A. and Entwistle, K.W. (1972). *Australian Journal of agricultural Research,* **23,** 457.

Martin, I.C.A. and Watson, P.F. (1976). *Theriogenology,* **5,** 29.

Rogers, B.J. and Morton, B.E. (1973). *Journal of Reproduction and Fertility,* **35,** 477.

Watson, P.F. (1975). *Veterinary Record,* **97,** 12.

Watson, P.F. and Martin, I.C.A. (1975a). *Australian Journal of biological Sciences,* **28,** 145.

Watson, P.F. and Martin, I.C.A. (1975b). *Australian Journal of biological Sciences,* **28,** 153.

60

STORAGE OF RAM SEMEN

G. COLAS and M. COUROT
I.N.R.A.—Station de Physiologie de la Reproduction, 37380, Nouzilly, France

SUMMARY

Storage of ram semen with maintenance of its fertilizing ability is possible for a short duration of less than 24 h, as liquid semen and for a longer time as deep frozen semen. Milk or egg yolk diluents can be used for preservation of liquid semen at $+15°C$ and $+4°C$ respectively. Owing to the large number of spermatozoa needed in a small volume per AI (125 to 500×10^6 in 0.2 ml), the dilution rate is low; consequently since the fertilizing capacity of ram semen decreases during storage, storage must not exceed 24 h. Deep freezing of ram semen is possible in a pelleted form or in PVC straws when diluted respectively in tris buffer-sugar-egg yolk-glycerol (dilution in one step), or in lactose-egg yolk-milk-glycerol (dilution in two steps) extenders. The duration of survival after thawing is shorter than with fresh semen therefore AI needs to be carried out precisely at the proper time during oestrus. The fertilizing ability of the semen varies according to the season of collection, at least for those breeds sensitive to photoperiod and it maintains this fertilizing ability for several years. Both methods of storage of ram semen are thus available for practical use in the field.

INTRODUCTION

Storage of ram semen with preservation of its fertilizing ability is of great practical interest and importance. As in cattle, the development of artificial insemination in sheep would allow larger use of the best sires for whatever trait is desired, or make possible the preservation of genes in a more basic approach to field problems. However, such developments have remained limited until the present time, due to specific difficulties encountered with ram semen, the storage of which is deleterious to its fertilizing capacity (Dauzier, Thibault and Wintenberger, 1954; Salamon and Robinson, 1962). According to the requirement, two methods for the preservation of ram semen are available: either a very short term storage of semen here referred to as chilled semen, or a long term storage as deep frozen semen. The methods call for different techniques with different practical applications and both will be presented here.

CHILLED SEMEN

For many years, people have attempted to extend or store ram semen but with little success, due to the peculiar property of ram semen which maintains the motility of the sperm cells for a longer period of time than their fertilizing capacity (Dauzier, Thibault and Wintenberger, 1954). Thus it is imperative to compare results of *in vitro* experiments with those of fertility trials. Moreover, because the anatomy of the cervix of the ewe limits the deposition of semen only in its entrance, with some successful passage through the cervix (Andersen, Aamdal and Fougner, 1973), it is necessary to inseminate a large number of spermatozoa (Salamon, 1962) in a small volume of diluent (Allison and Robinson, 1971). Thus ram semen must be stored at low dilutions. In the present review, the diluents, the temperature of storage and the dilution rate are considered in order to obtain the optimum conditions for preserved liquid ram semen.

(a) Diluents

Successful storage of ram semen, measured in terms of the preservation of its fertilizing ability has been obtained with two groups of diluents, the basal components of which are egg yolk or milk.

(i)   Milk extenders

Whole milk or skim milk have been used as extenders after heating in order to remove a protein which was found to be harmful to ram spermatozoa (Dauzier, Thibault and Wintenberger, 1954; Salamon, 1962; Jones, 1969). Antibiotics have always been added to the diluents. Catalase was shown to have no positive effect in milk extenders, whereas it is beneficial in egg yolk extenders (Colas *et al.,* 1968). At present, heated reconstituted skim milk plus antibiotics is the method employed for storing liquid ram semen for use in AI in France. It allows for a good preservation of fertilizing ability of ram semen for as long as 14-16 h at +15°C with a lambing rate of 65 to 75% following AI at one oestrus.

(ii)   Egg yolk extenders

These diluents are made of egg yoke and buffer plus sugars and antibiotics (Salamon and Robinson, 1962; Ozkoca, 1967; Colas *et al.*, 1968). With fractions of egg yolk, most of the protective activity was shown to be associated with large molecules of the supernatant light fraction (centrifugation 35,000 g, 30 min) which are probably proteins or lipoproteins (Watson and Martin, 1975). It appears that egg yolk extenders also allow ram semen to be well preserved but with lower fertility than milk at +15°C— e.g. lambing rate 66.4 *v.* 75.4% from 895 females.

(b)   Temperature of storage

Storage of spermatozoa requires the lowering of temperature to avoid exhaustion of the cells. With liquid semen, interactions have been observed between the diluent, the dilution rate and the temperature of cooling and storage. Thus ideally, one must choose the best method of dilution according to the temperature of storage desired. Milk extenders only allow cooling to +15°C rather than +4°C. After storage at +4°C, and warming to +35°C, the percentage of dead spermatozoa was higher than at +15°C (Fig. 1).

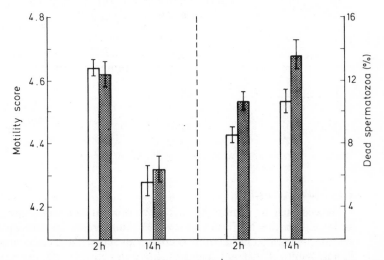

Figure 1—Effect of temperature and duration of storage on ram semen diluted in milk extender. (Open column   15°C; dotted column   4°C.)

(From Colas *et al.* 1968)

On the other hand, egg yolk protects ram spermatozoa during chilling as it reduces changes in the mitochondria and the acrosomes which are important for motility and fertility (Jones and Martin, 1973). It is likely that this protection operates by attachment of the egg yolk to the sperm plasma membrane (Watson, 1975).

(c)   Dilution rate

There is a highly significant linear relationship between lambing rate and total sperm number inseminated; a dose of $120\text{-}125 \times 10^6$ normal spermatozoa has been shown to be necessary for maximum fertility from AI of ewes in natural oestrus (Salamon, 1962). It has also been observed that a higher sperm number is necessary to fertilize ewes at a synchronized oestrus, due to modifications of transport and survival of spermatozoa in the female genital tract after hormonal treatment (Quinlivan and Robinson, 1969; Hawk and Conley, 1971; Colas *et al.,* 1973). In the latter case, the same level of fertility can be achieved with one instead of two AI at the induced oestrus (Table 1) provided the same total sperm number was inseminated at the right time and oestrus was highly synchronized by injection of PMSG at sponge removal.

TABLE 1

FERTILITY OF EWES INSEMINATED ONCE OR TWICE DURING A SYNCHRONIZED OESTRUS

No. AI	Time after AI (h)*	Lambing rate (%)	No. ewes
1	55	69.6	112
2	48 and 60	71.8	110

* Time is given from the end of treatment with progestagen (i.e. sponge removal and PMSG injection), until AI. All the ewes received a total of 500×10^6 spz. in one or two AI of liquid semen.

However, it is important not to increase the volume of the inseminated dose when more sperm cells are required as there is a better fertility when a lower volume of diluted semen is delivered into the female tract (Allison and Robinson, 1971; Colas *et al.,* 1973). Thus, the dilution rate must be low and the ram semen needs to be prepared and stored in a concentrated form, 625 to 2000×10^6 spz/ml. The survival of the sperm cells depends upon the diluent (Schindler and Amir, 1961) and the dilution rate (Amir *et al.,* 1973) —the higher the concentration, the lower and the shorter is the survival rate (Mampouya, 1973). But, whatever the diluent, the temperature of cooling and the dilution rate, there is a regression of fertility on time of storage. This suggests that the maximum duration of storage of liquid ram semen is less than 24 hours. Thus the best results of fertility are always obtained with storage for relatively short durations, namely a few hours.

SPERM FROZEN AT LOW TEMPERATURE

Studies on sperm storage at low temperature have been carried out over several years. However, the setting up of a freezing technique for ram semen has proved difficult and the literature shows that fertility is generally lower with frozen semen than with chilled semen. This is one reason why preservation of spermatozoa at $-196°C$ is not widely adopted and still is only used on a limited number of females.

Since the AI method is used on ewes both in natural oestrus and in oestrus which has been induced by progestagen and that the loss of fertility occurs in both cases (Salamon, 1971; Colas, 1975a), it is necessary to treat these cases separately.

(a) Ewes in natural oestrus

(normal oestrus or second oestrus after synchronization with intravaginal sponges)

The method is based on the use of a solution of tris, glucose, citric acid and glycerol added at low rates (1 : 2, sperm : diluent) to semen. Set up by Salamon and co-workers, this process gives a high concentration of spermatozoa prior to freezing and provides after thawing a large number of motile cells in a small volume. Diluted semen is then pelleted on dry ice and the frozen pellets are stored in liquid nitrogen.

Other workers have recourse to this type of storage; Fraser (1968), Kareta, Pilch and Wierzbowski (1972) and Naumov *et al.* (1974). With rare exceptions, the freezing method in pellets is the most currently used for ewes in natural oestrus.

Outside the properties of the diluents, the fertility rate seems to be related to two factors: timing of AI and number of motile spermatozoa per ewe.

(i) Timing of insemination

A study of the optimum conditions for sperm delivery in the Merino ewe has led Salamon (1971) to conclude (Table 2):

—The percentage of ewes lambing is better when spermatozoa are inseminated 15 to 17 h after the observed onset of oestrus.

—The inseminations carried out in early or in late oestrus result in a similarly lower fertility.

TABLE 2

EFFECT OF INTERVAL FROM DRAFTING* TO INSEMINATION ON FERTILITY OF EWES USING FROZEN SEMEN

No. of inseminations	Time of AI†(h)	% ewes lambing
1	1-3	42.9(77)
	15-17	51.3(80)
2	1-3 and 11-13	59.5(79)
	15-17 and 25-27	58.2(82)

* Drafting twice daily at 08.00h and 19.00h.
† Time of AI relative to time of drafting.
( ) No. of ewes inseminated.

In fact two inseminations always give a slightly higher level of fertility than one insemination (Salamon, 1971; Visser and Salamon, 1973). This is due to the fact that the time between onset of oestrus and ovulation is more variable in a ewe in natural oestrus (Cognié, Mariana and Thimonier, 1970) and that survival of spermatozoa after thawing is greatly reduced (Loginova and Zheltobrukh, 1968, 1972; Mattner, Entwistle and Martin, 1969; Lightfoot and Salamon, 1970). Thus, the chance of contact between the male and female gametes is improved by increasing the number of AI (Kareta, Pilch and Wierzbowski, 1972). However, the difference in fertility between a single and double insemination is generally small. So, one is left to ask if this difference can justify the increase of work caused by the second operation as Salamon has shown in recent trials (1976).

(ii)   Number of motile spermatozoa per ewe

The fertilization rate also depends upon the number of motile cells deposited during oestrus. It was with 360×10^6 motile spermatozoa that Visser and Salamon (1973) obtained the best results (Table 3).

TABLE 3

EFFECT OF THE NUMBER OF MOTILE SPERMATOZOA/EWE ON FERTILITY OF FROZEN SEMEN

No. of motile spermatozoa/ewe ($\times 10^6$)	No. AI / ewe	Ewes lambing (%)
90	1	25.3 (75)
180	1	32.0 (75)
	2	36.2 (69)
360	2	56.1 (66)

( ) No. of ewes inseminated.

(From Visser and Salamon 1973)

This number is very high in comparison to those commonly used when semen is only chilled, which reveals the loss of the quality of the gametes which undergo deep freezing.

(b)   Ewes treated with progestagens

This problem has been studied in Norway (Andersen, Aamdal and Fougner, 1972) and especially in France where the geographic breakdown of flocks does not allow AI in sheep unless used in conjunction with a process of oestrous synchronization. The technique can be summarized as follows: semen is added to a lactose-egg yolk solution (Nagase and Graham, 1964) and chilled to +4°C. It is then diluted again in a concentrated suspension of skim milk (450 mOsm) containing glycerol, the pH of which is adjusted to 6.6-6.7 (Colas, 1975b). Both dilutions are made in order to obtain, prior to freezing, a concentration of glycerol of 4% and a sperm density of 900×10^6/ml.

Diluted semen is packaged in straws (0.45 ml) suspended horizontally in the vapour above liquid nitrogen and then stored in the freezing liquid.

(i) Effect of PMSG

Figure 2 shows there is a close relationship between fertility from frozen semen and interval between onset of oestrus and AI. Thus, lambing rate is maximum when semen is deposited 12 h after oestrus is detected. When ewes are inseminated at a predetermined time after sponge removal (i.e. without oestrous detection), it is necessary that oestrous synchronization be as precise as possible so that most of the ewes are inseminated at the optimum time.

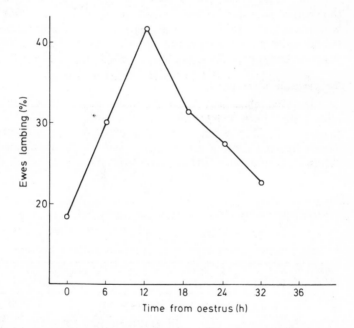

Figure 2—Influence of interval from onset of oestrus to first AI on fertility with frozen semen (Lacaune breed—sexual season).

For this reason PMSG is used in conjunction with the FGA intravaginal sponge treatment (Colas, 1975a). Table 4 shows that PMSG has a marked effect on sheep reproduction when semen is frozen.

TABLE 4

REPRODUCTIVE PERFORMANCES FOLLOWING INSEMINATION* WITH FROZEN-THAWED SEMEN — EFFECT OF PMSG

PMSG (I.U.)	Conception rate (%)	Prolificacy (%)	Lambing† (%)
0	20.6 (131)**	126.0 (27)	26.0 (131)
400	43.6 (133)	158.0 (58)	69.0 (133)

( ) No. of ewes; † % lambs/ewe inseminated.
  * ewes were treated with FGA (40 mg) and artificially inseminated (2 AI ewe) during the breeding season.
  ** P < 0.01.

(From Colas 1975a)

Under these conditions, recent data (Colas, 1976; unpublished) has shown that ewes can be inseminated, in both the anoestrous or oestrous season and give satisfactory results, with only one insemination per oestrus (55 hours after PMSG injection) provided that the number of motile spermatozoa per ewe is $350\text{-}400 \times 10^6$.

(ii)   Season of the year

In the same flocks and during the same season (spring and autumn) the reproductive performances of ewes inseminated with Ile-de-France and Berrichon du Cher semen frozen in autumn or in spring has been studied. Semen frozen in spring always gave lower results than semen frozen in autumn (i.e. in the breeding season) (Table 5).

TABLE 5

INFLUENCE OF THE SEASON OF FREEZING OF SEMEN ON THE REPRODUCTIVE PERFORMANCES OF EWES *

Season of freezing	Conception rate (%)	Lambing† (%)	Fecundity (%)
I. — AI in Spring			
Spring	41.0 (56) (P < 0.05)	165.2 (23)	67.8 (56)
Autumn	57.5 (120)	169.8 (69)	97.5 (120)
II. — AI in Autumn			
Spring	51.1 (45)	165.0 (23)	84.4 (45)
Autumn	60.9 (69)	173.8 (42)	105.8 (69)

* Ewes treated with FGA + PMSG and inseminated (2 AI/female) without oestrous detection. Within each AI period, comparisons are made in the same flocks.
† % lambs/ewe inseminated.   (From Colas and Brice 1976)

These data indicate that at least for breeds sensitive to photoperiod (e.g. Ile-de-France and Berrichon du Cher), semen quality varies according to the time of year. The question arises now whether the decline of the fertilizing ability of the sperm cell is due to a modification of characteristics of seminal plasma or to a loss of quality of the gamete itself. Preliminary results would tend to indicate that the second hypothesis is the more likely.

(c)   Influence of long term storage on fertility of frozen semen

Ram spermatozoa frozen in the breeding season and stored for several years does not seem to be affected either in its survival (Patt and Nath, 1969; Salamon and Visser, 1974) or in its fertilizing ability (Salamon and Visser, 1974; Colas and Brice, 1976) (Table 6).

However, to obtain more precise information on this problem, it will be necessary to study the effects of long term storage on spermatozoa collected at different periods of the year, especially in spring.

TABLE 6

INFLUENCE OF LONG TERM STORAGE ON FERTILITY FROM FROZEN SEMEN

Period of storage	Ewes lambing (%)
2 weeks	54.4 (68)
5 years	52.9 (70)

( ) No. of ewes inseminated (1 or 2 AI/ewe) at the second oestrus after synchronization.
(From Salamon and Visser 1974)

(d)   Examination of ram spermatozoa after freezing

In the case of fresh semen, an estimation of motility of raw semen or of the per cent of motile cells after dilution gives a rather good idea of its fertility.

After freezing, the relationship between the fertilizing ability and the values observed in vitro is more difficult to establish.

Different tests in vitro are proposed for predicting the fertility of frozen ram semen. The most common among them, the recovery rate immediately after thawing is very imprecise, at least for certain diluents, for it does not take into account the survival ability of the gametes which is known to be reduced after freezing.

Alternatively, the viability of the gametes during incubation (+38°C) after thawing gives a fairly precise idea of the quality of the semen. As shown in Table 7, fertility and fecundity of FGA+ PMSG-treated ewes are related to the per cent of motile spermatozoa 0, 1 and 3 hours after incubation. Furthermore, the test of survival at 1 and 3 h is sufficient since there is

a significant correlation between per cent of motile spermatozoa at 0 and
1 h (r = 0.66).

TABLE 7

RELATIONSHIP BETWEEN SURVIVAL OF FROZEN RAM SPERM
AFTER INCUBATION (+38°C) AND ITS FERTILIZING ABILITY

	Motile spermatozoa	
% Survival	< 40-35-20%*	> 40-35-20%*
Conception Rate (%)	50.8 (59)	61.6 (188)
Lambing (%)	86.4 (59)	101.0 (188)

( ) No. of ewes treated with FGA + PMSG and artificially inseminated (2 AI/ewe).
 * Survival is quoted as per cent of motile spermatozoa at 0, 1 and 3 h after thawing
respectively.
                                                                (From Colas and Brice 1975)

Amongst the morphological tests, only those based on the changes of the
acrosome are successfully used in the ram. Smorag and Kareta (1974) in
fact have shown that increasing the number of unimpaired spermatozoa
improved percentage of ewes lambing.

CONCLUSION

From the data presented here it can be seen that many problems have
still to be solved in the storage of ram semen. At a temperature of + 15°C or
+4°C, it is necessary to improve the survival and fertilizing ability of highly
concentrated semen stored for more than one day. It is also necessary to
carry out further work on freezing technology in order to reduce the unfavourable effect of the freezing process on the cell and consequently increase its fertilizing ability.

However, the high fertility rates achieved now with unfrozen spermatozoa stored for one day is enabling AI to be used more widely especially
when oestrus is induced by hormonal treatment (in France: 35,000 ewes inseminated in 1972 and 85,000 in 1975).

Finally, one is led to ask what is the future of both storage methods. In
other words, will it be possible to abandon the use of liquid ram semen in
the next few years in favour of frozen semen?

It is clear that the rejection rate of ejaculates required before and after
freezing and the high numbers of spermatozoa required to ensure fertilization, partly reduce the efficiency of the ram as measured by number of
ewes/ejaculate. However, even with better techniques it is doubtful if it
will be possible in all cases to reach a fertility level as high as that recorded
with chilled semen. Also the loss of quality which occurs during the non-
breeding season of the ram will probably limit the collection for long term

stored semen to the sexual season. Thus, from an economic point of view, AI with frozen semen remains expensive. The choice of the method will be dependent upon the purpose required and it is likely that both methods will continue to be used, each of them in fields where the different motives warrant their use.

ACKNOWLEDGEMENT

The skilful assistance of Mr L. Cahill for the English correction is gratefully acknowledged.

REFERENCES

Allison, A.J. and Robinson, T.J. (1971) *Australian Journal of Biological Science* **24** : 1001.

Amir, D., Schindler, H., Eyal, E. and Lehrer, A.R. (1973) *Annales de Biologie Animale, Biochimie, Biophysique* **13** : 1.

Andersen, K., Aamdal, J. and Fougner, J.A. (1973) *Zuchthygiene* **8** : 113.

Cognié, Y., Mariana, J.M. and Thimonier, J. (1970) *Annales de Biologie Animale, Biochimie, Biophysique* **10** : 15.

Colas, G. (1975a) *Annales de Biologie Animale, Biochimie, Biophysique* **15** : 317.

Colas, G. (1975b) *Journal of Reproduction and Fertility* **42** : 277.

Colas, G. and Brice, G. (1975) *Journées de la Recherche Ovine Caprine en France* : 277.

Colas, G. and Brice, G. (1976) *VIIIth International Congress of Reproduction and Artificial Insemination, Krakow* Vol. III: 977.

Colas, G., Brice, G. and Guerin, Y. (1974) *Bull. Tech. Inform. Minist. Agric.* **274**: 795.

Colas, G., Dauzier, L., Courot, M., Ortavant, R. and Signoret, J.P. (1968) *Annales de Zootechnie* **17** : 47.

Colas, G., Thimonier, J., Courot, M. and Ortavant, R. (1973) *Annales de Zootechnie* **22** : 441.

Dauzier, L., Thibault, C. and Wintenberger, S. (1954) *Annales d'Endocrinologie* **15**(3) : 341.

Fraser, A.F. (1968) *VIth International Congress of Reproduction and Artificial Insemination, Paris* **2** : 1033.

Hawk, H.W. and Conley, H.H. (1971) *Journal of Animal Science* **33** : 255.

Kareta, W., Pilch, J. and Wierzbowski, S. (1972) *VIIth International Congress of Animal Reproduction and Artificial Insemination, Munich* **2** : 1479.

Jones, R.C. (1969) *Australian Journal of Biological Science* **22** : 983.

Jones, R.C. and Martin, I.L.A. (1973) *Journal of Reproduction and Fertility* **35** : 31.

Lightfoot, R.J. and Salamon, S. (1970) *Journal of Reproduction and Fertility* **22** : 385.

Loginova, N.V. and Zheltobrukh, N.A. (1968) *VIth International Congress of Animal Reproduction and Artificial Insemination, Paris* **2** : 1077.

Loginova, N.V. and Zheltobrukh, N.A. (1972) *VIIth International Congress of Animal Reproduction and Artificial Insemination, Munich* **2** : 1507.

Mampouya, C. (1973) Thesis. Universiy of Clermont-Ferrand.

Mattner, P.E., Entwistle, K.W. and Martin, I.C.A. (1969) *Australian Journal of Biological Science* **22** : 181.

Nagase, N. and Graham, E.F. (1964) *Vth International Congress on Animal Reproduction and Artificial Insemination, Trento* **4** : 387.

Naumov, N., Miljkovic, V., Atanasov, L.J., Mrvos, G., Minhajlovski, P., Tanev, G., Zasov, M. and Lazarov, J. (1974) *Proceedings of First Yugoslavia Congress on Animal Reproduction and Artificial Insemination, Ohrio* **2-4-VI** : 171.

Ozkoca, A. (1967) *Lalahan Zoot. Arart. Enstit. Dergisi* **7** : 131.

Patt, J.A. and Nath, J. (1969) *Cryobiology* **5** : 385.

Quinlivan, T.D. and Robinson, T.J. (1969) *Journal of Reproduction and Fertility* **19** : 73.

Salamon, S. (1962) *Australian Journal of Agricultural Research* **13** : 1137.

Salamon, S. (1971) *Australian Journal of Biological Research* **24** : 183.

Salamon, S. (1976) Personal communication.
Salamon, S. and Robinson, T.J. (1962) *Australian Journal of Agricultural Research* **13** : 271.
Salamon, S. and Visser, D. (1974) *Journal of Reproduction and Fertility* **37** : 433.
Schindler, H. and Amir, D. (1961) *Journal of Agricultural Science,* Cambridge **56** : 183.
Smorag, Z. and Kareta, W. (1974) *Medyc. Weteryn* **30** : 689.
Visser, D. and Salamon, S. (1973a) *Australian Journal of Biological Science* **26** : 513.
Watson, P.F. (1975) *Journal of Reproduction and Fertility* **42** : 105.
Watson, P.F. and Martin, I.C.A. (1975) *Australian Journal of Biological Science* **28** : 145.

61

FERTILITY AFTER NON-SURGICAL INTRA-UTERINE INSEMINATION WITH FROZEN-PELLETED SEMEN IN EWES TREATED WITH PROSTAGLANDIN $F_2\alpha$

Y. FUKUI and E.M. ROBERTS

School of Wool and Pastoral Sciences, University of N.S.W., Australia

SUMMARY

Two experiments were conducted to investigate aspects of non-surgical intra-uterine insemination with frozen-pelleted ram semen.

In Experiment 1, 126 out of 190 (66.3%) ewes treated with Prostaglandin $F_2\alpha$ ($PGF_2\alpha$) on day seven to 11 of the oestrous cycle showed oestrus within five days after treatment. Ewes in oestrus were inseminated once with either 0.1 ml of fresh-undiluted semen or 0.3 ml of frozen-thawed semen. Intra-uterine insemination without surgery was successful in 42 (45.7%) out of 92 ewes from both $PGF_2\alpha$ and control groups. Levels of fertilization were estimated following laparotomy and egg recovery. Percentage fertilized of recovered ova were as follows: normal insemination with fresh-undiluted semen 87.5% (14/16) in untreated ewes and 76.5% (13/17) in $PGF_2\alpha$-treated; intra-uterine insemination with frozen-thawed semen 71.4% (10/14) in untreated ewes and 64.7% (11/17) in $PGF_2\alpha$-treated.

In Experiment 2, intra-uterine insemination with frozen-thawed semen stored for 14 months was performed over a period of seven days at second or third oestrus in ewes returning after insemination following treatment with $PGF_2\alpha$ or at natural oestrus for untreated sheep. The total number of ewes involved in the experiment was 182. Intra-uterine insemination was attempted in 129 ewes; 55 of these (42.6%) were inseminated successfully using the intra-uterine technique and the remaining 74 ewes were inseminated deep-cervically. Fresh-undiluted semen was deposited in 53 ewes by normal insemination.

Lambing rates (percentage of ewes lambing of inseminated ewes alive) were as follows: 64% for normal insemination with fresh-undiluted semen; 19.7% for deep-cervical insemination with frozen-thawed semen; 50.9% for intra-uterine insemination with frozen-thawed semen. There was no significant difference in lambing rate between the normal method with fresh-undiluted semen and intra-uterine method with frozen-thawed semen.

INTRODUCTION

It has been demonstrated that Prostaglandin $F_{2\alpha}$ when injected into cattle is luteolytic and causes the animals to return to oestrus and ovulate within two to four days (Lauderdale, 1972; Liehr, Marion and Olson, 1972;

Louis, Hafs and Morrow, 1972; Rowson, Tervit and Brand, 1972).

In sheep, Hearnshaw, Restall and Gleeson (1973) reported that $PGF_2\alpha$ infusion in ewes on day eight of the oestrous cycle induced oestrus and a fresh ovulation. Hawk (1973) showed that of 50 ewes given two 5 mg doses three to four hours apart, or 15 mg of $PGF_2\alpha$ in a single dose on day nine of the cycle, 43 were in oestrus within two or three days after treatment. Otake, Kikuma, Nomoto, Domei and Nakahara (1975) also reported that 12 out of 16 ewes came into oestrus following three subcutaneous injections totalling 12-17 mg $PGF_2\alpha$ on two consecutive days between day three and 13 of the cycle and 47% were pregnant compared to 56% of control animals.

More recently, Haresign (1976) obtained a comparable lambing rate (69%) in ewes mated naturally after treatment with a double injection of a $PGF_2\alpha$ analogue (ICI 80996) compared with control untreated ewes (72%). Fairnie, Cumming and Martin (1976) reported on the fertility of a synchronized oestrus with either a single or double injection of $PGF_2\alpha$ analogue. However, it would appear that the use of AI with frozen ram semen at an oestrus induced by means of $PGF_2\alpha$ or its analogue has not previously been reported.

The successful revival of ram spermatozoa after deep-freezing has not been accompanied by full fertility as in the bull. Nevertheless, lambing rates of 55-67% have been obtained (Colas and Brice, 1970; Salamon and Lightfoot, 1970; Andersen and Aamdal, 1972; Andersen, Aamdal and Fougner, 1973; Salamon and Visser, 1974; Visser and Salamon, 1973, 1974). Several workers have reported that frozen-semen introduced surgically into the uterus of the ewe resulted in a high fertilization rate, but often a very low percentage of embryos survived (Lightfoot and Salamon , 1970b; Killeen and Moore, 1970b, 1971). However, Andersen *et al.* (1973) demonstrated a method of non-surgical intra-uterine insemination in sheep and this method did not result in a high incidence of early embryonic loss. Further development of this insemination technique has been essential to allow the use of frozen semen under practical conditions. This part of the present study (Experiment 2) has already been presented by Fukui and Roberts (1976).

The purpose of the study reported here was to investigate results of non-surgical intra-uterine insemination with frozen-pelleted semen at the first synchronized oestrus after treatment with $PGF_2\alpha$.

A similar experiment was conducted at subsequent oestrous periods (2nd and 3rd) following $PGF_2\alpha$ treatment.

MATERIALS AND METHODS

Two experiments were conducted in May 1975 on the University of New South Wales Field Station (Experiment 1) and on a private property (Experiment 2), at Hay in the south-west of New South Wales.

The design and timetable of the experiments are shown in Table 1.

TABLE 1

DESIGN OF THE EXPERIMENTS

Experiment 1 — Fertilization rates	
Period of experiment	12-24th May, 1975
Comparison:	
1. Treatment	PGF$_2\alpha$ v. Control
2. Semen	Fresh-undiluted v. frozen-thawed.
3. AI techniques	* Normal (fresh-undiluted) v. normal (frozen-thawed) v. deep-cervical (frozen-thawed) v. intra-uterine (frozen-thawed).
No. of ewes	375
Experiment 2 — Non return and lambing data	
Period of experiment	3rd-9th May, 1975
Comparison:	
AI techniques	Normal (fresh-undiluted) v. deep-cervical (frozen-thawed) v. intra-uterine (frozen-thawed).
No. of ewes	182

* Posterior cervical insemination.

(a) Sheep

Three hundred and seventy-five Merino ewes were used for Experiment 1 and 182 for Experiment 2. The latter ewes had returned to service after insemination. Each ewe was identified by serially numbered ear tags and colour branding for the different groups. For each experiment, 15 vasectomized rams which had been electro-ejaculated to confirm their status, were used.

(b) Prostaglandin F$_2\alpha$ treatment

For Experiment 1, a single intramuscular injection of 16 mg PGF$_2\alpha$ Trimethamine salt (PGF$_2\alpha$) in 1 ml of physiological saline was performed in ewes between day seven and 11 of the oestrous cycle. A total of 190 out of 375 ewes were treated with PGF$_2\alpha$ and the remaining 185 ewes were used as a control group.

(c) Semen

Semen was collected from two mature Merino rams for Experiment 1 and from a mature Merino ram for Experiment 2. All ejaculates were examined immediately after collection and only those with a good initial motility were used. For Experiment 1, the semen from the two rams was pooled. Semen was diluted 1 : 2 (v/v) with a diluent containing Tris (hydroxymethyl) amino methane, Tris (360 mM)-glucose (33.3 mM)-citric acid (113.7 mM). The diluted semen contained 18% egg-yolk and 6% glycerol. Processing of freezing, pelleting and thawing procedures were performed

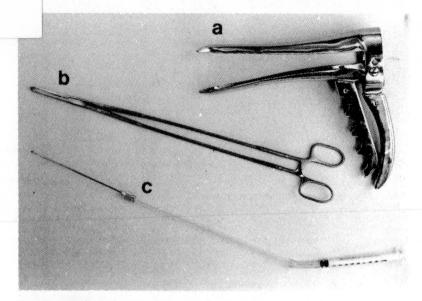

Figure 1—Equipment for intra-uterine insemination.
(a) Duck-billed Speculum
(b) Forcep
(c) Insemination Pipette attached with a ball-tipped needle

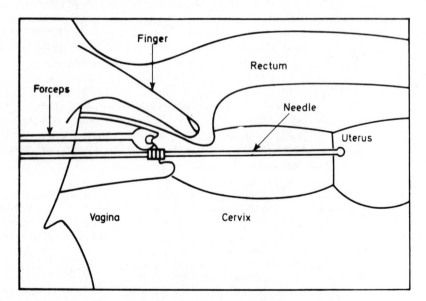

Figure 2—Illustration of the technique of intra-uterine insemination.

according to the method of Visser and Salamon (1973, 1974). The pelleted semen was stored in liquid nitrogen for 7-10 months (Experiment 1) or 14 months (Experiment 2). After thawing in dry, clean test tubes, the percentage of motile spermatozoa was examined before insemination with a minimum level of 40% motility being acceptable.

(d) Artificial insemination

For Experiment 1, the ewes treated with $PGF_2\alpha$ and non-treated ewes, were run with 15 harnessed vasectomized rams and were drafted for incidence of oestrus four times daily (0800, 1300, 1700 and 2300 h) for five successive days after treatment. Ewes in oestrus were inseminated once between 10 and 23 h after the onset of oestrus. Insemination doses were 0.1 ml of fresh-undiluted semen containing a mean of 365×10^6 motile spermatozoa and 0.3 ml of frozen-thawed semen containing a mean of 359×10^6 motile sperm cells.

For Experiment 2, ewes in oestrus were drafted off at 0700 h and 1500 h daily for seven days (3rd-9th May). A single insemination was carried out with either fresh-undiluted or frozen-thawed semen at 10 to 18 h after heat detection. The insemination doses were 0.1 ml containing a mean of 384×10^6 motile spermatozoa for fresh-undiluted semen, and 0.3 ml containing a mean of 312×10^6 motile sperm cells for frozen-thawed semen.

In both experiments, the conventional method of insemination into the mouth of the cervix with fresh-undiluted semen was performed with the aid of a duck-billed speculum and headlight. In the case of insemination using frozen-thawed semen, non-surgical intra-uterine insemination was carried out with a method similar to that of Andersen *et al.* (1973). Semen was deposited deep-cervically (about 2-4 cm) in ewes where penetration through the full length of the cervical canal, was not possible.

The equipment and the technique used in intra-uterine insemination are illustrated in Figures 1 and 2, respectively. The duck-billed speculum was inserted gently into the vagina and the mouth of the cervix located using a headlight. The site was fixed by means of special long forceps (30 cm long) and an insemination pipette attached to a ball-tipped 17 gauge hypodermic needle of 10 cm length introduced through the external orifice into the cervical canal. After withdrawal of the speculum, the method was similar to that of Andersen *et al.* (1973).

The conventional method of insemination with frozen-thawed semen was also performed in Experiment 1.

(e) Recovery of ova

The reproductive tracts were exposed by laparotomy under 2-3 ml of 2% Xylocaine local anaesthesia between 47 and 65 hours after AI. Ova were recovered *in vivo* by flushing the Fallopian tube of the same side with fresh ovulation points, with 2 ml of 0.9% physiological saline.

Data shown as follows were recorded:

Number of corpora lutea
Cell stages of ova
Number of spermatozoa on the zona pellucida

For the confirmation of cell stages of ova and counting of number of spermatozoa on the zona pellucida, all ova were isolated from the flushing medium then mounted and slightly compressed between a clean glass slide and coverslip held apart by four strips of Vaseline. The uncleaved eggs were stained further for the presence of pronuclei using the method of Mattner (1963).

(f)   Non-return and lambing rate

Non-return to service was checked by observation of raddle marks on the 21st day from the last date of insemination. Udder examination, according to the method of Dun (1963), was conducted at 160 days after the last date of insemination.

RESULTS

(a)   Synchronization of oestrus and ovulation

In Experiment 1, 126 out of 190 ewes treated with $PGF_2\alpha$ showed oestrus within five consecutive days (66.3%). 56.3% of these treated ewes came on heat on the second and third day after injection. The pattern of the induced oestrus is shown graphically in Figure 3. Overall 23.8% of the ewes used as control animals showed oestrus over the five days. Therefore, theoretically, about 81% of the control ewes had been cycling during one complete oestrous cycle of 16.5 days.

Number of ewes with new corpora lutea/number of ewes examined were 112/120 (93.3%) in the $PGF_2\alpha$ group and 74/78 (94.9%) in control group.

(b)   Fertilization rate

In Experiment 1, 122 ewes in the $PGF_2\alpha$-treated and 79 ewes in the control group were inseminated. For each group, 107 ewes in the treated group and 73 ewes in the non-treated group were available for flushing ova. A total of 21 ewes from all groups were unfit to flush the Fallopian tubes due to adhesions between the uterus and oviducts, hydrosalpinx, and other abnormalities.

Results of ova examination are summarized in Table 2.

The percentages of successful intra-uterine insemination were 48.2% (27/56) in $PGF_2\alpha$-treated ewes and 41.7% (15/36) in control ewes.

It was found by analysis of variance that there was no significant difference between $PGF_2\alpha$-treated and control groups in both number of

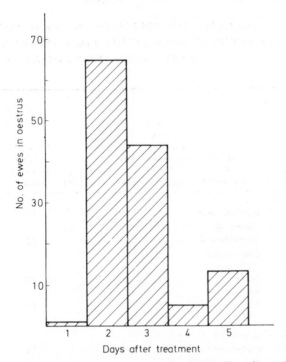

Figure 3—The pattern of synchronized oestrus after $PGF_2\alpha$ (Experiment 1).

ewes with fertilized ova as percentage of ewes with recovered ova, and number of fertilized ova as per cent of ova collected. The different insemination techniques including the type of semen resulted in significant differences ($P < 0.01$) in fertilization rate in both experiments. There was no significant difference between intra-uterine insemination and the other two methods with frozen-thawed semen in both $PGF_2\alpha$-treated ($\chi^2 = 4.05$ d.f. = 2) and control group ($\chi^2 = 3.04$ d.f. = 2), when a chi-square test was performed on numbers of fertilized ova as percentage of number of collected ova.

However there was no significant difference between intra-uterine insemination with frozen-thawed semen and the conventional method with fresh-undiluted semen for number of fertilized ova as per cent of number of ova collected in both $PGF_2\alpha$-treated (64.7% v. 76.5%; $\chi^2 = 0.57$ d.f. = 1) and control group (71.4% v. 87.5%; $\chi^2 = 1.2$ d.f. = 1). There was also no significant difference between treated and control group for number of collected ova as per cent of number of corpora lutea (71.1% v. 68.1%: $\chi^2 = 0.21$ d.f. = 1). The conventional insemination method with frozen-thawed semen in the control group resulted in a higher fertilization rate compared with the treated group (68.8% v. 34.6%: $P < 0.05$).

TABLE 2

FERTILIZATION FOLLOWING SEVERAL INSEMINATION TECHNIQUES IN EWES AT THE FIRST SYNCHRONIZED OESTRUS AFTER PGF$_2\alpha$ (EXPERIMENT 1)

Treatment	Type of AI and semen	No. ewes inseminated and available for flushing	No. of C.L.	No. ova recovered	No. ewes yielding recovered ova	No. of fertilized ova/(% recovered ova fertilized)	No. ewes yielding fertilized ova/(% ewes yielding fertilized ova)
PGF$_2\alpha$	Normal AI fresh-undiluted	22	23	17	16	13 (76.5)*	13 (81.3)**
	Normal AI frozen-thawed	29	34	26	22	9 (34.6)	9 (40.9)
	Deep-cervical AI frozen-thawed	29	35	25	20	10 (40.0)	9 (45.0)
	Intra-uterine frozen-thawed	27	29	17	16	11 (64.7)	10 (62.5)
	Total	107	121	86	74	43 (50.0)	41 (55.4)
Control	Normal AI fresh-undiluted	18	23	16	14	14 (87.5)	12 (85.7)
	Normal AI frozen-thawed	19	23	16	14	11 (68.8)	10 (71.4)
	Deep-cervical frozen-thawed	21	26	16	14	7 (43.8)	6 (42.7)
	Intra-uterine frozen-thawed	15	19	14	11	10 (71.4)	8 (72.7)
	Total	73	91	62	53	42 (66.7)	36 (67.9)

* as % recovered ova that were fertilized.
** as % ewes in which recovered ova were fertilized.

For comparison of number of spermatozoa attached to the zona pellucida of fertilized ova, numbers of spermatozoa were transformed to logarithms and tested by analysis of variance. There was no significant difference between treatment and control group, but a highly significant difference was found between the insemination techniques as well as type of semen ($P < 0.001$, Table 3).

Table 3 shows the distribution of cell stages in fertilized ova. Using a chi-square test, there was no significant difference between treated and control group for the four stages of development of ova ($\chi^2 = 2.40$ d.f. = 3); but significant differences were found between different techniques of insemination with either fresh-undiluted or frozen-thawed semen at 4-7 ($P < 0.001$), and at 8-15 cells stage ($P < 0.01$).

TABLE 3

OVA DEVELOPMENT AFTER INSEMINATION AT FIRST OESTRUS FOLLOWING PGF$_2\alpha$ TREATMENT (EXPERIMENT 1)

Treat-ment	Type of AI and semen	No. of ova at cell stages				Total No. fertil-ized ova	Mean No. of sperm on the zona pellucida
		1 cell	2-3 cells	4-7 cells	8-15 cells		
PGF$_2\alpha$	Normal AI with fresh-undiluted	3	0	7	3	13	30
	Normal AI with frozen-thawed	4	1	1	3	9	2
	Deep-cervical AI with frozen-thawed	3	1	3	3	10	2
	Intra-uterine AI with frozen-thawed	4	0	1	6	11	24
	Total	14	2	12	15	43	16
Control	Normal AI with fresh-undiluted	2	2	6	4	14	24
	Normal AI with frozen-thawed	1	2	3	5	11	9
	Deep-cervical AI with frozen-thawed	2	1	4	0	7	2
	Intra-uterine AI with frozen-thawed	4	0	0	6	10	9
	Total	9	5	13	15	42	13

TABLE 4

FERTILITY FOLLOWING THREE DIFFERENT INSEMINATION TECHNIQUES (EXPERIMENT 2)

Type of AI and semen	No. ewes insemin-ated	Non-Return		Lambing	
		No. ewes	Percentage	No. ewes	Percentage
Normal AI with fresh-undiluted	53	43	81.1	32	64.0*
Intra-uterine AI with frozen-thawed	55	38	69.1	28	50.9
Deep-cervical AI with frozen-thawed	74	33	44.6	13	19.7†

* 3 ewes were missing at lambing
† 8 ewes were missing at lambing

(c)   Non-return and lambing rate

Results of Experiment 2 are summarized in Table 4.

Using a chi-square test of independence it was shown that the non-return rates and lambing rates for the three different techniques of insemination were significantly different (P< 0.001). There was also a highly significant difference between deep-cervical and intra-uterine insemination with frozen-thawed semen for lambing rate (19.7% v. 50.9% : P< 0.001). However, there was no significant difference between the conventional method with fresh-undiluted semen and the intra-uterine method using frozen-thawed semen, for non-return rates ($\chi^2 = 1.83$ d.f.= 1).

DISCUSSION

The percentage of ewes showing synchronized oestrus after $PGF_2\alpha$ (Expt. 1) was unexpectedly low (66.3%), although $PGF_2\alpha$ treatment was given to ewes between day seven and 11 of the oestrous cycle when there would have been a functional corpus luteum in the ovary. However, it should be remembered that only about 81% of the control ewes had had functional ovarian cycles. The percentage synchronized was higher than the figure (54%) of Fairnie, Cumming and Martin (1976) who gave a single injection of $PGF_2\alpha$ analogue (ICI 80996) in ewes on day 10 of the cycle, but lower than the results of Otake *et al.* (1975). It has been confirmed that the peak of induced oestrus was the second and third day after treatment, when 84.9% of 126 ewes which showed oestrus did so during those two days.

Although overall it was shown that fertilization rate in $PGF_2\alpha$-treated ewes was lower than in control ewes (50% v. 66.7%), these figures were not significantly different. In the present experiment, the percentage of ewes with fertilized ova was 59.1% with fresh-undiluted semen and 37.0% with frozen-thawed semen by intra-uterine insemination as compared to control (66.7% for fresh-undiluted and 53.3% for frozen-thawed semen with the method of intra-uterine insemination).

In both Experiments 1 and 2, intra-uterine insemination without surgery resulted in satisfactory fertilization (Table 2) and lambing rates (Table 4). There was no statistical difference in fertilization rates between the conventional insemination with fresh-undiluted semen and the intra-uterine method with frozen-thawed semen (76.5% v. 64.7% for $PGF_2\alpha$ and 87.5% v. 71.4% for control). Also in Experiment 2, it was confirmed that fertility was much higher in ewes inseminated directly into the uterus than in ewes inseminated with the deep-cervical method (50.9% v. 19.4%). It appears that the intra-uterine technique has given increased opportunity for the deposited spermatozoa to pass to the Fallopian tubes without any damage or loss of fertilizing capacity during passage of the cervical barrier. It also has to be remembered that all ewes in Experiment 2 returned to service after an earlier synchronization and therefore may have been a less fertile sample of ewes.

From Table 3 it has been shown that eggs fertilized by spermatozoa inseminated into the uterus developed earlier than those fertilized by sperm cells inseminated either by the conventional or deep-cervical method. This may have been due to rapid transport of sperm. Shelton and Moore (1967) found that ova recovered from progestagen-treated ewes following AI tended to be at an earlier stage of development than ova recovered from normal ewes. Quinlivan and Robinson (1967) also found that the accumulation of spermatozoa in the Fallopian tubes of the synchronized oestrous ewes, particularly following AI, is delayed. The deposition of sperm cells into the uterus without surgery might be able to overcome ova abnormalities due to late fertilization in progestagen-treated ewes.

The rams used in both experiments had had extremely high numbers of spermatozoa per ejaculate with excellent motility (4-7×10^9 spermatozoa/ml and 85-90% motility). This has given a high number of motile sperm cells per inseminating dose (average 325×10^6 per 0.1 ml of fresh-undiluted and 336×10^6 for 0.3 ml of frozen-thawed semen). Thus, 68.8% of 16 recovered ova were fertilized by the conventional insemination with frozen-thawed semen in the control group but not in $PGF_2\alpha$ group (34.6%).

With frozen semen, Colas (1972) and Visser and Salamon (1973, 1974) indicated that a double insemination containing 180×10^6 motile spermatozoa might be necessary to obtain over 50% of lambing rate by means of the conventional technique of insemination. From the present study, it has been shown that 50% lambing rate would be possible from a single non-surgical intra-uterine insemination with frozen-thawed semen containing about 300×10^6 motile spermatozoa. However, the number of motile sperm cells per dose could be decreased to half or a quarter according to the report of Andersen *et al.* (1973) who obtained high conception rate in non-synchronized ewes (89%) and in synchronized ewes (54%) with about 150×10^6 of total numbers of spermatozoa per dose.

Since Andersen *et al.* (1973) have demonstrated the technique of non-surgical intra-uterine insemination, it may be necessary to consider the site of deposition of semen in the ovine reproductive tract.

In cattle, intra-uterine insemination is technically easy except in some heifers. Lasley and Bogart (1943); Holt (1946); Weeth and Herman (1951); Adler (1960) and Moller, MacMillan and Shannon (1972) have reported that the intra-uterine method results in a significantly or slightly better fertility than deep-cervical or vaginal insemination methods. However, it has been generally accepted in sheep that non-surgical insemination into the uterus is impossible (Grant, 1933; Gun, 1936).

Many workers have performed intra-uterine insemination by surgery in sheep (Salamon and Lightfoot, 1967; Mattner *et al.*, 1969; Lightfoot and

Salamon, 1970a, b and Killeen and Moore, 1970a, b, 1971). Lightfoot and Salamon (1970b) reported that intra-uterine insemination with laparotomy resulted in high fertilization rates using frozen-thawed semen (90.5%), but high embryonic mortality occurred following uterine insemination as compared with following either the cervical or cervical traction methods of insemination (47% v. 13% and 6%). However, Andersen *et al.* (1973) who performed intra-uterine insemination non-surgically, reported that early embryonic loss was not observed and the phenomenon could therefore be connected with the surgical interference.

In the present study, a comparison of the non-return with the lambing rates (Table 4) for each group does not indicate a greater incidence of early embryonic death in ewes inseminated into the uterus without surgery. However, investigations should be undertaken to determine whether or not early embryonic loss is affected by surgical procedures.

The percentages of successful penetration into the uterus was 45.7% in Experiment 1 and 42.6% in Experiment 2, while Andersen *et al.* (1973) succeeded in 136 out of 220 ewes (61.8%). The method is still time-consuming when compared with the conventional insemination technique and the operator needs some training. Andersen (1974, personal communication) stated that the time necessary for passing the catheter through the cervical canal varied between 15 seconds to several minutes depending on the shape and course of the canal.

For more successful penetration, studies of improved equipment, simple and accurate insemination techniques and the anatomy of the cervical canal in the ewe, deserve attention before a large-scale operation is commenced.

In conclusion, it was found from Experiment 1 that $PGF_2\alpha$ did not affect fertilization rate and ova development when compared with controls. A single non-surgical intra-uterine insemination with frozen-thawed semen resulted in comparable fertility with no significant differences to the conventional insemination with fresh-undiluted semen. Using the technique of intra-uterine insemination, a higher dilution rate may be possible. Further work is required to determine the minimum number of motile spermatozoa per dose for maximum fertility. This approach together with studies to increase the percentage of uterine penetration will be the substance of future work.

ACKNOWLEDGEMENTS

The authors wish to thank Dr J.W. Lauderdale, the Upjohn Company, Kalamazoo, Michigan, USA for the supply of Prostaglandin $F_2\alpha$; Mr F. Fysh for his generous provision of sheep and facilities (for Experiment 2) and Dr S. Salamon for advice on techniques of freezing pelleted ram semen.

REFERENCES

Adler, H.C. (1960) *Acta Veterinaria Scandinavica* **1** : 105.
Andersen, K. and Aamdal, J. (1972) *World Review of Animal Production* **8** : 77.
Andersen, K., Aamdal, J. and Fougner, J.A. (1973) *Zuchthygiene* **8** : 113.

Colas, G. (1972) *Proceedings of VIIth International Congress of Animal Reproduction and Artificial Insemination* **2** : 925.

Colas, G. and Brice, G. (1970) *Annales de Zootechnie* **19** : 353.

Dun, R.B. (1963) *Australian Journal of Experimental Agriculture and Animal Husbandry* **3** : 228.

Fairnie, I.J., Cumming, I.A. and Martin, E.R. (1976) *Proceedings of the Australian Society of Animal Production* **XI** : 133.

Fukui, Y. and Roberts, E.M. (1976) *Proceedings of VIIIth International Congress of Animal Reproduction and Artificial Insemination* Vol. IV, 991.

Grant, R. (1933) *Transactions at the Royal Society of Edinburgh* **58** : 1.

Gun, R.M.C. (1936) *Bulletin of Commonwealth Scientific and Industrial Research of Australia* No. **94** : 125.

Haresign, W. (1976) *British Society of Animal Production* **22** : 137.

Hawk, H.W. (1973) *Journal of Animal Science* **37** : 314.

Hearnshaw, H., Restall, B.J. and Gleeson, A.R. (1973) *Journal of Reproduction and Fertility* **32** : 322.

Holt, A.F. (1946) *Veterinary Record* **58** : 309.

Killeen, I.D. and Moore, N.W. (1970a) *Australian Journal of Biological Sciences* **23** : 1271.

Killeen, I.D. and Moore, N.W. (1970b) *Australian Journal of Biological Sciences* **23** : 1279.

Killeen, I.D. and Moore, N.W. (1971) *Journal of Reproduction and Fertility* **24** : 63.

Lasley, J.F. and Bogart, R. (1943) *Research Bulletin, Missouri Agricultural Experiment Station* **376** : 1.

Lauderdale, J.W. (1972) *Journal of Animal Science* **35** : 246.

Liehr, R.A., Marion, G.B. and Olson, H.H. (1972) *Journal of Animal Science* **35** : 247.

Lightfoot, R.J. and Salamon, S. (1970a) *Journal of Reproduction and Fertility* **22** : 385.

Lightfoot, R.J. and Salamon, S. (1970b) *Journal of Reproduction and Fertility* **22** : 399.

Louis, T.M., Hafs, H.D. and Morrow, D.A. (1972) *Journal of Animal Science* **35** : 247.

Mattner, P.E. (1963) *Australian Journal of Biological Sciences* **16** : 877.

Mattner, P.E., Entwistle, K.E. and Martin, I.C.A. (1969) *Australian Journal of Biological Sciences* **22** : 181.

Moller, K., MacMillan, K.L. and Shannon, P. (1972) *Proceedings of VIIth International Congress of Animal Reproduction and Artificial Insemination* **2** : 1437.

Otake, M., Kikuma, T., Nomoto, S., Domei, I. and Nakahara, T. (1975) *Japanese Journal of Animal Reproduction* **20** : 132.

Quinlivan, T.D. and Robinson, T.J. (1967) In "The Control of the Ovarian Cycle in Sheep" (Sydney University Press) p.117.

Rowson, L.E.A., Tervit, R. and Brand, A. (1972) *Journal of Reproduction and Fertility* **29** : 145.

Salamon, S. and Lightfoot, R.J. (1967) *Nature (London)* **216** : 194.

Salamon, S. and Lightfoot, R.J. (1970) *Journal of Reproduction and Fertility* **22** : 409.

Salamon, S. and Visser, D. (1974) *Journal of Reproduction and Fertility* **37** : 433.

Shelton, J.N. and Moore, N.W. (1967) In "The Control of the Ovarian Cycle in the Sheep" (Sydney University Press) p.59.

Visser, D. and Salamon, S. (1973) *Australian Journal of Biological Sciences* **26** : 513.

Visser, D. and Salamon, S. (1974) *Australian Journal of Biological Sciences* **27** : 423.

Weeth, H.J. and Herman, H.A. (1951) *Journal of Dairy Science* **34** : 195.

62

STUDIES IN ARTIFICIAL INSEMINATION OF SHEEP IN WESTERN AUSTRALIA

J.H. FIRTH

Muresk Agricultural College, Western Australian Institute of Technology

SUMMARY

Fertility following insemination of fresh semen extended to 100×10^7 or 200×10^7/ml, with either heat-treated cow milk or Tris-based diluents, was indistinguishable for a 0.1 ml inseminate dose. Insemination 8 h after a single morning draft resulted in higher fertility than 2 h. Chilled semen up to 24 h and deep frozen semen gave satisfactory results.

INTRODUCTION

Recent interest in artificial insemination (AI) of sheep in Western Australia (WA) has stimulated the extension of the technique in this State. Lightfoot and Smith (1968) and Knight and Lindsay (1975) reported lambing to a single ram service as 35% and 46% when several rams serve a single ewe, which highlights WA's infertility.

Salamon (1962) reported that 120×10^6 spermatozoa were required for optimal lambing in NSW. Lightfoot and Salamon (1970a) obtained 70% and 49% lambing using a double insemination of 160×10^6 spermatozoa fresh and frozen semen respectively. There have been no reports of the effects of lower sperm numbers, use of chilled semen, stage of oestrus or single inseminations on resultant lambing to AI in WA.

The results for chilled and frozen semen have been conflicting, as reviewed by Lightfoot (1969). Salamon and Robinson (1962b) and Rabocev (1965) found a highly significant decline in fertility with increasing days of storage, and that egg yolk-glucose-citrate was a better diluent than cow milk. Lapwood, Martin and Entwistle (1972) obtained a very rapid decline in fertility after storage at 5°C for 24 h. With frozen semen, Visser and Salamon (1974) obtained 25.3% and 32% fertility for a single insemination with 90 and 180×10^6 spermatozoa and 36.2% and 56.1% when a double insemination of each dose was given. Colas (1975) obtained 68.3% lambing with a double insemination of 450×10^6 spermatozoa. This report examines a recently developed technique for freezing semen and the resultant lambing to a single insemination with a dose of 100×10^6 motile spermatozoa.

MATERIALS AND METHOD

The semen, collected by artificial vagina, was diluted at 30°C to a sperm concentration of 150×10^7/ml in tests 2-4, whereas the sperm concentration was varied in tests 1 (Table 2) and 5 (Table 5). The diluents used were heat-treated fresh cow milk (Salamon and Robinson, 1962a) and the Tris-based diluents described below (see Table 1). The initial diluent used with the frozen semen was glucose (246 mM), L-Glutamic acid (34 mM), NaHCO$_3$ (59.5 mM) and egg yolk (20% v/v).

TABLE 1

TRIS-BASED DILUENTS

Component	mM of component in diluent		
	Test No.		
	1 & 4*	5 (chilling)	5 (freezing)
Tris	300	360	400
Fructose	27.75	33.3	
EDTA	20	20	
Citric acid·†	90	108	120
Egg yolk (v/v) %	14	20	20
Glycerol (v/v) %			12

* Maxwell and Salamon (unpublished data) pers. comm.
† to adjust pH of diluent to 7.0

For liquid storage (tests 4 and 5), the diluted semen was cooled to 5°C in 2.5-3 h, and stored at this temperature for up to 54 h. In test 4, single and double inseminations were performed and the second insemination used semen that was 6 h older than that used for the first insemination. In test 5, only a single insemination was performed and the semen was 20 h old.

The frozen semen (test 5) was prepared by the two-step method, and semen of good initial motility and concentration was used. The semen was initially diluted at 30°C with glucose-egg yolk (1 : 3/4, semen : diluent) and cooled to 5°C in 2.5 h. Tris-based diluent, also at 5°C was then added to the cooled portion (1 : 3/4, semen : diluent), such that the final dilution rate was 1 : 1 1/2 (semen : total diluent), with a sperm concentration of 180-200 × 10^7 spermatozoa/ml and a glycerol concentration of 240 mg/10^9 spermatozoa. The diluted semen was equilibrated at 5°C for 2.5 h, frozen by the pellet method (0.15 ml) on dry ice and stored in liquid nitrogen. A sample of the frozen semen was thawed and subjected to a 6 h incubation test. Only semen with at least 40% recovery of motile spermatozoa and satisfactory incubation results was used for insemination. The frozen pellets were thawed in large, dry tubes (6/tube) and a single 0.1-0.13 ml inseminate dose was used, depending on recovery rates, such that each dose contained 100 × 10^6 motile spermatozoa.

Teaser rams, previously checked by electro-ejaculation, were introduced into a flock of mature Merino ewes 13 days prior to commencement of the

AI programme. They were harnessed and raddled the night before commencing drafting. Ewes from the first draft were excluded from the tests. The oestrous ewes were drafted off once daily (a.m.) and the insemination method was as described by Lightfoot and Salamon (1970b). The time difference between drafting and insemination was 2 h in tests 1 and 2, or 8 h in tests 2-4, and 8 h in test 5. The vagino-cervical mucus was scored (m.s.) on a scale of 1-5 (Restall, 1961).

Tests 1-4 were performed in February 1975 and test 5 in December 1975 on properties in the Northam area of WA. The ewes were grazed on Geraldton based subclover pastures and in test 5 supplementary feeding with lupin grain was started 5 days before the AI programme commenced.

RESULTS

Test 1

The lambing data for the two diluents and two sperm concentrations are presented in Table 2. The mean lambing was 55.2% (160/290) and there were no differences between treatments.

TABLE 2

PERCENTAGE OF EWES LAMBING FOLLOWING INSEMINATION WITH FRESH DILUTED SEMEN

No. of motile spermatozoa/ 0.1 ml inseminate ($\times 10^6$)	Type of diluent	
	Milk	Tris
100	57.3% (43/75)	57.3% (43/75)
200	50.0% (41/82)	56.9% (33/58)

Test 2

In the second test, lambing results for ewes inseminated 1-2 h or 8-9 h after drafting were 41.8% (16/39) and 57.9% (22/38). The differences were not statistically significant.

Test 3

The results of Test 3 are presented in Table 3.

TABLE 3

FERTILITY OF EWES WITH VARIOUS MUCUS SCORES

Mucus score* at drafting	(A) 1-2	(B) 3-5	(C) 1-2	(D) 1-2
Number & time of insemination	Single (a.m.)	Single (a.m.)	Double (a.m. & p.m.)	Single† (p.m.)
% ewes lambing	37.1	53.2	64.9	63.6
	(13/35)	(25/47)	(24/37)	(21/33)
% N.R.	63	55	68	65

* Assessed 1-2 h after drafting.   †6-7 h after assessment of m.s.
Lambing results for B, C and D significantly greater than A (P< 0.05)

Test 4

The results of test 4 are presented in Table 4.

TABLE 4

PERCENTAGE OF EWES LAMBING AFTER INSEMINATION WITH LIQUID STORED SEMEN

No. of Inseminations	Time of Storage (h)		
	0 (A)	24-30 (B)	48-54 (C)
Single	44.9 (22/49)	51.7 (30/58)	34.1 (14/41)
Double	61.1 (33/54)	48.8 (20/41)	40.9 (9/22)
Overall	53.4 (55/103)	50.5 (50/99)	36.5 (23/63)

Semen stored 48-54 h significantly decreased fertility (P< 0.05), whereas lambing to semen stored 0 and 24-30 h was statistically indistinguishable. Differences between single and double insemination were not statistically significant.

Test 5

The results of test 5 are presented in Table 5.

TABLE 5

EWES LAMBING TO FRESH, CHILLED AND FROZEN SEMEN

Type of Semen	No. Inseminated	No. Lambing	% lambing
Fresh	106	67	63.2
Chilled (20 h)	97	54	55.7
Frozen (2 months)	95	50	52.6

There was no statistically significant difference when using 100×10^6 *motile* spermatozoa whether fresh, after chilling, or deep freezing diluted semen.

DISCUSSION

Salamon (1962) and Entwistle and Martin (1972) report satisfactory lambing following inseminations of ewes with 100×10^6 spermatozoa. Below this sperm number, fertility was found to decline. Salamon and Robinson (1962a) found that cow milk, skim milk and egg yolk-glucose-citrate were each suitable diluents for fresh semen. The results of test 1 in WA support these findings and suggest that the Tris-based extender is a suitable diluent for fresh semen. These findings have importance for drier regions of WA, where there is difficulty in obtaining fresh cow milk. By using the Tris-based diluent, semen can be effectively extended to a concentration of 100×10^6 spermatozoa per inseminate so more ewes can be covered by the same rams.

Restall (1961) found that ewes inseminated at mucus scores 1-2 and 3-4 resulted in 55.4% and 65.6% lambing. These results were supported by Salamon (1971) who, in two tests, inseminated 1-2 h or 15-17 h after detection (drafting) of oestrous ewes, resulting in 42.9% and 51.3% and in the second test, 69.1% and 75.7% ewes lambing. If the ewes were inseminated again 10 h later, the average increases were 13-15% and 6.2% if ewes were early or late in oestrus respectively. Salamon concludes that most gain from the second insemination could be obtained when the first is performed at early oestrus. Entwistle and Martin (1972) obtained similar findings, the differences being slightly larger. The results of the second and third test support these findings, though two findings of importance are noted. Column D shows conclusively that the increased lambing over early oestrous ewes is due to a better-timed insemination, rather than larger sperm numbers or double insemination. Column A (Table 3) points out the inaccuracy of using non-return (NR) as an assessment of lambing percentage. The differences obtained in NR and lambing (63% and 37%) are conflicting, as all other results were accurately predicted by NR. Knight and Lindsay (1973) studied paddock mating in WA and found that ewes served by one ram only had a NR of 65%, but only 30% lambed. Also, it has been noticed (personal observation) that ewes mating to one ram only, do so at the beginning of oestrus. Hence, both results seem to suggest that embryonic losses occur due to the time difference between insemination and ovulation.

A partial budget was performed to determine the most economic method of insemination in WA. It was found to be a single p.m. insemination of morning drafted ewes, though this does not result in the highest lambing.

Salamon and Robinson (1962b), Rabocev (1965), Jones, Martin and Lapwood (1969) and Entwistle and Martin (1972) reported decreased fertility following insemination of liquid-stored semen. The general conclusions from these papers were that egg yolk-glucose-citrate buffers were suitable for chilled storage, and sperm concentration was vitally important; in fact many more spermatozoa per inseminate were required for satisfactory lambing. The results of chilling and storage in tests 4 and 5 support these findings, though indicating the reduction in fertility with 100×10^6 motile spermatozoa is not as great with the Tris-based diluent.

In the fourth test, Column B differs from the other results mainly because there was no effect of double insemination. This was due to a combination of two factors. Firstly, the single and double inseminations were performed on ewes of m.s. 3-5 and 1-2 respectively. Secondly, the stored semen used for the second insemination was stored six hours longer than the first, simulating the field situation where semen would only be available daily. The results of the fifth test, where the semen had been transported 140 km, and stored 20 h, gave satisfactory results with 100×10^6 motile spermatozoa, when ewes were inseminated by the "single p.m." method. These results justify more research into the use of Tris-based diluents for chilled storage of ram semen.

The studies of frozen semen date back to Spallanzani in 1776 (cited by Luyet and Gehenio, 1940), and recent research has made great progress into its use. Most workers have used high dilution rates and thawing solutions for pellet-frozen semen. Lightfoot and Salamon (1970a) obtained 49% lambing with a double insemination with 160×10^6 spermatozoa. Salamon (1971) obtained 47.1% and 59% for single and double inseminations with 160×10^6 spermatozoa. These both required centrifugation of the thawed semen before insemination, so as to establish a sufficient cervical population. This was suggested by Mattner, Entwistle and Martin (1969) and Lightfoot and Salamon (1970b) to be related to the concentration of spermatozoa, rather than total spermatozoa in the inseminate dose. Visser and Salamon (1973) found that thawed concentrated semen was as efficient as re-concentrated thawed semen, when the same number of spermatozoa were inseminated. This fact has been supported by the results of the fifth test, as well as "in vitro" studies (unpublished). The semen in this study was more concentrated, and the inseminate dose was related to number of motile spermatozoa, rather than total number. It appears, therefore, that the frozen semen was both of adequate concentration and sufficiently high viability to create the presence of a desirable cervical population, previously having been described as necessary for satisfactory lambing performance.

Colas (1975) obtained 68.3% and 71.3% lambing for frozen and fresh semen respectively, using a double insemination of 450×10^6 spermatozoa. The semen was frozen in straws, which have previously been found unsuitable for WA and Merinos (personal observations). The results of the fifth test compare favourably with the best results from elsewhere, and have the added advantage of serving more ewes per ejaculate of ram semen. These results suggest that the freezing and thawing techniques described are satisfactory, and adaptable to the field situation due to the small number of pellets required and the thawed semen does not require re-concentration. A single insemination gave encouraging results, the inseminate volume was manageable and easily incorporated to the field, allowed handling large numbers of ewes and reduced the need for technicians assisting the inseminator.

The results of the fourth and fifth tests suggest that both chilled and frozen semen could be effectively used in an AI programme in WA. Further research into the effects of larger numbers of motile spermatozoa, whether a result of increased volume, double insemination, increased concentration or improved diluents, as well as increasing the frequency of drafting, need to be performed to evaluate conclusively the efficiency and potential of frozen and chilled semen.

REFERENCES

Colas, G. (1975) *Journal of Reproduction and Fertility* **42** : 277.

Entwistle, K.W. and Martin, I.C.A. (1972) *Australian Journal of Agricultural Research* **23** : 467.

Jones, R.C., Martin, I.C.A. and Lapwood, K.R. (1969) *Australian Journal of Agricultural Research* **20** : 141.

Knight, T.W. and Lindsay, D.R. (1973) *Australian Journal of Agricultural Research* **24** : 579.

Lapwood, K.R., Martin, I.C.A. and Entwistle, K.W. (1972) *Australian Journal of Agricultural Research* **23** : 457.

Lightfoot, R.J. (1969) PhD Thesis. University of Sydney.

Lightfoot, R.J. and Salamon, S. (1970a) *Journal of Reproduction and Fertility* **22** : 399.

Lightfoot, R.J. and Salamon, S. (1970b) *Journal of Reproduction and Fertility* **22** : 385.

Lightfoot, R.J. and Smith, J.A.C. (1968) *Australian Journal of Agricultural Research* **19** : 1029.

Luyet, B.J. and Gehenio, P.M. (1940) Life and Death at Low Temperatures (Biodynamica, Normandy, Missouri).

Mattner, P.E., Entwistle, K.W. and Martin, I.C.A. (1969) *Australian Journal of Biological Science* **22** : 181.

Rabocev, B.K. (1965) *Ovtsevodstov* **11 (No. 9)** : 14.

Restall, B.J. (1961) *Proceedings of a Conference: Artificial Breeding of Sheep in Australia,* University of New South Wales, p.67.

Salamon, S. (1962) *Australian Journal of Agricultural Research* **13** : 1137.

Salamon, S. (1971) *Australian Journal of Biological Science* **24** : 183.

Salamon, S. and Robinson, T.J. (1962a) *Australian Journal of Agricultural Research* **13** : 52.

Salamon, S. and Robinson, T.J. (1962b) *Australian Journal of Agricultural Research* **13** : 271.

Visser, D. and Salamon, S. (1973) *Australian Journal of Biological Science* **26** : 513.

Visser, D. and Salamon, S. (1974) *Australian Journal of Biological Science* **27** : 423.

63

ORGANIZATION OF ARTIFICIAL BREEDING PROGRAMMES OF SHEEP IN WESTERN AUSTRALIA

I.J. FAIRNIE

School of Veterinary Studies, Murdoch University, Murdoch, Western Australia

SUMMARY

Artificial insemination (AI) of sheep provides a way of making maximum effective use of elite rams and ewes. The main problems associated with large scale AI programmes at the moment are the transportation of rams and semen, the low fertility of ewes in the early part of the mating season, predicting the daily numbers of ewes in oestrus for AI and the optimal use of labour and other resources during the AI programme. Synchronization of ovulation can play a part in overcoming some of these problems in the short term, and the development of a practical on-farm system for using frozen semen will be a major development in the long term. However, both these techniques could add considerably to the costs of a programme. The problem of planning programmes to maximize efficiency and to keep costs low is one that will always be part of any large scale sheep AI programme.

INTRODUCTION

AI is used to gain maximum benefits from elite stock. In Western Australia (WA) the largest user of AI, the Australian Merino Society, uses AI to increase productivity in terms of wool and meat (Shepherd 1976). In recent years a number of stud breeders have used AI to make better use of rams purchased at considerable expense. In general terms rams are housed close to where AI is carried out, semen is harvested from them as required and AI continues over a 16 to 18 day period to give all ewes a chance to come into oestrus. The basic techniques currently used for AI have been described by Martin and Fairnie (1976).

The success of an AI programme can be measured in a number of ways—acceptable fertility (about 50% ewes lambing to one insemination), the cost of each lamb born, the lift in productivity. Probably most importantly, the flock owner must feel financially and mentally satisfied.

This chapter outlines factors which currently limit the success of AI programmes and describes methods of overcoming them.

LIMITATIONS IN AI PROGRAMMES

(a) Rams

The current practice in co-operative breeding programmes is to move a group of rams from district to district throughout the "joining" season. In

WA this extends from October through to March and the one group of rams can be used at up to seven district centres over this period. In the absence of a reliable system for long or short term storage of semen, semen has to be harvested as frequently as necessary and used almost immediately after extension.

Semen quality is often poor for the first three or four days on arrival at the next district centre. Semen quantity is reduced during periods of continual hot weather. Heat treated fresh cows-milk is the commonly used extender and the search for a reliable substitute continues as it becomes more uncommon for farms to keep their own dairy cow.

Within each district there may be a number of properties which request AI and so far no satisfactory way has been found to service properties other than where the rams are housed. The variable fertility reported for chilled transported semen (Firth 1979) makes it an unattractive proposition to many farmers.

(b)   Ewes

Lightfoot (1972) has reviewed the factors influencing the recorded low fertility of Merino ewes in WA. Prior to February many ewes fail to mate until some 15 to 20 days after the introduction of rams, a phenomenon known as the "teaser" effect. During this part of the "joining" season it can be expected that ewes with no prior exposure to rams will not demonstrate behavioural oestrus whilst an AI programme is operating.

The release of vasectomized "teaser" rams at least two weeks prior to AI stimulates these ewes to commence cycling (Figure 1).

It can be seen from Figure 1 that "teasing" of ewes causes some degree of synchronization of oestrus in two peaks spaced five to seven days apart, although this effect seems less obvious with the Tambellup property, February 1974. It can be seen that on some properties, 33% of ewes to be inseminated ("A") had been marked by "teaser" rams within five days of AI commencing but overall this point is reached at a variable time (three to six days). If ewes were coming into oestrus at a regular rate, 33% should be marked by about the sixth day of AI, and 50% ("B") by the eighth or ninth day. In Figure 1 it appears that this second point is reached commonly at the eighth day and only earlier on one property. Some farmers have attempted to use this natural apparent synchronization of oestrus to shorten their AI programme. However, it can be seen from this data that this "teaser" effect mainly serves to make predictions of the daily numbers of oestrus ewes unreliable.

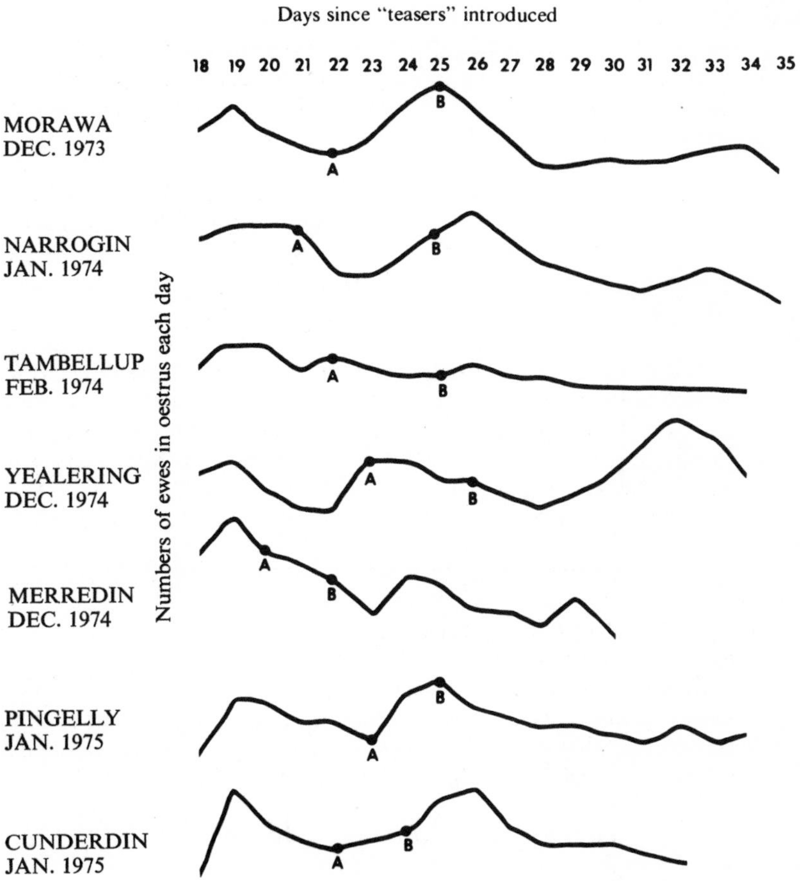

Figure 1—Daily incidence of oestrus in ewes as assessed by rump raddle marks from "teaser" rams. (A is point by which 33% ewes were inseminated and B 55%).

Recent work by C.M. Oldham (unpublished) has demonstrated that the teaser effect is only a temporary one and that after two to three months, ewes will return to their normal seasonal oestrous activity even in the continual presence of "teaser" rams (Figures 2 and 3).

It is necessary to use "teasers" in the joining season prior to February to stimulate most ewes to demonstrate oestrus during the AI programme. If the "teaser" effect is only temporary and the breeder also wishes to follow up his AI programme with natural paddock mating (usually giving a six week programme in all) this could leave only about two to three weeks for "teasing" the ewes prior to AI commencing.

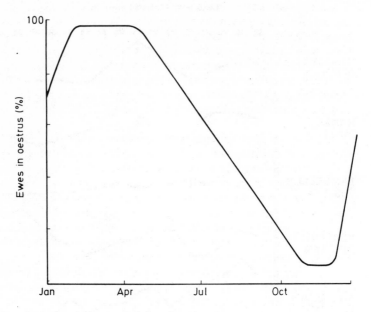

Figure 2—Seasonal oestrous activity of merino sheep.

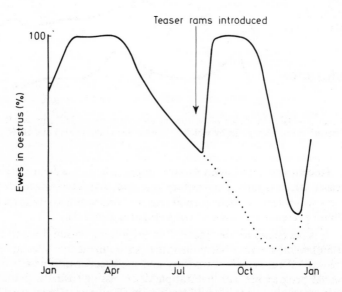

Figure 3—Stimulatory influence of teaser rams on the oestrous activity of merino ewes in the "non-breeding" season.

Reference has been made above to the variable fertility recorded from the use of chilled semen on properties other than that where the rams are housed. Differences in fertility within the range 25% to 55% ewes lambing have been recorded in ewes inseminated on the same day with the same semen but on different properties. There are also differences in fertility between days on the same property within the range 45% to 60% under normal conditions of AI. This effect becomes exaggerated under synchronization of oestrus regimes using exogenous hormones (Fairnie, Cumming and Martin 1976a).

Because of the difficulties experienced with the preparation and transport of chilled semen, some districts have decided to bring ewes from other properties onto the property where the rams are housed. Transporting ewes from their home property to the rams is costly in labour and fuel, as well as feed if the ewes are agisted on the property where AI is being carried out. Again there can be differences in fertility between flocks of ewes held on the one property which may be related to transport of the ewes to and from the property, changes in nutrition or behavioural disturbances when ewes from different properties are run together for AI.

Some farmers have preferred to transport oestrous ewes daily to the rams for AI and collect their ewes from the previous day's insemination. In this case ewes are transported twice 24 hours apart but do not mix with other ewes even during actual insemination. Ewes treated in this manner have comparable fertility to similar ewes which stay on the property where the rams are housed.

(c) Resource Utilization

Rams

Because of the difficulty of predicting when the majority of ewes in a district will have been inseminated, most AI programmes are planned to continue for a 16 to 18 day period. This restricts the use of a group of rams to a maximum of seven districts during the joining period, and means that many elite ewes in other districts are not able to be inseminated and the rams are under-utilized. This would tend to restrict overall progress in programmes such as those undertaken by the Australian Merino Society.

Labour

Many farmers prefer to have a constant maximum daily throughput of ewes. However, the unpredictability of the "teaser" effect, both in time and daily numbers of oestrous ewes, means that programmes are deliberately scaled down to ensure that on only one or two days will the labour force be fully extended. Thus on other days (Figure 1) manpower will be slightly or grossly under-utilized.

(d)   Costs

Moncrieff (personal communication) has constructed a model based on a population of 2,000 ewes to be inseminated under a number of systems over a 16 day period. He has estimated the annual increased production necessary to cover the costs of AI under a number of circumstances. At the lowest level with labour supplied at no cost it is necessary to achieve at least a 0.4% increase in annual production. At the other end of the scale with current costs for labour and an inseminator, a 3.0% increase in production is necessary. In addition if the farmer has to pay to bring an inseminator to his property this adds costs equivalent to a 0.1% increase in production for every 10 km the inseminator has to travel. Similar costs are also incurred if the farmer transports oestrous ewes daily to the property where the rams are housed.

(e)   Co-ordination

Co-ordination becomes necessary in co-operative breeding schemes using AI. It involves timetabling the transport of rams between districts, attempting to meet breeders' requirements concerning when AI will be undertaken and planning strategies such as the release of "teaser" rams, the use of synchronizing agents and feeding of chemical regimes to improve ovulation rates etc. Keeping individual farmers to the overall plan constitutes a major problem of organization.

OVERCOMING THE LIMITATIONS

(a)   Storage of Ram Semen

A practical on-farm system for the use of frozen semen with subsequent acceptable fertility will overcome most of the limitations outlined above. Rams will not have to be transported from district to district nor will they have to remain in a district for 16 to 18 days.

The need for co-ordination of the programme from property to property will only relate to the availability of labour and inseminators. Breeders will have more choice about when their elite ewes are to be inseminated.

Frozen semen will mean that more individual farmers will undertake their own AI programme as they have in recent times with cattle. Trained skilled personnel are required for collection and processing of semen. The actual skill required for insemination is easily learnt and no differences in fertility between inseminators have been demonstrated in the current large scale AI programmes in WA.

Whilst frozen semen would mean reduced labour costs to the farmer, it has been estimated (Hambly, personal communication) that the costs associated with freezing and storage of ram semen could add between 50 cents to one dollar to the cost of each ewe inseminated. Using Moncrieff's model this would require a compensatory increase in production to between 3 and 4% per annum.

The present system for the use of AI requires rams to travel from district to district during the "joining" season. Work is in progress to determine the effects of transporting rams between properties on semen quality and fertility. Semen quantity can probably be maintained with the feeding of lupins before and during the "joining" season.

(b) Synchronization of Ovulation

This topic has been extensively reviewed elsewhere (Cumming 1979). Synchronization regimes offer a way of making better use of resources such as rams and labour, and Lightfoot, Croker and Marshall (1979) have suggested that synchronization of ovulation allows a more strategic and economic use of hyperovulatory agents such as lupins. Recent work by Findlay, Cumming and Fairnie (1977) has shown that an injectible FSH/LH releasing hormone (Hoe 776—Hoechst) can be used in conjunction with synchronization regimes to lift ovulation rates. In conjunction with synchronization programmes, it is also feasible to conduct embryo transfers from elite superovulated ewes to recipient ewes to make even better use of elite rams.

Fairnie, Cumming and Martin (1976b) reported on the use of three agents for synchronization of ovulation—Cloprostenol, a prostaglandin $F_2\alpha$ analogue (ICI 80996), Cronolone, a progestagen intravaginal sponge and a progesterone implant ("Silestrus"—Abbott). They concluded that, overall, Cronolone sponges were the most satisfactory for use in current AI programmes.

Whatever method is used for synchronization, it involves further costs to the farmer. Using Moncrieff's model again, and assuming a cost of 50 cents per ewe treated, a throughput of 500 ewes per day and two days extra work to implement the synchronization programme, the required lift in annual production in the flock to cover costs is increased to 3.0% at the lowest level (from 0.4%) and increased to 3.8% at the highest level (from 3.0%).

Thus although labour and other resources may be used more efficiently, the cost of synchronization treatments are sufficiently high to require a further lift in production to justify them.

Although hard to justify to the individual breeder, in a large scale co-operative breeding programme the use of synchronization regimes allows the use of elite rams in other districts and gives the opportunity to further lift the productivity of the co-operative flocks overall.

(c) Co-ordination

Until such time as a reliable system for using frozen semen is developed it is going to be necessary to continue to co-ordinate the movement of rams between properties. This is best done with having one overall director of the programme with regional directors co-ordinating at the district level. It is imperative that all farmers involved know and understand what is required of them. Table 1 lists the items required for attention before and during the programme.

TABLE 1

CHECK LIST OF REQUIREMENTS FOR A
PROPERTY ARTIFICIAL INSEMINATION PROGRAMME

Item	When Required
A.	Before AI Programme
Order harnesses and raddles* for "teaser" rams	90 days
Vasectomize rams	80 days
Test "teasers" for fertility	30 days
Select ewes for AI	30 days
Check yards and drafting facilities	30 days
Release unharnessed teasers (1% of ewes)	14 days
Pick up AI rams	5 days
Control dust in yards**	1 day
Harness and release teasers (Rate 3% of ewes)	16 hours
B.	During AI Programme
Drafting ewes	Daily before 0800 hours
Check harnesses and raddles	First day, then every 3 or 4 days
Draft teasers out from ewes	Last day
Replace raddles with new colour	Last day
Put harnesses on follow up rams or teasers to assess non-return rates	Last day
(Check raddles in 8 days time)	

 * Allow four raddles per teaser, two of each of two colours.
** Use water daily, or a one only application of sump oil at the rate of 6 litres/square metre (1½ gallons/square yard).

CONCLUSION

It can be seen that until frozen semen of acceptable fertility is available AI programmes must rely on the transport of rams from district to district. To make maximum effective use of these rams, synchronization of oestrus regimes can assist in reducing the amount of time elite rams spend on each property, and the time required of the farmers in each district to implement each programme.

REFERENCES

Cumming, I.A. (1979) In: *Sheep Breeding*. Ed. by Tomes, G.J., Robertson, D.E. and Lightfoot, R.J. p. 403. London: Butterworths.

Fairnie, I.J., Cumming, I.A. and Martin, E.R. (1976a) *Proceedings of Australian Society of Animal Production* **11** : 133.

Fairnie, I.J., Cumming, I.A. and Martin, E.R. (1976b) *Proceedings of 53rd Annual Conference Australian Veterinary Association*, p. 186.

Findlay, J.K., Cumming, I.A. and Fairnie, I.J. (1977) *Animal Breeding Abstracts* **45** : 2811.

Firth, J.H. (1979) In: *Sheep Breeding*. Ed. by Tomes, G.J., Robertson, D.E. and Lightfoot, R.J. p. 547. London: Butterworths.

Lightfoot, R.J. (1972) *Journal of Agriculture Western Australia* **13** : 102.

Lightfoot, R.J., Croker, K.P. and Marshall, T. (1979). In: *Sheep Breeding*. Ed. by Tomes, G.J., Robertson, D.E. and Lightfoot, R.J. p. 451. London: Butterworths.

Martin, E.R. and Fairnie, I.J. (1976) *Proceedings of Australian Society of Animal Production* **11** : 1P.

Shepherd, J.H. (1979) In: *Sheep Breeding*. Ed. by Tomes, G.J., Robertson, D.E. and Lightfoot, R.J. p. 235. London: Butterworths.

64

ARTIFICIAL INSEMINATION OF SHEEP IN THE SOVIET UNION

N.A. JHELTOBRUCH

All-Union Research Institute for Sheep Breeding, Stavropol, USSR

Artificial insemination of farm livestock is the most progressive method of animal improvement and requires suitable techniques for the collection, storage and use of semen. The practical value of AI in sheep production is that its application permits the more widespread use of valuable ram sires for flock improvement.

The founder of the method of artificial insemination was the notable Russian biologist Ilia Ivanovich Ivanov. He was the first to develop an industrially acceptable technique of semen collection and insemination, developed some original devices and instruments and proved the practicability of AI in sheep. He conducted experiments on the storage of semen, the use of diluents and carefully monitored the quality of offspring from ewes artificially inseminated.

The first experiment on artificial insemination involved 4,700 ewes, and was conducted in 1928. The number of inseminated ewes has since increased, and now in the USSR 42–44 million ewes are inseminated every year, which is 72–76 per cent of all ewes in the country. In the most developed sheep regions 90–95 per cent of ewes are artificially inseminated.

It is possible to say, without exaggeration, that all progress achieved in sheep breeding in the USSR in the last 25 years is connected to a great extent with the wide use of artificial insemination. It is thanks to this method that it has been possible to create in so short a period a number of fine-fleeced breeds. The Kasachskiy Archaro Merinos breed, for example, was produced as a result of artificial insemination of native sheep with semen taken from wild Archaro rams at slaughter. This technique and its subsequent development have played a very important part in the development of sheep breeding in the USSR.

CURRENT METHODS

The largest part of the sheep population in the USSR is found on large-scale farms carrying from 3,000–60,000 sheep. Ewes on such farms are kept in flocks of 600-800. A ewe flock is the main structural unit in which the work on artificial insemination is organized. With rare exceptions, all ewes on such farms are inseminated artificially. In almost all of the sheep regions in Russia there is a clearly marked seasonality of reproduction, which limits the periods used for artificial insemination. For this reason inseminations are usually conducted in the autumn months.

For selection of ewes on heat 8-10 aproned teaser rams are put with each flock during the insemination period. After inseminations have been completed these rams are used for normal mating of ewes which failed to conceive to AI.

Selection of ewes on heat is conducted once daily, early in the morning. The interval between selection of ewes on heat and time of insemination is defined by such physiological processes as heat duration and time of the beginning of ovulation. Since heat in the majority of ewes continues for about 35 hours on average, daily single heat tests ensures selection of practically all ewes on heat.

In order that those ewes which are selected late in the oestrous period will be fertilized, it is necessary to carry out the insemination, immediately after finishing the heat test. There are some animals which come into heat only shortly before selection, and for this reason single inseminations may turn out to be unsatisfactory because the semen dies before ovulation occurs. Such ewes are better inseminated later, or a repeat insemination given 12–24 hours later. Some ewes are therefore inseminated twice in order to increase their fertility.

Fresh semen, as well as semen stored for up to 24 hours at +2 to 4°C is used, although about 70 per cent of ewes are inseminated using fresh undiluted semen. When fresh semen is used, one or two rams, which are kept on the artificial insemination station, are used for each ewe flock. Every ejaculate is used within 20–30 minutes of collection.

For more effective use of valuable rams, semen is sometimes transported to the ewe flocks. Semen transport is organized by the State artificial insemination stations or by central stations on the farms. On these stations only the best rams are kept, semen from which is used for insemination of ewes in the appointed flocks. Collected semen is diluted 2–3 times by a glucose-citrate-egg yolk medium and is cooled to +2 to 4°C. Semen transported from State stations may be stored for 12–24 hours before use, but by carefully organized transport between farms, storage time is often reduced to 2–3 hours. When using fresh semen conception rates are 75–80 per cent although when stored semen is used conception rates fluctuate from 50 to 60–65 per cent.

All semen for artificial insemination is collected by artificial vagina. Semen collected using electro-ejaculators is not used on a large scale because it is labour-consuming, and also of much poorer quality. Fresh semen is used in 0.05 ml doses, containing 120–150 million live sperm. When using stored semen the dose is increased to 0.1–0.15 ml.

It is only possible to achieve good results if rams produce good quality semen, and the ewes are in good body condition. In this connection, during the preparation period which continues for $1\frac{1}{2}$ months, great attention is given to creating conditions for normal spermatogenesis and oogenesis. Rams are given good quality feed and are kept grouped together. When preparing ewes for mating, a very important factor is the timely weaning of lambs which must be completed not later than $1\frac{1}{2}$ months before beginning artificial insemination. It is known that green pasture fodders stimulate ovarian function, increasing the ovulation rate. Quality and composition of feed has a great influence on the reproductive performance of ewes.

Up to the end of the preparation period estimations are made of the reproductive capacity of the rams. Rams are trained to use the artificial vagina—this is very important for young rams which are being mated for the first time. Simultaneously, the quality of the semen produced is recorded. Usually every ejaculate is checked for volume, semen concentration and activity. In cases where the use of stored semen is necessary the suitability of semen for cooling is also estimated. Normally, only those rams which show a high degree of sexual activity and give semen of good quality are used.

All selective breeding work in the USSR is carried out by State Breeding Stations, a network of which embraces all administrative districts. In accordance with defined State breeding plans, these stations supply and distribute rams of appropriate breeds and productivity lines, supervise ram progeny and performance testing, take stock of results of artificial insemination and generally organize the selective breeding work of the region. Priority is given to State Breeding Stations for the purchase of the best rams from the leading studs. With regard to the new large flocks that have recently developed it has been necessary to develop new systems of AI. One of the requirements of these flocks is a reduction in the spread of lambing with the aim of getting young stock all of the same age and at the same time. This requirement is achieved in one of two ways. Firstly, during the period of artificial insemination, all stock (some 3,000–5,000 ewes) are formed into one large flock and the ewes are inseminated over a 5 to 8 day period. Lambing in such flocks usually continues for 10–12 days. Alternatively heat synchronization is used in the separate flocks within the farm.

In some cases the method of oestrous cycle synchronization has obvious advantages. At present methods of regulating reproductive function in ewes are widely practised in the Karakul breed. Ovulation rates are increased by treating ewes before insemination with pregnant mare's serum gonadotrophin (PMSG) on a large cycle. This is a labour-consuming method, but injection of PMSG in Karakul ewes increases their fertility by an average of 30–40 per cent.

In recent years intensive investigations have been conducted to develop methods of stimulating reproduction during seasonal anoestrus. This is used widely in Karakul sheep and also to obtain caractultcha from ewes needing culling from the flock. Usually in ewe flocks every year there are about 20 per cent of ewes unfit for further breeding. These are culled after fattening and used for meat production. Prior to fattening, such ewes are inseminated in the spring after hormonal treatment to induce ovulation and are killed on day 130–135 of pregnancy. In this way such ewes produce not only meat, but also the valuable caracultcha lambs. It is important to note that the effectiveness of out of season breeding is not yet sufficiently high. Treatment with progestagens and PMSG results in oestrus in 80 per cent of ewes, but because of low fertilization, conception rates are only 40–50 per cent. Nevertheless, the gain from the additional lamb production covers all expenditure on such treated animals.

UNSOLVED PROBLEMS, AND PERSPECTIVES IN DEVELOPMENT

As is known, the main advantage of artificial insemination consists of increased selection pressure and therefore the more intensive use of the most valuable rams. Even now, with the well organized sheep breeding work as it exists, these advantages are not fully exploited. Experiments over many years in the USSR have shown that, by using fresh semen for insemination, the load on one ram during the season is usually not more than 400–500 ewes. This figure can be increased several times by inseminating ewes with stored semen, although in this case, it is still only possible to inseminate 1,000–2,000 ewes with semen from the best rams.

The development of an effective method of long-term storage of ram's semen in a frozen state would have great advantages. Besides increasing the number of offspring taken from valuable rams, a method of prolonged storage of semen would offer wide possibilities for technological improvement, for example, the introduction of new genotypes, the more rapid multiplication of new breeds.

However, in spite of investigations in many countries, including the USSR, this problem has not yet been solved. The satisfactory results published so far have either not been confirmed by further experiments, or the suggested methods to increase fertility (e.g. intrauterine insemination) are technically difficult to perform in the field situation.

Methods for an earlier estimation of productivity and ram reproductive capacity have great significance in increasing selection intensity. The existing methods of genetic assessment by progeny testing takes too long. Usually, assessment begins when the rams are approximately $1\frac{1}{2}$ years old and is complete when breeding rams are three years old. Thus, the time left for using the rams is reduced by almost half. With no satisfactory method for long term storage of ram semen, it is possible to increase the rams' breeding life only by reducing the progeny testing period. For this reason a means of early estimation of productivity (at the age of 4–6 months) both of rams and their progeny is needed.

To reduce expenditure on keeping young rams, early estimation of their reproductive capacity is required as well as early genetic assessment. The most significant indicators of reproductive capacity are sexual activity, volume and concentration of the semen and its fertilizing capacity. One solution to this problem may be through a study of heritability of reproductive capacity in the ram and the development of basic tests for selection of desirable types of young animals. Simple and reliable methods of objective estimation of the insemination capacity of ram's semen may be of great significance.

Technology of artificial insemination also requires further perfection, as the process at present is very labour-consuming.

There is also a need for significant modernization of the instruments used in artificial insemination. In the first instance, the instruments of semen insertion require perfection, owing to the fact that by the usual cervical method it is necessary to use at least 100 million live sperm for every ewe. The effectiveness of this method of insemination in a number of cases turns out to be very low. It is considered that a possible solution may be a method of deep-cervical or intra-uterine insemination. Development of a method of deep insertion of the semen may eventually show a definite influence on increasing the fertilizing capacity of frozen ram semen.

The range of questions which require solution is not limited to those listed above. Successful solution of these will favour perfection of the technology of artificial insemination, but further work is required to increase the effectiveness of genetic selection.

65

THE EFFECT OF FEEDING A HIGH PROTEIN SUPPLEMENT BEFORE JOINING ON TESTICULAR VOLUME OF RAMS

D.R. LINDSAY, P.B. GHERARDI and C.M. OLDHAM

University of Western Australia, Nedlands, Western Australia

SUMMARY

Merino rams were fed a high protein supplement of lupin grain for 8 weeks before joining. The testicular volume of the supplemented rams increased significantly (22.5 ml per week) but that of control rams did not change significantly (2.7 ml per week).

At joining selected rams from both groups were allocated to various mating treatments based on (a) the farmer's normal ram-ewe ratio and (b) the amount of testicular tissue per 100 ewes. During joining the mean testes volume of all rams fell at a constant rate (40 ml per week) irrespective of ram-ewe ratio or pre-joining supplementation. There were no significant differences in relative return rates of ewes among ram treatments. The body weights of the rams changed in the same direction as testicular volume both before and during joining, but not to the same degree.

INTRODUCTION

Knight (1973) has shown that in Merino rams spermatozoa are produced by the testes at a relatively constant rate of about 20×10^6 spermatozoa/g testis per day. This means that rams with big testes produce more spermatozoa per day than rams with small testes. Knight (1974) has suggested that a rational criterion for selecting ram-ewe ratios at joining might be based on the grams of testicular tissue per 100 ewes. It has recently been shown in this laboratory (Lindsay, 1976) that the testes of rams are very labile and respond rapidly to changes in nutrition—individual rams being capable of doubling the size of their testes after consuming feed supplemented by the high protein grain, lupin, for as little as 8 weeks. Such rapid growth of the testis is accompanied by equivalent growth and output from the tubules (Oldham *et al.*, 1978) so that the capacity of a ram to produce spermatozoa can be increased by high protein supplementation.

Pasture conditions in south Western Australia in December, January and February when most sheep flocks are joined, are invariably dry and deteriorating in quality (Rossiter and Ozanne, 1970). It seems likely, therefore, that supplementation of rams in the field during this period should induce testicular growth and increase the capacity of individual rams to produce spermatozoa. The experiment described here was designed to test this hypothesis.

MATERIALS AND METHODS

(a) Sheep

Identical experiments were carried out on three commercial properties in the Pingelly district in south Western Australia. Maiden ewes were eliminated from the experimental flocks from each property and the remaining ewes were divided randomly into four groups of approximately 300 ewes each which were joined in separate paddocks. Joining dates for the three properties were 26th January, 9th February and 13th February.

(b) Ram treatments

Eight weeks before the proposed date of joining the ram flock on each farm was divided randomly into two. One group acted as unsupplemented controls, the other received 500 g of lupins per day for 4 weeks and then 1000 g per day for the remaining 4 weeks until joining. The grain was offered whole, on the ground, every second day. All supplementary feeding ceased at joining.

From 8 weeks before, until the end of joining the rams were weighed fortnightly and the size of their testes was measured by palpation and comparison with a calibrated orchidometer, a series of testis-shaped beads ranging in volume from 50-400 ml.

(c) Joining treatments

Rams were selected from the supplemented and control groups and were allocated to flocks of 300 ewes on the following basis:

> Group (1) Control rams mated at the usual ram proportion for the farm (mean 2.3%).
>
> Group (2) Supplemented rams mated at the same proportion as (1).
>
> Group (3) Supplemented rams mated at a proportion so that the combined volume of their testes was equivalent to that of the rams in (1) (mean 1.3%).
>
> Group (4) Control rams mated at the same proportion as in (3).

(d) Returns to service

The rams were fitted with "Sire Sine" harnesses and crayons (Radford *et al.,* 1960) and the colours of the crayons were changed every 14 days, at which times markings were recorded. From the data "relative return rates" (the proportion of ewes marked in one marking period that were marked in the next period) were calculated. From the "relative return rates" a comparison was made among flocks of the relative capacity of ewes to conceive during a 2-week period.

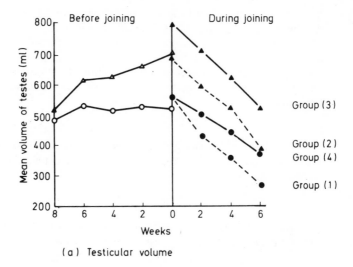

(a) Testicular volume

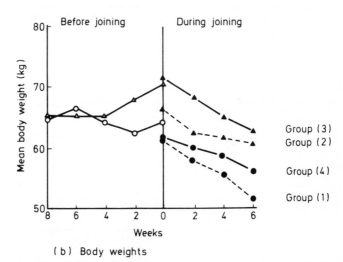

(b) Body weights

Figure 1—Changes in mean testicular volume and body weight of rams before and during joining. (▵—▵) lupin supplemented rams. (○—○) control rams. (See text for description of joining treatment groups).

RESULTS

(a) Changes in testicular volume

The testicular volume of rams supplemented with lupins increased linearly and significantly on each farm. The combined results for all three farms are shown in Figure 1. The average weekly gain in testicular volume for the lupin fed animals was 22.5 ml. The gain of 2.7 ml per week in the control rams was not significant. After mating, the volume of the testes of all rams fell at a constant rate of 40 ml per week regardless of their pre-joining feeding or ratio of rams to ewes at joining. The body weights of the rams changed in the same direction as their testicular volumes but not to the same degree.

(b)   Return rates

The combined results for the three properties are shown in Table 1. There were no significant differences between treatment groups in relative return rates.

TABLE 1

RELATIVE RETURN RATES* OF EWES MATED TO SUPPLEMENTED AND CONTROL RAMS AT RAM-EWE RATIOS BASED ON NUMBERS OR TESTICULAR VOLUME

	Ewes served	Ewes returning	Relative return rate (%)
Group 1 Control rams 2.3%	758	137	18
Group 2 Supplemented rams 2.3%	806	143	18
Group 3 Supplemented rams same testicular volume as (1) 1.3%	739	138	19
Group 4 Control rams same proportion as (3) 1.3%	693	108	16

* The proportion of ewes marked in one marking period that were marked in the next period.

DISCUSSION

The results of this experiment highlight the lability of testicular tissue and its sensitivity to the nutritional regime of the rams. There is equivocation in the literature as to the influence of nutrition and nutritional components such as energy and protein on spermatogenesis in rams, but the measurements of spermatogenesis used have generally been complicated and probably less meaningful than simple measurement of testicular size

(Lindsay, 1976). If the role of rams under field conditions is to produce sufficient viable spermatozoa for optimum fertilization then the enhancement of total sperm production per ram is an important animal production criterion. Differences in nutrition, except in extreme cases, do not appear to affect either libido (Mattner and Braden, 1975) or semen quality (Moule, 1963). The main response therefore to differences in nutrition of rams is in the quantity of semen produced.

During joining the volume of the testis of rams falls rapidly. This is probably a response to reduce intake of food as their live weight also falls during joining. Knight (1973) and Knight, Lindsay and Gherardi (1976) have demonstrated that testicles become smaller even in the absence of a change in live weight but that the main influence is probably nutritional. What is interesting from a practical viewpoint is that the rate at which rams lose testicular volume is not influenced by the original volume at the start of joining or the ratio of rams to ewes. This means that rams with superior daily sperm production apparently maintain that superiority throughout joining.

The practical implications of these findings are that it should be possible to use a lower proportion of rams following supplementary feeding than if the animals are run on natural feed. It is clear from the initial experiments reported here that, even in the control animals, a sufficiently low proportion of rams to ewes was not reached to show differences in fertility of the flocks. It can be stated nonetheless, that no depression in fertility was apparent when rams whose testis size totalled 700 ml were joined at a rate of 1.3 rams per 100 ewes.

ACKNOWLEDGEMENTS

We are grateful to the participating farmers, Messrs P. Stewart, J. Hughes and C. Trott. Also to Mr H. Armstrong for technical assistance. This project was supported by a grant from the Australian Meat Research Committee.

REFERENCES

Knight, T.W. (1973) PhD Thesis, University of Western Australia.
Knight, T.W. (1974) In "Sheep Fertility: Recent Research and its Application in Western Australia" **1** : 64.
Knight, T.W., Lindsay, D.R. and Gherardi, S.G. (1976) Unpublished data.
Lindsay, D.R. (1976) *Proceedings of the Australian Society of Animal Production* **11** : 217.
Mattner, P.E. and Braden, A.W.H. (1975) *Australian Journal of Experimental Agriculture and Animal Husbandry* **15** : 330.
Moule, G.R. (1963) *Australian Veterinary Journal* **39** : 299.
Oldham, C.M., Adams, N.R., Lindsay, D.R. and MacKintosh, J.B. (1976). Unpublished data.
Oldham, C.M., Adams, N.R., Gherardi, P.B., Lindsay, D.R. and MacKintosh, J.B. (1978) *Australian Journal of Agricultural Research* **29**: 173.
Radford, H.M., Watson, R.H. and Wood, G.F. (1960) *Australian Veterinary Journal* **36** : 57.
Rossiter, R.C. and Ozanne, P.G. (1970) In "Australian Grasslands" ed. R.M. Moore (Australian National University Press, Canberra).

INDEX

AI,
  fertility after, 533
  intrauterine, 533
  limitations of, 560
  organization of, 555
  progress of, 245, 565
  use of, 547
Age,
  effect on puberty, 380
  effect on response to lupins, 367
  effect on semen production, 497
Australian sheep breeding schemes, 13
Australian sheep industry, 1

Behaviour,
  abnormal, 484
  effect on management, 481
  grazing, 482
  mating, 473, 482
Breed,
  choice of, 153, 162, 181
  classification, 69
  comparisons, 75, 172
  dual purpose, 181
  effects on fertility, 341, 342
  effects on ovulation rate, 351
  effects on testis growth, 498
  exotic, 65, 153, 159
  multiplication, 157
  new, 189
  performance, 289
  sampling, 154
  testing, 163
Breeding programmes, 13, 56, 79
Breeding schemes, 205, 215, 221, 235,
    251, 269, 275
  advantages of, 231
  development of, 224
  factors affecting, 225
  future of, 228
  results of, 223, 233
  selection in, 250
  structure of, 234, 235
  theory of, 205

Carcass traits, 177, 195

Clover disease, 375
Coloured wool fibres, inheritance of, 96
Conception rate,
  after prostaglandins, 451
  effect of progestagens, 428, 440
  factors affecting, 443
Controlled breeding, 423, 439
  development of, 423
  factors affecting, 435
  improvements of, 440
Co-operative breeding schemes, 251, 275
  genetic selection in, 280
Cross-breeding, 30, 58, 107, 171

Dual purpose breeds, 181

Embryo,
  culture of, 457
  development, 316
  mortality, 315, 327, 335
    diagnosis of, 321, 327
    effect of ovulation rate, 335
    effect of stress, 319
    effect of temperature, 319
    physiology of, 323
  storage, 457
  transfer, 457
Environment,
  adaptation to, 132
  effects on production, 131, 133
Exotic breeds, 65, 153
  multiplication of, 157
  need for, 160
  use of, 153, 159

Fat,
  growth of, 195
  partition of, 200
  sex differences, 196
Feed conversion efficiency, 115
Fertility, 288, 341
  after AI, 533
  effect of breed, 341
  effect of clover disease, 373
  effect of lupins, 357, 363

577